Metallocene-based Polyolefins

Volume Two

June 27, 00

To Our friend Bob Nowack

Best Chu

[signature] Swopper

Wiley Series in Polymer Science

Series Editor:
John Scheirs
ExcelPlas Australia
PO Box 163
Casula, NSW 2170
AUSTRALIA

Modern Fluoropolymers
High Performance Polymers for Diverse Applications

Polymer Recycling
Science, Technology and Applications

Forthcoming titles:

Dendritic Polymers

Polymer–Clay Nanocomposites
Unique Polymers with a Dispersed Nanophase

Metallocene-based Polyolefins

Preparation, properties and technology

Volume Two

Edited by

JOHN SCHEIRS
ExcelPlas Australia, Casula, NSW, Australia

and

W. KAMINSKY
Institut für Technische und Makromolekulare Chemie,
Universität Hamburg, Germany

WILEY SERIES IN POLYMER SCIENCE

John Wiley & Sons, Ltd
Chichester · New York · Weinheim · Brisbane · Singapore · Toronto

Other Wiley Editorial Offices

John Wiley & Sons, Inc., 605 Third Avenue,
New York, NY 10158-0012, USA

WILEY-VCH Verlag GmbH, Pappelallee 3,
D-69469 Weinheim, Germany

Jacaranda Wiley Ltd, 33 Park Road Milton,
Queensland 4064, Australia

John Wiley & Sons (Asia) Pte Ltd, Clementi Loop #02-01,
Jin Xing Distripark, Singapore 129809

John Wiley & Sons (Canada) Ltd, 22 Worcester Road,
Rexdale, Ontario M9W 1L1, CANADA

Library of Congress Cataloging-in-Publication Data
Metallocene-based polyolefins : preparation, properties, and
technology / edited by J. Scheirs and W. Kaminsky.
p.   cm. — (Wiley series in polymer science)
Includes bibliographical references and index.
ISBN 0-471-98086-2 (set : alk. paper)
1. Polyolefins—Synthesis.   2. Metallocenes.   3. Metallocene
catalysts.   I. Scheirs, John.   II. Kaminsky, W. (Walter), 1941–   .
III. Series.
TP1180.P67M48   1999
668.4'234—dc21   99-21151
                                                                CIP

British Library Cataloguing in Publication Data

A catalogue record for this book is available from the British Library

ISBN 0 471 99911 3 (Vol. 1)
ISBN 0 471 99912 1 (Vol. 2)
ISBN 0 471 98086 2 (Set)

Typeset in Times by Techset Composition Ltd, Salisbury, Wiltshire
Printed and bound in Great Britain by Biddles Ltd, Guildford, Surrey
This book is printed on acid-free paper responsibly manufactured from sustainable forestry, in which at least two trees are planted for each one used for paper production.

Contents

I   MECHANISTIC ASPECTS OF METALLOCENE CATALYSTS

II   CYCLOOLEFINS, DIENES AND FUNCTIONALIZED OLEFIN POLYMERIZATION

VI RHEOLOGY AND PROCESSING OF METALLOCENE-BASED
POLYMERS

VII APPLICATIONS OF METALLOCENE-BASED POLYOLEFINS

Contributors

V. Almquist
Borealis AS
N-3960 Stathelle
NORWAY

M. Arndt-Rosenau
Institut für Technische und
Makromolekulare Chemie
Universität Hamburg
Bundesstrasse 45
D-20146 Hamburg 13
GERMANY

E. A. Benham
Research and Development
Phillips Petroleum Company
Bartlesville, OK 74004
USA

S. Betso
Dow Chemical Company
Bachtobelstrasse 3
8810 Horgen
SWITZERLAND

R. W. Bohmer
Research and Development
Phillips Petroleum Company
Bartlesville, OK 74004
USA

L. W. Brenek
Research and Development
Phillips Petroleum Company
Bartlesville, OK 74004
USA

J. M. Carella
Department of Materials Science &
INTEMA
Facultad de Ingeneria
Universidad Nacional de Mar del Plata
Avenida Juan B. Justo 4302
7600 Mar del Plat
ARGENTINA

M. C. Carter
Research and Development
Phillips Petroleum Company
Bartlesville, OK 74004
USA

L. Cavallo
Dipartmento di Chimica
Universitá di Napoli
Via Mezzocannone 4
80134 Naples
ITALY

M. C. Chen
Exxon Chemical Company
Polymers Group
P.O. Box 5200
Baytown, TX 77522
USA

C. Y. Cheng
Polypropylene Department
Exxon Chemical Company
Baytown Polymers Center
P.O. Box 5200
Baytown, TX 77522
USA

Y. W. Cheung *Wilson Cheung*
Polyolefins Research
Dow Chemical Company
Building B1470-D
2301 N. Brazosport Blrd
Freeport, TX 77541
USA

T. C. Chung
Polymer Science Program
Department of Materials Science and
Engineering
Pennsylvania State University
University Park, PA 16802
USA

P. Corradini
Dipartmento di Chimica
Università di Napoli
Via Mezzocannone 4
80134 Naples
ITALY

J. DeGroot
Polyethylene and INSITE Technology
R&D Dow Plastics
Building B-1607
Dow Chemical Company
Freeport, TX 77541
USA

L. Deng
University of Calgary
Calgary
Alberta
CANADA

J. A. Dibbern
DuPont Dow Elastomers L.L.C.
2301 N. Brazosport Blvd
Freeport, TX 77541
USA

C. Diehl   *Charles Diehl*
Polyolefins Research
Dow Chemical Company
Building B1470-D
2301 N. Brazosport Blvd
Freeport, TX 77541
USA

J. D. Domine
Exxon Chemical Company
Polyolefins
P.O. Box 5200
Baytown, TX 77522
USA

J. M. Dionisio
Research and Development
Phillips Petroleum Company
Bartlesville, OK 74004
USA

A. Eckstein
Institut für Makromolekulare Chemie
Albert-Ludwigs Universität Freiburg
Stefan-Meier-Str. 23
D-79104 Freiburg 1
GERMANY

D. L. Embry
Research and Development
Phillips Petroleum Company
Bartlesville, OK 74004
USA

D. R. Fahey
Research and Development
Phillips Petroleum Company
Bartlesville, OK 74004
USA

A. M. Fatnes
Borealis AS
N-3960 Stathelle
NORWAY

A. Follestad
Borealis AS
N-3960 Stathelle
NORWAY

C. Friedrich
Institut für Makromolekulare Chemie
Albert-Ludwigs Universität Freiburg
Stefan-Meier-Str. 23
D-79104 Freiburg 1
GERMANY

A. Giarusso
Dipartimento di Chimica Industriale
e Ingeneria Chimica
Politecnico di Milano
Piazza Leonardo da Vinci 32
20133 Milan
ITALY

G. Guerra
Dipartimento di Chimica
Università di Salerno
Baronissi
84081 Salerno
ITALY

M. J. Guest   *Martin Guest*
Polyolefins Research
Dow Chemical Company
Building B1470-D
2301 N. Brazosport Blvd
Freeport, TX 77541
USA

R. W. Hankinson
Research and Development
Phillips Petroleum Company
Bartlesville, OK 74004
USA

J. C. Haylock
Montell Polyolefins
912 Appleton Road
Elkton, MD 21921
USA

J. J. Hemphill
DuPont Dow Elastomers L.L.C.
2301 N. Brazosport Blvd
Freeport, TX 77541
USA

T. Ho
The Dow Chemical Company
Building B-1470
2301 N. Brazosport Blvd
Freeport, TX 77541
USA

S. M. Hoenig
Polyolefins Research
Dow Chemical Company
Building B1470-D
2301 N. Brazosport Blrd
Freeport, TX 77541
USA

J. Janzen
Polymer Physics, Polymers &
Materials Division
Corporate Technology
Phillips Petroleum Company
Bartelsville, OK 74004
USA

L. T. Kale
Dow Chemical Company
Building B-1470
2301 N. Brazosport Blvd
Freeport, TX 77541
USA

W. Kaminsky
Institut für Technische und
Makromolekulare Chemie
Universität Hamburg
Bundesstrasse 45
D-20146 Hamburg 13
GERMANY

H. Knuuttila
Borealis Polymers Oy
P.O. Box 330
Fin-06101 Porvoo
FINLAND

K. Koch
Polyethylene and INSITE Technology
R&D
Dow Plastics
Building B-1607
Dow Chemical Company
Freeport, TX 77541
USA

J. Lastovica
Polyethylene and INSITE Technology
R&D Dow Plastics
Building B-1470
Dow Chemical Company
Freeport, TX 77541
USA

M. K. Laughner
DuPont Dow Elastomers L.L.C.
2301 N. Brazosport Blvd
Freeport, TX 77541
USA

A. Lehtinen
Borealis Polymers Oy
P.O. Box 330
FIN-06101 Porvoo
FINLAND

C. Y. Lin
Exxon Chemical Company
Polyolefins
P.O. Box 5200
Baytown, TX 77522
USA

B. Löfgren
Polymer Science Centre
Helsinki University of Technology
P.O. Box 356
FIN-02151 Espoo
FINLAND

P. M. Margl
University of Calgary
Calgary
Alberta
CANADA

J. M. Martin
The Dow Chemical Company
Building B-1470
2301 N. Brazosport Blvd
Freeport, TX 77541
USA

A. K. Mehta
Exxon Chemical Company
Polyolefins Group
P.O. Box 5200
Baytown, TX 77522
USA

I. S. Melaaen
Borealis AS
N-3960 Stathelle
NORWAY

L. Moore
Research and Development
Phillips Petroleum Company
Bartlesville, OK 74004
USA

R. Mülhaupt
Institut für Makromolekulare Chemie
Albert-Ludwigs Universität Freiburg
Stefan-Meier-Str. 23
D-79104 Freiburg 1
GERMANY

A. Orscheln
Research and Development
Phillips Petroleum Company
Bartlesville, OK 74004
USA

S. J. Palackal
Research and Development
Phillips Petroleum Company
Bartlesville, OK 74004
USA

A. Penlidis
Institute for Polymer Research
Department of Chemical Engineering
University of Waterloo
Waterloo, ON N2L 3G1
CANADA

R. A. Phillips
Montell Polyolefins
912 Appleton Road
Elkton, MD 21921
USA

L. Porri
Dipartimento di Chimica Industriale
e Ingeneria Chimica
Politecnico di Milano
Piazza Leonardo da Vinci 32
20133 Milan
ITALY

R. C. Portnoy
Exxon Chemical Company
Polyolefins
P.O. Box 5200
Baytown, TX 77522
USA

L. M. Quinzani
PLAPIQUI (UNS–CONICET)
Camino La Carrindanga km 7
8000 Bahía Blanca
ARGENTINA

G. Ricci
Instituto di Chimica delle
Macromolecole – CNR
Via E. Bassini 15
20133 Milan
ITALY

D. C. Rohlfing
Polymer Physics, Polymers &
Materials Division
Corporate Technology
Phillips Petroleum Company
Bartelsville, OK 74004
USA

H. Salminen
Borealis Polymers Org.
P.O. Box 330
FIN-06101 Porvoo
FINLAND

K. Sehanobish
Polyolefins Research
The Dow Chemical Company
Building B-1470
2301 N. Brazosport Blvd
Freeport, TX 77541
USA

J. Seppälä
Polymer Science Centre
Helsinki University of Technology
P.O. Box 356
FIN-02151 Espoo
FINLAND

G. Di Silvestro
Departimento di Chimica Organica e
Industriale
Università di Milano
Via Venezian 21
20133 Milan
ITALY

J. B. P. Soares
Institute for Polymer Research
Department of Chemical Engineering
University of Waterloo
Waterloo, ON N2L 3GI
CANADA

J. Stewart
Research and Development
Phillips Petroleum Company
Bartlesville, OK 74004
USA

C. E. Stouffer
Research and Development
Phillips Petroleum Company
Bartlesville, OK 74004
USA

F. Stricker
Institut für Makromolekulare Chemie
Albert-Ludwigs Universität Freiburg
Stefan-Meier-Str. 23
D-79104 Freiburg 1
GERMANY

A. M. Sukhadia
Research and Development
Phillips Petroleum Company
Bartlesville, OK 74004
USA

M. B. Welch
Research and Development
Phillips Petroleum Company
Bartlesville, OK 74004
USA

W. M. Whitte
Research and Development
Phillips Petroleum Company
Bartlesville, OK 74004
USA

M. D. Wolkowicz
Montell Polyolefins
912 Appleton Road
Elkton, MD 21921
USA

T. K. Woo
University of Calgary
Calgary
Alberta
CANADA

S. Wu
Polyolefins Research
Dow Chemical Company
Building B-1470
2301 N. Brazosport Blvd
Freeport, TX 77541
USA

A. Zambelli
Dipartimento di Chimica
Università degli Studi di Salerno
Via S. Allende
84080 Baronissi (Salerno)
ITALY

T. Ziegler
University of Calgary
Calgary
Alberta
CANADA

Series Preface

The Wiley Series in Polymer Science aims to cover topics in polymer science where significant advances have been made over the past decade. Key features of the series will be developing areas and new frontiers in polymer science and technology. Emerging fields with strong growth potential for the twenty-first century such as nanotechnology, photopolymers, electro-optic polymers etc. will be covered. Additionally, those polymer classes in which important new members have appeared in recent years will be revisited to provide a comprehensive update.

Written by foremost experts in the field from industry and academia, these books place particular emphasis on structure–property relationships of polymers and manufacturing technologies as well as their practical and novel applications. The aim of each book in the series is to provide readers with an in-depth treatment of the state-of-the-art in that field of polymer technology. Collectively, the series will provide a definitive library of the latest advances in the major polymer families as well as significant new fields of development in polymer science.

This approach will lead to a better understanding and improve the cross fertilization of ideas between scientists and engineers of many disciplines. The series will be of interest to all polymer scientists and engineers, providing excellent up-to-date coverage of diverse topics in polymer science, and thus will serve as an invaluable ongoing reference collection for any technical library.

John Scheirs
June 1997

Preface

This book provides an overview of metallocene catalysts and metallocene-based polyolefins. The manufacture of polyolefins by metallocene catalysts represents a revolution in the polymer industry. Polymerization of olefin monomers with single-site metallocene catalysts allows the production of polyolefins (such as polyethylene or polypropylene) with a highly defined structure and superior properties. Furthermore, the structure of these metallocene catalysts can be varied to 'tune' the properties of the polymer. Thus, for the first time it is possible to tailor carefully the properties of large-volume commodity polymers such as polyethylene and polypropylene.

Metallocene polymerization catalysts generally have a constrained transition metal (usually a group 4b metal such as Ti, Zr or Hf) which is sandwiched between cyclopentadienyl ring structures to form a sterically hindered site. Variations on this theme include ring functionalization with various alkyl or aromatic groups, ring bridging with either a Si or C atom and metal coordination to either an alkyl group or halogen atom.

While the term metallocene classically described compounds with π-bound cyclopentadienyl ring structures, today's catalysts are better described as being 'single-site catalysts'. They differ from traditional olefin polymerization catalysts by the fact that the catalytically active metal atom is generally in a constrained environment and thereby only allows single access by monomers. By confining the polymerization reaction to a single site instead of the conventional multiple sites, these catalysts permit close control over comonomer placement, side-chain length and branching. Such single-site catalysts allow precise control over polymer design because the polymer grows by a single mechanism rather than multiple routes. Metallocene catalysts can yield a polymer with almost perfect regularity and tacticity. This enables a range of different polyolefins each with a well-defined structure to be manufactured in one reactor configuration. Such polymers are also characterized by very narrow polydispersities and low extractables (one-fifth that of conventional polyethylenes).

Metallocene-based resins offer improved strength and toughness, enhanced optical and sealing properties and increased elasticity and cling performance. Metallocene-based polyolefins are having most impact in applications such as film packaging for products such as baked goods, meat and poultry, frozen foods, snack foods, cheese wraps and shrink film. Metallocene catalysts allow the manufacture of highly flexible and tenacious films. Traditionally cling and stretch films used to stabilize pallet loads contain a cling additive such as polyisobutylene which limit their recyclability. However new metallocene-catalyzed polyolefin films can achieve remarkable stretch properties without the need of such additives. Flexible plastics based on new metallocene polyolefins can also replace plasticized PVC in many applications without requiring the addition of environmentally-damaging phthalate plasticizers. In addition to these areas, metallocene penetration is occurring in applications such as adhesives, disposable medical packaging, flexible foam products and wire and cable insulation.

Metallocene technology has matured significantly in the last 3 years. There has been no text published to date which covers these developments. Polyolefin technology is growing faster than any other polymer technology and metallocenes are at the centre of this frenetic activity. It has been estimated that within 10 years, 50% of polyolefins will be made by the metallocene route. In addition, copolymerization of ethylene with styrene opens the door to a new series of commodity polymers. Metallocenes can also polymerize bulky monomers such as norbornene, so a range of polymers await that hitherto were not commercially viable or possible.

This book gives comprehensive coverage to all areas of metallocene technology–catalyst structure, comonomer incorporation, polymerization mechanisms and conditions, reactor configurations, special properties, comparison with conventional polyolefins, rheological and processing behavior and fields of application.

John Scheirs

December 1998

Brief Historical Perspective

Metallocene catalysts for ethylene polymerization were reported as long ago as 1957 by Breslow and Newburg [1]. These were based on bis(cyclopentadienyl)titanium dichloride activated with an aluminium alkyl (AlEt$_3$). Such catalysts however, had no commercial viability because their polymerization activity was far inferior to comparable Ziegler–Natta catalysts based on titanium tetrachloride and AlEt$_3$.

The field was relatively dormant for almost two decades, then 18 years after his original discovery Breslow [2] reported that a substantial increase in the ethylene polymerization rate of bis(cyclopentadienyl)titanium dichloride–dimethylaluminium chloride could be achieved through the addition of small amounts of water to the catalyst. Shortly thereafter, Sinn, Kaminsky and coworkers [3] showed that a titanium-based metallocene mixed with trimethylaluminium became a highly active polymerization catalyst when water was added to the aluminium alkyl cocatalyst before reaction with the metallocene. They attributed the enormous increase in catalytic activity accompanying the addition of water to the formation of an alumoxane—a chain of alternating oxygen and aluminium atoms to which methyl groups are bound [now commonly referred to as methylalumoxane (MAO)]. Methylalumoxane is a hydrolysis product of trimethylaluminium. The bulky non-coordinating MAO anion stabilizes the metallocene cation to create the active site. In 1980, Sinn, Kaminsky *et al.* [4] took methylalumoxane that they synthesized separately and added this to a bent zirconocene (a zirconium atom bound to two chlorine atoms and sandwiched between two cyclopentadienyl rings) to yield a system that gave unprecedented catalytic activity (10^6 g PE/g Zr.h.bar).

Unfortunately the prototype zirconium metallocene-MAO system developed by Kaminsky was unsuitable for commercial production of polyethylene. In addition the catalyst exhibited very poor activity for propylene polymerization and offered no stereoselectivity. In 1984 John Ewen (then at Exxon) first demonstrated activity and molecular weight improvements for polyethylene by modifying the structure of Kaminsky's catalyst. Ewen also made the very first stereoselective metallocene catalysts for producing isotactic polypropylene [5, 6]. This stereoselective poly-propylene catalyst was a combination of Kaminsky's MAO and a Brintzinger

Figure 1 Professor Walter Kaminsky in his laboratory where he discovered highly active, metallocene polymerization catalysts based on zirconocenes in combination with methylalumoxane

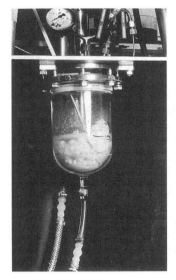

Figure 2 Kaminsky's early polymerization reactor for metallocenes research. While studying zirconocenes in combination with triethyl and methyl aluminium he accidentally discovered that water condensed in sample tube led to an unexpected yield in the amount of polyethylene produced

Figure 3 John Ewen made the very first stereoselective metallocene catalysts which produced isotactic polypropylene. He is also responsible for the discovery of *ansa* zirconium metallocenes for the production of highly syndiotactic polypropylene

metallocene called *rac*-ethylidenebis(indenyl)titanium dichloride. This so-called *ansa* metallocene was selected because tying of the cyclopentadienyl rings together with a two carbon 'handle' made it more rigid (*ansa* literally means bent handle in Latin). The main feature of these catalysts is that the cyclopentadienyl ligands are connected by a bridge, hence they could not turn against each other and this made the overall molecule more rigid.

Around the same time Kaminsky, Brintzinger and coworkers produced highly isotactic polypropylene by using a zirconium metallocene [7]. A couple of years later Ewen produced yet another *ansa* zirconium metallocene with bilateral symmetry (typical of life: i.e., right and left handed) and predictably it produced highly syndiotactic (i.e. alternating pendant methyl groups), high molecular weight polypropylene when combined with MAO [8]. The syndiotactic polymer was more transparent then isotactic polypropylene but less stiff.

In the late 1980s, James Stevens and coworkers at Dow made the remarkable discovery that titanium catalysts based on certain monocyclopentadienyl (monoCp) metallocenes in which a donor ligand stabilizes the metal center are much better than those based on the corresponding bis-cyclopentadienyl metallocenes [9]. These catalysts have become known as 'constrained geometry catalysts' with one key

Figure 4 Some of the persons responsible for the development of Dow's constrained geometry catalysts for polyolefin polymerization. Back row from left: Kurt Swogger (Vice President of Dow Polyolefins and INSITE™ Technology), James Stevens (catalyst development), Brian Kolthammer (process development), Bill Knight (materials science), Che-I Kao (process and polymer fundamentals). Front row from left: Jackie de Groot (applications development), Steve Chum (product development).

advantage being very high activities for ethylene and alpha-olefin copolymerizations. This catalyst technology plus new materials science and process capability were trademarked by the Dow Chemical Company as INSITE™ Technology. Such catalysts have a silicon atom bonded to one of the carbons of a cyclopentadienyl ring coordinated to a titanium atom. The silicon atom is also bonded to an amino nitrogen which is coordinated to the titanium. The 'short-leash' silicon–nitrogen bond pulls on the cyclopentadienyl ring and opens up the bond angle between the ring and the other ligands making the metal site more accessible to longer chain monomers. In the 1990s, several new families of polymers were developed based on INSITE™ technology.

John Scheirs, April, 1999

REFERENCES

1. Breslow, D. S. and Newburg, N. R., *J. Am. Chem. Soc.*, **79** 5072 (1957).
2. Long, W. and Breslow, D. S., *Liebigs Ann. Chem.*, 463 (1975).
3. Andresen, A., Cordes, H. G., Herwig, J., Kaminsky, W., Merck, A., Mottweiler, R., Pein, J., Sinn, H. And Vollmer, H. J., *Angew, Chem, Int. Ed. Engl*, **15**, 630 (1976).
4. Sinn, H. J., Kaminsky, W., Vollmer, H.-J. and Woldt, R., *Angew. Chem., Int. Ed. Engl.*, **19**, 390 (1980).
5. Ewen, J. A., *J. Am. Chem. Soc.*, **106**, 6355 (1984).
6. Ewen, J. A. and Welborn, H. C., Eur. Patent Appl. 0,129,368 (1984); Ewen, J. A., *Stud. Surf. Sci. Catal.*, **25** 271 (1986).
7. Kaminsky, W., Kulper, K., Brintzinger, H. H. and Wild, F. R., *Angew. Chem., Int. Ed., Engl.*, **24**, 507 (1985).
8. Ewen, J. A., Jones, R. L., Razavi, A. and Ferrara, J. D., *J. Amer. Chem. Soc.*, **110**, 6255 (1988).
9. Stevens, J. C., *Stud. Surf. Sci. Catal.*, **89**, 277 (1994).

About the Editors

John Scheirs

Dr John Scheirs was born 1965 in Melbourne and studied applied chemistry at the University of Melbourne. His PhD thesis was on the mechanism and fate of chromocene catalysts in the gas phase polymerization of high-density polyethylene. This work led to an understanding of how the polymerization catalysts affect polymer morphology and thermooxidative stability. Subsequently he worked with silica-supported polymerization catalysts based on chromocene and silyl chromate for the commercial production of high-density polyethylene at Kemcor Australia (an Exxon and Mobil venture). He has also been involved with metallocene-based packaging films and their characterization by thermal analysis. Dr Scheirs has authored over 50 scientific papers including eight encylopedia chapters, a book on polymer recycling and has given presentations at ACS, IUPAC and ANTEC symposia.

Walter Kaminsky

Professor Walter Kaminsky was born 1941 in Hamburg and studied chemistry at the University of Hamburg. His thesis was in the field of metallocene chemistry. Since 1979 he has occupied the role of Full Professor of Technical and Macromolecular Chemistry at the University of Hamburg.

He is currently supervising a group of 20 students and scientists in the field of metallocene/MAO chemistry and a group in the field of pyrolysis of plastic waste and scrap tyres. His past experience includes discovering a highly active, soluble metallocene catalyst system for the polymerization of olefins and for the copolymerization of ethylene with cyclic olefins. Other research interests are in the fields of chemical engineering, polymer recycling by pyrolysis and macromolecular chemistry.

He has published more than 200 papers and books and holds 20 patents. He has also organized several international symposia in the field of olefin polymerization and metallocene catalysis.

In 1991 he received, together with Hans H. Brintzinger, the Heinz Beckurts Prize for the isotactic polymerization of propylene with metallocene catalysts and in 1995, together with Hans H. Brintzinger and Hansjörg Sinn, the Alwin Mittasch Medal for Metallocene Catalysts. In 1997 he received the Carothers Award of the American Chemical Society (Delaware Section), and in the same year the Walter Ahlström Prize in Helsinki, Finland. Since 1997 he has been an Honorary Member of the Royal Society of Chemistry in London, and Honorary Professor of the Zheijiang University in China since 1998. In 1999 he received the Benjamin Franklin Medal for Chemistry as well as the SPE ANTEC Outstanding Achievement Award.

PART I
Mechanistic Aspects of Metallocene Catalysts

1

Molecular Modeling Studies on Stereospecificity and Regiospecificity of Propene Polymerization by Metallocenes

PAOLO CORRADINI AND LUIGI CAVALLO
Università di Napoli, Naples, Italy

GAETANO GUERRA
Università di Salerno, Baronissi (Salerno), Italy

1 INTRODUCTION

It is generally accepted that the stereospecificity of Ziegler–Natta catalysts is mainly due to non-bonded interactions at the catalytic site. In the 1960s Cossee and his group [1–3] and subsequently Allegra [4] and Corradini, Guerra and co-workers [5–8], made some evaluation of the non-bonded interactions for the proposed catalytic models, for heterogeneous isospecific catalysis. The detailed structure of these models, based on crystallochemical considerations relative to possible terminations of crystals of $TiCl_3$, was, of course, only hypothetical. However, the proposed polymerization mechanism [3] and the derived mechanism of enantio-selectivity, which involves the growing chain orientation [5–8], is in good agreement with a large number of experimental facts for the first generation ($TiCl_3$-based) and for the high-yield $MgCl_2$-supported catalysts.

The stimulating opportunity of applying molecular modeling studies, in particular non-bonded interactions analysis, to much better defined catalytic model sites was given by the discovery of new stereospecific homogeneous catalysts based on metallocenes in combination with alkyl-Al-oxanes [9–15]. In fact, the well described

Metallocene-based Polyolefins Edited by J. Scheirs and W. Kaminsky

crystal structures of precursor metallocenes can be used to model in a reasonable manner the coordination of the π-ligands in the catalytic models. Remarkable contributions to the comprehension of different aspects of the stereospecific polymerization mechanisms have been obtained thanks to molecular modeling studies conducted by several research groups (by molecular mechanics [16–27] and quantum mechanics [28–44] techniques) on metallocene-based catalytic systems.

In this review, molecular modeling contributions to the elucidation of even fine details relative to stereospecificity and regiospecificity of metallocene catalytic systems, which mainly come from molecular mechanics studies, are outlined as follows:

Section 2. A brief description of the generally assumed polymerization mechanism and of the elements of chirality for stereospecific olefin polymerization is presented.

Section 3. The enantioselectivity mechanism involving a 'chiral orientation of the growing chain' is presented for model sites of homogeneous catalytic systems based on metallocenes. The discovery of this enantioselectivity mechanism, never considered before, in our opinion constitutes the main contribution of molecular modeling to the comprehension of stereospecific polymerizations. The experiments, often designed to verify the possible validity of this mechanism, are also briefly reviewed.

Section 4. The contributions of molecular modeling to rationalizing the stereospecific behavior of catalytic systems based on metallocenes of different symmetries and in different experimental conditions are described.

Section 5. The ability of molecular modeling to predict also detailed stereostructures of chain segments close to isolated regioirregularities, that is, to predict enantioselectivities of regioirregular insertions, is described.

Section 6. Recent molecular mechanics studies which are able to rationalize the observed dependence of the regiospecificity of these catalytic systems on the type of stereospecificity and on ligand substituents are reviewed.

2 POLYMERIZATION MECHANISM AND ELEMENTS OF CHIRALITY

The basic assumptions about the polymerization mechanism, common to all the molecular modeling studies, are as follows:

(i) the mechanism is monometallic and the active center comprises a transition metal–carbon bond [2,3,45,46];
(ii) the mechanism is in two stages: coordination of the olefin to the catalytic site, followed by insertion into the metal–carbon bond through a *cis*-opening of the double bond [3,47,48].

As a consequence, the generally assumed models of the olefin-bound intermediates are metal complexes containing a π-coordinated propene molecule, a σ-coordinated alkyl group (simulating the growing chain) and the π-coordinated ligands of the precursor metallocenes. Geometries of coordination of the π-ligands, similar to those observed in the crystal structures of the precursor metallocenes, have generally been assumed as starting points for energy minimizations.

Non-bonded interaction calculations for these models have suggested that the coordination of a further ligand is unlikely, and have forced to assume (with the hypothesis of a metal oxidation number of IV) cationic models such as those of Figures 1 and 2 [16,18].

The cationic character of the active site, proposed as long ago as 1961 [49], has more recently been confirmed by the synthesis of a wide series of Group 4 metallocene cations able to polymerize ethene and propene without any aluminum cocatalyst [50–53]. The finding by Ewen *et al.* [13] that ethylenebis(1-indenyl)ZrCl$_2$/methylalumoxane and ethylenebis(1-indenyl)ZrCH$_3^+$B(C$_6$F$_5$)$_4^-$ produce isotactic polypropylene with the same microstructural defects is the best available proof of the cationic nature of the active species.

In order to indicate the terminology used, the elements of chirality in the Ziegler–Natta polymerization will be briefly recalled.

First, upon coordination a prochiral olefin, such as propene, gives rise to non-superimposable *si* and *re* coordinations [54]. The nomenclature *si*, *re* (defined for specifying heterotopic half spaces) [54] is used instead of the nomenclature *R*, *S* (defined for double or triple bonds π-bonded to a metal atom [55,56]) in order to avoid confusion with the symbols *S* and *R* used for other chiralities at the same catalytic site. Since the name of a fixed enantioface of a 1-olefin depends on the bulkiness of the substituent in position 1, the use of the *si*, *re* nomenclature can be confusing when different monomers are considered. For the case of propene, which

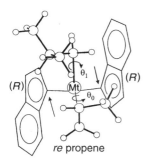

re propene

Figure 1   A model catalytic complex comprising an isopropylbis(1-indenyl) ligand, a propene molecule *re* coordinated and an isobutyl group (simulating a growing primary polypropene chain). The chirality of coordination of the bridged π-ligand is (*R,R*), labeled according to the absolute configurations of the bridgehead carbon atoms which are marked by arrows. The coordinated bridged π-ligand presents a local C$_2$ symmetry axis and the two coordination positions which are available for monomer and growing chain are homotopic. For the sake of clarity, for the bridged π-ligand only the C–C bonds are sketched and the carbon atoms of the methyl groups of the isopropyl bridge are omitted. The dihedral angles θ_0 and θ_1 (see text), used in the energy plots in Figures 6 and 3, respectively, are also indicated

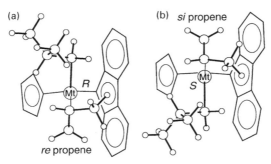

Figure 2 Model catalytic complexes comprising an isopropyl(cyclopentadienyl-fluorenyl) ligand, a propene molecule and an isobutyl group all coordinated. No chirality of coordination of the bridged π-ligand exists. The coordinated bridged π-ligand presents a local C_s symmetry plane and two coordination positions which are available for monomer and growing chain are enantiotopic. Enantiomeric minimum energy pre-insertion intermediates for the models with (a) *R* and (b) *S* chirality at the metal atom are shown. For the sake of clarity, for the bridged π-ligand only the C–C bonds are sketched and the carbon atoms of the methyl groups of the isopropyl bridge are omitted. A regular alternation of insertion steps based on intermediates such as those in (a) and (b) assure the syndiospecific behavior of the model (see Scheme 2)

is the only monomer considered in this review, the *si* and *re* coordinations correspond to the *S* and *R* coordinations, respectively. According to the described mechanism, the isotactic polymer is generated by a large series of insertions of all *si* or all *re* coordinated monomers, while the syndiotactic polymer would be generated by alternate insertions of *si* and *re* coordinated monomers.

A second element of chirality is the configuration of the tertiary carbon atom of the growing chain nearest to the metal atom.

A third element of chirality is the chirality of the catalytic site which, in particular, can be of two different kinds:

(i) The chirality arising from coordinated ligands, other than the alkene monomer and the growing chain. For the case of metallocenes with prochiral ligands, it is possible to use the notation (*R*) or (*S*), in parentheses, according to the Cahn–Ingold–Prelog rules [56,57] extended by Schlögl [58]. For instance, the (*R,R*) chirality of coordination of the isopropylbis(1-indenyl) ligand, which is labeled according to the absolute configurations of the bridgehead carbon atoms (marked by arrows), is shown in Figure 1.

(ii) An intrinsic chirality at the central metal atom, which for tetrahedral or assimilable to tetrahedral situations, can be labeled with the notation *R* or *S*, by the extension of the Cahn–Ingold–Prelog rules as proposed by Stanley and Baird [59]. This nomenclature has been used by us to distinguish configurationally different olefin-bound intermediates which may arise by exchanging the relative positions of the growing chain and of the incoming monomer. For instance, two models with intrinsic chirality at the central metal atom R and S are shown in Figure 2(a) and (b), respectively, for the case of a metallocene with an isopropyl(cyclopentadienyl-9-fluorenyl) ligand. For the case of the models with C_1 symmetric metallocenes, we will mainly use the more mnemonic notation according to which the relative disposition of the ligands that presents the coordinated monomer in the more (less) crowded region is referred to as 'inward (outward) propene coordination' [24].

As will be shown in detail in the following, one or both of these kinds of chirality at the catalytic site can be present in the models. For the case of model complexes in which two carbon polyhapto ligands are tightly connected through chemical bonds (the so-called bridge), and which we shall hereafter call 'stereorigid', only the chirality of kind (ii) can change during standard polymerization reactions.

In the simpler cases, the discrimination between the two faces of the prochiral monomer may be dictated by the configuration of the asymmetric tertiary carbon atom of the last inserted monomer unit (chain-end stereocontrol) as well as by the chirality of the catalytic site (chiral-site stereocontrol). The distribution of steric defects along the polymer chain may be indicative of which kind of stereocontrol is operating. For instance, for the case of prevailingly isotactic polyolefins, Bernoullian statistics in the (m and r) diad distributions have been shown [60] to be consistent with chain-end stereocontrol, whereas non-Bernoullian distributions originate from chiral site stereocontrol [61,62].

In molecular modeling studies [16–27], it is assumed that the enantioselectivity of stereospecific polymerizations is connected with the energy differences between diastereoisomeric situations which originate from a combination of two or more of the above elements of chirality.

In particular, our molecular mechanics calculations were aimed: (i) at the recognition of the 'stable' geometries of the relevant diastereoisomers; (ii) at choosing, among these, the geometries which are as near as possible to the transition states of the insertion reactions; and (iii) at making comparative estimates of the corresponding activation energies. In this context, we intend as 'stable' geometries, the geometries of minimum conformational energy, under constraints such as those indicated in the following paragraphs.

Geometries of intermediates, corresponding to the coordination stage, are considered to be sufficiently close to the transition states, and considered as suitable conformers of pre-insertion intermediates, only if the insertion can occur through a 'least nuclear motion' [3,63–65]. This corresponds to geometries of the catalytic intermediates for which:

(a) the double bond of the olefin is nearly parallel to the metal-growing chain bond;
(b) the first C—C bond of the chain is nearly perpendicular to the plane defined by the double bond of the monomer and by the metal atom ($50° < \theta_1 < 100°$ rather than $\theta_1 \approx 180°$, which is also consistent with the formation of an adjuvant α-agostic interaction; see the discussion at the end of Section 3).

Moreover, olefin-bound intermediates for which the methyl group of the propene and the second carbon atom (and its substituents) of the growing chain are on the same side with respect to the plane defined by the Mt–C (Mt = metal) bonds (e.g. $\theta_1 \approx +60°$ and $-60°$ for the *re* and *si* coordinated monomer, respectively) are assumed to be less suitable for successive monomer insertion. In fact, the insertion paths starting from these intermediates involve large non-bonded interactions [5,8,65,66].

In several cases, starting from the energy minima located in this way, models corresponding to other significant points of the insertion reaction paths, possibly closer to the transition states, have also been considered [65,66]. In particular, in this review, results of molecular mechanics calculations relative to pseudo-transition states have been also reported. For these pseudo-transition states, the geometry of the olefin and of the first carbon atom of the growing chain has been set equal to that determined by Ziegler and co-workers [67] for the insertion reaction

$$[Cp_2ZrCH_3]^+ + CH_2{=}CH_2 \longrightarrow [Cp_2ZrCH_2CH_2CH_3]^+$$

3 ENANTIOSELECTIVITY. MECHANISM OF THE CHIRAL ORIENTATION OF THE GROWING CHAIN

Different kinds of homogeneous catalysts, based on Group 4 metallocene/ methylalumoxane systems, have been discovered [9,15]. Depending on the kind of the metallocene π ligands, these systems present completely different stereo-specific behaviors.

For instance, catalytic systems comprising the metallocene stereorigid ligand ethylenebis(1-indenyl) or ethylenebis(4,5,6,7-tetrahydro-1-indenyl) polymerize propene to isotactic polymer with a non-Bernoullian distribution of steric defects, consistent with chiral-site stereocontrol [9,14,15].

Catalytic systems comprising the metallocene stereorigid ligand isopropyl(cyclopentadienyl-9-fluorenyl) instead polymerize propene to syndiotactic polymer, and the distribution of steric defects is again consistent with prevailingly chiral-site stereocontrol [12].

Catalytic systems comprising as metallocene ligands two cyclopentadienyl non-stereorigidly connected can produce, at low temperatures, isotactic polymer with a Bernoullian distribution of steric defects, consistent with chain-end stereocontrol [9,68].

A necessary (but not sufficient) prerequisite for models of stereospecific catalytic systems is the enantioselectivity of each monomer insertion step. The possible origin of the enantioselectivity, for models of several kinds of catalytic systems, has been investigated through molecular mechanics analyses.

Let us first consider models relative to pre-insertion intermediates such as those described in the previous section. For these models, the internal energy is minimized with respect to the relevant dihedral angles determining the positions of the atoms near to the catalytic center. These dihedral angles (see Figure 1) are defined as in our previous publications. In particular, θ_0 is associated with rotations of the olefin around the axis connecting the metal to the center of the double bond, while θ_1 is associated with the rotation around the bond between the metal atom and the first carbon atom of the growing chain. At θ_0 near $0°$ the olefin is oriented in a way suitable for primary insertion, while θ_0 near $180°$ corresponds to an orientation suitable for secondary insertion. θ_1 near $0°$ corresponds to the conformation having the first C−C bond of the growing chain eclipsed with respect to the axis connecting the metal atom to the center of the double bond of the olefin. For instance, the situation depicted in Figure 1 corresponds to $\theta_0 \approx 0°$ and $\theta_1 \approx -60°$.

Figure 3(a) and (b) plot as a function of θ_1 the optimized energy for the catalytic site models of Figures 1 and 2(a), respectively. These models coordinate to a zirconium atom a propene molecule, an isobutyl group (simulating a primary growing chain) and a stereorigid π-ligand, which can be an (R,R) coordinated rac-isopropylbis(1-indenyl) or an isopropyl(cyclopentadienyl-9-fluorenyl) with R chirality to the metal.

The starting points for the energy optimizations are conformations with $\theta_0 = 0°$. Whatever the energy, the absolute value of θ_0 for the optimized conformations is never greater than $20°$. Hence these models simulate pre-insertion intermediates for primary insertion of propene into a primary polypropylene growing chain. The full and dashed lines refer to re and si coordinated propene, respectively.

The absolute energy minima labeled a in Figure 3 correspond to $\theta_1 \approx -60°$, for the re monomer coordination, and are sketched in Figures 1 and 2(a), respectively. These models minimize the interactions between the growing chain (at $\theta_1 \approx -60°$) and the methyl of the propene monomer (re coordinated). Therefore, these are assumed to be pre-insertion intermediates suitable for the re-monomer primary insertions.

Minimum energy situations labeled b in Figure 3, corresponding to similar θ_1 values but opposite monomer enantiofaces, are slightly higher in energy (1–2 kcal/mol). However, since the methyl group of the propene and the second carbon atom (and its substituents) of the growing chain are on the same side with respect to the plane defined by the Zr−C(olefin) bonds, the corresponding models are assumed to be unsuitable for the successive monomer insertion (see below).

The minimum energy situations labeled c in Figure 3, corresponding to the opposite monomer enantioface, minimize the interactions between the methyl group of the propene monomer and the growing chain ($\theta_1 \approx +60°$ for si coordinated monomer). Therefore, as discussed in previous papers [5,18,65,66], they are also assumed to be suitable for the successive insertion reaction. However, these models are strongly disfavored by repulsive interactions of the growing chain with one of the six-membered rings of the ligands.

In our framework, the energy differences $E_c - E_a$ gives an approximation of the enantioselectivity of primary ($\Delta E_{enant,p}$) insertion step.

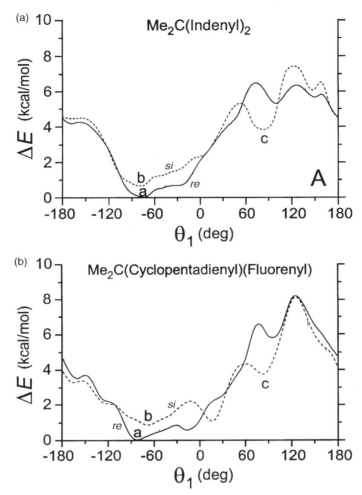

Figure 3 The optimized energies plotted as a function of θ_1 for model complexes comprising (a) (R,R)-coordinated isopropylbis(1-indenyl) ligand, i.e. **4** in Scheme 1, and (b) isopropyl(cyclopentadienyl-9-fluorenyl) ligand, i.e. **25** in Scheme 1, for the R chirality at the metal atom. The full and dashed lines refer to re and si coordinated propene, respectively. In both cases the chirality of the site favors negative values of θ_1, which, in turn, favor the re coordinations of propene. The models of pre-insertion intermediates corresponding to the absolute energy minima, labeled a in parts (a) and (b), are sketched in Figures 1 and 2(a), respectively. For the model complex comprising the isopropyl(cyclopentadienyl-9-fluorenyl) ligand, for the S chirality at the metal atom the chirality of the site favors positive values of θ_1, which, in turn, favor the si coordination of propene [the corresponding minimum energy pre-insertion intermediate is shown in Figure 2(b)]

This analysis indicates that the enantioselectivity of these models is not due to direct interactions of the π-ligands with the monomer, but to interactions of the π-ligands with the growing chain, determining its chiral orientation ($\theta_1 \approx -60°$ preferred to $\theta_1 \approx +60°$) which, in turn, discriminates between the two prochiral faces of the propene monomer [16–18].

Molecular mechanics calculations have also been performed on pseudo-transition states, constructed as described in the previous section, originated from the pre-insertion intermediates labeled a, b and c in Figure 3. The pseudo-transition states originated from the pre-insertion intermediates labeled b, presenting the methyl group of the propene and the second carbon atom (and its substituents) of the growing chain on the same side with respect to the plane defined by the Zr$-$C(olefin) bonds, are always of higher energy. An additional kind of evaluation of the enantioselectivities for the primary insertions ($\Delta E^{\neq}_{\mathrm{enant,p}}$) is hence obtained by energy differences between pseudo-transition states following pre-insertion intermediates labeled a and c in Figure 3. The pseudo-transition states corresponding to the pre-insertion intermediates labeled a in Figure 3(a) and (b) [sketched in Figures 1 and 2(a)] are sketched in Figure 4(a) and (b), respectively.

The enantioselectivities as evaluated from pre-insertion intermediates and pseudo-transition states ($\Delta E_{\mathrm{enant,p}}$ and $\Delta E^{\neq}_{\mathrm{enant,p}}$, respectively) are collected in Table 1, for stereorigid models based on the π-ligands sketched in Scheme 1. In Table 1, the local symmetry of the coordinated bridged π-ligand (C_2, C_1 or C_s), the possible chirality of coordination of the π-ligand (which does not change during the polymerization reaction) and, for C_1 or C_s local symmetries, the possible chirality at the central metal atom (which generally is inverted at each insertion step; see the next section) are also indicated. It is apparent that the values of the

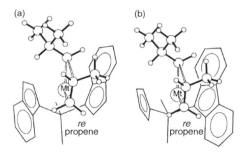

Figure 4   Minimum energy pseudo-transition states for primary propene insertion for model complexes with (a) isopropylbis(1-indenyl) ligand (isospecific), for the (R,R) coordination, and (b) isopropyl(cyclopentadienyl-9-fluorenyl) ligand (syndiospecific), for the R chirality at the metal atom. The corresponding pre-insertion intermediates, labeled a in Figures 3(a) and (b), are sketched in Figures 1 and 2(a), respectively

Table 1   Calculated non-bonded energy contributions (kcal/mol) to the enantioselectivity for pre-insertion intermediates ($\Delta E_{\text{enant,p}}$), and for pseudo-transition states ($\Delta E^{\neq}_{\text{enant,p}}$). The calculation method is described in Ref. 27

Ligand	Metal atom	Chirality of coordination of the π-ligand	Propene coordination position	$\Delta E_{\text{enant,p}}$	$\Delta E^{\neq}_{\text{enant,p}}$	Favored propene enantioface	$E_{\text{out}} - E_{\text{inw}}$
C_2 symmetric ligands							
1	Zr	(R,R)	—	0.1		si	
2	Zr	(R,R)	—	3.3		re	
3	Zr	(R,R)	—	3.7		re	
4	Zr	(R,R)	—	3.8	2.3	re	
5	Zr	(R,R)	—	3.6		re	
6	Zr	(R,R)	—	0.0		—	
7	Zr	(R,R)	—	0.0		—	
8	Zr	(R,R)	—	3.7		re	
9	Zr	(R,R)	—	4.4		re	
10	Zr	(R,R)	—	4.9	3.0	re	
11	Zr	(R,R)	—	4.8	3.2	re	
12	Zr	(R,R)	—	0.1	0.3	re	
13	Zr	(R,R)	—	4.0	2.4	si	
14	Zr	(R,R)	—	5.3	3.1	re	
15	Zr	(R,R)	—	5.9		re	
16	Zr	(R,R)	—	5.5		re	
17	Zr	(R,R)	—	5.6		re	
18	Zr	(R,R)	—	4.3		re	
18	Ti	(R,R)	—	5.6		re	
19	Zr	(R,R)	—	5.7	4.7	re	
19	Ti	(R,R)	—	7.4		re	
20	Zr	(R,R)	—	0.4		re	
21	Zr	(R,R)	—	6.0		re	
22	Zr	(R,R)	—	6.1		re	
23	Zr	(R,R)		1.9		re	
24	Zr	(R,R)		4.7		re	
C_s symmetric ligands:							
25	Zr		R	3.7	2.1	re	
26	Zr		R	3.2		re	
27	Zr		R	2.3		re	
28	Zr		R	4.8		re	
C_1 symmetric ligands:							
29	Zr	(R)	$R \equiv$ inward	4.6		re	2.9
			$S \equiv$ outward	1.1		si	
30	Zr	(R)	$R \equiv$ inward	7.6		re	4.6
			$S \equiv$ outward	2.1		re	
31	Zr	(R)	$S \equiv$ inward	0.1		re	−0.9
			$R \equiv$ outward	3.8		re	
32	Zr	(R)	$S \equiv$ inward	0.3		re	−0.9
			$R \equiv$ outward	2.6		re	
33	Zr	(R)	$R \equiv$ inward	6.3		re	3.1
			$S \equiv$ outward	2.2		re	
34	Zr	(R)	$R \equiv$ inward	4.4		re	3.4
			$S \equiv$ outward	1.5		si	

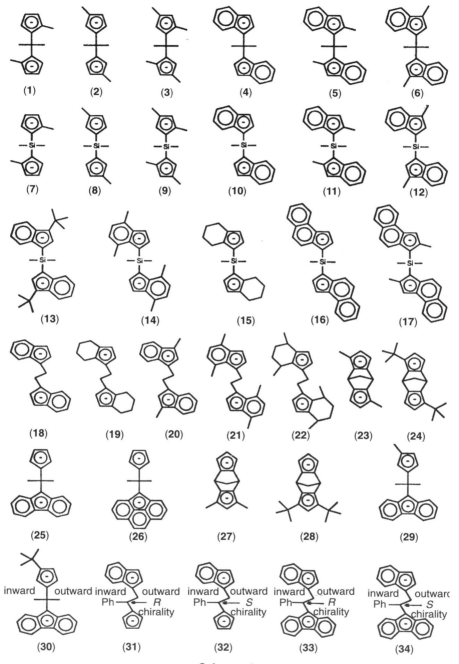

Scheme 1

enantioselectivities obtained by the two above-described methods ($\Delta E_{enant,p}$ and $\Delta E^{\neq}_{enant,p}$) present similar trends for different catalytic models.

The molecular modeling studies relative to both pre-insertion intermediates and pseudo-transition states indicate that, for all metallocenes reported in Scheme 1 (independently of their structure and symmetry), when a substantial enantioselectivity is calculated for primary monomer insertion, this is mainly due to non-bonded energy interactions of the methyl group of the chirally coordinated monomer with the chirally oriented growing chain.

According to analogous molecular mechanics analyses [25,69], this enantioselectivity mechanism would operate also for catalytic systems with oscillating stereocontrol, leading to atactic–isotactic stereoblock polymers [70,71], such as those based on two unbridged 2-phenylindenyl ligands [71].

This mechanism of enantioselectivity is in agreement with the available experimental data. In this respect, we recall that a relevant experimental proof of the remarkable role played by the growing chain in determining the steric course of the insertion reaction was given by Zambelli and co-workers by the observation of the stereospecificity in the first step of polymerization. In fact, ^{13}C NMR studies of the polymer end groups have shown that in the first step of polymerization, when the alkyl group bonded to the metal is a methyl group, the propene insertion is essentially non-enantioselective, whereas when the alkyl group is an isobutyl group the first insertion is enantioselective as the successive insertions. This holds for both heterogeneous [72] and homogeneous Ziegler–Natta catalysts [73]. The results of molecular mechanics calculations on the proposed metallocene models relative to possible initiation steps are in perfect agreement with these experimental findings [72,73], and also rationalize [74] the partial enantioselectivity of a catalytic system based on the ethylenebis(1-indenyl) ligand for 1-butene insertion at an $Mt-CH_3$ bond [73].

This enantioselective mechanism also accords with the elegant analysis and optical activity measurements by Pino et al. [75] on the saturated propene oligomers obtained with this kind of catalyst (under suitable conditions), proving that the re insertion of the monomer is favored in the case of (R,R) chirality of coordination of the ethylenebis(1-indenyl) ligand.

Moreover, similar studies on simultaneous deuteration and deuterooligomerization of 1-olefins using catalysts based on (R,R)-ethylenebis(4,5,6,7-tetrahydro-1-indenyl)zirconium derivatives have shown that in the dimerizations and oligomerizations of 1-olefins (propene, 1-pentene, 4-methylpentene) the R enantioface of the olefin is predominantly involved, whereas in the deuterations of 1-olefins (1-pentene, styrene) the S enantioface is favored [76]. These results confirm that the growing chain plays a primary role in the enantioface discrimination for the stereospecific polymerization catalyses. On the other hand, the presence of opposite enantioselectivities in deuteration and oligomerization has also been rationalized by molecular mechanics calculations analogous to those previously described, thus constituting a further valuable support to the proposed catalytic models and enantioselective mechanism [77].

More recent results by Gilchrist and Bercaw [78] relative to deuteration and deuterodimerization experiments on isotopically chiral 1-pentene are also in agreement with a mechanism involving a chiral orientation of the growing chain.

In the final part of this section, the possible origin of the lower enantioselectivity (ca 2 kcal/mol) [79,80] for the chain end-controlled catalytic systems based on metallocenes including two cyclopentadienyl rings is discussed. In this case, no chirality of coordination for the aromatic ligands or chirality at the central metal atom exists. The two possible diastereomeric pre-insertion intermediates for *re* and *si* coordinations of the monomer for the case of an *si* chain (that is, a growing chain in which the last monomeric unit has been obtained by a *cis*-addition of an *si* coordinated monomer molecule) suitable for isotactic and syndiotactic propagations, are shown in Figure 5(a) and (b), respectively.

According to our calculations, these models of pre-insertion intermediates are geometrically similar to those found for stereorigid metallocenes. In fact, not only do they present the methyl group of propene and the second carbon atom (and its substituents) of the growing chain on opposite sides with respect to the plane defined by the Zr−C bonds, but also θ_0, θ_1 and the other dihedral angles of growing chain are similar to those found for the stereorigid metallocenes such as those in Figures 1 and 2. However, in this case the two situations corresponding to $\theta_1 \approx +60°$ or $-60°$ [Figure 5(a) and (b), respectively] are energetically similar [65].

The chain end control of the enantioselectivity for these poorly isospecific systems could be related to easier insertion paths starting from models for isospecific propagation [Figure 5(a)]. In fact, according to molecular modeling studies relative to other significant points of the insertion paths, possibly closer to the transition

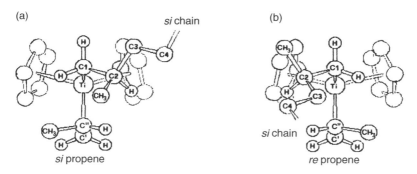

Figure 5 The two minimum energy diastereoisomeric pre-insertion intermediates for a model site comprising two cyclopentadienyl ligands, a propene molecule and a primary growing chain. The growing chain is labeled a *si*-chain, since the chirality of its tertiary carbon atom closest to the metal has been obtained by a primary insertion of an *si*-coordinated propene. For the model corresponding to isospecific propagation (a) the chain (atoms C3, C4,...) points away from the olefin, whereas for the model corresponding to syndiospecific propagation (b) it points towards the olefin

states, to obtain acceptable insertion paths starting from the intermediate for syndiotactic propagation [Figure 5(b)] the positions of a larger number of nuclei of the growing chain should be changed in the rate-determining step [65].

It is worth noting that the proposed pre-insertion intermediates, such as those in Figures 1, 2 and 5, are suitable for the formation of an α-agostic bond between the metal and the hydrogen atom on the opposite side with respect to the incoming monomer [81]. In fact, the transition state for the insertion step in non-chiral scandium-based [82] and chiral zirconium-based [83] catalysts has been found to be α-agostic stabilized.

According to our models, the formation of an α-agostic bond is adjuvant to but not necessary for the enantioselectivity of the monomer insertion. In fact, the conformation with $|\theta_1| \approx 60°$, which discriminates between re and si monomer enantiofaces, is often imposed by the chiral environment, even in the absence of a possible α-agostic bond. Moreover, as described in detail in Ref. 23, the observed enantioselectivity of the primary propene insertion into a secondary growing chain $[Mt-CH(CH_3)CH_2 \cdots]$, for the case of catalytic system based on the ligands **18** and **19**, has been rationalized on the basis of pre-insertion intermediates which do not involve α-agostic bonds.

4 STEREOSPECIFICITY

4.1 CHAIN MIGRATORY INSERTION MECHANISM

For a given catalytic model, the enantioselectivity of each insertion step does not assure its stereospecificity. In fact, the possible presence, in addition to the kind, of stereospecificity depends on possible differences between stereostructures of transition states of two successive insertion steps. These stereostructures are related to the best geometry of coordination of the growing chain at the site in the absence of the monomer molecule (as at the end of each polymerization step) and also to the stability of such geometry before the coordination of a new monomer molecule.

Already according to simple quantum mechanical considerations by Lauher and Hoffman [84] for d_0, d_1, d_3 or low-spin d_2 complexes of the kind MtCp$_2$L, the best coordination of L is not the one (most symmetrical and presumably sterically most favorable) along the symmetry axis which relates the two bent cyclopentadienyl rings. This is confirmed by more recent and complete calculations by Ziegler and co-workers relative to homogeneous $[Cp_2ZrCH_3]^+$ and $[CpSiH_2NHTiCH_3]^+$ systems [34]. According to these calculations, energy minimum situations would correspond to angular deviations $\alpha \approx 60°$ from the symmetry axis (see Scheme 2), and additional stabilization of geometries with $\alpha \neq 0°$ at the unsaturated metal center could occur because of a γ-agostic interaction [34], analogous to that postulated by Brookhart and Green [85]. It is worth noting that in the case the two cyclopenta-

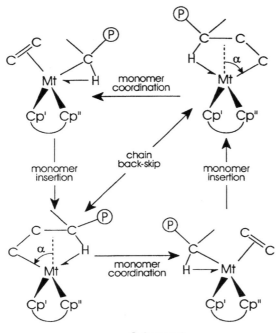

Scheme 2

dienyl ligands Cp′ and Cp″ in the cation $(MtCp'Cp''L)^+$ are different, a geometry with $\alpha \neq 0°$ makes the metal atom chiral.

On the basis of the chain migratory insertion mechanism [1–3,12,17] reported in Scheme 2, at the end of each polymerization step the growing chain occupies the coordination site previously occupied by the olefin monomer. Hence if the ligand rearrangement rate is much lower than the insertion rate of a new monomer molecule, many consecutive polymerization steps correspond to models obtained by exchanging the relative positions of the growing chain and of the incoming monomer. This is extremely relevant for the stereospecific behavior of the models since, if a chirality at the central metal atom [i.e. of type (ii)] exists, it is inverted in most polymerization steps, as shown in Scheme 2.

In the framework of the chain migratory insertion mechanism, the stereospecific behavior of the model sites depends on the relationship between the two situations obtained by exchanging, in the coordination step, the relative positions of the growing chain and of the incoming monomer.

Depending on the local symmetry of the π-coordinated bridged ligand, these two situations can be (i) identical, (ii) enantiomeric or (iii) diastereomeric, as follows.

(i) *Identical*, i.e. the two available coordination positions are homotopic [86]. These model sites, if the insertion step is enantioselective, are consequently isospecific. Model sites presenting a C_2 symmetry axis, which locally relates the atoms which are relevant to the non-bonded interactions with the monomer and the growing chain, were proposed by Allegra [4] many years ago for heterogeneous catalysis on $TiCl_3$.

Analogous models based on metallocenes with a C_2 symmetry axis relating the atoms of the stereorigid π ligand (such as **1–24** in Table 1) account for the isospecific behavior of corresponding metallocene-based catalytic systems [9,16,23]. In fact, the calculated enantioselectivity values compare well with the observed tacticities of polypropylene samples obtained by corresponding catalytic systems.

For instance, by considering **1–22** as bridged and substituted bis(cyclopentadienyl) ligands, substituents in position 2 (and 2′) are not sufficient to give enantioselectivity (**1** and **7**), substituents in position 3 (and 3′) give substantial enantioselectivities (**2–5**, **8–11**, **13–22**), while substituents of similar encumbrances in positions 3 and 4 (and 3′ and 4′) again produce poor enantioselectivities (**6** and **12**]). These computational results, qualitatively similar to those already presented in Ref. 87, agree well with the experimental findings relative to catalytic systems based on analogous C_2 symmetric π-ligands [13,88,89].

Of course, the enantioselectivity, and hence the isospecificity, of the catalytic models strongly depend on the encumbrance of the π-ligand, increasing along the series 3-methylcyclopentadienyl (e.g. 3.7 kcal/mol for **8**), indenyl (e.g. 4.9 kcal/mol for **10**), 4,7-dimethylindenyl (e.g. 5.3 kcal/mol^{-1} for **14**) and tetrahydroindenyl (e.g. 5.9 kcal/mol for **15**). This is in good qualitative agreement, for instance, with the percentage of mmmm pentads evaluated for polypropylene samples obtained for different catalytic systems in strictly similar conditions by Resconi and co-workers [90].

The calculation results listed in Table 1 are also able to account for the influence of the bridge on the isospecificity. In fact, in agreement with the experimental results (see, e.g., Ref. 90), the calculated enantioselectivities for a given π-ligand are generally smaller for the case of the $-C(CH_3)_2-$ (**2–5**) than for the $-Si(CH_3)_2-$ bridge (**8–11**), whereas for the case of the $-CH_2-CH_2-$ bridge the enantioselectivities are often intermediates (e.g. **4**, **10** and **18**).

It is worth noting that the calculations on models with a 3-*tert*-butyl-1-indenyl ligand (**13**) predict a substantial enantioselectivity but in favor of the opposite monomer enantioface [*si* and *re* for (R,R) and (S,S) coordinated π-ligand, respectively].

(ii) *Enantiomeric*, i.e. the two available coordinate positions are enantiotopic [86]. These model sites, if the insertion step is enantioselective, are consequently syndiospecific. Model sites based on metallocenes with a C_s symmetry plane relating the atoms of the stereorigid π-ligand (such as **25–28** in Table 1) [17,91] account for the syndiospecific behavior of corresponding metallocene-based catalytic systems [12,13,92].

It is worth noting that the lower syndiospecificity of catalytic systems based on **26**, with respect to those based on **25** [13], is accounted for by these calculations. This is easily rationalized in the framework of the enantioselective mechanism which imposes on the growing chain (both in the pre-insertion intermediate and in the transition state) a chiral orientation towards the cyclopentadienyl ligand and favors the insertion of the propene enantioface presenting the methyl group closer to the bulkier π-ligand [as shown in Figures 2 and 4(b), respectively]. In fact, the two additional aromatic carbon atoms which are present in **26**, with respect to **25**, generate direct interactions with the propene coordinated with the right enantioface, thus reducing the enantioselectivity.

(iii) *Diastereomeric*, i.e. the two available coordination positions are diastereotopic [86]. These model sites would be prevailingly isospecific (or syndiospecific) if these two situations are enantioselective in favor of the same (or opposite) monomer enantioface, and tendentially hemi-isospecific [93] if only one of the two situations is enantioselective.

The calculated enantioselectivities relative to the C_1 symmetric models of Table 1 qualitatively agree with the stereospecific behavior of corresponding catalytic systems. For instance, the calculations relative to the model complexes with ligands **30** and **29** account well for the isospecificity and hemi-isospecificity of corresponding catalytic systems [13,94]. The calculations are also able to account qualitatively for the tacticities of catalytic system based on ligands differing only for the chirality at the tertiary carbon atom of the bridge [95]. As described in detail in Ref. 24, this chirality imposes the chirality of the ethylene bridge conformation (λ or δ for S or R, respectively) which, for crowded complexes (such as those based on **33** and **34**), has a substantial influence on the stereospecific behavior of propene polymerization. In fact, both models **33** and **34** are enantioselective for the inward monomer coordination, but present a reduced enantioselectivity for the propene outward coordination. However, this enantioselectivity is in favor of the same or of the opposite monomer enantioface, respectively. This is in good qualitative agreement with measured tacticities of corresponding catalytic systems (46% vs 5% of mmmm pentads, respectively) [95]. According to the calculations, the influence of the bridge conformation chirality on the enantioselectivity would also be mainly mediated by the chiral orientation of the growing chain [24].

For non-crowded complexes, the influence of the ethylene bridge conformation (and hence of the chirality of the tertiary carbon atom of the bridge) on the stereospecific behavior is poor (cf. calculations for ligands **31** and **32** in Table 1). This is in qualitative agreement with measured tacticities of corresponding catalytic systems (39% vs 34% of mmmm pentads, respectively) [95].

In summary, characterizations of stereosequences in polymers obtained by catalytic systems based on well characterized metallocene complexes have produced a general acceptance of the chain migratory insertion mechanism and of models of type (i), (ii) and (iii).

4.2 POSSIBLE BACK-SKIP OF THE GROWING CHAIN

Several experimental facts, relative to the behavior of different metallocene-based catalytic systems in propene polymerization, can be rationalized by considering a disturbance of the chain migratory insertion mechanism. This perturbation can be essentially ascribed to a kinetic competition between the monomer coordination in the olefin-free state and a back-skip of the growing chain to the other possible coordination position (see Schemes 2 and 3).

   The possible occurrence of back-skip of the chain for catalytic systems based on C_2 symmetric metallocenes would not change the chirality of the transition state for the monomer insertion and hence would not influence the corresponding polymer stereostructure.

   On the contrary, this phenomenon would invert the chirality of the transition state for the monomer insertion for catalytic systems based on C_s symmetric metallo-

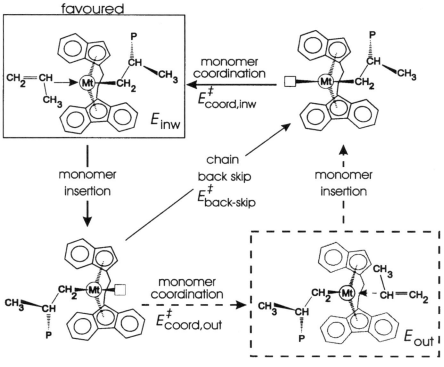

Scheme 3

cenes. In fact, it has been invoked to rationalize typical stereochemical defects (isolated m diads) in syndiotactic polypropylenes [13,17,92]. This mechanism of stereoerrors formation has been confirmed by their increase in polymerization runs conducted with reduced monomer concentrations [96]. In fact, it is reasonable to expect an increase in the frequency of the back-skip of the chain by reducing the monomer concentration and hence the frequency of monomer insertion.

For catalytic systems based on C_1 symmetric metallocenes, a driving force for the occurrence of the back-skip phenomenon, related to the energy difference between the two diastereomeric situations obtained by exchanging the relative positions of monomer and chain, has to be considered.

The energy differences between minima corresponding to diastereomeric pre-insertion intermediates with different chiralities at the central metal atom $(E_{out} - E_{inw})$ for several catalytic models with C_1 symmetric metallocenes are listed in the last column of Table 1. It is worth noting that, when substantial energy differences between minimum energy diastereomeric intermediates are present, lower energies correspond to the monomer coordination in the (more crowded) inward coordination position [24].

Of course, the probability of occurrence of a back-skip of the chain, in the olefin-free state, is only indirectly dependent on this energy difference. In fact, it is dependent on the difference between the activation energy for the chain back-skip [$E_{back-skip}^{\neq}$ in Scheme 3] and the activation energy for the formation of the high energy olefin-bound intermediate [$E_{coord,out}^{\neq}$ in Scheme 3]. However, since the degree to which empirical force fields can be used for prediction of transition states is not well established and since the activation energy $E_{coord,out}^{\neq}$ is expected to increase with increasing E_{out}, for the sake of simplicity we take $E_{out} - E_{inw}$ as a semi-quantitative evaluation of the driving force for the back-skip of the chain [24].

In particular, it is reasonable to expect that, for the models with large $E_{out} - E_{inw}$ values (**29**, **30**, **33** and **34**), the growing chain in successive coordination steps can occupy frequently the (less crowded) outward coordination position, leaving the inward position free for the monomer coordination. For all these models, the lower energy diastereomer with inward monomer coordination is more enantioselective than the higher energy diastereomer with outward monomer coordination.

The results in Table 1 are in good qualitative agreement with several experimental results. In particular, the appreciable probability of two successive additions at the enantioselective catalytic face and the near zero probability of two successive additions at the non-enantioselective catalytic face, observed for the essentially hemi-isospecific catalytic system based on **29** [13,92], are rationalized in terms of the lower energies for the single minimum of the enantioselective steps with respect to the two minima of the non-enantioselective steps [24]. The increased stereo-regularity at decreasing monomer concentration (i.e. increasing the frequency of the back-skip of the chain) for catalytic systems based on **33** and **34** (not shown by systems based on **31** and **32**) [95] can be explained on similar grounds [24].

5 ENANTIOSELECTIVITIES OF REGIOIRREGULAR INSERTIONS

Molecular mechanics analyses of the kind described in previous sections are able to rationalize not only the enantioselectivity (and stereospecificity) of regioregular primary insertion steps but also the enantioselectivities relative to occasional secondary monomer insertions and relative to primary insertions following these secondary insertions [23].

For instance, Figure 6(a) plots as a function of θ_0 the optimized energies for the model complex **4**, which is a π-ligand with C_2 symmetry, with (R,R) chirality of coordination of the bridged π-ligand. Let us recall that, in our framework, energy minima for $\theta_0 \approx 0°$ or $180°$ can correspond to pre-insertion intermediates suitable for primary and secondary monomer insertions, respectively.

The energy minima labeled a, b and c in Figure 6(a), corresponding to $\theta_0 \approx 0°$, are coincident with those labeled with the same letters in the energy plot in Figure 3(a). As already cited, the absolute minimum energy (labeled a) corresponds to the pre-insertion intermediate for propene primary insertion with *re* enantioface, which is sketched in Figure 1.

As for situations possibly suitable for secondary propene insertion ($\theta_0 \approx 180°$), the optimized energy for the *si* monomer coordination, labeled **e** in Figure 6(a), is higher by nearly 2 kcal/mol with respect to the absolute minimum. The corresponding model sketched in Figure 7(a) is considered suitable for *si*-monomer secondary insertion. Much higher is the energy corresponding to $\theta_0 = 180°$ for the *re* monomer coordination [situation labeled d in Figure 6(a)] owing to the repulsive interactions of the methyl group of propene with the six-membered rings of one of the indenyl ligands.

In summary, there is a substantial enantioselectivity of this isospecific C_2 symmetric catalytic model for the lower energy (and experimentally observed) primary monomer insertion and the enantioselectivity would also be higher for the higher energy (experimentally detected) secondary monomer insertion. Anyhow, it is worth noting that the enantioselectivity of the isospecific model site is in favor of opposite monomer prochiral faces, for primary and secondary insertions [23].

This result is in perfect agreement with the observed microstructure of polypropylene chains obtained by isospecific catalytic systems including the aforementioned and also analogous bridged π-ligands [93,94]. It is also worth noting that the non-bonded interactions generating enantioselectivity are those between the methyl substituent of the coordinated propene and the chiral oriented growing chain, in the case of primary insertion, and between the propene methyl group and one of the six-membered rings of the π-ligand, in the case of secondary insertion [23].

For the case of catalytic model sites based on metallocenes **18**, a detailed molecular mechanics analysis has been conducted also for the case of primary or secondary propene insertions on secondary polypropylene chains (for which the last propene insertion has been secondary) [23]. According to this analysis, the enantioselectivities observed after an occasional secondary monomer insertion are

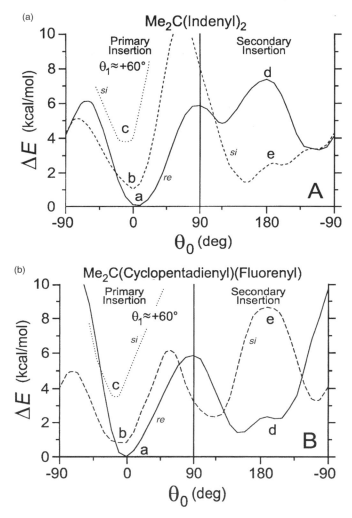

Figure 6 The optimized energies plotted as a function of θ_0 for the model complexes comprising the (a) (R,R)-coordinated isopropylbis(1-indenyl) ligand and (b) isopropyl(cyclopentadienyl-9-fluorenyl) ligand for the R chirality at the metal atom. The full and dashed lines refer to re- and si-coordinated propene, respectively. The dotted lines are parts of the optimized energy curves obtained by requiring, for the si-coordinated monomer, that the methyl group of propene and the second carbon atom (and its substituents) of the growing chain are on opposite sides with respect to the plane defined by the Zr−C bonds (i.e. requiring $\theta_1 \approx +60°$). The minimum energy pre-insertion intermediates for primary insertion ($\theta_0 \approx 0°$), labeled a in parts (a) and (b), are sketched in Figures 1 and 2(a), respectively. The minimum energy pre-insertion intermediates for secondary insertion ($\theta_0 \approx 180°$), labeled e and d in parts (a) and (b), are sketched in Figure 7(a) and (b), respectively

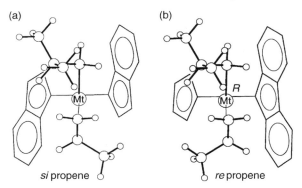

Figure 7 Pre-insertion intermediates for secondary propene insertion into a primary polypropene chain for the case of (a) isospecific model complex based on the (*R,R*)-coordinated isopropylbis(1-indenyl) ligand and (b) syndiospecific model complex based on the isopropyl(cyclopentadienyl-9-fluorenyl) ligand for the *R* chirality at the metal atom. The enantioselectivity of the isospecific model site is in favor of opposite monomer prochiral faces, for primary and secondary insertions [compare Figure 1 with Figure 7(a)]. The enantioselectivity of the syndiospecific model site is in favor of the same monomer prochiral face for primary and secondary insertions [compare Figure 2(a) with Figure 7(b)]

easily accounted for in the framework of the mechanism of the 'chiral orientation of the growing chain'. In fact, substituting the usual primary growing chain with a secondary growing chain reduces the bulkiness of the substituents of the second carbon atom of the chain (which is secondary carbon for the secondary chain, but is tertiary carbon for the primary chain). Correspondingly, the energy differences between conformations with positive and negative values of θ_1 (which determines the enantioselectivity, in the framework of our model; see above) is reduced, leading to a less pronounced enantioselectivity [23]. These results are able to rationalize the probability distributions of stereochemical configurations of regioirregular units in isotactic polymer samples prepared in the presence of corresponding catalytic systems [97,98].

As will be discussed in the following section, syndiospecific catalytic systems based on metallocenes are highly regioregular. As a consequence, their enantio-selectivity in possible regioirregular insertions has not been experimentally inves-tigated. However, an analysis of the enantioselectivity of possible secondary propene insertions on syndiospecific catalytic models based on C_s symmetric metallocenes is reported here, owing to its relevance for the rationalization of the dependence of regiospecificity on the type of stereospecificity (see Section 6.1) [27].

As an example, Figure 6(b) plots as a function of θ_0 the optimized energies for the model complex **25**, which is a π-ligand with C_s symmetry, for the polymerization step involving an R chirality at the metal atom.

The energy minima labeled a, b and c in Figure 6(b), corresponding to $\theta_0 \approx 0°$, are coincident with those labeled with the same letters in the energy plot in Figure

3(b). The absolute minimum energy (labeled a) corresponds to the pre-insertion intermediate for propene primary insertion with *re* enantioface which is sketched in Figure 2(a).

As for situations possibly suitable for secondary propene insertion ($\theta_0 \approx 180°$), the optimized energy for the *re* monomer coordination, labeled d in Figure 6(b), is higher by nearly 2 kcal/mol with respect to the absolute minimum. The corresponding model sketched in Figure 7(b) is considered suitable for *re*-monomer secondary insertion. Much higher is the energy corresponding to $\theta_0 = 180°$ for the *si* monomer coordination [situation labeled e in Figure 6(b)] owing to repulsive interactions of the methyl group of propene with a six-membered ring of the fluorenyl ligand.

In summary, there is a substantial enantioselectivity of this syndiospecific C_s symmetric catalytic model for the lower energy (and experimentally observed) primary monomer insertion and also for the higher energy (experimentally undetected) secondary monomer insertion. Anyhow, it is worth noting that the enantioselectivity of the syndiospecific model site is in favor of the same monomer prochiral face, for primary and secondary insertions (contrary to the isospecific model sites) [27].

6 REGIOSPECIFICITY

The regiospecificity of metallocene-based catalytic systems is generally very high. In fact, regioirregularities cannot be detected by standard NMR characterization methods in most of the polypropylene samples obtained by these catalytic systems.

In particular, highly regioregular polymer samples are obtained by catalytic systems based on titanocenes and also by syndiospecific or aspecific catalytic systems based on zirconocenes and hafnocenes [12,27,68,75]. In contrast, isotactic polypropylene samples from catalytic systems based on zirconocenes and hafnocenes contain isolated secondary (2,1 insertions, usually 0–2%) propene units and isolated 3,1 propene units (arising from unimolecular isomerization of 2,1 units, 0–5%) in the isotactic sequences of primary propene insertions [97–105]. Regioirregularities have been observed not only for isospecific catalytic systems based on C_2-symmetric π-ligands but also for isospecific catalytic systems based on C_1-symmetric π-ligands, such as **30** in Scheme 1 [106].

For isospecific catalysts based on *ansa*-zirconocenes and hafnocenes, regioirregularities are reduced when the C_2 symmetric ligand contains at each cyclopentadienyl ligand, in addition to the enantioselectivity determining substituents in positions 4 (β), alkyl groups in position 2 (α) [27,107,108]. Moreover, regioirregularities are substantially absent for catalytic systems based on C_2 symmetric ligands with a *tert*-butyl group in position 3 (β') [109]. On the other hand, for isospecific catalysts based on *ansa*-zirconocenes, regioirregularities increase when the indenyl ligands are substituted in both positions 4 and 7 [110]. The last effect is particularly strong (leading to amounts of regioirregularities larger than 20%) for the case of the catalytic system based on $C_2H_4(4,7\text{-Me}_2\text{-H}_4\text{Ind})_2ZrCl_2$ [105,111]. Some

Table 2 Calculated non-bonded energy contributions (kcal/mol) to the regiospecificity [111] for pre-insertion intermediates (ΔE_{regio}), and for pseudo-transition states ($\Delta E_{\text{regio}}^{\neq}$) compared with observed percentages of regioirregularities

Ligand	Alkyl substitution	Regioirregularities (%)	Ref.	ΔE_{regio}	$\Delta E_{\text{regio}}^{\neq}$
10	None	0.4_8	109	2.0	2.4
11	2-Methyl	0.3_3	27	3.0	3.1
14	4,7-Dimethyl	1.8_4	27	1.2	1.5
13	3-*tert*-butyl	<0.1	108	4.1	4.0

data relative to catalytic systems based on the same bisindenyl skeleton (**10**) are collected in Table 2 to show the dependence of the overall amount of regioirregularities on alkyl substitution in different positions.

Molecular mechanics calculations similar to those described in the previous sections can be used to evaluate energy differences between catalytic models (pre-insertion intermediates and pseudo-transition states) suitable for primary and secondary insertions. In the framework of the assumed mechanism, this energy difference gives a rough estimate of the non-bonded energy contribution to the regiospecificity of the insertion reaction.

Although the main contribution to the energy differences determining the regiospecificity could be electronic in most cases, the analysis of the non-bonded energy contribution has allowed us to rationalize the dependence of regiospecificity on some of the above-cited factors, i.e. type of stereospecificity and the nature and position of the π-ligand alkyl substituents.

6.1 TYPE OF STEREOSPECIFICITY

As reported previously, whereas syndiospecific and aspecific zirconocene based catalytic systems are highly regiospecific, isospecific systems produce substantial amounts of regioirregular monomeric units, independently of the nature of the π-ligands and of the bridge between them (unless suitable substitutions of the ligands are involved). This dependence of the degree of regiospecificity on the symmetry rather than on the nature of the π-ligand is, of course, not easy to rationalize by invoking differences in the electronic contributions to the regiospecificity.

On the other hand, the results discussed in the previous section indicate that the non-bonded contribution to the energy differences between secondary and primary pre-insertion intermediates, for zirconocene-based catalytic models, is poorly dependent on the symmetry of the π-ligands and hence on their stereospecificity ($E_d - E_a$ and $E_e - E_a \approx 2$ kcal/mol in Figure 6(a) and (b), respectively). A similar energy difference has been also calculated for the analogous C_{2v} symmetric model for aspecific propene polymerization (based on two bridged cyclopentadienyl ligands) [27].

However, for the enantioselective model complexes (syndio- and isospecific), the energy difference between secondary and primary pre-insertion intermediates, for a given chirality of coordination of the monomer, changes considerably with the symmetry of the π-ligands. In particular, for the syndiospecific (C_s symmetric) model complexes the non-bonded energy contribution to the regiospecificity is large for the enantioface which is wrong for the primary insertion [$E_e - E_b \approx 8 \, \text{kcal/mol}$ in Figure 6(a)] whereas, for isospecific (both C_2 and C_1 symmetric) model complexes this contribution is large for the right enantioface [$E_d - E_a \approx 8 \, \text{kcal/mol}$ in Figure 6(b)].

For generic aspecific, syndiospecific and isospecific model complexes, schematic plots of the internal energy versus the reaction coordinate, for both primary and secondary insertions, are sketched in Figure 8(a), (b) and (c), respectively. The minima at the centers and at the ends of the energy curves correspond to olefin-free intermediates including a growing chain with n and $n + 1$ monomeric units, respectively. Movements from the central minima towards the left and the right correspond to possible reaction pathways leading to primary and secondary (subscripts p and s in the following) insertions, respectively. For the enantioselective complexes, the reaction pathways for monomer enantiofaces which are right and wrong (subscripts r and w in the following) for primary insertion are different, and are indicated by full and dashed lines, respectively. For each pathway, the two energy barriers correspond to the coordination and insertion steps.

The energy minima between the energy barriers for the monomer coordination and insertion correspond to olefin-bound intermediates of the kind simulated by our molecular mechanics calculations (Figures 3 and 6). The possible dissociation of the monomer coordinated with the wrong enantioface can lead back to the olefin-free intermediate or, directly, to the olefin-bound intermediate with the right enantioface (through some isomerization mechanism, for which the monomer does not leave the coordination sphere of the metal).

The pre-insertion intermediates of our molecular mechanics analysis (when different from the coordination intermediates) would correspond to situations closer to the transition state for the insertion reactions. For an easier comparison, the labels (a–e) used for coordination and pre-insertion intermediates in Figures 6(a) and (b) are also reported close to the schematic energy plots in Figure 8(b) and (c), respectively.

For the sake of simplicity, minimum energy pathways that according to our calculations are expected to be similar are assumed to be identical, independently of the symmetry (stereospecificity) of the catalyst. However, the plots for the syndiospecific [Figure 8(b)] and isospecific [Figure 8(c)] models are different since, as discussed previously, the enantioselectivities for primary and secondary insertions are in favor of the same or opposite monomer enantiofaces, respectively.

In this framework, the lower regiospecificity of isospecific catalytic systems can be rationalized by assuming that the activation energy for the rotation of the

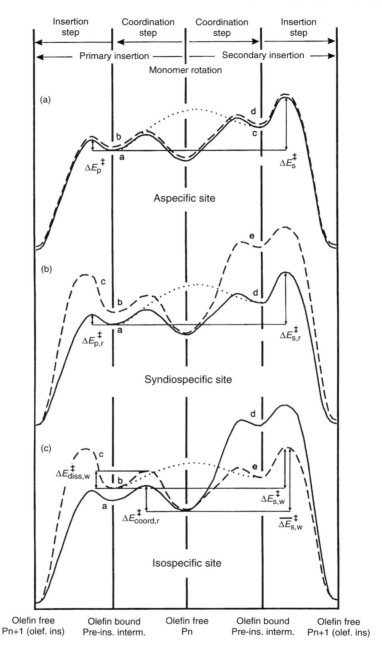

coordinated monomer around θ_0 between the orientations suitable for the primary and secondary insertions (schematically shown by dotted lines in Figure 8) is in general lower than (or comparable to) the activation energy for the secondary insertion [27].

For syndiospecific model complexes, since their enantioselectivity is in favor of the same monomer enantioface for both primary and secondary insertions, when coordination of the monomer with the wrong enantioface for the primary insertion occurs [situation b in Figures 6(b) and 8(b)], the most probable event is the dissociation of the coordinated monomer. With low probability, also possible is the primary insertion of the wrong enantioface which introduces a stereoirregularity in the polymer chain. Secondary insertions are expected to be essentially absent [see the high energy of the situation e in Figures 6(b) and 8(b)]. With the assumption of a low energy barrier for the monomer rotation around θ_0, the regioselectivity would be simply determined by the differences between the activation energies for the secondary and primary insertions of the more suitable enantioface ($\Delta E_{s,r}^{\ddagger} - \Delta E_{p,r}^{\ddagger}$) (and independent of the energy barrier for the monomer coordination). Moreover, the regiospecificity is expected to be high and similar to that of the corresponding aspecific catalytic complex [27].

For isospecific model complexes, since their enantioselectivity is in favor of opposite monomer enantiofaces for primary and secondary insertions, when the coordination of the monomer with the enantioface unsuitable for the primary insertion occurs [situation b in Figures 6(a) and 8(c)] in addition to the dissociation of the coordinated monomer and to a low-probability primary insertion (generating the stereoirregularities), a low-probability secondary insertion (generating the regio-

---

Figure 8 Schematic plots of the internal energy versus the reaction coordinate, for both primary and secondary insertions, for generic (a) aspecific, (b) syndiospecific and (c) isospecific model complexes. The minima at the centers and at the ends of the energy curves correspond to olefin-free catalytic intermediates including a growing chain with n and $n + 1$ monomeric units, respectively. Movements from the central minima towards the left and the right correspond to possible reaction pathways leading to primary and secondary insertions, respectively. For the enantioselective complexes [(b), (c)] the reaction pathways for monomer enantiofaces being right and wrong for primary insertion are different, and are indicted by full and dashed lines, respectively. The two energy barriers encountered for each pathway correspond to the coordination and insertion steps. The energy minima between the energy barriers for the monomer coordination and insertion correspond to olefin-bound catalytic intermediates of the kind simulated by our molecular mechanics calculations (e.g. Figures 3 and 6). In particular, the labels (a)–(e) close to the curves of parts (b) and (c) correspond to the coordination and pre-insertion intermediates of Figure 6(b) and (a), respectively. The dotted lines indicate the rotation of the coordinated monomer around θ_0 whose activation energy is assumed to be lower than (or comparable to) the activation energy for the secondary insertion. The activation energies, which in this framework are relevant to the regiospecificity, are also indicated

(a)

(b)

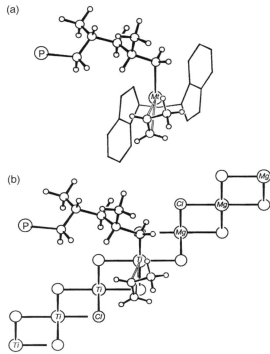

Figure 9 Minimum energy pre-insertion intermediates suitable for primary propene insertion for isospecific models (a) of homogeneous catalytic systems based on the (*R,R*)-coordinated isopropylbis(1-indenyl) ligand and (b) of heterogeneous catalytic systems based on Ti chloride supported on $MgCl_2$; the catalytic model site is located on a lateral termination of a structural layer and presents a Δ chirality. In both models the methyl groups of propene and the second carbon atom (and its substituents) of the growing chain are on opposite sides with respect to the plane defined by the Zr–C bonds (i.e. $\theta_1 \approx -60°$ and *re* monomer)

irregularities) would also be possible. This is due to the fact that the barrier for the dissociation of the coordinated monomer is expected to be non-negligible with respect to the activation energy for the secondary insertion. Hence for these isospecific model complexes, the amount of regioirregularities in the polymer chains would be not determined (as for the cases of aspecific and syndiospecific model complexes) by the differences between the activation energies for the secondary and primary insertions. Instead, it would be related to the difference between the activation energies for the dissociation of the monomer (coordinated with the wrong enantioface) and the activation energy for its secondary insertion ($\Delta E_{s,w}^{\neq} - \Delta E_{diss,w}^{\neq}$).

6.2 LIGAND SUBSTITUENTS

The calculated non-bonded energy contributions to the regiospecificity for pre-insertion intermediates and pseudo-transition states (ΔE_{regio} and $\Delta E^{\neq}_{regio}$, respectively) for *ansa*-zirconocenes based on the same bisindenyl skeleton (**10**), unsubstituted or alkyl-substituted in different positions, are reported, for instance, in Table 2 and compared with the overall amount of regioirregular insertions experimentally observed for corresponding catalytic systems.

First, it is worth noting that the calculated non-bonded energy contributions to the regioselectivities for pre-insertion intermediates and for pseudo-transition states (ΔE_{regio} and $\Delta E^{\neq}_{regio}$) are generally similar.

Results such as those shown in Table 2 indicate that non-bonded energy interactions are able to account for the increases and decreases of regiospecificity experimentally observed after methyl substitution in position 2 and dimethyl substitutions in positions 4 and 7, respectively. In particular, strong interactions of the methyl group of the coordinated propene occur with the 2-methyl substituent in models suitable for secondary insertion whereas, they occur with the 4-methyl substituent in models suitable for primary insertion [112].

The calculated non-bonded energy contribution to the regiospecificity for the model based on ligand **10** strongly increases after 3-butyl substitution (model **13**). This is due to non-bonded interactions between the methyl group of the coordinated propene and methyl groups of the *tert*-butyl substituent, which are stronger in models suitable for secondary insertion than in models suitable for primary insertion. This accounts for the very high regiospecificity of corresponding iso-specific catalytic systems, which is comparable to those observed for syndiospecific and aspecific *ansa*-zirconocene-based catalytic systems [109].

7 CONCLUSION

Through the examination of models for homogeneous Ziegler–Natta polymerization of olefins, we have seen that a general pattern in the similarity of behaviors seems to emerge for the enantioselectivity mechanisms, indicating the importance of the non-bonded interactions. For all the different models proposed for the considered catalytic systems, the chiral orientation of the first $C-C$ bond of the growing chain, imposed by the chirality of the site, seems to be crucial in determining the enantioselectivity. This mechanism of enantioselectivity, which was also proposed long ago for models for the heterogeneous Ziegler–Natta catalytic, agrees well with a large number of experimental facts, some of which result from experiments designed to test its validity.

Models corresponding to calculated minimum energy pre-insertion intermediates suitable for primary propene insertion for a metallocene catalyst (based on **4**) and for a catalytic site located on a lateral termination of a structural layer (TiCl$_3$ supported on $MgCl_2$ are compared in Figure 9(a) and (b), respectively. The similarity in the positions and relative orientations of growing chain and incoming monomer is apparent.

Non-bonded energy interactions are also able to rationalize not only the stereo-specificities observed for different metallocene-based catalytic systems (isospecific, syndiospecific, hemi-isospecific) but also the origin of particular stereodefects and their dependence on monomer concentration as well as stereostructures associated with regioirregular insertions.

Non-bonded energy analysis has allowed to also rationalize the dependence of regiospecificity on the type of stereospecificity and also on nature and position of alkyl substituents of the π-ligands, although the main contribution to the regio-specificity could be electronic in most cases.

This final section reports also some example, mainly related to research conducted in our group, of other contributions of the molecular mechanics to the comprehen-sion of various aspects of the mechanisms of the Ziegler–Natta catalysis.

Simple molecular mechanics calculations have been able to rationalize the occurrence of different kinds of chain terminations for some homogeneous Zieg-ler–Natta catalysts. In particular, the occurrence of β-hydrogen abstraction for the complex with two Cp rings and of β-methyl abstraction for the complex with two Cp* rings [113,114] have been qualitatively explained by analysis of the non-bonded interactions predominant in the two cases [114]. Moreover, the evaluation of non-bonded energy interactions on model catalytic sites based on benz[e]indenyl and 2-methylbenz[e]indenyl [115] has allowed us to account for the experimental finding, for corresponding catalytic systems, that the β-hydrogen transfer to the monomer is inhibited by the presence of 2-methyl substituents [108,116].

Molecular mechanics techniques have been also used by us to elucidate the origin of the diastereoselectivity observed for the cyclization step in the cyclopolymeriza-tion of a non-conjugated diene (1,5-hexadiene) with homogeneous catalysts based on zirconocene/methylalumoxane systems [117], assuming model catalytic sites analogous to those proposed for polymerization of α-olefins. In particular, model sites with two Cp rings or with two Cp* rings have been compared and the opposite diastereoselectivities of the two corresponding systems [that is, the achievement of poly(methylene-1,3-cyclopentane) prevailingly *trans* or *cis*, respectively] have been rationalized [118].

Recently, molecular modeling studies have also made some contribution to the comprehension of the mechanism of the homogeneous stereospecific polymerization catalysis of conjugated dienes [119].

Further contributions and developments of molecular modeling are expected for the future. A more precise knowledge of the transition states relative to the insertion and chain termination reactions not only could allow for more quantitative predic-tions relative to stereospecific and regiospecific behaviors, but also could lead to semi-quantitative predictions of average molecular masses and of relative amounts of chemically different chain terminations. Molecular modeling studies can also make relevant contributions to the rationalization of the large variations of the relative reactivities of 1-olefins for copolymerizations with different metallocene-based catalytic systems. So far, only a molecular mechanics study of the relative reactivities of ethene and propene in heterogeneous $TiCl_3$ based catalysts has been published [120]. Moreover, several aspects relative to the stereospecificity of polymerization of bulky olefins with metallocene-based catalytic systems remain to be clarified. Let us cite, for instance, the unexpected isotacticity of the polymer samples obtained in the polymerization of C_3-branched 1-olefins by (typically syndiospecific) catalytic systems based on C_s symmetric ligands [121,122] (see also Chapter 3).

The last decade has been an exciting period in the field of Ziegler–Natta polymerizations, and very elegant experimental and theoretical studies have allowed a deep comprehension of homogeneous catalytic systems. It is highly desirable that molecular modeling will be used to shed light also on the much less defined, but industrially relevant, heterogeneous catalytic systems [123].

8 ACKNOWLEDGMENTS

We are grateful to our co-workers Dr G. Moscardi and Dr M. Toto. We thank Professor V. Busico and Professor M. Vacatello of the University of Naples and Dr L. Resconi of Montell Polyolefins for useful discussions. The financial support of the Ministero Università e della Ricerca Scientifica e Tecnologica of Italy, of the Consiglio Nazionale delle Ricerche and of Montell Polyolefins is gratefully acknowledged.

9 REFERENCES

1. Cossee, P., *Tetrahedron Lett.*, **17**, 12 (1960).
2. Cossee, P., *J. Catal.*, **3**, 80 (1964).
3. Cossee, P., in Ketley, A. D. (Ed.), *The Stereochemistry of Macromolecules*, Marcel Dekker, New York, 1964, Vol. 1, Chapt. 3.
4. Allegra, G., *Makromol. Chem.*, **145**, 235 (1971).
5. Corradini, P., Barone, V., Fusco, R. and Guerra, G., *Eur. Polym. J.*, **15**, 133 (1979).
6. Corradini, P., Barone, V., Fusco, R. and Guerra, G., *J. Catal.*, **77**, 32 (1982).
7. Corradini, P., Barone, V. and Guerra, G., *Macromolecules*, **15**, 1242 (1982).
8. Corradini, P., Barone, V., Fusco, R. and Guerra, G., *Gazz. Chim. Ital.*, **113**, 601 (1983).
9. Ewen, J. A., *J. Am. Chem. Soc.*, **106**, 6355 (1984).
10. Ewen, J. A., in Keii, T. and Soga, K. (Eds), *Catalytic Polymerization of Olefins. Studies in Surface Science and Catalysis*, Vol. 25, Elsevier, New York, 1986, p. 271.
11. Ewen, J. A., Haspeslagh, L., Atwood, J. and Zhang, H., *J. Am. Chem. Soc.*, **109**, 6544 (1987).
12. Ewen, J. A., Jones, R. L., Razavi, A. and Ferrara, J. D., *J. Am. Chem. Soc.*, **110**, 6255 (1988).

13. Ewen, J. A., Elder, M. J., Jones, R. L., Haspeslagh, L., Atwood, J. L., Batt, S. G. and Robinson, K., *Makromol. Chem. Symp.*, **48/49**, 253 (1991).
14. Kaminsky, W., Külper, K., Brintzinger, H. H. and Wild, F., *Angew. Chem., Int. Ed. Engl.*, **24**, 507 (1985).
15. Kaminsky, W., *Angew. Makromol. Chem.*, **145/146**, 149 (1986).
16. Corradini, P., Guerra, G., Vacatello, M. and Villani, V., *Gazz. Chim. Ital.*, **118**, 173 (1988).
17. Cavallo, L., Corradini, P., Guerra, G. and Vacatello, M., *Macromolecules*, **24**, 1784 (1991).
18. Corradini, P. and Guerra, G., *Prog. Polym. Sci.*, **16**, 239 (1991).
19. Hortmann, K. and Brintzinger, H. H., *New J. Chem.*, **16**, 51 (1992).
20. Castonguay, L. and Rappé, *J. Am. Chem. Soc.*, **114**, 5832 (1992).
21. Kuribayashi, H. K., Koga, N. and Morokuma, K., *J. Am. Chem. Soc.*, **114**, 8687 (1992).
22. Hart, J. and Rappé, A., *J. Am. Chem. Soc.*, **115**, 6159 (1993).
23. Guerra, G., Cavallo, L., Moscardi, G., Vacatello, M. and Corradini, P., *J. Am. Chem. Soc.*, **116**, 2988 (1994).
24. Guerra, G., Cavallo, L., Moscardi, G., Vacatello, M. and Corradini, P., *Macromolecules*, **29**, 4834 (1996).
25. Pietsch, M. A. and Rappé, A., *J. Am. Chem. Soc.*, **118**, 10908 (1996).
26. van der Leek, Y., Angermund, K., Reffke, M., Kleinschmidt, R., Goretzki, R. and Fink, G., *Chem. Eur. J.*, **3**, 585 (1997).
27. Guerra, G., Longo, P., Cavallo, L., Corradini, P. and Resconi, L., *J. Am. Chem. Soc.*, **119**, 4394 (1997).
28. Jolly, C. A. and Marynick, D. S., *J. Am. Chem. Soc.*, **111**, 7968 (1989).
29. Koga, N. and Morokuma, K., *J. Phys. Chem.*, **94**, 5454 (1990).
30. Kuribayashi, H. K., Koga, N. and Morokuma, K., *J. Am. Chem. Soc.*, **114**, 2359 (1992).
31. Blomberg, M. R. A., Siegbahn, P. E. M. and Svensson, M., *J. Phys. Chem.*, **96**, 9794 (1992).
32. Prosenc, M., Janiak, C. and Brintzinger, H. H., *Organometallics*, **11**, 4036 (1992).
33. Mohr, R., Berke, H. and Erker, G., *Helv. Chim. Acta*, **76**, 1389 (1993).
34. Woo, T., Fan, L. and Ziegler, T., *Organometallics*, **13**, 2252 (1994).
35. Sini, G., MacGregor, S. A., Eisenstein, O. and Teuben, J. H., *Organometallics*, **13**, 1049 (1994).
36. Bierwagen, E. P., Bercaw, J. E. and Goddard, W. A., III, *J. Am. Chem. Soc.*, **116**, 1481 (1994).
37. Weiss, H., Ehrig, M. and Ahlrics, R., *J. Am. Chem. Soc.*, **116**, 4919 (1994).
38. Meier, R. J., Doremaele, G. H. J. V., Iarlori, S. and Buda, F., *J. Am. Chem. Soc.*, **116**, 7274 (1994).
39. Fusco, R. and Longo, L., *Macromol. Theory Simul.*, **3**, 895 (1994).
40. Lohrenz, J. C. W., Woo, T., Fan, L. and Ziegler, T., *J. Organomet. Chem.*, **497**, 91 (1995).
41. Yoshida, T., Koga, N. and Morokuma, K., *Organometallics*, **14**, 746 (1995).
42. Fan, L., Harrison, D., Woo, T. and Ziegler, T., *Organometallics*, **14**, 2018 (1995).
43. Lohrenz, J. C. W., Woo, T. K. and Ziegler, T., *J. Am. Chem. Soc.*, **117**, 12793 (1995).
44. Margl, P. M., Lohrenz, J. C. W., Blöchl, P. E. and Ziegler, T., *J. Am. Chem. Soc.*, **118**, 4434 (1996).
45. Zambelli, A. and Tosi, C., *Fortschr. Hochpolym.-Fortschr.*, **15**, 31 (1974).
46. Boor, J., Jr, *Ziegler–Natta Catalysis and Polymerizations*, Academic Press, New York, 1979.
47. Natta, G., Farina, M. and Peraldo, M., *Chim. Ind. (Milan)*, **42**, 255 (1960).
48. Zambelli, A., Giongo, M. G. and Natta, G., *Makromol. Chem.*, **112**, 183 (1968).
49. Dyachkovsky, F. S., Shilova, A. J. and Shilov, E., *J. Polym. Sci., Part C*, **16**, 2333 (1967).
50. Eisch, J. J., Piotrovsky, A. H., Brownstein, S. K., Gabe, E. J. and Lee, F. J., *J. Am. Chem. Soc.*, **107**, 7219 (1985).

51. Jordan, R., Bajgur, C., Willet, R. and Scott, B., *J. Am. Chem. Soc.*, **108**, 7410 (1986).
52. Bochmann, M., Jaggar, A. and Nicholls, J., *Angew. Chem., Int. Ed. Engl.*, **29**, 780 (1990).
53. Xinmin, Y., Stern, C. and Marks, T. J., *J. Am. Chem. Soc.*, **113**, 3623 (1991).
54. Hanson, K. R., *J. Am. Chem. Soc.*, **88**, 2731 (1966).
55. Corradini, P., Paiaro, G. and Panunzi, A., *J. Polym. Sci., Part C*, **16**, 2906 (1967).
56. Cahn, R. S., Ingold, C. and Prelog, V., *Angew. Chem., Int. Ed. Engl.*, **5**, 385 (1966).
57. Prelog, V. and Helmchem, G., *Angew. Chem., Int. Ed. Engl.*, **21**, 567 (1982).
58. Schlögl, K., *Top. Stereochem.*, **1**, 39 (1966).
59. Stanley, K. and Baird, M. C., *J. Am. Chem. Soc.*, **97**, 6598 (1975).
60. Bovey, F. A. and Tiers, G. V. D., *J. Polym. Sci.*, **44**, 173 (1960).
61. Sheldon, R. A., Fueno, T., Tsunetsugu, T. and Furukawa, J., *J. Polym. Sci., Part B*, **3**, 23 (1965).
62. Doi, Y. and Asakuru, K., *Makromol. Chem.*, **176**, 507 (1975).
63. Hine, J., *J. Org. Chem.*, **31**, 1236 (1966).
64. Hine, J., *Adv. Phys. Org. Chem.*, **15**, 1 (1977).
65. Venditto, V., Guerra, G., Corradini, P. and Fusco, R., *Polymer*, **31**, 530 (1990).
66. Venditto, V., Guerra, G., Corradini, P., and Fusco, R., *Eur. Polym. J.*, **27**, 45 (1991).
67. Woo, T. K., Fan, L. and Ziegler, T., *Organometallics*, **13**, 432 (1994).
68. Zambelli, A., Ammendola, P., Grassi, A., Longo, P. and Proto, A., *Macromolecules*, **19**, 2703 (1986).
69. Cavallo, L., Guerra, G. and Corradini, P., *Gazz. Chim. Ital.*, **126**, 463 (1996).
70. Chien, J. C. W., Llinas, G. H., Rausch, M. D., Lin, G. Y., Winter, H. H., Atwood, J. L. and Bott, S. G., *J. Am. Chem. Soc.*, **113**, 8569 (1991).
71. Coates, G. W. and Waymouth, R. M., *Science*, **267**, 217 (1995).
72. Zambelli, A., Sacchi, M. C., Locatelli, P. and Zannoni, G., *Macromolecules*, **15**, 211 (1982).
73. Longo, P., Grassi, A., Pellecchia, C. and Zambelli, A., *Macromolecules*, **20**, 1015 (1987).
74. Cavallo, G., Guerra, G., Oliva, L., Vacatello, M. and Corradini, P., *Polym. Commun.*, **30**, 000, (1989).
75. Pino, P., Cioni, P. and Wei, J., *J. Am. Chem. Soc.*, **109**, 6189 (1987).
76. Pino, P., Galimberti, M., Prada, P. and Consiglio, G., *Makromol. Chem.*, **191**, 1677 (1990).
77. Cavallo, G., Guerra, G., Vacatello, M. and Corradini, P., *Chirality*, **3**, 299 (1991).
78. Gilchrist, J. H. and Bercaw, J. E., *J. Am. Chem. Soc.*, **118**, 12021 (1996).
79. Resconi, L., Abis, L. and Franciscono, G., *Macromolecules*, **25**, 6814 (1992).
80. Erker, G. and Fritze, C., *Angew. Chem., Int. Ed. Engl.*, **31**, 199 (1992).
81. Brookhart, M., Green, M. L. H. and Wong, L., *Prog. Inorg. Chem.*, **36**, 1 (1988).
82. Piers, W. E. and Bercaw, J. E., *J. Am. Chem. Soc.*, **112**, 9406 (1990).
83. Kraudelat, H. and Brintzinger, H. H., *Angew. Chem., Int. Ed. Engl.*, **29**, 1412 (1990).
84. Lauher, J. W. and Hoffman, R., *J. Am. Chem. Soc.*, **98**, 1729 (1976).
85. Brookhart, M. and Green, M. L. H., *J. Organomet. Chem.*, **250**, 395 (1983).
86. Mislow, K. and Raban, M., *Top. Stereochem.*, **1**, 1 (1967).
87. Cavallo, L., Corradini, P., Guerra, G. and Vacatello, M., *Polymer*, **32**, 1329 (1991).
88. Mise, T., Miya, S. and Yamazaki, H., *Chem. Lett.*, 1853 (1989).
89. Röll W., Brintzinger, H. H., Rieger, B. and Zolk, R., *Angew. Chem., Int. Ed. Engl.*, **29**, 279 (1990).
90. Several data are collected in Ref. 27, Table 1.
91. Cavallo, L., Corradini, P., Guerra, G. and Vacatello, M., *Organometallics*, **15**, 2254 (1996).
92. Bercaw, J., paper presented at the 211th National Meeting of the American Chemical Society, New Orleans, LA, March 24–28, 1996.
93. Farina, M., Di Silvestro, G. and Sozzani, P., *Macromolecules*, **26**, 946 (1993).

94. Ewen, J. A., *Macromol. Symp.*, **89**, 181 (1995).
95. Rieger, B., Jani, G., Fawzi, R. and Steiman, M., *Organometallics*, **13**, 647 (1994).
96. Ewen, J. A., Elder, M. J., Jones, R. L., Curtis, S. and Cheng, H. N., in Keii, T. and Soga, K. (Eds), *Catalytic Olefin Polymerization*, Kodansha, Tokyo, 1990, p. 439.
97. Grassi, A., Zambelli, A., Resconi, L., Albizzati, E. and Mazzocchi, R., *Macromolecules*, **21**, 617 (1988).
98. Mizuno, A., Tsutsui, T. and Kashiwa, N., *Polymer*, **33**, 254 (1992).
99. Soga, K., Shiono, T., Takemura, S. and Kaminsky, W., *Makromol. Chem., Rapid Commun.*, **8**, 305 (1987).
100. Cheng, H. and Ewen, J., *Makromol. Chem.*, **190**, 1931 (1989).
101. Tsutsui, T., Ishimaru, N., Mizuno, A., Toyota, A. and Kashiwa, N., *Polymer*, **30**, 1350 (1989).
102. Tsutsui, T., Mizuno, A. and Kashiwa, N., *Makromol. Chem.*, **190**, 1177 (1989).
103. Rieger, B., Mu, X., Mallin, D., Rausch, M. and Chien, J., *Macromolecules*, **23**, 3559 (1990).
104. Busico, V. and Cipullo, R., *J. Organomat. Chem.*, **497**, 113 (1995).
105. Spaleck, W., Antberg, M., Aulbach, M., Bachmann, B., Dolle, V., Haftka, S., Küber, F., Rohrmann, J. and Winter, A., in Fink, G., Mülhaupt, R. and Brintzinger, H. H. (Eds), *Ziegler Catalysts*, Springer, Berlin, 1995, p. 83.
106. Razavi, A., Vereeke, D., Petres, L., Den Davw, K., Nafpliotis, L. and Atwood, J. L., In Fink, G., Mühlhaupt, R. and Brintzinger, H. H. (Eds), *Ziegler Catalysts*, Springer, Berlin, 1995, p. 111.
107. Röll, W., Brintzinger, H. H., Rieger, B. and Zolk, R., *Angew. Chem., Int. Ed. Engl.*, **29**, 279 (1990).
108. Stehling, U., Diebold, J., Kirsten, R., Röll, W., Brintzinger, H. H., Jüngling, S., Mühlhaupt, R. and Langhauser, F., *Organometallics*, **13**, 964 (1994).
109. Resconi, L., Piemontesi, F., Nifant'ev, I. E. and Ivichenko, P. V., *PCT Int. Appl.*, W096 22 995, to Montell.
110. Resconi, L., Piemontesi, F., Camurati, I., Balboni, D., Sironi, A., Moret, M., Rychlicki, H. and Ziegler, R., *Organometallics*, **15**, 5046 (1996).
111. Resconi, L. and Moscardi, G., *Polym. Prepn.*, **38**, 832, (1997).
112. Toto, M., Cavallo, L., Corradini, P., Guerra, G. and Resconi, L., *Macromolecules*, **31**, 3431, (1998).
113. Resconi, L., Piemontesi, F., Franciscono, G., Abis, L. and Fiorani, T., *J. Am. Chem. Soc.*, **114**, 1025 (1992).
114. Eshuis, J. J. W., Tan, Y. Y., Meetsma, A., Teuben, J. H., Renkema, J. and Evens, G. G., *Organometallics*, **11**, 362 (1992).
115. Cavallo, L. and Guerra, G., *Macromolecules*, **29**, 2729 (1996).
116. Jüngling, S., Mühlhaupt, R., Stehling, U., Brintzinger, H. H., Fisher, D. and Langhauser, F., *J. Polym. Sci., Part A: Polym. Chem.*, **33**, 1305 (1995).
117. Resconi, L. and Waymouth, R., *J. Am. Chem. Soc.*, **112**, 4953 (1990).
118. Cavallo, L., Guerra, G., Corradini, P., Resconi, L. and Waymouth, R., *Macromolecules*, **26**, 260 (1993).
119. Guerra, G., Cavallo, L., Corradini, P. and Fusco, R., *Macromolecules*, **30**, 677 (1997).
120. Guerra, G., Pucciariello, R., Villani, V. and Corradini, P., *Polym. Commun.*, **19**, 2699 (1986).
121. Borriello, A., Busico, V., Cipullo, R., Chadwick, J. C. and Sudmeijer, O., *Macromol. Rapid Commun.*, **17**, 589 (1996).
122. Oliva, L., Longo, P. and Zambelli, A., *Macromolecules*, **29**, 6383 (1996).
123. Cavallo, L., Guerra, G. and Corradini, P., *J. Am. Chem. Soc.*, **120**, 2428 (1998).

2

Chirotopicity of Metallocene Catalysts and Propene Polymerization

GIUSEPPE DI SILVESTRO
Università di Milano, Milan, Italy

1   INTRODUCTION

As a curious coincidence, in the early 1980s two papers appeared, both related to new aspects of the synthesis and microstructure of polypropene. Farina *et al.* [1] announced the preparation of a new stereoisomer of polypropene possessing an equilibrium between ordered and disordered stereocenters, hemiisotactic polypropene (hit-pp). Sinn and Kaminsky [2] reported the activation of some metallocene complexes for the polymerization of α-olefins by methylalumoxane (MAO). The work of Farina's group was a classical academic paper about research on new polymeric stereoisomers and the synthetic approach used was not suitable for industrial production. However, the structure of the hemiisotactic polypropene was recognized by Ewen [3] in a sample obtained with a metallocene catalyst possessing some spatial restrictions near the reactive catalytic sites. As a consequence, the new stereoisomer of polypropene can be prepared, if necessary, in large quantity by a more accessible process than inclusion polymerization.

Metallocene-based Polyolefins Edited by J. Scheirs and W. Kaminsky
© 2000 John Wiley & Sons Ltd

2 CHIROTOPICITY CONCEPT IN INCLUSION POLYMERIZATION IN PERHYDROTRIPHENYLENE (PHTP) ADDUCTS. SYNTHESIS OF HEMIISOTACTIC POLYPROPENE

The control of the microtacticity of polymers prepared by new catalysts has been interpreted with different models [4]; in this chapter we will refer only to the model discussed by Farina, because it uses a stereochemical base common to different fields of chemistry (organometallic, organic and macromolecular stereochemistry). No arbitrary assumptions are inserted in the model and each statistical parameter relates to a definite stereochemical situation. In some respects the model is complicated, but its results are founded on stereochemical bases.

The proposed model is based on the chirotopicity concept [5] (from the Greek, local chirality) whose first application in polymer chemistry was in the inclusion polymerization of prochiral diene monomers performed on the adducts with the *anti-trans, anti-trans, anti-trans* isomer of PHTP [6]. Inclusion polymerization is a topochemical reaction promoted by γ-radiation that transforms a long sequence of monomers inserted in the channel of the PHTP–monomer adduct into macromolecules isolated from each other by the channel walls.

In some papers inclusion compounds are described as tubulates; this word can suggest, for the including space, a cylindrical symmetry; however, several results obtained in inclusion polymerization of PHTP adducts the need for a different symmetry of the reactive space: 'a tube with helicoidal symmetry'.

In a helix all points possess the same chirality, they are homotopic; as a consequence, the synthesis of isotactic polypentadienes can be seen as a hard copy of the homotopic relationships existing between the reactive polymerization sites [7]. The 'm' sequences of isotactic polymers are, in fact, a copy of 'm' relations of adjacent points of the helix; in other words, the polymer is obtained under stereochemical control of local chirality, the chirotopicity. The same phenomenon in the Ziegler–Natta mechanism is reported as catalytic site control [3].

The first synthesis of hemiisotactic polypropene was performed by a complex procedure: first, 2-methylpentadiene is polymerized in its PHTP adduct; the isotactic polymer is hydrogenated by the diimine 'generated *in situ*'. This reaction does not epimerize the adjacent chirotopic carbon atoms.

The number of permitted sequences in hit-pp is greatly reduced with respect an atactic polymer and the NMR spectrum was interpreted at a very high level. The methyl region of the ^{13}C NMR spectrum registered at room temperature was interpreted at the nonad/undecad level and some tridecads were recognized. Octads and decads were recognized in the methylene region [8].

In a later section the statistical approach to Ziegler–Natta polymerization will be illustrated. We will stress here the importance of the discovery of hemiisotactic polypropene for demonstration of the Arcus–Cossee [9] mechanism (Scheme 1). The presence of alternating ordered and disordered stereogenic carbons is incompatible

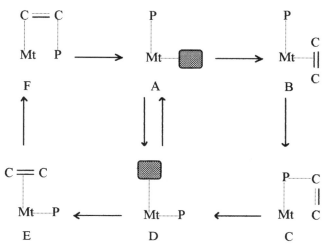

Scheme 1 Arcus–Cossee alternating mechanism (pathway A–B–C–D–E–F–A) and Ewen 'retention mechanism' (pathway A–B–C–D–A)

with any mechanism where a stereocontrolling reaction step does not succeed an aspecific one and vice versa. The synthesis of hemiisotactic polypropene is consistent with the conformational model presented by Corradini and co-workers [10], where it was assumed that the catalyst structure influences the propagating chain conformation and hence the stereochemistry of monomer insertion. In the case of the hemiisotactic polypropene obtained by Ewen [3], in the stereospecific polymerization step the chain is bonded at the less crowded part of the catalyst and assumes an almost rigid conformation; propene coordination occurs with a defined face (*re* or *si*). In the non-stereospecific step, the chain is in the more crowded part of the catalyst and no monomer face selection occurs owing to the presence of two almost equal energy conformations of the growing chain.

3 CLASSIFICATION OF RIGID METALLOCENE CATALYSTS ACCORDING TO THE SYMMETRY

The scheme proposed by Corradini's group [10] uses conformational analysis in order to interpret the microstructure of the growing chain; however, the effects of the experimental conditions can hardly be interpreted by conformational calculations if, as an example, the solvent effect on the polymer microstructure is to be accounted for. The original aim of our group was to develop a scheme based on rigorous stereochemical grounds useful to relate both structural variations in catalyst structures and polymerization conditions (solvent, temperature, etc.) in term of energy differences. As an expected result, it should be possible to optimize the catalyst structure and polymerization conditions with respect to polymer micro-structure.

The starting point was Ewen's experimental evidence that stereorigid catalysts are able to control the stereochemical pathway of polymerization. From our point of view it is not important how the conformational rigidity between two ligands is obtained or the exact nature of the ligand. Obviously, the presence of substituents modifies the reactivity and or the stability of complexes under the polymerization conditions [11].

We considered that if symmetry relationships exist between the catalytic sites, the same relationships must be present between the energy levels accessible to the chain in the rate-determining step as in Corradini's model.

For metallocene complexes classification, the general formula L_2MtX_2 is used, where Mt is the transition metal, L_2 is a bidentate ligand or two monodentate ligands bonded to the metal and X are halogen atoms or organic groups such as methyl or phenyl; the two ligands may or may not have the same structure or they can constitute the two moieties of the bidentate ligand. We consider the substitutions in the outside portion of the catalyst to be non-influential with respect to the reactive catalytic sites.

The activation reaction of the complex with methylalumoxane (MAO) [2] or other electrophilic reagents in the presence of the monomer produces the cationic center represented as $(L_2MtMP)^+$, where the monomer (M) is η^2-coordinated to the metal atom and the growing chain end (P) is σ-bonded. If we assume a tetrahedral coordination for the metal atom in the original complex and in the active center, we can draw the corresponding Fisher projections (Figure 1).

Metallocene catalysts can be divided into five classes (and no more than five). This number is obtained by considering all the possible conditions for chirotopicity of the catalytic sites X. If they are not chirotopic (e.g. they are bisected by a mirror plane) we have two possibilities: the two sites are equal or different. If, on the other hand, they are chirotopic, three possibilities exist: the two catalytic sites are homotopic (equal), enantiotopic (mirror image of each other) or diastereotopic (different) [4e].

The symmetry relationships between the X atoms and the local symmetry of the metal atom are deduced from the actual structure of the catalyst, or, more simply,

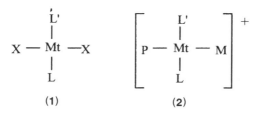

Figure 1   Fisher projection of metallocene complex **(1)** and of the active center **(2)**.

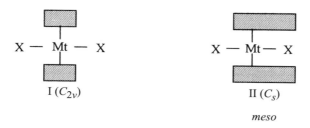

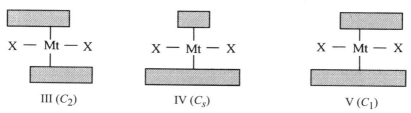

Scheme 2 Classification of metallocene catalysts according to the symmetry

from the naive representation of Scheme 2 where the shaded areas represent the shape requirement of the ligands with respect to the L_1-Mt-L_2 plane.

The five symmetry classes for all stereorigid catalysts or for each instantaneous conformation of a mobile complex are shown in Scheme 2. Table 1 reports the symmetry relationships in rigid metallocene catalysts or in an instantaneous conformation of mobile complexes.

Table 1 Symmetry relationships in rigid metallocene catalysts and predicted polymer microstructure according to the alternating mechanism

Parameter	Catalyst				
	I	II	III	IV	V
Symmetry of the L_2MtX_2 complex	C_{2v}	C_s	C_2	C_s	C_1
Chirotopicity of atoms X in L_2MtX_2	No	No	Yes	Yes	Yes
Relationship between atoms X in L_2MtX_2	Equal	Different	Homotopic	Enantiotopic	Diastereotopic
Chirotopicity of atom Mt in $[L_2MtMP]^+$	No	No	Yes	Yes	Yes
Stereogenicity of atom Mt in $[L_2MtMP]^+$	No	Yes	No	Yes	Yes
Polymer structure predicted for catalyst control and alternating mechanism	Atactic	Atactic	Isotactic	Syndiotactic	n.p.[a]

[a] Non-predictable.

In the first two classes the X atoms are achirotopic: in class I they are related by two orthogonal planes of symmetry and, as a consequence, the two sites are identical. In class II the X atoms lie on the plane of symmetry and the two sites are different. The difference between class I and II can be better understood if we use the asymmetric submolecular unit concept [12]: in class I the X groups occupy a generic position and are related by a symmetry operation whereas in class II they are in a special position because they lie on the symmetry operator.

In classes III, IV and V the two sites are chirotopic: in class III a twofold rotation axis relates the two sites so they are homotopic; in class IV the two X atoms occupy the two specular positions respect to the vertical mirror plane of symmetry, so they are enantiotopic; in class V no elements of symmetry exist and the two sites are diastereotopic.

4 SYNTHESIS OF METALLOCENES WITH CHIRAL GROUPS AND DEFINED SYMMETRY

The same classification made according to symmetry can be extended to metallocenes containing chiral substituents such as that synthesized by Erker [13]. In this case, metallocene can be represented with its Fisher projection or as a plane with four sectors (A–D) (Figure 2). Axes of this last representation are centered on the metal atom and perpendicular to each other. Symmetry relationships between sectors determine the class of symmetry of the catalysts.

We studied the classification of metallocene catalysts having 1–4 substituents R; we analyzed only symmetric substitution respect to the metal [14]; Figure 3 reports possible ways for preparation of a class IV catalyst; in the figure + and − represent the absolute configuration of stereocentres. This analysis can be used for a rational

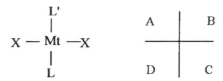

Figure 2 Fisher projection of a metallocene with chiral substituents

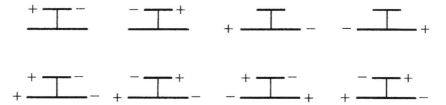

Figure 3 The eight structures of metallocenes with chiral substituents permitted for Class IV

synthesis of 3,4-cyclopentadienyl, indenyl or 5,6-fluorenyl derivatives. Non-symmetric substitution produces only C_1 symmetric molecules (class V). However, the classification is more general and can be applied to other kinds of molecular structures. Microstructural results for polypropene obtained with soluble optically active catalysts can be interpreted according to the catalyst symmetry [14b].

5 PROPENE POLYMERIZATION AND STATISTICAL GENERATION OF CHAIN MICROSTRUCTURE

In the last row of Table 1, we reported the expected microstructure of polypropene when the polymerization occurs under strict control of the catalytic site and the alternating Arcus–Cossee [9] mechanisms is operating (Scheme 1: pathway A–B–C–D–E–F–A).

The monomer insertion occurs through a four-center cooperative mechanism between the chain, the metal atom and the monomer. At each step of the polymerization a new σ-bond is formed between the metal and one of the unsaturated atoms of the coordinated monomer; at the same time the polymeric chain binds to the other olefin carbon through a *cis*-1,2 insertion. During the growth process the chain lies alternately in the two catalytic sites and apparently it moves from one site to the other (alternating mechanism; see pathway A–B–C–D–E–F–A in Scheme 1) and as a consequence the symmetry relationships between the sites must be reflected in the energy levels accessible to the chain. Demonstration of the validity of the mechanism is outside the scope of this chapter but, in our opinion, syntheses of syndio- and hemiisotactic polypropene support the mechanism at least when polymerizations are performed at low temperatures and in non-polar solvents [15].

Syndiotactic polypropenes are obtained by using metallocene catalysts belonging to class IV; the enantiomorphic relationship between the two catalytic sites (in polymeric terms the two catalytic sites are in a reversed relation as in an r dyad) is perfectly reproduced by the enantiomorphic chirality of successive chirogenic carbons in the chain (successive carbons possess reversed chiralities). In the synthesis of hit-pp the evidence for the alternating mechanism is stronger: the presence of the perfectly alternating ordered and disordered stereogenic carbons can be explained only by a polymerization mechanism in which a regular succession of reaction steps is present, one with steric control analogous to that observed in the syndiotactic polymer while the other lacks steric control. This situation holds only if the reaction occurs on the two diastereomeric sites present in class V catalysts [4e].

As can be observed, stereoregular polymers can be obtained only if the catalytic sites are chirotopic; it is not important whether they are homo-, enantio- or diastereotopic. The synthesis of isotactic polymers through the use of class III catalysts is not evidence for the alternating mechanism; the homotopic relationship between the catalytic sites (the two catalytic sites maintain the chirality as in the polymer the successive carbons generate m dyads) does not allow us to distinguish between the alternating mechanism and growth on the same catalytic site. The presence of steric errors in isotactic polymers is evidence for the presence of polymerization processes different from the pure alternating type.

The insertion of the monomer is 1–2 (primary insertion); however, secondary or 1–3 insertions can occur, as revealed by head-to-head and tail-to-tail junctions and $CH_2-CH_2-CH_2$, sequence respectively.

Ewen et al. [16] proposed a more complex model: in the 'retention mechanism' (referred to as 'isomerization without insertion' and also as 'skipped insertion' corresponding to the A–B–C–D–A pathway in Scheme 1, the growing chain and the coordinated monomer remain in the same position in successive propagation steps. Evident analogies exist between the alternating mechanism and the well known S_N2 mechanism in bimolecular substitution reactions where inversion occurs; the skipped insertion can be related to the S_Ni process (internal nucleophic substitution) where retention of configuration is observed. The presence of 1–3 insertions should be related to an S_N1 mechanism where substitution with racemization at the stereocenter, elimination reactions or [2,3] sigmatropic rearrangements of the monomeric unit are present.

6   PROPENE POLYMERIZATION AND STATISTICAL GENERATION OF CHAIN MICROSTRUCTURE

The Arcus–Cossee scheme and the classification of metallocene catalysts according to the local symmetry of catalytic sites were the basis of a probability approach to the analysis of polymer microstructure from ^{13}C NMR spectra at the pentad level.

In the general case the chirality of the catalytic site of metallocene catalysts excludes the use of symmetric Bernoulli or Markov chains, which are used in radical or ionic processes performed in an achiral environment [7]. This approach was first applied to hemiisotactic polypropene by our group and by Fink [11b].

A general scheme of chain generation in the presence of a class V metallocene has to consider a very complex situation: the two diastereotopic sites react with different kinetics both in a regular alternating reaction and in a skipped insertion, and the stereochemical control can be due to the catalytic site chirality and/or to the ultimate stereogenic carbon atom.

From these premises we developed four probability schemes of increasing complexity, including Bernoulli and Markov chains which alternate regularly (alternating mechanism) or follow in a statistical way (mixing of alternating and retention mechanisms) [4e].

The diastereotopic nature of the two catalytic sites of class V metallocenes and the hypothesis of different reactivity need two different sets of probability parameters: the probability of having a stereogenic carbon with a given configuration, say 1, is expressed by parameter a (for asymmetric Bernoulli processes) or by p_{i1} (for asymmetric Markov chains) when reaction occurs at a given site, and by b or q_{i1} when the other site is involved.

The quantitative expressions of different stereosequences were developed as the product of the conditional probability of the considered stereosequence and the unconditional probability of the first event at the beginning of the sequence. All the possible ways of formation of each stereosequence have to be evaluated; at the pentad level 1024 (2^{10}) different ways were considered.

The schemes were developed for class V catalysts, but they can be reduced to a simpler form on the basis of the symmetry of the catalyst considered.

A short presentation is reported here. Interested readers can refer to the original paper [4e].

6.1 MODEL A

The model consists of pure alternating mechanism reactions: the steric control derives from the catalytic site. The two Bernoulli asymmetric processes require two probability parameters (a and b) of forming stereogenic carbons of 1 (or 0) configuration in either catalytic site. The effect of symmetry makes $b = 1 - a$ and $b = a$ in class IV and III catalysts, respectively. For class I and II $a = b = 0.5$. The synthesis of atactic polypropene in the presence of biscyclopentadienyl-ZrCl$_2$ is accounted for by the presence in the catalyst of a horizontal plane of symmetry which makes the two re and si monomer faces completely equivalent [2]. Stereo-control cannot be expected in the absence of other controlling effects.

6.2 MODEL B

Model B also consists of pure alternating reactions, but the chain-end configuration takes part in steric control. The effect of the two different sources of stereocontrol (catalytic site and chain-end configuration) makes the probability parameter of a given configuration (say 1) dependent on the configurations of both the catalytic site and last-entered monomeric unit (0 or 1). In class V we need four independent probability parameters, two (p_{11} and p_{00}) related to a given site and two (q_{11} and q_{00}) related to the other; p and q refer to the site at which the last monomeric unit forms.

The conditions $q_{11} = p_{11}$ and $q_{00} = p_{00}$ (class III symmetry) converts the system into a Markov single-site model used for isospecific polymerization with a hetero-geneous catalyst. The scheme discussed by Ewen [3] coincides with the class IV model, where $q_{00} = p_{11}$ and $q_{11} = p_{00}$.

In class II, $p_{00} = p_{11} = p_m$ and $q_{00} = q_{11} = q_m$; in this case two Bernoulli symmetric processes (expressed in m and r) alternate with one another. In class I $q_m = p_m$ and the distribution follows the Bernoulli symmetric trial expressed in m and r.

6.3   MODEL C

In this model the Bernoulli parameters a and b are defined as in model A; c is the probability of successive insertion (skipped insertion) at the site related to the b parameter and d is the analogous parameter for the catalytic site related to a.

6.4   MODEL D

This last model works like model C with the combined influence of the chain-end configuration and of chirotopicity of the catalytic site at various stages of the polymerization mechanism. The class V model D contains eight parameters, four of them related to the probability of the formation of a stereogenic carbon of a given configuration and coinciding with previously discussed parameters of model B, and the other four parameters relating to the occurrence of consecutive additions at the same catalytic site. Consecutive addition is assumed depending on the configuration of the last stereogenic carbon in the chain. Parameters I and K are the probabilities of occurrence of two consecutive insertions at the site related to parameter q when the last carbon has configuration 1 or 0, respectively, and J and L have the same meaning but refer to the other catalytic site. I, J, K and L take into account the influence of 'right' or 'wrong' configuration of the chain-end on the probability of the occurrence of the control phenomenon.

6.5   TEST OF MODELS AT PENTAD LEVEL

The method used to test models compares the expected sequence concentrations, according to a set of probability parameters of a model, with the real polymer microstructure and changes the parameter set until a minimum is reached in the Hamilton [17] agreement factor. The polymer structure is usually determined by ^{13}C NMR spectroscopy; in the case of polypropene, although sequences as long as a tridecad can be recognized, quantitative sequence determination only at the pentad level is generally used. Eight out of ten pentads are individually resolved in the usual experimental conditions, because two of them (mmrm and rmrr) merge into a single peak.

A preliminary check of the correctness of pentad concentrations is made by considering the stoichiometric relationships.

Because of the merging of two pentads in the usually high-temperature registered spectra, only one relationship can be used:

$$\text{mmmr} + 2\,\text{rmmr} + \text{rrrm} + 2\,\text{mrrm} - \text{xmrx} - 2\,\text{mmrr} = 0 \tag{1}$$

where xmrx ($=$ mmrm $+$ rmrr) is the unresolved peak.

In some literature data, the sum of pentad concentrations differs from 100; however this difference does not greatly influence the validity of the test of the models. More important is the difference in the pentad relationship evaluated according to equation (1): a high value of this difference limits the validity of the experimental data and the confidence of the test with complex models, e.g. class V model D. As discussed later, the stoichiometry test between tetrads and pentads should be used for a better confidence test of experimental concentration.

The agreement factor (Af), according to Hamilton, is calculated by using the equation

$$AF = \sqrt{\frac{\sum (I_{obs} - I_{calc})^2}{\sum (I_{obs})^2}} \qquad (2)$$

where I_{obs} and I_{calc} are, at each step of the iteration, the observed and calculated intensities according to the model under test.

Applications of this method to literature data have already been reported [18] and will not be discussed here.

7 EVALUATION OF SOME ENERGY PARAMETERS RELATED TO MICROSTRUCTURAL DEFECTS

In 1969, Farina [19] interpreted the microstructure of a polymer as a result of 'wrong' vs 'right' reactions on a growing center of an ionic polymerization performed with chiral initiators. The growth process was described, according to Fueno and Furukawa [4a], as an asymmetric Markov chain and the probability parameters are a measure of difference in activation energy of parallel reactions at reactive centers. The same approach was proposed by Farina *et al.* [4e] for metallocene-promoted polymerization due to the analogous stereochemical bases of the model so far illustrated and was applied to literature data on syndiotactic polypropenes. However, the relationships used are a particular case of a more general method.

We report equations for class V models B and C because they are, in our experience, very useful for relating polymer microstructure to experimental polymerization conditions.

7.1 CLASS V MODEL B. ENERGY PARAMETERS

In this model, four different steric control sources are present and, as a consequence, four defective insertion mechanisms can be devised: (a) defective insertion related to the stereospecific catalytic site, (b) defective insertion related to the chain-end bonded to the stereospecific catalytic site, (c) defective insertion related to the non-stereospecific catalytic site and (d) defective insertion related to the chain-end bonded to the non-stereospecific catalytic site.

The corresponding Arrhenius equations are reported:

$$\exp(\Delta G_1/RT) = (p_{10}p_{00}/p_{11}p_{01})^{1/2} \qquad (3)$$

where $\Delta G_1 = $ free energy excess due to the defective insertion related to the stereospecific catalytic site.

$$\exp(\Delta G_2/RT) = (p_{10}p_{01}/p_{11}p_{00})^{1/2} \tag{4}$$

where $\Delta G_2 = $ free energy excess due to the defective insertion related to the chain-end bonded to the stereospecific catalytic site.

$$\exp(\Delta G_3/RT) = (q_{10}q_{00}/q_{11}q_{01})^{1/2} \tag{5}$$

where $\Delta G_3 = $ free energy excess due to the defective insertion related to the non-stereospecific catalytic site.

$$\exp(\Delta G_4/RT) = (q_{10}q_{01}/q_{11}q_{00})^{1/2} \tag{6}$$

where $\Delta G_4 = $ free energy excess due to the defective insertion related to the chain-end bonded to the non-stereospecific catalytic site.

7.2 CLASS V MODEL C. ENERGY PARAMETERS

In model C, steric defects can come from the lack of steric control by the two diastereotopic sites and/or skipped insertion at the two sites; four different free energy excess equations can be written as due to (a) defective insertion related to the stereospecific catalytic site, (b) defective insertion related to the non-stereospecific catalytic site, (c) defective insertion related to the skipped insertion related to the non-stereospecific catalytic site and (d) defective insertion related to the skipped insertion related to the stereospecific catalytic site.

The correspond Arrhenius equations are (as subscript we use the letter used as the probability parameter):

$$\exp(\Delta G_a/RT) = a/(1-a) \tag{7}$$

where $\Delta G_a = $ free energy excess due to the defective insertion related to the stereospecific catalytic site.

$$\exp(\Delta G_d/RT) = d/(1-d) \tag{8}$$

where $\Delta G_d = $ free energy excess due to the defective insertion related to the skipped insertion related to the stereospecific catalytic site.

$$\exp(\Delta G_b/RT) = b/(1-b) \tag{9}$$

where $\Delta G_b = $ free energy excess due to the defective insertion related to the non-stereospecific catalytic site.

$$\exp(\Delta G_c/RT) = c/(1-c) \tag{10}$$

where $\Delta G_c = $ free energy excess due to the defective insertion related to the skipped insertion related to the non-stereospecific catalytic site.

The symmetry of classes III and IV reduces the number of free energy excess equations as in the same way as discussed for model B and C probability parameters [18].

8 PROPENE POLYMERIZATION WITH CLASS III CATALYSTS

Isotactic polypropene has been a key point in macromolecular stereochemistry and one of the most important industrial polymers. However, in the present context, the synthesis of isotactic polypropene is of less importance than that of hemiisotactic and syndiotactic polypropene.

Computer simulations for class III catalysts, changing the probability parameters, revealed that no or only small variations in pentad concentrations are produced; as a consequence, a reliable test for the application of the proposed scheme cannot be made with isotactic polymers. Better results are expected for class IV and V catalysts.

We will discuss later the case of isotactic polypropene synthesized using class V catalysts.

9 PROPENE POLYMERIZATION WITH CLASS IV CATALYSTS

The synthesis of syndiotactic polypropenes is one of the most interesting results of the metallocene-catalyzed polymerization of propylene as it permitted the understanding of the mechanism of polymerization in the presence of metallocenes and the effect of some modifications of the structure of the catalyst and of the polymerization conditions. The polymer itself possesses interesting properties in the solid state: its melting-point is near that of the isotactic polymer, and polymorphism has been described which seems to be related to the nature and the concentration of steric defects [20].

As discussed before, the studies with class IV are the most interesting cases for a stereochemical and statistical analysis of propylene polymerization with a metallocene catalyst; in this case Ewen's idea that decreasing the mobility of ligands should increase the steric control of polymerization is completely demonstrated. The effect on the polymer microstructure of the symmetry of chirotopicity of catalytic sites is also evident.

For syndiospecific polymerization, Ewen et al. [16] discussed the application of the enantiomorphic-site model they had proposed for isotactic polymerization.

Some of our selected experimental results are reported in Table 2 (items 1–8). The related probability and energy parameters according to model C are reported in Table 3. The catalysts used are cyclopentadienylfluorenyldiphenylmethane-$ZrCl_2$ for syndiotactic polypropene (items 1–8) and 3-methylcyclopentadienylfluorenyldimethylmethane-$ZrCl_2$ for hemiisotactic polypropene (items 9–14), both activated with methylallumoxane (MAO).

Table 2 Polymerization conditions and pentad concentrations of syndiotactic (1–8) and hemi-isotactic (9–14) polypropenes[a]

Item	T (°C)	P (bar)	Solvent[b]	mmmm	mmmr	rmmr	mmrr	xmrx	rmrm	rrrr	rrrm	mrrm
1	−20	7.0	Toluene	0	0	0	0	0	0	100	0	0
2	0	4.0	Toluene	0	0	0.54	1.52	0	0	95.91	2.03	0
3	20	1.7	Toluene	0	0.65	1.01	2.43	1.45	0	91.01	3.21	0.23
4	45	2.8	Toluene	0	0.80	3.23	9.94	0.91	0.34	72.99	9.23	2.55
5	60	4.2	Toluene	0	0	1.97	4.51	7.86	0	74.9	8.58	2.18
6	20	1.7	Heptane	0	0	0.91	2.05	1.25	1.25	88.35	4.99	1.21
7	20	1.7	DCB	0.34	1.54	3.50	15.14	3.25	0.39	54.17	18.61	3.06
8	20	1.7	DCM	0	0.42	1.88	4.46	19.56	5.35	46.00	20.37	1.97
9	−20	1.0	Toluene	13.46	12.08	6.83	24.67	2.44	1.97	20.64	12.34	5.58
10	60	4.2	Toluene	17.13	14.25	4.20	21.79	6.63	2.27	13.47	12.49	7.77
11	−10	1.0	DCB	16.64	13.10	6.35	23.60	2.55	0.76	18.77	11.94	6.27
12	60	4.2	DCB	19.16	13.09	4.47	15.84	14.21	6.89	7.92	11.69	6.68
13	−40	1.0	DCM	18.58	13.77	5.60	23.52	4.23	1.90	13.53	12.10	6.74
14	20	1.7	DCM	32.94	15.59	2.55	19.03	5.78	1.18	7.94	7.21	7.78

[a] xmrx = mmrm + rmrr.
[b] DCB = o-Dichlorobenzene; DCM = dichloromethane.

Table 3 Statistical parameters and energy values (ΔG, cal/mol) according to model C for syndiotactic (1–8) and hemiisotactic (9–14) polypropenes (polymerization conditions as in Table 2)

Item	a	d	b	c	ΔG_a	ΔG_d	ΔG_b	ΔG_c
1	—	—	—	—	—	—	—	—
2	0.992(1)	0.000(8)	0.007(9)	0.000(8)	2622	3888	2622	3888
3	0.987(2)	0.006(9)	0.012(8)	0.006(9)	2528	2888	25282	888
4	0.393(3)	0	0.060(7)	0	1731	—	1731	—
5	0.976(6)	0.041(3)	0.023(4)	0.041(3)	2468	2081	2468	2081
6	0.984(6)	0.010(9)	0.015(4)	0.010(9)	2421	2624	2421	2624
7	0.889(1)	0.007(7)	0.110(9)	0.007(7)	1212	2829	1212	2829
8	0.996(5)	0.173(5)	0.003(5)	0.173(5)	3299	909	3299	909
9	0.972(5)	0	0.460(4)	0	1792	—	80	—
10	0.988(3)	0.140(9)	0.475(1)	0.079(2)	2936	1195	66	1623
11	0.984(2)	0.037(7)	0.477(5)	0	2160	1664	47	—
12	0.885(7)	0	0.585(8)	0.231(2)	1355	—	229	795
13	0.969(4)	0.003(4)	0.471(9)	0.037(1)	1599	2631	52	1506
14	0.959(1)	0.187(6)	0.631(1)	0	1836	853	310	—

As a general trend we can observe a decrease in the stereocontrolling capability with increase in reaction temperature and/or solvent polarity (items 1–5 and 3, 7, 8). Obviously, because of the symmetry of the system we have $b = 1 - a$ and $c = d$. What is particularly interesting is that in item 8 the value of ΔG_d is lower than in all the other cases. Hence we can conclude that in a highly polar solvent, such as methylene chloride, the probability of the defect connected with two consecutive insertions on the same site is particularly high. This appears logical if we consider the status of ion pairs of the catalytic system during the polymerization process. A high polar solvent creates a greater separation of the ion pairs, allowing all possible defective reactions such as skipped insertion and epimerization.

10 USE OF MODELS IN HEMIISOTACTIC POLYPROPENES

The most important case for our discussion is hemiisotactic polypropene because it is of great importance in the elucidation of the polymerization mechanism with metallocene catalysts. On the other hand, hemiisotactic polypropene is a fortuitous result of a controlling process regularly alternating to a non-controlling process. As we observed in the classification of catalysts, the stereochemical result in the presence of a class V catalyst cannot be predicted. In fact, symmetry is very useful for class I–IV catalysts but cannot predict the relative reactivity or stereocontrol efficiency of diastereotopic sites.

Some of our selected experimental results are reported in Table 2 (items 9–14). The analysis according to model C is reported in Table 3 (items 9–14).

It is useful to recall the theoretical pentad concentration of the pure hemiisotactic polymer [equation (11)]. As the ratios between pentads are peculiar to the hemiisotactic structure, deviations from these values can be easily interpreted according the polymerization alternating mechanism or as being due to the lack of control of the catalytic site or to the skipped insertion.

$$\text{mmmm : mmmr : rmmr : mmrr : xmrx : rmrm : rrrr : rrrm : mrrm}$$
$$= 3 : 2 : 1 : 4 : 0 : 0 : 3 : 2 : 1 \quad (11)$$

As for the syndiospecific catalyst, we can see that the stereocontrolling capability of the stereospecific site decreases with increased temperature and solvent polarity. For the aspecific site, no evident effect is observed.

As for the case of catalysts belonging to class IV, there is a greater difference in value between ΔG_a and ΔG_d when the reaction is conducted in methylene chloride. Hence the effect of the separation of the ion pair due to a highly polar solvent is observed in an increased probability of defective reactions connected with the ion-pair separation. The effect is not evident at very low temperature (item 13) and the kinetics of all defective reactions are almost small.

11 STATISTICAL RELIABILITY OF MODEL TESTS. USE OF EVEN SEQUENCES IN STATISTICAL MODELS

A test of a model is strictly related to (a) the complexity of the problem under study, (b) the number and the physical meaning of the parameters of the model used and (c) the number of experimental data (related to the number of parameters) and their reliability. There is no doubt that the study of the microstructure of a polypropene in quantitative terms is difficult: the ^{13}C NMR spectrum is normally measured at high temperature in solvents such as dichloro- or trichlorobenzene or $C_2D_2Cl_4$. Measuring a spectrum under quantitative conditions is a time-consuming experiment and high-field spectrometers are not generally used in this kind of work. NMR spectra are measured with a 200–400 MHz spectrometer. These facts limit the spectral resolution and the methyl region is commonly integrated only at the pentad level. Assignment of the ^{13}C NMR spectrum of hemiisotactic polypropene at the nonad or undecad level was obtained at room temperature because the hemiisotactic polymer is readily soluble in these conditions [8,21]; isotactic and syndiotactic polypropene are insoluble at the required concentration. Hence the obtained data are a compromise between spectral resolution, integration of pentads and run time. The use of high-field spectrometers decreases several of the problems discussed here.

The statistical data in Table 3 were obtained at the minimum of the Hamilton factor and by working with differences in parameters of less than 10^{-5} units. However, the original values of the pentad concentration do not support this level of accuracy. In a previous paper [22] we used a map of Af in order to study the effect on Af on the difference in the parameters. Such maps can be used to evaluate the difference in parameters compatible with the accuracy in the integration of the NMR spectrum; in fact, for every point on the curve with the same Af value, we can evaluate the corresponding pentad concentrations and compare them with the experimental values. This very simple method was confirmed by a study on the structure of the errors in models [23]. As a practical consequence of this study, we took great care with the integration of the NMR spectrum. We averaged at least five independent integrations to obtain 'experimental pentad concentrations'. A different result of this study allowed us to rationalize the difficulty we observed in the test of polypropene obtained with class V catalysts with bulky substituents [24]. In this case the bulk substituent precludes reaction at the aspecific site and the chain must move back to the stereospecific catalytic site; an isotactic polymer is produced. The case studied by Razavi [24] looks very similar to the synthesis of pure syndiotactic polypropene at $-20\,^{\circ}$C at a high monomer pressure (item 1, Table 2). Also in this last case we cannot evaluate the probability parameters, being at the border of the polymerization mechanism considered in the models.

A second point was made in the reliability of the tests: as we have already indicated, pentad concentrations must be checked for stoichiometric relationships. This is a general problem, as was demonstrated by Bovey [25] in his classical book and by Tulleken and co-workers [26] in a paper devoted to stoichiometric relationships in vinyl polymers.

Recently we proposed an approach through the use of stoichiometric relationships between tetrads and pentads in order to increase the number of independent data and the reliability of the experimental set of pentad concentrations [27]. This allows us to use model D, which requires eight parameters to be optimized.

We applied this method to a hemiisotactic polypropene obtained at 20 $^{\circ}$C in toluene (the ^{13}C NMR spectrum is reported in Figure 4). Hemiisotactic polypropene is a good test because chemical assignment of the NMR spectrum is excellent also

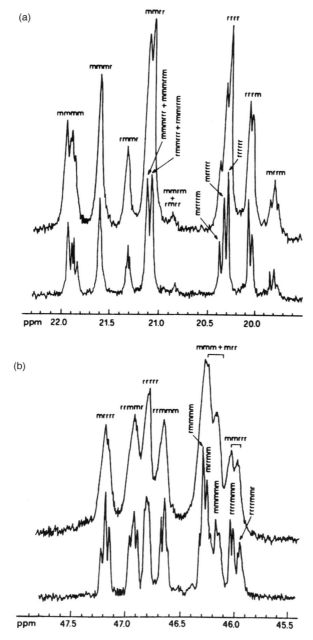

Figure 4 (a) methyl region and (b) methylene region of the ^{13}C NMR spectrum of hemiisotactic polypropene (toluene, 20 °C. 1.7 atm) registered at (top) 300 and (bottom) 600 MHz.

for long sequences. The variation of chemical shifts with temperature is well known for the methyl region; for the methylene region we used the assignment published by Farina and co-workers [8].

Tetrads and pentads give rise to a set of 16 data, sufficient to test model D with a good confidence level.

All even and odd experimental stereosequences, obtained from the 300 MHz spectrum in Figure 4, were optimized by a best-fitting procedure according to all relationships among tetrads and pentads. The optimized set was not far from the experimental one, indicating that the previous integration was good. Statistical analysis gave the following results: $p_{11} = 1$; $p_{00} = 0.035$; $q_{11} = 0.44$; $q_{00} = 1$; $J = 0.04$; $L = 0$; $I = 0.007$; $K = 0$.

As expected, we see that one site is highly stereocontrolling ($p_{11} = 1$), whereas the opposite one is aspecific (q_{00} close to 0.5). However, we can also observe that the other parameter related to an aspecific site (q_{11}) is 1 and not close to 0.5. This means that when the last monomeric unit has absolute configuration 1, the monomer insertion at the aspecific site has a probability close to 100% of entering with absolute configuration 1. In contrast, when the last monomeric unit has absolute configuration 0, the related parameter is close to 0.5. A strong effect of the chain-end configuration is thus observed. However, this effect does not influence the probability of skipped insertion (J, L, I and K are always close to zero).

12 CONCLUSION

We stress that a stereochemical-based model allows a simpler interpretation of polypropene microstructure obtained using soluble metallocene catalysts and tests of different polymerization models.

The effect of ion-pair separation due to the increased solvent polarity and/or polymerization temperature can be easily measured without changing the statistical expressions of the stereosequences.

This approach can be extended to stereosequences of any length if supported by more resolved spectra. The use of all stoichiometric relationships, particularly among even and odd stereosequences, allows us to obtain a set of sequence concentrations for a deeper investigation of the polymerization mechanism.

13 ACKNOWLEDGMENTS

Financial support from CNR (Italian National Research Council), Progetto Finaliz-zato Chimica Fine 2 and MURST (Italian Ministry of University and Scientific and Technological Research) is gratefully acknowledged. The author thanks all those who contributed to this work, particularly Dr Stefano Ambrosio, Dr Nicola Caronzolo and Dr Alberto Terragni. A special tribute is devoted to Professor

Mario Farina, whose interest in macromolecular stereochemistry was the basis of this and many other studies.

14 REFERENCES

1. Farina, M., Di Silvestro, G. and Sozzani, P., *Macromolecules*, **15**, 1451 (1982).
2. Sinn, H. and Kaminsky, W., *Angew. Chem.*, **92**, 396 (1980).
3. Ewen, J. A., *J. Am. Chem. Soc.*, **106**, 6355 (1984).
4. (a) Fueno, T. and Furukawa, J., *J. Polym. Sci., Part A*, **2**, 3681 (1964); (b) Cheng, H. N., Babu, G. N., Newmark, R. A. and Chien, J. C. W., *Macromolecules*, **25**, 6980 (1992); (c) Zambelli, A., Locatelli, P., Provasoli, A. and Ferro, D. R., *Macromolecules*, **13**, 267 (1980); (d) Itabashi, Y., Chujo, R. and Doi, Y., *Polymer*, **25**, 1640 (1984); (e) Farina, M., Di Silvestro, G. and Terragni, A., *Macromol. Chem. Phys.*, **196**, 353 (1994).
5. Mislow, K. and Raban, M., *Top. Stereochem.*, **1**, 1 (1967).
6. Farina, M., *Makromol. Chem., Suppl.*, **4**, 21 (1981).
7. Farina, M., Audisio, G. and Natta, G., *J. Am. Chem. Soc.*, **89**, 5071 (1967).
8. Di Silvestro, G., Sozzani, P., Savarè, B. and Farina, M., *Macromolecules*, **18**, 928 (1985).
9. (a) Arcus, C. L., *J. Chem. Soc.*, 2801 (1955); (b) Cossee, P., in Ketley, A. D. (Ed.), *The Stereochemistry of Macromolecules*, Marcel Dekker, New York, 1967, Vol. I, Chapt. 3.
10. Cavallo, L., Guerra, G., Vacatello, M. and Corradini, P., *Macromolecules*, **24**, 1784 (1991).
11. (a) Kaminsky, W., Kulper, K., Brintzinger, H. H. and Wild, F. R. W. P., *Angew. Chem., Int. Ed. Engl.*, **24**, 507 (1985); (b) Fink, G., *Makromol. Chem. Macromol. Symp.*, **66**, 157 (1993).
12. Farina, M. and Morandi, C., *Tetrahedron*, **30**, 1819 (1974).
13. Erker, G., *J. Am. Chem. Soc.*, **115**, 4590 (1993).
14. (a) Di Silvestro, G., Terragni, A. and Galbiati, A., in *XII Convegno A.I.M.*, *Palermo*, 1995, p. 519 (available from A.I.M., c/o Department of Chemistry and Industrial Chemistry, University of Pisa, via Risorgimento 35, 56216, Pisa, Italy); (b) Cesarotti, E., Terragni, A. and Di Silvestro, G., in *XI FECHEM, Parma,* 1995, p. 87 (available from Tipolitografia Benedettina, Parma, Italy) (full papers are in preparation).
15. Farina, M., Di Silvestro, G. and Terragni, A., *Macromol. Chem. Phys.*, **195**, 353 (1995).
16. Ewen, J. A., Elder, M. J., Jones, R. L., Curtis, S. and Cheng, H. N., in Keii, T. and Soga, K. (Eds), *Catalytic Olefin Polymerization*, Elsevier, New York, 1980, p. 439.
17. Hamilton, W. C., *Acta Crystallogr.*, **18**, 502 (1965).
18. Di Silvestro, G., Sozzani, P. and Terragni, A., *Macromol. Chem. Phys.*, **197**, 3209 (1996).
19. Farina, M., *Makromol. Chem.*, **122**, 237 (1969).
20. Balbontin, G., Dainelli, D., Galimberti, M. and Paganetto, M. G., *Makromol. Chem.*, **193**, 693 (1992).
21. Schilling, F. C. and Tonelli, A. E., *Macromolecules*, **13**, 270 (1980).
22. Farina, M., Di Silvestro, G. and Sozzani, P., *Macromolecules*, **26**, 946 (1993).
23. Terragni, A., Thesis, Scuola di Specializzazione G. Natta, Politecnico di Milano (1996).
24. Razavi, A., personal communication.
25. Bovey, F. A., *High Resolution NMR of Macromolecules*, Academic Press, New York, 1972.
26. Van Der Burg, W. M., Chadwick, J., Sudmeijer, O. and Tulleken, H. J. A. F., *Makromol. Chem. Theory Simul.*, **2**, 385 (1993).
27. Di Silvestro, G., Terragni, A. and Ambrosio, S., in *Atti XIII Convegno A.I.M.*, *Genova*, 1997, p. 317 (available from A.I.M., c/o Department of Chemistry and Industrial Chemistry, University of Pisa, via Risorgimento 35, 56126 Pisa, Italy).

3

Stereospecific Polymerization of Chiral α-Olefins

ADOLFO ZAMBELLI
Università di Salerno, Baronissi (Salerno), Italy

1 INTRODUCTION

Soon after the discovery of the isotactic-specific polymerization of propene, in the presence of heterogeneous catalytic systems such as $TiCl_3-AlR_2Cl$ (R = hydrocarbyl) [1], Natta, Pino and, later, Ciardelli and co-workers began to investigate the polymerization of chiral α-olefins with the aim of obtaining some information about the mechanism of the steric control of the polyinsertion and, possibly, of resolving racemic α-olefins, by means of the just-discovered Ziegler catalysts, mimicking enzymatic resolution [2–6].

Both goals were, at least partially, achieved. By polymerizing racemic α-olefins branched on carbon 3 (*rac*-3-methyl-1-pentene and *rac*-3,7-dimethyl-1-octene), in the presence of the same heterogeneous catalysts able to promote isotactic-specific polymerization of propene and the other α-olefins, they obtained crystalline, essentially isotactic, polymers. In fact, the stereochemical sequence of the monomer units was prevailingly isotactic, not only with reference to the backbone substituted carbons, but also with respect to the configuration of the substituents. The resulting polymer could be resolved (by chromatography on an optically active support) into optically active fractions [3,7,8].

Partial resolution of the racemic monomers was also achieved, by using as polymerization promoter a heterogeneous Ziegler–Natta catalyst 'asymmetrized' by addition of optically active ingredients [6,8–10]. Under these conditions, one enantiomer polymerizes faster than the other and the final result is an optically active polymer, together with optically active residual unreacted monomer.

Metallocene-based Polyolefins Edited by J. Scheirs and W. Kaminsky
© 2000 John Wiley & Sons Ltd

Partial resolution of *rac*-3-methyl-1-pentene was also achieved, e.g., by copoly-merization with optically active 3,7-dimethyl-1-octene. Even in this case, not only the copolymer is optically active, but also the unreacted 3-methyl-1-pentene recovered after the polymerization [11]. The absolute configuration of the unreacted 3-methyl-1-pentene is opposite to that of the optically active 3,7-dimethyl-1-octene comonomer [5,8].

From these experimental results, the authors concluded that the active species promoting isotactic-specific polymerization are 'intrinsically' chiral, being able not only to control the stereochemistry of the insertion of prochiral α-olefins, but also to select between mirror related chiral monomers. The same authors termed 'stereo-selective' the polymerization of racemic monomers in the presence of Ziegler–Natta catalysts, considered as racemic mixtures of chiral active species able to select the enantiomeric monomers, according to their configuration, and 'stereoelective' the polymerization performed in the presence of 'asymmetrized' Ziegler–Natta catalysts (uneven mixtures of mirror related chiral active species, promoting preferential polymerization of one of the enantiomers of the racemic monomers). Copolymer-ization of racemic monomers with an optically active comonomer was also termed 'stereoelective'.

2 RELEVANT STEREOCENTERS

Coordination of the monomer to metallic centers, *cis* addition to the double bond and insertion of the coordinated monomer into the metal–carbon bond of the active site are common features of all α-olefin coordination polymerizations investigated up to now. Depending on the catalyst, the monomer insertion can be primary (almost always) or secondary (in two cases [12]):

primary insertion [1,2]:

$$Mt-CH_2 - CH(R) \cdots + CH_2{=}CH(R) \longrightarrow Mt-CH_2-CH(R)-CH_2-CH(R) \cdots$$

secondary insertion [2,1]:

$$Mt-CH(R)-CH_2 \cdots + CH_2{=}CH(R) \longrightarrow Mt-CH(R)—CH_2—CH(R)—CH_2 \ldots$$

The insertion of the monomer may involve migration of the chain or not, depending on the particular catalyst (it is likely that chain migratory insertion is general but, in the presence of particular catalysts, it is followed by back-skipping of the chain to the preferred enantiotopic coordination site of the catalyst before coordination of a new molecule of monomer [15]). In principle, the stereochemistry of the polyinsertion is controlled by all the stereocenters that, being close enough to the reactive metal–carbon bond, more or less affect the energy of the transition state. The simplest mechanism controlling the stereochemistry of the polyinsertion of α-olefins is 1,3 asymmetric induction (either *like* or *unlike*). The relevant stereo-centers are the enantioface of the coordinated monomer and the substituted carbon of the last unit of the growing chain end.

When the metal atom of the active site is also chiral, the stereocenters to be possibly considered become three. If, in addition, also the α-olefin is chiral, the more or less relevant stereocenters become five: the metal, the substituted skeletal carbon of the last unit of the growing chain end and the chiral substituent and the two stereocenters of the coordinated monomer diastereoface. Consequently, the possible stereoisomers of the transition state become 2^5 (i.e. 2^4 diastereomeric pairs of mirror related enantiomers).

3 DIASTEREOSELECTIVITY

While the faces of prochiral α-olefins are mirror related, the faces of chiral α-olefins are diastereotopic. As a consequence, they should show a different reactivity when attacked by any reagent, including the active species promoting coordination polymerization, and afford diastereoisomeric products. Actually, when polymerization of *rac*-3-methyl-1-pentene is performed in the presence of either heterogeneous [16] or homogeneous coordination catalysts [17], one can easily observe that the initiation step is diastereoselective because the reactive metal–carbon bond of the active sites (e.g. a metal–$^{13}CH_3$) preferentially attacks the *like* monomer diaster-eofaces and produces mostly *erythro* end groups that can be easily identified by ^{13}C NMR, as shown in Figures 1 and 2.

It is worth noting that similar results have been observed by using different catalytic systems such as (1) $\delta TiCl_3-Al(CH_3)_3$ (heterogeneous, isotactic specific) [1,16], (2) *rac*-ethylenebis(1-indenyl)ZrCl$_2$–MAO (homogeneous, isotactic-specific) [17,18] and (3) isopropyl(cyclopentadienyl)(1-fluorenyl)ZrCl$_2$–MAO (homo-geneous, promoting syndiotactic polymerization of propene) [17,19].

For catalytic system 1 the stereochemical structure of the end groups formed by terminating polymerization by hydrolysis of the reactive metal–carbon bonds:

$$Mt-CH_2-\overset{\overset{\textstyle CH(CH_3)C_2H_5}{|}}{CH}\cdots \quad \xrightarrow{H_2O} \quad Mt-OH +CH_3-\overset{\overset{\textstyle CH(CH_3)C_2H_5}{|}}{CH}\cdots$$

was also investigated and was again that expected from faster reaction of the *like* diastereoface [16].

Therefore, it seems reasonable to conclude that the diastereoselectivity is about the same during the propagation and the initiation steps, at least for catalysts 1 and 2, producing polymers of very similar structure. In fact, the ^{13}C NMR spectra of the polymers produced in the presence of catalysts 1 and 2 are almost identical, showing

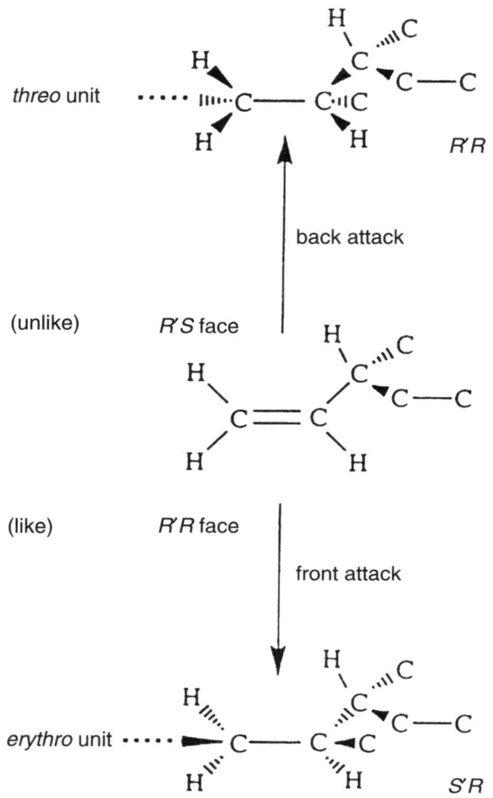

Figure 1 The front attack of Mt–CH$_3$ on the like faces is faster than the back attack and produces *erythro* monomer units

that they are copolymers of *erythro* and *threo* units in proportion similar to the *erythro/threo* ratios observed on the end groups. The copolymers are able to crystallize and Pino and co-workers [2–5] succeeded in achieving the optical resolution of the polymers obtained in the presence of catalyst 1. Thus the reaction products are racemic mixtures of mirror-related coisotactic copolymer macromolecules of uneven content of mirror-related comonomers.

Surprisingly, the polymer obtained in the presence of catalyst 3, although showing a ^{13}C NMR spectrum distinctly different from the others, is also crystalline, and the diffraction spectrum is almost the same as observed previously for the polymers obtained with heterogeneous catalysts and having a prevailingly isotactic structure, with reference to the configuration of the backbone substituted carbons (see Figure 3) [17].

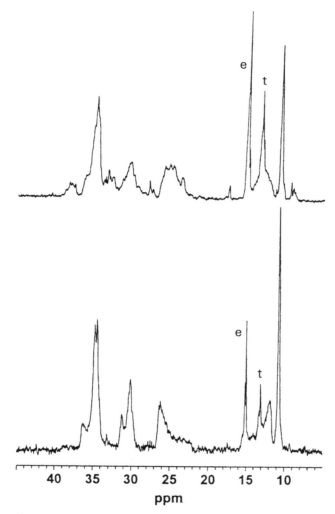

Figure 2 ^{13}C NMR spectra of poly-3-methyl-1-pentene obtained in the presence of (a) catalyst 2 and (b) catalyst 3; e and t (*erythro* and *threo*) are the resonances of the enriched methyls of the end groups resulting from insertion in Mt $-^{13}$ CH$_3$ bonds

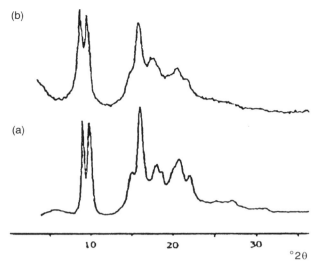

(b)

(a)

10   20   30

°2θ

Figure 3 X-ray powder diffraction spectra (Cu Kα radiation) of poly-(*R,S*)-3-methyl-1-pentene obtained in the presence of metallocene catalysts (a) 2 and (b) 3

As discussed in Section 5, catalysts 1 and 2 promote isotactic-specific polymerization of propene whereas catalyst 3 promotes syndiotactic-specific polymerization of propene.

4   STEREOSELECTIVE POLYMERIZATION

All the heterogeneous catalytic systems used by Natta, Pino, Ciardelli and co-workers for promoting the polymerization of chiral α-olefins produce isotactic polymers of propene through a stereochemical mechanism dominated by the fixed chirality of the metal stereocenters, that are highly enantioelective, towards the mirror related faces of the monomer [20–24]. These chiral active species are expected to be also enantioselective towards the mirror related diastereofaces of chiral α-olefins and, consequently, the resulting macromolecules should be more or less optically active, depending on how much different are the reactivities of the like diastereofaces in comparison with the unlike ones.

This is the simplest rationalization of the 'stereoselective' polymerization of α-olefins branched on C-3 (3-methyl-1-pentene and 3,7-dimethyl-1-octene) in the presence of isotactic-specific catalysts, either heterogeneous, or based on metallocene precursors of C_1 or C_2 symmetry [18,25]. Polymerization of chiral α-olefins branched further than C-3 is much less stereoselective [26].

5 STEREOELECTIVE POLYMERIZATION

Preferential polymerization of one enantiomer of, e.g., 3-methyl-1-pentene could be expected when polymerizing the racemate in the presence of a dissymmetric isotactic-specific catalyst. Pino and co-workers achieved this goal by modifying, with optically active ingredients, heterogeneous catalysts, promoting isotactic polymerization of propene. In this way, by polymerizing rac-3-methyl-1-pentene, rac-3,7-dimethyl-1-octene and rac-4-methyl-1-hexene, after partial conversion, they recovered optically active polymers together with optically active monomer leftovers.

The observed enantiomeric excesses were modest and the authors pointed out that, most probably, the enantiomeric excesses of the active species inside the catalysts, were also very modest. However it is also worth noting that even an optically pure catalyst, 100% enantioface-selective, would produce polymers with an enantiomeric purity limited by the only partial diastereoface selectivity that, e.g. for 3-methyl-1-pentene, is about 8 : 2 in favor of the like faces [27]. Ciardelli et al. also achieved preferential polymerization of one enantiomer by copolymerizing rac-3-methyl-1-pentene with optically active 3,7-dimethyl-1-octene. The catalyst was heterogeneous, without chiral ingredients, and the more reactive enantiomer of 3-methyl-1-pentene was that with the same absolute configuration as the optically active comonomer. Similar results were obtained by copolymerization of rac-3,7-dimethyl-1-octene with optically active 3-methyl-1-pentene [11].

The elegant experiments reported by Ciardelli are particularly complex terpolymerizations involving mirror-related active sites, the two enantiomers of one chiral monomer and one enantiomer of a further chiral monomer.

In the simpler case of copolymerizing the two enantiomers of, e.g., 3-methyl-1-pentene in the presence of mirror-related active sites, the copolymerization equation was solved and the reactivity ratios for the two enantiomers were determined [28]. The $r_1 \times r_2$ (see Scheme 1) product is > 1, showing that the stereocenters of the last

$$C_{(sS)} \quad
\begin{array}{l}
+ M_S \xrightarrow{K_{(sS)S}} C_{(sS)} \\[1em]
+ M_R \xrightarrow{K_{(sS)R}} C_{(sR)}
\end{array}$$

$$C_{(sR)} \quad
\begin{array}{l}
+ M_S \xrightarrow{K_{(sR)S}} C_{(sS)} \\[1em]
+ M_R \xrightarrow{K_{(sR)R}} C_{(sR)}
\end{array}$$

Scheme 1 C_s is the chiral active site producing monomer units with an S configuration of the backbone-substituted carbon. $C_{(sS)}$ is the same active site bonded to a monomer unit with an S configuration substituent. $C_{(sR)}$ is the same active species bonded to a monomer unit with an R substituent. The Ks are defined as usual: $r_1 = K_{(sS)}S/K_{(sS)}R$; $r_2 = K_{(sR)}R/K_{(sS)}R$

unit of the growing chain end are by no means irrelevant as that of the propylene unit during isotactic-specific polymerization seems to be. As a consequence, at least in some cases, polymerization of α-olefins with a very bulky substituent, in the presence of chiral active species, might experience (dual) control of the stereochemistry of the insertion both (i) from the metal stereocenter and (ii) from the stereocenter(s) of the last unit of the growing chain end [29–31].

Stereoelective polymerization of 4-methyl-1-hexene has been also performed in the presence of an optically pure chiral catalyst obtained from $C_2H_4(Ind)_2$ Zr-di-*o*-acetyl (*R*)-mandelate and MAO [32]. The highest optical purity observed in the monomer left over was 17%, much more than any observed previously when using 'asymmetrized' heterogeneous catalysts, but still severely limited by the incomplete diastereoface selectivity of the catalyst.

6 C_s SYMMETRIC METALLOCENES

Coordination polymerization catalysts prepared by using stereorigid C_s symmetric metallocene precursors promote the syndiotactic-specific polymerization of propene [14]. The chain migratory insertion mechanism implies that the configuration of the chiral active species [e.g. the isopropylidene(cyclopentaidenyl)(1-fluorenyl)Zr-alkyl cation (see Fig. 4)] is inverted at each monomer insertion step.

The mentioned catalyst also promotes syndiotactic-specific polymerization of higher linear α-olefins and of 4-methyl-1-pentene [33].

Polymerization of (+)-(*S*)-4-methyl-1-hexene affords a 'pseudosyndiotactic' polymer [34], i.e. a polymer with a sequence of diastereomer units with alternating configurations of the substituted backbone carbons and identical configurations of the substituents (by definition a true syndiotactic polymer implies a sequence of mirror-related units).

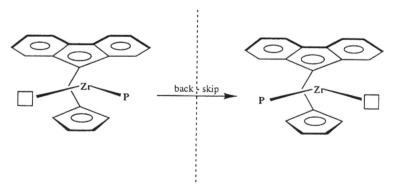

Figure 4   Considering the stereocenter of P, the two active species of the figure are diastereoisomeric

Polymerization of *rac*-4-methyl-1-hexene affords a polymer with alternating backbone-substituted carbon configurations and a sequence of the configurations of the substituents more or less at random [34]. These results are as expected and confirm that the reactivities of the monomer diastereofaces are little different owing to the distance of the chiral carbon of the substituent, two bonds apart from the unsaturation.

Surprisingly, the just mentioned catalyst promotes prevailing isotactic-specific polymerization of prochiral 3-methyl-1-butene [35] and, from *rac*-3-methyl-1-pentene, one obtains a copolymer of the mirror-related monomers with a prevailing isotactic sequence of the backbone-substituted carbons [17].

It is apparent that for α-olefins with very bulky substituents, the possibility of dual steric control should be given serious consideration.

1,3-Like asymmetric induction from the last unit of the growing chain end would reinforce the steric control coming from the metal stereocenter for catalysts based on C_1 or C_2 symmetric stereorigid metallocene precursors, whereas it would be conflicting for catalysts based on C_s symmetric metallocenes and strong enough to prevail.

A further possibility, not conflicting with the previous one, could be that owing to the low polymerization rate observed for 3-branched α-olefins, after the insertion of each monomer unit the resulting catalytic species isomerizes to the most stable diastereoisomer before coordination of a further monomer (back skipping of the chiral chain to the preferred enantioface of the catalyst). The relevance of the carbon stereocenters of the last unit of the growing chain end agrees with the 'stereoelective' copolymerization reported by Ciardelli and seems to be confirmed by the fact that (+)-(S)-3-methyl-1-pentene, unlike the racemate, does not produce any solid polymer in the presence of the catalyst considered here [17].

Finally, ^{13}C NMR analysis, while showing that insertion in the Mt–^{13}CH$_3$ bond is diastereoselective in favor of the *like* diastereofaces, suggests that this does not happen during propagation steps in the presence of the C_s metallocene [17].

It is unfortunate that 3-branched α-olefins, either prochiral or chiral, have never been polymerized in the presence of polymerization catalysts based on C_2 and C_1 symmetric metallocenes. Optical resolution of the polymers of racemic olefins obtained in the presence of metallocene catalysts has never been attempted.

7 CONCLUSION

The stereochemical polymerization mechanism for propene and linear α-olefins is relatively simple because 1,3-asymmetric induction does not play a relevant role when a stereorigid metal stereocenter is present on the active species.

When the size of the olefin substituent increases, 1,3-asymmetric induction can no longer be neglected and can reinforce or oppose the steric control by the metal stereocenter [29–31]. For chiral 3-branched α-olefins even the chiral carbon of the substituent plays a considerable role in determining the stereochemistry of the polyinsertion.

The very low polymerization rate of these bulky α-olefins suggests that polymerization could be, to a certain extent, under thermodynamic, rather than kinetic, control.

8 REFERENCES

1. Natta, G., Pino, P., Corradini, P., Danusso, F., Mantica, E., Mazzanti, G. and Moraglio, G., *J. Am. Chem. Soc.*, **77**, 1708 (1955).
2. Pino, P., Lorenzi, G. P. and Lardicci, L., *Chim. Ind. (Milan)*, **42**, 712 (1960).
3. Pino, P., Ciardelli, F., Lorenzi, G. P. and Natta, G., *J. Am. Chem. Soc.*, **84**, 1487 (1962).
4. Pino, P., *Adv. Polym. Sci.*, **4**, 393 (1965).
5. Pino, P., Oschwald, A., Ciardelli, F., Carlini, C. and Chiellini, E., in Chien, J. W. (Ed.), *Coordination Polymerization*, Academic Press, New York, 1975, p. 25.
6. Pino, P., Fochi, G., Piccolo, O. and Giannini, U., *J. Am. Chem. Soc.*, **104**, 7381 (1982).
7. (a) Pino, P., Montagnoli, G., Ciardelli, F. and Benedetti, E., *Makromol. Chem.*, **93**, 158 (1966). (b) Pino, P., Ciardelli, F. and Montagnoli, G., *J. Polym. Sci., Part C*, **16**, 3265 (1968).
8. Ciardelli, F., Carlini, C., Montagnoli, G., Lardicci, L. and Pino, P., *Chim. Ind. (Milan)*, **50**, 860 (1968).
9. Carlini, C., Nocci, R. and Ciardelli, F., *J. Polym. Sci., Polym. Chem. Ed.*, **15**, 767 (1977).
10. Masi, E., Menconi, F., Altomare, A., Ciardelli, F., Michelotti, M. and Solaro, R., in Soga, K. and Terano, M. (Eds), *Catalyst Design for Tailor-Made Polyolefins*, Kodansha, Tokyo, 1994.
11. (a) Ciardelli, F., Carlini, C. and Montagnoli, G., *Macromolecules*, **2**, 296 (1969); (b) Ciardelli, F., Carlini, C., Altomare, F., Menconi, F. and Chien, J. C. W., in Quirk, R. P. (Ed.), *Transition Metal Catalyzed Polymerizations*, Cambridge University Press, New York, 1988.
12. (a) Zambelli, A., Tosi, C. and Sacchi, C., *Macromolecules*, **5**, 649 (1972). (b) Pellecchia, C., Mazzeo, M. and Pappalardo, D., *Macromol. Rapid Commun.* **19**, 651 (1998).
13. Zambelli, A. and Tosi, C., *Adv. Polym. Sci.*, **15**, 32 (1974).
14. Ewen, J. A., *J. Am. Chem. Soc.*, **106**, 6355 (1984).
15. Pellecchia, C., Zambelli, A., Oliva, L. and Pappalardo, D., *Macromolecules*, **29**, 6990 (1996).
16. Zambelli, A., Ammendola, P., Sacchi, M. C., Locatelli, P. and Zannoni, G., *Macromolecules*, **16**, 341 (1983).
17. Oliva, L., Longo, P. and Zambelli, A., *Macromolecules*, **29**, 6383 (1996).
18. Kaminsky, W., Kulper, K., Brintzinger, H. H. and Wild, F. R. W. P., *Angew. Chem., Int. Ed. Engl.*, **24**, 507 (1985).
19. Ewen, J. A., Jones, R. L. and Razavi, A., *J. Am. Chem. Soc.*, **110**, 6255 (1988).
20. Zambelli, A., *NMR Basic Principles and Progress*, Springer, Berlin, 1971, Vol. 4, p. 101.
21. Zambelli, A., Bajo, G. and Rigamonti, E., *Makromol. Chem.*, **179**, 1249 (1978).
22. Wolfsgruber, C., Zannoni, G., Rigamonti, E. and Zambelli, A., *Makromol. Chem.*, **176**, 2765 (1965).
23. Longo, P., Grassi, A., Pellecchia, C. and Zambelli, A., *Macromolecules*, **20**, 1015 (1987).
24. Zambelli, A., Sacchi, M. C., Locatelli, P. and Zannoni, G., *Macromolecules*, **15**, 211 (1982).

25. Ewen, J. A. and Elder, M. J., in Fink, G. and Mulhaupt, H. H. (Eds), *Ziegler Catalysts*, Springer, Berlin, 1995, p. 99.
26. Pino, P., Guastalla, G., Rotzinger, B. and Mulhaupt, R., in Quirk, R. P. (Ed.), *Transition Metal Catalyzed Polymerizations. Alkenes and Dienes*, Harwood Academic, New York, 1983, p. 435.
27. Zambelli, A., Ammendola, P. and Sivak, A. J., *Macromolecules*, **17**, 461 (1984).
28. (a) Zambelli, A., Ammendola, P., Longo, P. and Grassi, A., *Gazz. Chim. Ital.*, **117**, 579 (1987); (b) Zambelli, A., Proto, A. and Longo, P., in Fink, G. and Mulhaupt, H. H. (Eds), *Ziegler Catalysts*, Springer, Berlin, 1995, p. 218.
29. Fueno, T. and Furukawa, J., *J. Polym. Sci., Part A*, **2**, 3681 (1964).
30. Fueno, T., Shelden, R. A. and Furukawa, J., *J. Polym. Sci., Part A*, **3**, 1269 (1965).
31. Farina, M., *Makromol. Chem.*, **122**, 237 (1969).
32. Chien, J. C. W., Vizzini, J. C. and Kaminsky, W., *Makromol. Chem., Rapid Commun.*, **13**, 479 (1992).
33. Albizzati, E., Resconi, L. and Zambelli, A., *Eur. Pat. Appl.* 387609 (1990).
34. Zambelli, A., Grassi, A., Galimberti, M. and Perego, G., *Makromol. Chem. Rapid Commun.*, **13**, 467 (1992).
35. Borriello, A., Busico, V., Cipullo, R., Chadwick, J. C. and Sudmeyer, D., *Makromol. Chem. Rapid Commun.*, **17**, 589 (1996).

4

Computational Modeling of Single-site Olefin Polymerization Catalysts

TOM K. WOO, LIQUN DENG, PETER M. MARGL AND
TOM ZIEGLER
University of Calgary, Calgary, Alberta, Canada

1 INTRODUCTION

Computational chemistry or molecular modeling involves simulating chemical reactions and processes at the atomic level within the virtual space of a computer. This type of modeling is evolving into an invaluable research approach in all areas of chemistry. It is already an integral part of the drug design process in the pharmaceutical industry and, with the explosive growth of computer technologies that is currently occurring, computational chemistry will undoubtedly find utility in the commercial study of transition metal catalysis. Although molecular modeling will never completely replace experiments performed in the laboratory, performing these experiments on computers will become more and more cost effective as laboratory costs continue to rise, the price of computers continues to drop and the computational methodologies continue to improve. In this chapter we will briefly introduce some of the computational methodologies that we have used to model metallocene-based catalysts and we will show, by way of examples, how computational molecular modeling can be used in aiding the design of new single-site olefin polymerization catalysts.

Metallocene-based Polyolefins Edited by J. Scheirs and W. Kaminsky
© 2000 John Wiley & Sons Ltd

2   COMPUTATIONAL METHODOLOGIES

Computational techniques have been used to predict anything from NMR spectra to the average shape of a solvated protein at 298 K. To study catalytic processes at the molecular level, it is necessary to model reaction energetics, dynamics and geometries. The methodologies we have used to study the energetics and structure of olefin polymerization catalysts range from sophisticated high-level quantum mechanical (QM) methods, simple molecular mechanics (MM) schemes and the combination of the two (combined QM/MM). In this section, these methodologies will be briefly outlined with emphasis on when and how they can be applied to study transition metal-based catalysis.

2.1   QUANTUM MECHANICAL METHODS

Perhaps the most direct way of determining the energetics and structure of a molecular system is to solve the quantum mechanical Schrödinger equation. This is a computationally demanding process and consequently only small molecular systems can be treated. The results, however, can be very accurate, in some cases attaining experimental accuracy. Other properties, such as transition state structures, reaction barriers and reaction profiles cannot be easily determined experimentally and are best calculated by computational means. Quantum mechanical methods can also provide a detailed picture of the electronic structure of a molecular system, which can yield invaluable insights into the nature of the system. Elementary reaction steps in homogeneous catalysis have been investigated with increasing success by quantum mechanical methods over the past decade [1].

2.1.1   Density Functional Theory

The Schrödinger equation can be solved to varying degrees of accuracy by a variety of different methods. The most widely used scheme is the *ab initio* Hartree–Fock method [2–4]. Almost every small organic molecule has been studied theoretically by this technique and consequently there is an ever increasing number of references to these types of calculations in scientific papers. Another quantum mechanical (QM) method, known as density functional theory [5] or DFT, is quickly emerging as the preferred methodology to deal with transition metal complexes and large-sized systems. The reason for this is that it has been shown to be faster and generally more accurate than Hartree–Fock-based methods for these types of systems [6]. Many metallocene-based and related single-site catalysts have been examined by DFT [7–23] and currently most quantum mechanical studies of these catalysts are performed with DFT.

The theoretical rigor of QM-based methods affords several advantages. In terms of studying olefin polymerization catalysts, the most significant feature of QM-based methods is that chemical reactions can be modeled at the molecular level. Thus, reaction intermediates and transition states can be studied and detailed reaction profiles can be elicited. From this, important kinetic and thermodynamic information of the catalytic processes can be determined. At the density functional level, reaction barriers can be estimated to within 15 kJ/mol [20,24]. Furthermore, high-level QM methods such as DFT are general and do not require parameterization to experimental results.

The primary disadvantage of QM-based methods is that their theoretical rigor also demands extensive computational resources. At the present time, detailed and expedient studies of molecular systems are limited to systems of under 100 atoms. For this reason, calculations are generally performed on truncated model systems which can be a severe approximation to the real system. For example, in modeling single-site catalysts systems, the solvent and counterion are generally ignored. Another issue to consider is that the results are difficult to interpret. For this reason, a well trained computational chemist is required to perform and interpret the calculations.

2.2 MOLECULAR MECHANICAL METHODS

Molecular mechanics is fundamentally different from QM-based methods, in that there is no attempt to solve the Schrödinger equation. There are no wavefunctions or molecular orbitals in molecular mechanics, in fact the electronic system is not treated explicitly (its effects, of course, are felt). Instead of blanketing the nuclei with a complicated electron density, in order to determine how much a geometric deformation increases or decreases the potential energy of a molecular system, the potential energy surface is determined from a set of very simple mathematical functions that are fitted to reproduce experimental results. In this case, simplicity is a virtue because molecular mechanics methods can effectively treat large macromolecules such as proteins or instantly treat small systems such as metallocenes.

One way of describing molecular mechanics is that it treats molecules as a set of balls connected together by springs. Each type of bond (e.g. C—C or C—H bond) is represented with a different kind of spring with a specific stiffness and equilibrium distance. If a bond is stretched, Hooke's law is assumed which results in a restoring force and a rise in energy of the system. In this way, each type of bond possesses a unique potential surface that is characteristic of the bond's natural length and strength. This is at the heart of molecular mechanics, the idea that a particular structural feature such as a C—H distance is essentially the same whether it is in butane or in DNA. In general, this is also true for bond angles, bond torsions, strain energies and so on. Thus, the structure and energy of large molecules can be formulated in terms of empirical parameters derived from the elementary features of smaller, well known molecules. Together, the sum of all the energy terms for the bond stretches (E_b), angles (E_θ), dihedrals (E_ϕ) and non-bonded interactions (E_{nb}) of a molecule forms the potential energy surface:

$$E_T = \sum E_b + \sum E_\theta + \sum E_\phi + \sum E_{nb}$$

The functional forms of the various energy terms and the parameters contained within them make up what is called a force field. A force field is generally highly parameterized in order to provide reasonable agreement with experiment. Furthermore, the parameters are often fitted to a specific group or type of molecular system and therefore force fields are generally designed to treat specific classes of molecules. For example, the AMBER [25] and CHARMM [26] force fields are designed to treat proteins and nucleic acids whereas the MM2 [27] force field is designed to treat small organic and main group inorganic molecules.

For studying transition metal-based catalytic processes, molecular mechanics methods possess two serious limitations. First, bond breaking and formation of covalent bonds cannot be treated and therefore chemical reactions cannot be simulated accurately. Second, force fields that can effectively deal with the complicated and varied bonding schemes of transition metals are only currently being developed [28,29]. These limitations confine molecular mechanics studies of transition metal catalysts to the qualitative regime. Studies of metallocene catalysts by molecular mechanics was pioneered by Corridini, Cavallo and co-workers [30–33]. These studies have contributed significantly to our understanding of the control of stereospecific α-olefin polymerization by metallocene catalysts. Attempts to apply molecular mechanics to obtain quantitative or semi-quantitative results have met with limited success [14,34–38].

2.3 COMBINED QM/MM METHOD

As outlined in previous sections, computational modeling of organometallic catalysts at the quantitative or semi-quantitative level necessitates a high-level quantum mechanical treatment because lower level molecular mechanics methods cannot accurately treat transition metals, nor can they properly simulate the bond breaking and forming processes in the elementary reaction steps. However, a quantum mechanical study often involves a stripped-down model system that only vaguely resembles the true system. If large ligand systems are involved, they are most often neglected in high-level calculations with the hope that they do not substantially influence the nature of the reaction mechanisms. Unfortunately, the surrounding ligand system, solvent or matrix can often play a critical mechanistic role. One dramatic example of this is that of the recently developed Ni(II) Brookhart polymerization catalyst [39,40]. Without an extended ligand system the catalyst acts only as a dimerization catalyst. However, by attaching an extended and sterically demanding ligand system, Brookhart and co-workers were able to transform the poor polymerization catalyst into a commercially viable material [see 41].

One reasonable approach to constructing a more sophisticated computational model which approximates these often neglected effects is the combined quantum mechanics and molecular mechanics (QM/MM) method [42–44]. In this hybrid method, part of the molecule, such as the active site, is treated quantum mechanically while the remainder of the system is treated with a molecular mechanics force field (Figure 1). This allows extremely large, transition metal-based systems that are out of the reach of pure QM calculations to be studied in an efficient and detailed manner. The key feature of the QM/MM method is that the QM calculation is performed on a truncated 'QM model' [Figure 1(b)] of the active site, where the large ligands have been removed and replaced with capping hydrogen atoms. Then the effects of the attached ligands are incorporated to form the potential surface of the whole system where the QM and MM regions interact with one another via steric and electrostatic potentials.

It should be noted that many modeling studies include separate QM calculations and MM calculations [22,34,37,38]. In these studies, a small model system is often optimized at the QM level. The QM geometry is then used in a molecular mechanics calculation where the extended ligand system is added and the QM geometry is frozen. This type of QM then MM calculation is distinct from a combined QM/MM calculation. Furthermore, it has been demonstrated that the QM then MM calculation can fail where the combined QM/MM calculation can provide accurate results [45].

Although the combined QM/MM methodology was first published in 1976 [46], it has only recently received mainstream attention [47]. For this reason, applications to olefin polymerization catalysis are limited [45,48–50] but the results are proving to be promising.

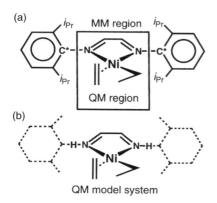

Figure 1 Example of the QM/MM partitioning in an Ni diimine olefin polymerization catalyst. (a) represents the 'real' system and (b) represents the corresponding model QM system for which the electronic structure calculation is performed. The hydrogen atoms cap the electronic system and are termed 'dummy' atoms. The dummy atoms correspond to 'link' atoms in the real system that are labeled C* in (a)

3 APPLICATIONS OF MOLECULAR MODELING TO SINGLE-SITE OLEFIN POLYMERIZATION CATALYSTS

3.1 MECHANISTIC INSIGHTS THROUGH COMPUTATIONAL MODELING

In this section we intend to demonstrate how important mechanistic insights can be procured from computational studies through the example of Brookhart's Ni(II) diimine olefin polymerization catalyst. In this study, both pure QM [45] and combined QM/MM [45] calculations were performed, providing a clear understanding of the effect of the extended ligand system in the catalyst.

Ni(II) diimine-based single site homogeneous catalysts of the type $(ArN=C(R)-C(R)=NAr)Ni-R'^+$ have emerged as promising alternatives to both traditional Ziegler–Natta and metallocene catalysts for olefin polymerization [39–41]. Brookhart and co-workers have shown that these catalysts are able to convert ethylene efficiently into high molecular weight polymers with a controlled level of polymer branching. In this polymerization system the bulky aryl groups play a crucial role, since without the bulky substituents the catalyst acts only as a dimerization catalyst owing to the favorability of the β-elimination chain termination process. From the structure of the catalyst, it is evident that the bulky aryl substituents partially block the axial coordination sites of the Ni center. It is probably this steric feature which impedes the termination of the insertion process, thereby promoting the intrinsically poor polymerization catalyst into a commercial viable material.

With the intention of examining, in detail, the role of the bulky substituents in the Brookhart polymerization catalyst, we have performed both pure QM calculations on the system without the bulky ligands and combined QM/MM calculations on the 'real' system. In the QM/MM model, the bulky R = Me and Ar = $2,6\text{-}C_6H_3(^iPr)_2$ groups are treated by the AMBER95 molecular mechanics potential whereas the Ni diimine core including the growing chain and monomer are treated by a density functional potential. Figure 1 shows the full catalyst system examined, where the carbon atoms with asterisks represent the link atoms at the QM/MM boundary which are replaced by dummy hydrogen atoms in the QM model system.

Using the prescription of Maseras and Morokuma [44], we combined the AMBER95 [25] molecular mechanics force field with the Amsterdam Density Functional (ADF) [51,52] program system. The QM system was calculated at the non-local density functional level with Becke's [53] 1988 exchange and Perdew's [54,55] 1986 correlation functionals. Full computational details are provided elsewhere [45].

Three processes are believed to dominate the polymerization chemistry of the catalyst system, namely, chain propagation, chain termination and chain branching, as shown in Figure 2. The propagation commences from an olefin π-complex which has been determined experimentally to be the catalytic resting state. Insertion of the

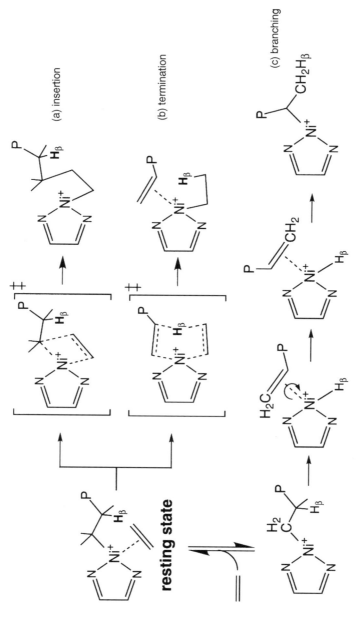

Figure 2 Proposed reaction mechanisms of (a) insertion, (b) chain termination and (c) chain branching mechanisms for the Brookhart Ni diimine olefin polymerization catalyst

olefin into the $M-C_\alpha$ bond forms a metal alkyl cation. Uptake of the monomer returns the system to the resting state. Chain termination occurs via monomer-assisted β-elimination, either in a fully concerted fashion as illustrated in Figure 2(b), or in a multistep associative mechanism as implicated by Johnson et al. [39]. The unique short-chain branching is proposed to occur via an alkyl chain isomerization process as sketched in Figure 2(c). In this proposed process, β-hydride elimination first yields a putative hydride olefin π-complex. Rotation of the π-coordinated monomer followed by rotation yields a secondary carbon unit and therefore a branching point. It is important to note that the branching process commences from the Ni–alkyl cation and not the resting state complex. Since monomer pressure affects the branching rate, the equilibrium between the free Ni–alkyl and the resting state is believed to influence the branching rate strongly. Experimentally the relative magnitudes of the free energy barriers are ordered such that propagation < branching < termination.

Table 1 compares the reaction barriers, $\Delta H^\ddagger$, for the propagation, branching and termination processes for both the pure QM model with no ligands modeled and the QM/MM model. Without the bulky ligands, the termination barrier is approximately 30 kJ/mol less than the insertion (propagation) barrier, suggesting that the system would be a poor polymerization catalyst. This is in agreement with experiment, where the Ni and Pd diimine systems without the bulky ligands are used as dimerization and *oligermization* catalysts [56]. When the bulky ligands are included, the termination and insertion barriers reverse their orders with a dramatic increase in the termination barrier. The termination transition state which is shown in Figure 3(a) has both axial coordination sites of the Ni occupied. As proposed by Brookhart

Table 1 Comparison of calculated and experimental barriers

	Reaction barrier (kJ/mol)		
	Insertion	Branching	Termination
Absolute:			
Pure QM[a] ($\Delta H^\ddagger$)	70.3	53.6	40.6
QM/MM ($\Delta H^\ddagger$)	55.3	64.0	77.9
Experimental[b] ($\Delta G^\ddagger$)	42–46	—	—
Relative to insertion:			
QM/MM ($\Delta\Delta H^\ddagger$)	0.0	8.8	22.6
Experimental[c] ($\Delta\Delta G^\ddagger$)	0.0	5.4[d]	23.4[e]

[a] Ref. 11.
[b] M. Brookhart, Department of Chemistry, University of North Carolina at Chapel Hill, personal communication.
[c] Ref. 39.
[d] Assuming that all branches are methyl branches (methyl branches are experimentally observed to predominate). Applying Boltzmann statistics to this ratio at 273.15 K yields a $\Delta\Delta G$ of 5.4 kJ/mol.
[e] The weight-average molecular weight, M_w, of 8.1×10^5 g/mol provides an estimate for the ratio of termination events to insertion events of 1 : 28 900. Using Boltzmann statistics to this ratio gives a $\Delta\Delta G$ of 23.4 kJ/mol.

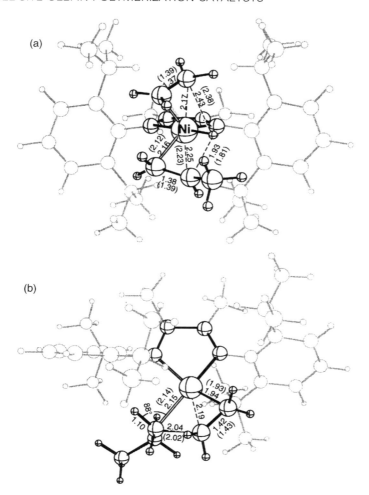

Figure 3 Optimized QM/MM (a) chain termination transition state via β-hydrogen transfer to the monomer and (b) insertion transition state. MM atoms are ghosted for clarity, and the dummy hydrogen atoms are omitted. Values in parentheses refer to the same geometric parameter found in the corresponding pure QM geometry. The pure QM model does not contain the bulky ligands. Distances are in ångstroms and angles in degrees

and co-workers [39,40], the bulky isopropyl groups on the aryl rings act to block the axial coordination sites, thereby dramatically increasing the termination barrier. In the insertion transition state which is sketched in Figure 3(b) the alkyl and olefin moieties lie in the coordination plane of the Ni center, removed from the bulky isopropyl groups. In this case there is little steric hindrance to the insertion process. In fact, the addition of the bulky aryl ligands actually reduces the insertion barrier. This stabilization of the insertion transition state with bulky ligands in the QM/MM model compared with that without the bulky ligands results from two effects. First, the resting state structure is destabilized by the bulky ligands, thereby lowering the insertion barrier by increasing the energy of the precursor. Additionally, there is the relaxation of the orientation of the aryl rings relative to the Ni diimine ring in the insertion transition state. Physically, there is a preference for the aryl rings and the diimine ring to be more parallel to maximize the π-bonding between the rings. When both axial coordination sites are occupied, as in the resting state and the termination transition state, the aryl rings are forced to be perpendicular to the diimine ring. In the insertion transition state where the axial sites are empty, the aryl rings can rotate away from the perpendicular orientations so as to enhance the π-bonding interaction between the rings, thereby stabilizing the insertion transition state. This concept helps rationalize the observed [39] increase in activity when the π-system of the diimine ring is extended. We conclude that the catalyst activity can be increased by enhancing the π-bonding interaction between the diimine rings and the aryl rings.

The chain branching mechanism originally proposed by Brookhart and co-workers [39] involves a discrete olefin hydride intermediate as sketched in Figure 2(c). Our calculations implicate a concerted pathway since no stable olefin hydride complex could be located. (The optimized QM/MM chain branching transition state is not displayed.) With the hybrid model, the chain isomerization barrier is calculated to be 64 kJ/mol, which is only slightly increased from 53.6 kJ/mol in the pure QM model.

Table 1 reveals that the reaction barriers calculated from our combined QM/MM model are in good agreement with the experimentally determined free energy barriers, both in relative and in absolute terms. This contrasts the results of the pure QM study where the bulky ligands were not modeled and even the order of the barrier heights was not reproduced. We note that there is excellent agreement between the calculated and experimental chain termination barrier relative to the insertion barrier. Specifically, the QM/MM relative barrier is $\Delta\Delta H^{\ddagger} = 22.6$ kJ/mol whereas the value determined experimentally is $\Delta\Delta G^{\ddagger} = 23.4$ kJ/mol. Our model therefore provides an accurate estimate of the polymer molecular weight, M_{w}. It is more difficult to compare directly the calculated barriers for chain branching with that of experiment because in our model we do not account for concentration effects. Figure 2(c) shows that the branching commences from the metal–alkyl complex and there is an equilibrium with the olefin-coordinated π-complex. Experimentally, the chain branching increases with decreasing monomer pressure, suggesting that the equilibrium between the π-complex and the metal–alkyl is crucial. To address this issue we are currently simulating the olefin capture process with the combined QM/MM *ab initio* molecular dynamics method [57].

As demonstrated in this example, computational chemistry offers a powerful method for eliciting mechanistic details of the catalysis. Such fundamental understanding of the chemistry is necessary for rational design and improvement of the catalyst systems. This example also demonstrates that nearly quantitative estimates of the reaction barriers and other properties, such as polymer molecular weights, can be made with the methodologies available. Consequently, effective screening of candidate catalysts can be performed with computational chemistry.

3.2 COMBINATORIAL-LIKE CHEMISTRY WITH MOLECULAR MODELING

In the pharmaceutical industry today, combinatorial chemistry is almost always part of the route towards the discovery of a new and effective drug [58]. Combinatorial synthesis of organometallic catalysts has not yet reached the maturity of that seen in the drug design industry. For this reason, screening the performance of a group of candidate catalysts is very laborious because each candidate is synthesized one at a time. Computational chemistry offers some unique potential since 'synthesis' on the computer is not as difficult a task as it is in the laboratory. On the computer, the chemist simply has to change one group or element to another and redo the calculation as opposed to performing a whole new synthesis. Thus computational chemistry is well suited to performing systematic studies of candidate catalysts in order to refine an existing system or to search out a new catalyst system.

We have recently undertaken a combinatorial-like approach to investigate systematically the intrinsic activity of d^0 transition metal complexes toward olefin polymerization [59,60]. We examined the chain propagation by ethylene insertion into the $M-C_2H_5$ bond for a number of d^0 $[L]M-C_2H_5^{(0,+,2+)}$ fragments, where $M = $ Sc(III), Y(III), La(III), Lu(III), Ti(IV), Zr(IV), Hf(IV), Ce(IV), Th(IV) and V(V), $L = NH-(CH_2)_2-NH^{2-}$ (1), $N(BH_2)-(CH_2)_2-(BH_2)N^{2-}$ (2), $O-(CH_2)_3-O^-$ (3), Cp_2^{2-} (4), $NH-Si(H_2)-C_5H_4^{2-}$ (5), $[(oxo)(O-(CH_2)_3-O)]^{3-}$ (6), $(NH_2)_2^{2-}$ (7), $(OH)_2^{2-}$ (8), $(CH_3)_2^{2-}$ (9), $NH-(CH_2)_3-NH^{2-}$ (10) and $O-(CH_2)_3-O^{2-}$ (11) (Figure 4). Only minimal or irreducible ligand structures were utilized since the aim of the study was to outline the influence of the metal and the first coordination sphere on the insertion energetics. The goal of the study was to provide insights into the necessary elements of selecting an olefin polymerization catalyst with an intrinsically low electronic insertion barrier.

All calculations were pure QM in nature and performed with the ADF density functional package. Systematic studies of this nature can currently be performed with modest computational facilities. Complete computational details are published elsewhere [59,60].

We first introduce some necessary terminology and nomenclature. For any given metal–ligand combination there is usually a preference in the metal–alkyl complex to adopt either a pyramidal or trigonal coordination as depicted in Figure 5. We have found that preferences for the two coordination types can be correlated with certain catalyst properties. Coordination of the monomer results in the formation of what is termed an olefin π-complex from which insertion proceeds. Figure 5 define two

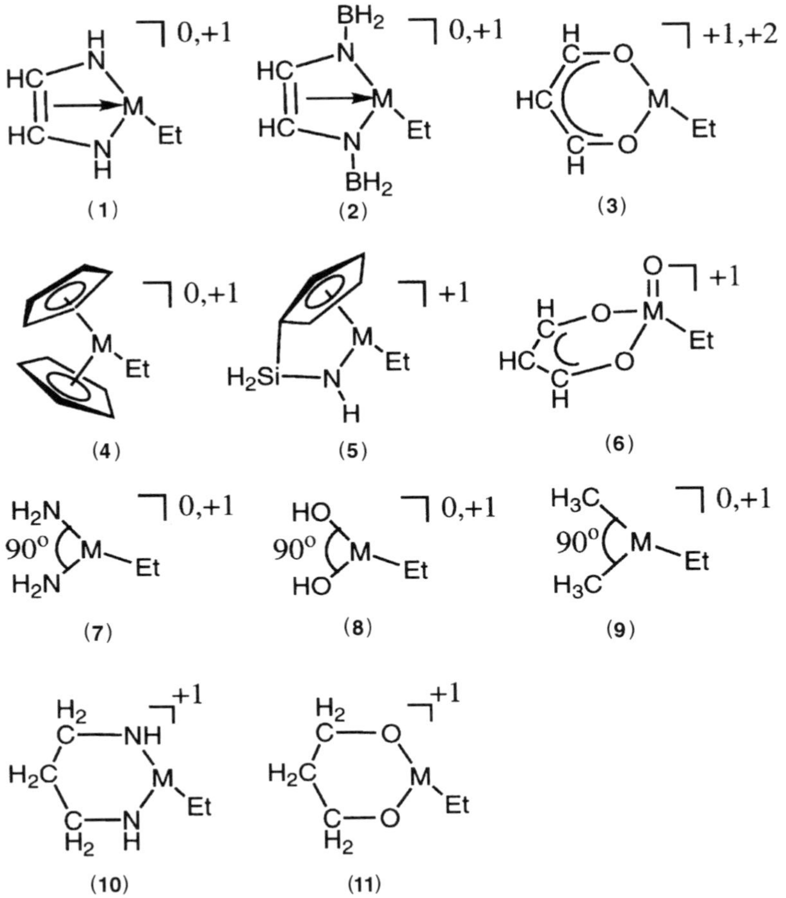

Figure 4 Ligand fragments used in systematic study of insertion barriers of various d^0 olefin polymerization catalysts

distinct types of insertion, frontside (FS) and backside (BS) insertion. The two are distinguished by whether the olefin inserts *syn* or *anti* to the β-agostic bond of the growing chain.

Plotted in Figure 6 are the insertion barriers of all metal–ligand combinations outlined in Table 2. It is apparent that the ethylene insertion barriers for all d^0 complexes are low and tend to be lower for light metals than for heavy metals. Analysis of the results is cumbersome and will not be detailed here. Instead, we provide a summary of the results in order to illustrate the kind of information that can be obtained from such a computational study. We have found (a) that olefin

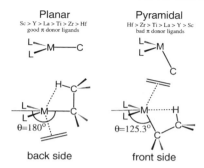

Figure 5 Schematic representation of the planar vs pyramidal configuration of the metal–alkyl complex and frontside vs back-side insertion process

insertion barriers for all d^0 complexes are generally small owing to a lack of metal d electrons which could fill an emerging carbon–carbon antibonding interaction in the insertion transition state. (b) Insertion barriers are smallest for metal–ligand combinations which have a high intrinsic aptitude for planar arrangement, such as light transition metals (Sc, Ti) ligated by good π-donor ligands such as amido ligands. (c) A large steric bulk facilitates insertion by favoring the insertion transition state if it restrains the metal–ligand framework to a planar arrangement. (d) Our results provide a means to manipulate insertion barriers by changing the metal, the donor atoms of the auxiliary ligands and the steric bulk of the auxiliary ligand. Steric modeling can be used to override intrinsic limitations imposed by the metal ion and the first coordination sphere of the auxiliary ligands.

The potential of systematic studies as illustrated above comes in extrapolating the results to original metal–ligand combinations which have not been synthesized and which may have more favorable patent possibilities. The results can even be used to

Table 2 Compound list for which ethylene insertion barriers are calculated and shown in Figure 6

Metal	Ligand framework (see Figure 4)
Sc(III)	1, 2, 3, 4, 7, 8, 9
Y(III)	1, 7, 8, 9
La(III)	1, 7, 8, 9
Ti(IV)	1, 2, 3, 4, 5, 6, 7, 8, 9, 10, 11
Zr(IV)	1, 3, 4, 6, 7, 8, 9, 10, 11
Hf(IV)	1, 4, 7, 8, 9, 10
V(V)	6
Lu(III)	1
Ce(IV)	1
Th(IV)	1
Ti(III)	7
Nb(III)	1

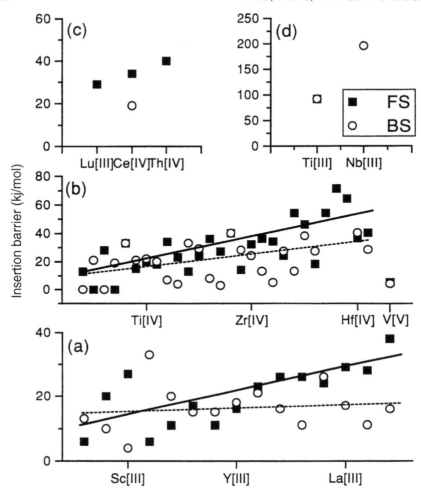

Figure 6 Barriers for the insertion of ethylene (y-axis) into the M–C$_\alpha$ bond of various d^0 catalyst systems (x-axis) in units of kJ/mol. Systems are grouped on the x-axis according to the central metal atom, with data points in the same sequence as they appear in Table 2. Circles and dashed lines refer to the BS insertion barrier, and squares and solid lines to the FS insertion barrier. (a) Barriers for compounds of the Sc triad; (b) the Ti triad and V; (c) lanthanides and actinides; and (d) d^1 and d^2 systems. Note the high activation barriers for the non-d^0 systems with Ti [29] and Nb [29] centers. The linear fit through the BS insertion barriers for the Ti triad (b, dashed line) shows an artificially enhanced slope since there are no BS insertion data points for [L]HfC$_2$H$_5^+$ (L = **7**, **8**, **9**)

extrapolate to complexes with non-zero d occupations. Efforts in our laboratory to study non-d^0 catalysts systematically are in progress.

3.3 A PRIORI *CATALYST DESIGN WITH MOLECULAR MODELING*

To date, computational modeling of metallocene catalysts (published in the open literature) has been confined to examining polymerization catalysts that have already been synthesized in the laboratory and shown to be promising. Although understanding how these existing catalysts function is valuable, the ultimate and more difficult challenge of molecular modeling is to design effective catalysts on the computer before time is spent on their synthesis in the laboratory. In this section we present our most recent efforts towards this goal.

An important development in the search for new single-site systems was the discovery of the first living Ziegler–Natta-type olefin polymerization catalyst by Scollard and McConville [61]. The catalyst is a Ti(IV) diamide system of the type $[ArNCH_2CH_2CH_2NAr]TiR^+$ where $Ar = 2,6\text{-}^iPr_2C_6H_3$ as shown in Figure 7. One intriguing characteristic of the catalyst system is that whereas the titanium system is a living polymerization catalyst, its zirconium analog produces only very low molecular weight oligomers with $n = 2–7$ [62]. In this study, it was our goal not only to understand the drastically different behavior of the Ti and Zr systems, but also to design a new ligand structure for the zirconium diamide system so as to boost its performance in terms of its activity and the molecular weight of the resulting polymer.

The computational methodology is similar to that of our examination of the Brookhart catalyst presented in Section 3.1. The full catalyst system is too large to be wholly treated at the DFT level. As a result, the full catalyst systems were examined with the combined QM/MM methodology. Figure 7 shows the full catalyst system where the carbon atoms with the asterisks represent atoms which are replaced with dummy hydrogen atoms in the model QM system. Full computational details are provided elsewhere [50].

The polymer length, measured by the polymer molecular weight (M_w), can be estimated by the difference in the rate of chain growth and chain termination. In terms of polymer lengths, dimerization catalysts and living olefin polymerization catalysts reside on opposite ends of the spectrum. With a dimerization catalyst, chain termination is more favorable than chain growth. In this way, once a monomer is added, the chain is immediately terminated so as to produce a dimer. On the other hand, in a living polymerization system the rate of chain termination compared with chain propagation is so insignificant that termination is not observed. In this way, the

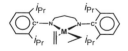

Figure 7 McConville-type catalyst. Atoms labeled with asterisks denote the atoms which are replaced by 'dummy' hydrogen atoms in the QM model system

polymer would appear to grow indefinitely. It is interesting that by simply changing the Ti to Zr, the McConville catalyst system is flipped from one end of the molecular weight spectrum to the other.

The combined QM/MM method was applied to the real McConville catalyst $[ArN(CH_2)_3NAr]MR^+$ (M = Ti, Zr; Ar = $2,6 - {^i}Pr_2C_6H_3$) with aryl rings attached to the chelating nitrogens. We calculate the barrier of termination to be 41 kJ/mol higher than the barrier of propagation for the titanium system. On the other hand, the difference in activation energy between propagation and termination is 0.4 kJ/mol for the zirconium complex. These computational results are in line with the findings by McConville that the titanium complex is a living olefin polymerization catalyst whereas the homologous zirconium complex is only able to oligomerize olefins. An analysis of the results showed that the isopropyl-substituted aryl rings in the titanium system are forced to stay perpendicular to the N−M−N plane in order to avoid the steric bulk of the diamide bridge. In this orientation the axial sites above and below the N−Ti−N plane are blocked and the termination transition state is destabilized. The steric interaction is reduced between the diamide bridge and the aryl rings in the zirconium system owing to the longer M−N bonds. As a result, the aryl rings can move out of the perpendicular position. Hence the isopropyl groups will be less efficient in retarding the chain termination by blocking the axial sites.

The lack of sufficient steric bulk in the zirconium diamide complex to retard the chain termination step led us to suggest a number of modified complexes based on the general principle. The modifications involved (see Figure 8) (i) increase of the steric bulk on the diamide bridge, **12** and **13**; (ii) increase of the steric bulk on the aryl rings, **14** and **15**; (iii) block one axial position by a hydrocarbon bridge, **16** and **17**. Results from the combined QM/MM calculations indicate that the suggested modifications, particularly those expressed in **15** and **17**, might be used to generate living olefin polymerization catalysts with much higher activities than the original

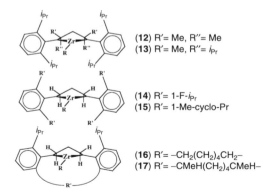

(12) R′= Me, R″= Me
(13) R′= Me, R″= i_{Pr}

(14) R′= 1-F-i_{Pr}
(15) R′= 1-Me-cyclo-Pr

(16) R′= $-CH_2(CH_2)_4CH_2-$
(17) R′= $-CMeH(CH_2)_4CMeH-$

Figure 8 Structures for the computationally screened catalysts based on the McConville diamine system

Table 3 Comparison of the catalytic capabilities of the new catalysts with the McConville and Brookhart catalysts for polymerization of ethylene

	Barrier (kJ/mol)			
Catalyst system	Insertion	Termination	$\Delta(\Delta E^{\ddagger})$ (kJ/mol)^a	Predicted M_w
Original McConville catalyst: [ArN(CH$_2$)$_3$NAr]MR^{+b}				
M = Ti, Ar = 2,6-^{i}Pr$_2$C$_6$H$_3$	39	81	41	5.6×10^8 (living)
M = Zr, Ar = 2,6-^{i}Pr$_2$C$_6$H$_3$	49	49	−0.4	dimer (2–7)
Modified McConville catalyst: [ArNCR′R′CH$_2$CR′R″NAr]ZrR$^+$				
R′ = R″ = CH$_3$ (**12**)	39	50	12	3.3×10^3
R′ = CH$_3$, R″ = ^{i}Pr (**13**)	12	28	16	1.8×10^4
[ArN(CH$_2$)$_3$NAr]ZrR$^+$				
Ar = 2,6-(CMe$_2$F)$_2$-C$_6$H$_3$ (**14**)	9	39	30	4.9×10^6
Ar = 2,6-(1-Me-cyclo-Pr)$_2$−C$_6$H$_3$ (**15**)	−7	30	37	1.0×10^8
[ArN(CH$_2$)$_3$NAr]ZrR$^+$ Ar = 1-^{i}Pr-6-R′−C$_6$H$_3$				
R′ = −CH$_2$(CH$_2$)$_4$CH$_2$− (**16**)	34	42	8	6.0×10^2
R′ = −CHMe(CH$_2$)$_4$CHMe− (**17**)	14	50	36	6.2×10^7
Brookhart catalyst: [ArNCH$_2$CH$_2$NAr]NiR$^+$ Ar = 2,6-^{i}Pr$_2$C$_6$H$_3$	55 (42–46)	78	23 (23)	5.8×10^4 (8.1×10^4)

^{a} $\Delta(\Delta E^{\neq})$: the difference in activation energy between insertion (chain propagation) and termination. Experimental values in parentheses.
^{b} The McConville catalysts [61,62]. The monomer used in the experiments is 1-hexene.

titanium-based diamide complex suggested by McConville. Table 3 details our computation estimates of the activities and molecule weights. Work by Piers and co-workers [63] is in progress to synthesize living zirconium-based diamide olefin polymerization catalysts similar to those suggested here.

4 CONCLUSION

Recent developments in molecular modeling combined with the rapid decrease in the cost of pure number crunching computing is transforming computational chemistry into a cost effective research tool for the refinement and design of new olefin polymerization catalysts. We believe that the mechanistic insights that modeling provides and its ability to perform large-scale systematic studies will lead to the *a priori* design of novel catalyst systems on the computer. Combined with experimental efforts, the utilization of computational chemistry in an industrial setting will provide a competitive edge in the fiercely competitive olefin polymerization business.

5 ACKNOWLEDGMENTS

This investigation was supported by the National Sciences and Engineering Research Council of Canada (NSERC), by the donors of the Petroleum Research Fund, administered by the American Chemical Society (ACS-PRF No. 31205-AC3), and by Novacor Research and Technology Corporation (NRTC) of Calgary. The generosity of the Izaak Walton Killam Memorial Foundation greatly appreciated by T.Z., L.D. and T.K.W. The authors are greatly indebted to the staff of NRTC for valuable discussions, especially Drs L. Fan, D. Harrison and J. McMeeking. They also thank Professors W. Piers, M. Brookhart and D. H. McConville for stimulating discussions.

6 REFERENCES

1. van Leeuwen, P. W. N. M., Morokuma, K. and van Lengthe, J. H., in Ugo, R. and James, B. R. (Eds), *Catalysis by Metal Complexes*, Kluwer, Dordrecht, 1995.
2. Raghavachari, K. and Anderson, J. B., *J. Phys. Chem.*, **100**, 12960 (1996).
3. Hehre, W. J., Radom, L., Schleyer, P. v. R. and Pople, J. A., *Ab Initio Molecular Orbital Theory*, Wiley, New York, 1986.
4. Szabo, A. and Ostlund, N. S., *Modern Quantum Chemistry*, McGraw-Hill, New York, 1982, Vol. 1.
5. Kohn, W., Becke, A. D. and Parr, R. G., *J. Phys. Chem.*, **100**, 12974 (1996).
6. Ziegler, T., *Chem. Rev.*, **91**, 651 (1991).
7. Woo, T. K., Fan, L. and Ziegler, T., *Organometallics*, **13**, 432 (1994).
8. Woo, T. K., Margl, P. M., Lohrenz, J. C. W., Blöchl, P. E. and Ziegler, T., *J. Am. Chem. Soc.*, **118**, 13021 (1996).
9. Woo, T. K., Margl, P. M., Blöchl, P. E. and Ziegler, T., *Organometallics*, **16**, 3454 (1997).
10. Woo, T. K., Fan, L. and Ziegler, T., *Organometallics*, **13**, 2252 (1994).
11. Deng, L., Margl, P. M. and Ziegler, T., *J. Am. Chem. Soc.*, **119**, 1094 (1997).
12. Fan, L., Krzywicki, A., Somogyvari, A. and Ziegler, T., *Inorg. Chem.*, **33**, 5287 (1994).
13. Fan, L., *et al.*, *Can. J. Chem.*, **73**, 989 (1995).
14. Fan, L., Harrison, D., Woo, T. K. and Ziegler, T., *Organometallics*, **14**, 2018 (1995).
15. Fan, L., Krzywicki, A., Somogyvari, A. and Ziegler, T., *Inorg. Chem.*, **25**, 4003 (1996).
16. Lohrenz, J. C. W., Woo, T. K. and Ziegler, T., *J. Am. Chem. Soc.*, **117**, 12793 (1995).
17. Lohrenz, J. C. W., Woo, T. K., Fan, L. and Ziegler, T., *J. Organomet. Chem.*, **497**, 91 (1995).
18. Margl, P., Lohrenz, J. C. W., Blöchl, P. and Ziegler, T., *J. Am. Chem. Soc.*, **118**, 4434 (1996).
19. Margl, P. and Ziegler, T., *Organometallics*, **15**, 5519 (1996).
20. Margl, P., Blöchl, P. and Ziegler, T., *J. Am. Chem. Soc.*, **118**, 5412 (1996).
21. Musaev, D. G., Froese, R. D. J., Svensson, M. and Morokuma, K., *J. Am. Chem. Soc.*, **119**, 367 (1997).

22. Cavallo, L. and Guerra, G., *Macromolecules*, **29**, 2729 (1996).
23. Froese, R. D. J., Musaev, D. G. and Morokuma, K., *J. Am. Chem. Soc.*, **119**, 7190 (1997).
24. Stanton, R. V. and Merz, K. M. J., *J. Chem. Phys.*, **100**, 434 (1993).
25. Cornell, W. D., *et al.*, *J. Am. Chem. Soc.*, **117**, 5179 (1995).
26. Brooks, B. B., *et al.*, *J. Comput. Chem.*, **4**, 187 (1983).
27. Allinger, N. L., *J. Am. Chem. Soc.*, **99**, 8127 (1977).
28. Allured, V. S., Kelly, C. M. and Landis, C. R., *J. Am. Chem. Soc.*, **113**, 1 (1991).
29. Rappé, A. K., Casewit, C. J., Colwell, K. S., and Skiff, W. M., *J. Am. Chem. Soc.*, **114**, 10024 (1992).
30. Corradini, P., Guerra, G., Vacatello, M. and Villani, V., *Gazz. Chim. Ital.*, **118**, 173 (1988).
31. Cavallo, L., Guerra, G., Oliva, L., Vacatello, M. and Corradini, P., *Polym. Commun.*, **30**, 16 (1989).
32. Guerra, G., Cavallo, L., Moscardi, G., Vacatello, M. and Corradini, P., *J. Am. Chem. Soc.*, **116**, 2988 (1994).
33. Cavallo, L., Guerra, G., Corradini, P., Resconi, L. and Waymouth, R. M., *Macromolecules*, **26**, 260 (1993).
34. Castonguay, L. A. and Rappé, A. K., *J. Am. Chem. Soc.*, **114**, 5832 (1992).
35. Hart, J. R. and Rappé, A. K., *J. Am. Chem. Soc.*, **115**, 6159 (1993).
36. Höweler, U., Mohr, R., Knickmeier, M. and Erker, G., *Organometallics*, **13**, 2380 (1994).
37. Kawamura-Kuribayashi, H., Koga, N. and Morokuma, K., *J. Am. Chem. Soc.*, **114**, 8687 (1992).
38. Woo, T. K., Fan, L. and Ziegler, T., in Fink, G., Mülhaupt, R. and Brintzinger, H. H. (Eds), *Ziegler Catalysts; Recent Scientific Innovations and Technological Improvements*, Springer, Berlin, 1996, p. 291.
39. Johnson, L. K., Killian, C. M. and Brookhart, M., *J. Am. Chem. Soc.*, **117**, 6414 (1995).
40. Johnson, L. K., Mecking, S. and Brookhart, M., *J. Am. Chem. Soc.*, **118**, 267 (1996).
41. Haggin, J., *Chem. Eng. News*, 6 (1996).
42. Singh, U. C. and Kollman, P. A., *J. Comput. Chem.*, **7**, 718 (1986).
43. Field, M. J., Bash, P. A. and Karplus, M., *J. Comput. Chem.*, **11**, 700 (1990).
44. Maseras, F. and Morokuma, K., *J. Comput. Chem.*, **16**, 1170 (1995).
45. Deng, L., Woo, T. K., Cavallo, L., Margl, P. M. and Ziegler, T., *J. Am. Chem. Soc.*, **119**, 6177 (1997).
46. Warshel, A. and Levitt, M., *J. Mol. Biol.*, **103**, 227 (1976).
47. Gao, J., in Lipkowitz, K. B. and Boyd, D. B. (Eds), *Reviews in Computational Chemistry*, VCH, New York, 1996, Vol. 7.
48. Froese, R. D. J., Musaev, D. G. and Morokuma, K., *J. Am. Chem. Soc.*, submitted for publication.
49. Musaev, D. G., Froese, R. D. J. and Morokuma, K., *Organometallics*, submitted for publication.
50. Deng, L., Ziegler, T., Woo, T. K., Margl, P. M. and Fan, L., *J. Am. Chem. Soc.*, submitted for publication.
51. Baerends, E. J., Ellis, D. E. and Ros, P., *Chem. Phys.*, **2**, 41 (1973).
52. Baerends, E. J. and Ros, P., *Chem. Phys.*, **2**, 52 (1973).
53. Becke, A., *Phys. Rev. A*, **38**, 3098 (1988).
54. Perdew, J. P., *Phys. Rev. B*, **34**, 7406 (1986).
55. Perdew, J. P., *Phys. Rev. B*, **33**, 8822 (1986).
56. Keim, W., *Angew. Chem., Int. Ed. Engl.*, **29**, 235 (1990).
57. Woo, T. K., Margl, P. M., Deng, L. and Ziegler, T., in preparation.
58. DeWitt, S. H. and Czarnik, A. W., *Acc. Chem. Res.*, **29**, 114 (1996).
59. Margl, P. M., Deng, L. and Ziegler, T., *Organometallics*, in press.
60. Margl, P. M., Deng, L. and Ziegler, T., *J. Am. Chem. Soc.*, submitted for publication.

61. Scollard, J. D. and McConville, D. H., *J. Am. Chem. Soc.*, **118**, 10008 (1996).
62. McConville, D. H., Department of Chemistry, University of British Columbia, personal communication.
63. Piers, W. E., Department of Chemistry, University of Calgary, personal communication.

PART II

Cycloolefins, Dienes and Functionalized Olefin Polymerization

5

Homo- and Copolymerization of Cycloolefins by Metallocene Catalysts

WALTER KAMINSKY AND MICHAEL ARNDT-ROSENAU
Institute for Technical and Macromolecular Chemistry, University of Hamburg, Germany

1   INTRODUCTION

Strained cyclic olefins such as cyclobutene, cyclopentene, norbornene, 1,4,5,8-dimethano-1,2,3,4,4a,5,8,8a-octahydronaphthalene (DMON) and 1,4,5,8,9,10-trimethano-1,2,3,4,4a,5,8,8a,9,9a,10,10a-dodecahydroanthracene (TMDA) (Figure 1) can be used as monomers and comonomers in a wide variety of polymers. They can be polymerized by ring opening polymerization (ROMP) featuring elastomeric materials [1] or by double bond opening (vinyl polymerization). Homopolymerization of cyclic olefins by double bond opening can be achieved by several transition metal catalysts, namely Ni [2,3], Pd [4–6] and metallocene catalysts [7] (Figure 2).

While polymerization of cyclic olefins with heterogeneous Ziegler–Natta catalysts is accompanied by ring opening [8], homogeneous metallocene, Ni or Pd catalysts promote vinyl polymerization. The polymers feature two chiral centers per monomer unit and therefore are ditactic. The cycloolefins may be divided into achiral, monocyclic and prochiral, polycyclic types. Polymerization of both types by chiral metallocenes may yield tactic, crystalline homopolymers [9]. The melting-points of these homopolymers are extremely high and in most cases decomposition occurs before melting. While atactic polymers can be dissolved in hydrocarbon solvents at least to some extent, tactic polymers are hardly soluble.

Metallocene-based Polyolefins Edited by J. Scheirs and W. Kaminsky
© 2000 John Wiley & Sons Ltd

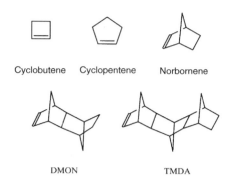

Figure 1 Cyclic olefin which can be polymerized using metallocene catalysts

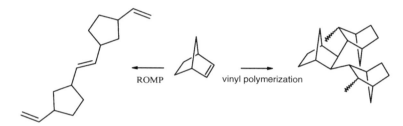

Figure 2 Mechanisms for the transition metal-catalyzed polymerization of norbornene

Cycloolefins may also be divided into two groups based on the possibility of undergoing β-hydrogen elimination reactions leading to chain termination or isomerizations of the growing chain. In contrast to norbornene and derivatives, the monocyclic monomers cyclobutene and especially cyclopentene may give β-hydrogen elimination reactions (Figure 3). A consequence is the unusual structure of

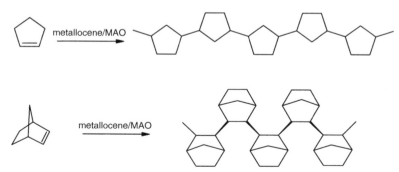

Figure 3 Polymerization of cyclopentene and norbornene by metallocene catalysts

poly(cyclopentene) containing *cis*- and *trans*-1,3-linkages of the monomer units [10,11].

Copolymers of norbornene and its derivatives with ethene (COC; cycloolefin copolymers) show a high glass transition temperature in combination with good chemical resistance and excellent transparency owing to their rigid cyclic monomer units [12]. Similar polymers have been produced by less active V catalysts or in a two-step process by hydrogenation of ROMP polymers (Figure 4) [13].

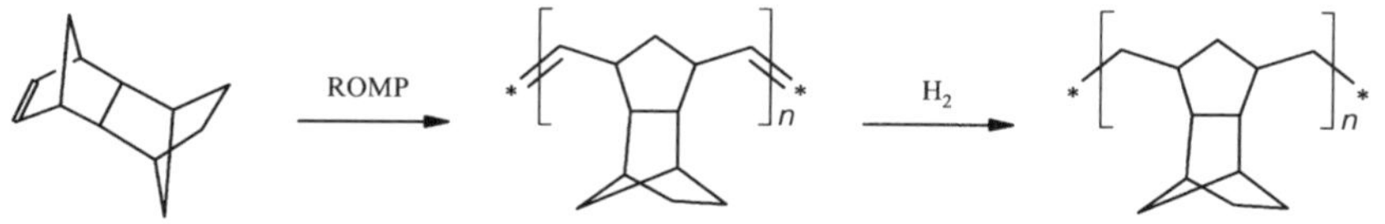

Figure 4 Generation of high T_g polymers from cycloolefins by ring opening polymerization of DMON and hydrogenation of the resulting polymer

2 HOMO- AND COPOLYMERIZATION OF CYCLOPENTENE

Cyclopentene was the first cyclic monomer to be polymerized by metallocene catalysts. Polymerization by *rac*-[En(Ind)₂]ZrCl₂/methylaluminoxane (MAO) at room temperature yields a semicrystalline polymer [9] which is hardly soluble in common organic solvents. Investigation of polymers produced at high catalyst concentrations shows that they have fairly low molecular weights which go along with melting over a broad range from 150 to 350°C. At low catalyst concentrations crystalline polymers with melting ranges from 200 to 395°C are produced. This has been shown by temperature-dependent wide-angle scattering using synchroton radiation [9f]. Crystalline reflexes are observed at $2\theta = 16$, 19.5 and 23.9°, corresponding to distances of 0.56, 0.46 and 0.37 nm. The plot (Figure 5) of the intensity of the reflex at 19.5° shows a decrease in intensity with increasing temperature for all signals. At 395°C the polymer is completely amorphous. After cooling of the sample, the reflexes appear again with even higher intensities. Therefore, a higher degree (>60%) of crystallinity is achieved by crystallization from the melt. The glass transition temperature of *cis*-1,3-poly(cyclopentene) is about 65°C [11a].

Owing to the low solubility of this semicrystalline material, the microstructure has been assigned by using oligomers and soluble, amorphic poly(cyclopentenes) which may be generated by using other metallocenes. Oligomerization of cyclopentene is possible by using low monomer/catalyst ratios or more conveniently by 'polymerization' of cyclopentene in the presence of hydrogen as a chain transfer agent. The structures of cyclopentene hydrooligomers have been elucidated by comparing them with model compounds [10] and 2D-NMR techniques [11]. The investigations have shown that all monomer units within poly(cyclopentene) are incorporated in a 1,3-

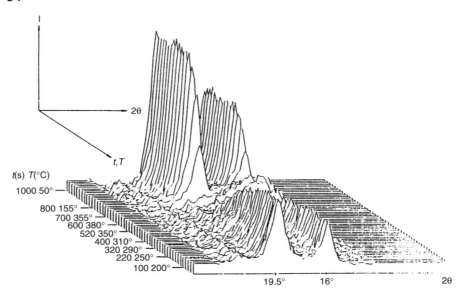

Figure 5 Temperature dependence of the wide-angle X-ray scattering (synchroton radiation) of poly(cyclopentene) produced by *rac*-[En(IndH$_4$)$_2$]ZrCl$_2$/MAO at 20 °C. The sample was heated from 50 to 400 °C in 600 s and cooled to 50 °C within 400 s

fashion. Surprisingly, in addition to *cis*-1,3-connected monomer units, *trans*-1,3 units are also detectable for some of the metallocenes. Both *cis*- and *trans*-1,3 incorporation have been proposed to result from isomerization of *cis*-1,2 inserted cyclopentene. The mechanism of the formation of *cis*-1,3-incorporated monomer units is similar to that proposed for the formation 1,3-enchained monomer units in polypropene (Figure 6). It involves a β'-hydrogen elimination, rotation of the resulting olefin around the Zr–olefin axis and reinsertion.

The mechanism of the formation of *trans*-1,3-linked monomer units has been extensively studied by Collins and co-workers using deuterated cyclopentene [10c]. The most plausible assumption is that also in the case of *trans* incorporation the first step is a β'-hydrogen elimination resulting in an olefin–hydride complex which isomerizes by migration of the transition metal to the other face of the olefin via a σ-C–H complex (Figure 7).

Collins and co-workers [10c] compared several catalysts for the polymerization of cyclopentene according to the structure of the resulting polymers (Table 1). Sterically more demanding ligands and elevated temperatures seem to favor the formation of *trans* structures.

Taking into account the stereo- and regioselectivity of the polymerization, four types of regular structures for poly(cyclopentenes) are possible (Figure 8), but until

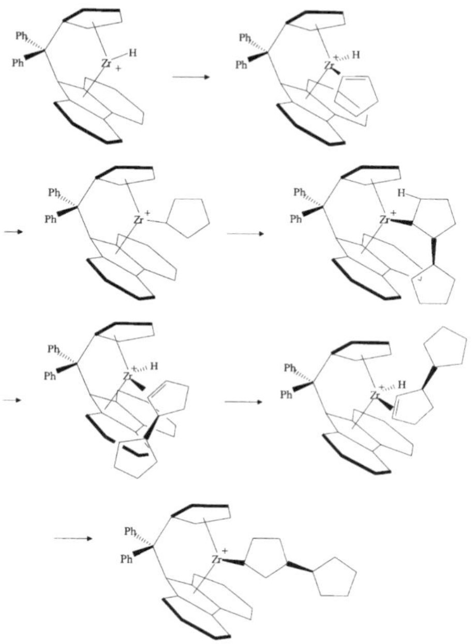

Figure 6 Mechanism of the formation of *cis*-1,3-enchained polycyclopentene shown for the polymerization catalyzed by [Ph$_2$C(Cp)(Flu)]ZrCl$_2$/MAO

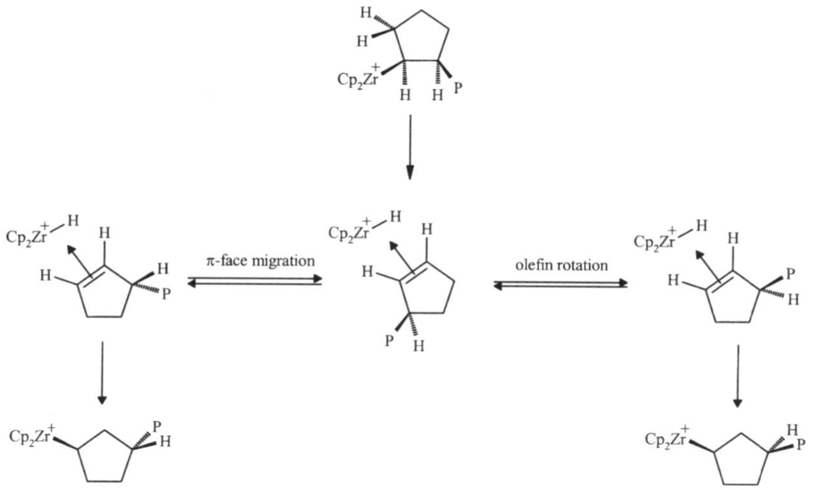

Figure 7 Schematic diagram of the isomerization processes in cyclopentene polymerization leading to the formation of *cis*- and *trans*-1,3-enchained configurational base units of the polymer

Table 1 Polymerization activity and microstructure of cyclopentene produced at 25 °C by metallocene/MAO catalysts [11a,12]

Catalyst	Productivity (kg/mol h)	*cis* (%)	M_n (g/mol)
rac-[En(IndH$_4$)$_2$]ZrCl$_2$	11	64	880
rac-[En(Ind)$_2$]ZrCl$_2$	6	>98	360
rac-[En(3-^{iso}PrCp)$_2$]ZrCl$_2$	5	34	1100
rac-[(CH$_2$)$_3$(Ind)$_2$]ZrCl$_2$	1	34	760
[En(Cp)$_2$]ZrCl$_2$	4	>99	460
Cp$_2$ZrCl$_2$	5	97	560
[Ph$_2$C(Flu)(Cp)]ZrCl$_2$	–	>99	–

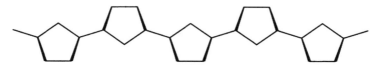

Erytho-disyndiotactic

Erytho-diisotactic

Threo-disyndiotactic

Threo-diisotactic

Figure 8 Microstructures of regio- and stereoregular poly(cyclopentenes)

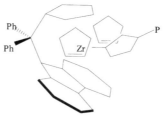

Figure 9 Simultaneous coordination of two monomers to [[Ph$_2$C(Cp)(Flu)] Zr(cyclopentyl)]$^+$. A possible explanation for the lack of stereoregularity observed in cyclopentene polymerization by this catalyst

now only the *erythro*-diisotactic polymer has been obtained as a crystalline material. This may have two reasons [11a]: (i) the isomerization process leading to the *cis* incorporation of the monomer can occur twice and thereby invert the configuration at an already *cis*-1,3-inserted monomer; and (ii) kinetic measurements indicate a reaction order of two for the polymerization of cyclopentene and therefore two monomers may be present at the active transition metal center, preventing the production of a highly syndiotactic poly(cyclopentene) from C_s-symmetric metallocenes such as [Me$_2$C(Cp)(Flu)]ZrCl$_2$ (Figure 9).

In contrast to homopolymerization, copolymerization of ethene and cyclopentene results in a *cis*-1,2-incorporation of the cyclic monomer into the polyethene chain [9a–d,11a]. Copolymerization parameters and investigations of the microstructure of ethene–cyclopentene copolymers show that the formation of cyclopentene blocks is unfavorable ($r_{CP} = 0$). Table 2 gives an overview of ethene–cyclopentene copolymerization by metallocene catalysts.

Table 2 Copolymerization parameter r_{ethene} for ethene–cyclopentene copolymerization as determined by non-linear regression from the copolymer composition (all parameters have been re-evaluated using this method)

Metallocene	Olefin	Temperature ($^\circ$C)	r_{Ethene}
rac-[En(Ind)$_2$]ZrCl$_2$	Ethene	−10	169
rac-[En(Ind)$_2$]ZrCl$_2$	Ethene	0	243
rac-[En(Ind)$_2$]ZrCl$_2$	Ethene	10	336
rac-[En(Ind)$_2$]ZrCl$_2$	Ethene	20	370
rac-[En(Ind)$_2$]HfCl$_2$	Ethene	30	45
rac-[MeSi(Ind)$_2$]ZrCl$_2$	Ethene	30	125
[Ph$_2$C(Cp)(Flu)]ZrCl$_2$	Ethene	30	85

^{a} The copolymerization parameter $r_{cyclopentene}$ was found to be zero because even at cyclopentene fractions above 99 mol% less than 40 % of cyclopentene was incorporated into the polymer chain. There was no evidence for blocks of cyclopentene although homopolymerization is possible. In some cases we already have shown that the microstructure of these copolymers is not that predicted from the first-order Markovian model used for evaluation of the parameters.

As can be seen from the copolymerization parameters, only a low content of cyclopentene is incorporated into the polymer chain even at molar fractions above 95% of cyclopentene in the reactor. Therefore, the copolymers are LLDPE-like, showing densities of 0.92–0.93 g/cm³ for copolymers containing 1.5–6 mol% cyclic monomer units. The crystallinity and melting-points of these copolymers at a given incorporation rate are lower than those of corresponding ethene–propene and ethene–hexene copolymers, reflecting the bulk of the comonomer.

Even more difficult is the incorporation of cyclopentene units into the chain of an α-olefin. Copolymerization of cyclopentene with propene yields copolymers showing a strong decrease in molecular weight with increase in the cyclopentene content [14], while attempts to copolymerize cyclopentene with pentene or hexene yield low molecular weight polypentenes (hexenes) showing cyclopent-2-en-1-yl and cyclopent-3-en-1-yl end groups [11a]. This indicates chain termination by β'-hydrogen elimination after insertion of cyclopentene.

The reactivity of cyclopentene in metallocene catalyzed polymerization may be explained by the scheme of reactions shown in Figure 10.

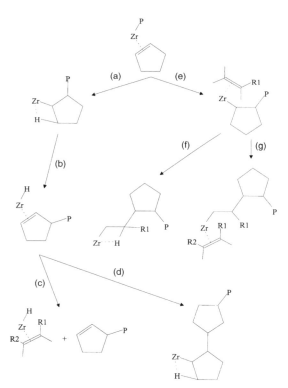

Figure 10   Schematic diagram summarizing the reactivity of cyclopentene in homo- and copolymerization with ethene and α-olefins

After *cis*-1,2 insertion of cyclopentene, the active species is stabilized by a β'-agostic interaction (a) or a small monomer (such as ethene or propene) which is able to approach the active site and insert (e, f, g) or the β'-hydrogen elimination takes place (b, in homopolymerization or with bulky comonomers). The hydride–olefin complex which results from the hydrogen elimination leads to chain termination if the olefin is replaced by a monomer (c) or undergoes isomerization reactions (rotation around the Zr–olefin bond; π-face migration) until the steric demand of the active species allows insertion of the next monomer unit (d). Chain termination is favored if higher α-olefins are present in the reaction mixture which are able to replace the polymer chain, whereas in case of cyclopentene which coordinates only weakly with transition metals isomerization is favored, resulting in *cis*- and *trans*-1,3-incorporation.

Similar effects are observed in propene polymerization (for a more detailed discussion and references, see Ref. 11c) after a regioirregular 2,1-insertion: in homopolymerization chain termination after 2,1-insertion is possible in addition to isomerization yielding a 1,3-enchained monomer unit, whereas in copolymerization with ethene, ethene is inserted after a 2,1-insertion. Isomerization reactions are also observed for propene (and butene) polymerization and oligomerization by metallocenes at low monomer concentration.

3 HOMO- AND COPOLYMERIZATION OF NORBORNENE AND NORBORNENE DERIVATIVES

Homopolymerization of norbornene and norbornene derivatives is also possible by using metallocene catalysts, although the activity is significantly lower. Table 3 gives an overview of the catalytic performance of several metallocenes [9e,15].

Norbornene is inserted into Zr–H and Zr–C bonds in as *cis-exo* orientation [11a,16]. Therefore, in contrast to cyclopentene, norbornene and its derivatives cannot undergo β-hydrogen elimination as one β-hydrogen is located on the *endo* side and *trans* to the Zr–C bond whereas elimination of the β'-hydrogen would result in the formation of an anti-Bredt olefin (Figure 11). Hence there is no possibility of relief of the steric stress accompanied by the incorporation of a disubstituted olefin into a Zr–C bond which already carries a 'branch' on the carbon atom. This explains why sterically less hindered metallocenes feature a higher activity (in fact ligand-free catalysts based on 'naked' nickel or palladium feature activities which are orders of magnitude higher).

Generally, polynorbornenes may be grouped into those soluble in toluene and those precipitating during the polymerization. WAXS and ^{13}C CP/MAS NMR investigations and high-temperature, high-resolution ^{13}C NMR studies of the soluble polymers confirm the different structures of the polymers and permit a further classification (Figure 12) [17].

Table 3 Comparison of several metallocene/MAO catalysts for the polymerization of norbornene at 25 °C in toluene ([Nor] = 2 M)

Catalyst	Catalyst (mol/l)	Al/ catalyst	Productivity (kg/mol h)	M_w (g/mol)	M_w/M_n
Cp$_2$ZrCl$_2$	7.1×10^{-4}	240	1.6	9.3×10^3	2.0
rac-[C$_2$H$_4$(Ind)$_2$]ZrCl$_2$	9.2×10^{-4}	560	0.8	Insoluble	
rac-[C$_2$H$_4$(H$_4$Ind)$_2$]ZrCl$_2$	1.0×10^{-3}	510	0.8	Insoluble	
rac-[Me$_2$Si(Ind)$_2$]ZrCl$_2^{d)}$	4.4×10^{-4}	410	0.7	Insoluble	
meso-[Me$_2$Si(Ind)$_2$]ZrCl$_2$	4.4×10^{-4}	470	0.7	Insoluble	
[Me$_2$C(Cp)(Flu)]ZrCl$_2$	5.6×10^{-4}	340	0.9	2.7×10^4	2.0
[Ph$_2$C(Cp)(Flu)]ZrCl$_2$	4.4×10^{-4}	410	2.2		

Hydrooligomerizations of norbornene have been used to assign the structures of the hydrodimers and -trimers investigated and correlated with the symmetry of the metallocene.

The polymerization of norbornene is shown to proceed by *cis-exo* insertion. From complexation with transition metals [18] and insertions into M−C bonds [19] it is well known that in comparison with the *endo* positions, the *exo* positions are more reactive. This can be explained in terms of electronic and steric effects [20].

Cis-exo insertion features *erythro*-ditactic polymers. While a 2,3-bis(*exo*)-disubstituted norbornane features an *R,S(S,R)* configuration at the configurational base unit, the *erythro*-ditacticities are represented by the configurations shown in Table 4.

Assuming the same mechanism as in the polymerization of α-olefins independent of the relative topicity the stereogenic C_s-symmetric metallocenes should feature *erythro*-disyndiotactic polymers, whereas C_2-symmetric metallocenes are expected to form *erythro*-diisotactic products. A single inversion of the relative topicity causes a *meso* (racemic) linkage in the product of the C_s-(C_2-)symmetric catalyst. The investigation of the hydrotrimers by ^{13}C NMR spectroscopy showed tactic products to be dominant in case of the bridged metallocene catalysts accompanied by the *meso,rac*- (*rac,meso*-)linked hydrotrimer (Figure 13).

Investigation of the dimer and trimer distribution indicated that polymerization of norbornene with rac-[Me$_2$Si(Ind)$_2$]ZrCl$_2$/MAO is highly stereospecific. From the formation of the *meso,meso*-trimer by the C_2-symmetric metallocene, an *erythro*-

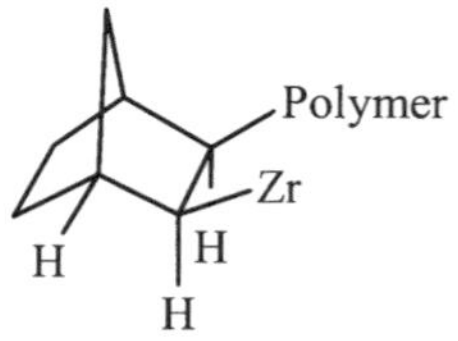

Figure 11 Owing to *cis-exo* insertion of norbornene into the growing polymer chain, no β-hydrogen elimination can occur

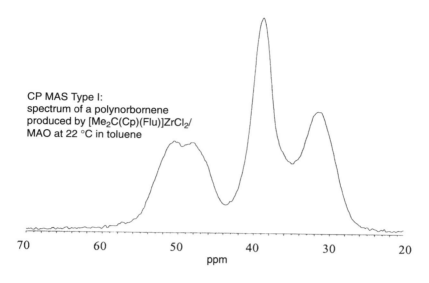

CP MAS Type I:
spectrum of a polynorbornene
produced by [Me$_2$C(Cp)(Flu)]ZrCl$_2$/
MAO at 22 °C in toluene

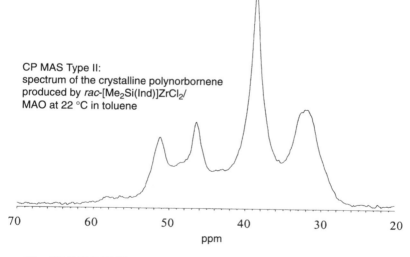

CP MAS Type II:
spectrum of the crystalline polynorbornene
produced by *rac*-[Me$_2$Si(Ind)]ZrCl$_2$/
MAO at 22 °C in toluene

Figure 12 CP/MAS NMR spectra of polynorbornenes produced by metallocene catalysts. While the polymer produced by [Me$_2$C(Cp)(Flu)]ZrCl$_2$/MAO is amorphous and soluble in toluene, that produced by *rac*-[Me$_2$Si(Ind)$_2$]ZrCl$_2$/MAO is insoluble and semicrystalline

Table 4 Configuration of the pseudoasymmetric centers and linkages in ditactic polymers

Type	Configuration of the centers and linkages
Erythro-diisotactic	$-(RS)(RS)(RS)(RS)(RS)(RS)(RS)-$
	m m m m m m m
Erythro-disyndiotactic	$-(RS)(SR)(RS)(SR)(RS)(SR)(RS)(SR)-$
	r r r r r r r
Erythro-atactic	$-(RS)(SR)(SR)(RS)(SR)(SR)(RS)(RS)-$
	r m r r m r m

rr-triad   mm-triad   mr-triad

Figure 13   Iso-, syndio- and atactic triad of poly(*cis-exo*-norbornene)

diisotactic microstructure of the polymer has been deduced, in agreement with the theory. The polynorbornene obtained from this catalyst reflects the high stereo-selectivity; it is semicrystalline. The WAXS of these polymers feature crystalline reflexes while those of less stereoregular polynorbornenes show two amorphous halos (Figure 14).

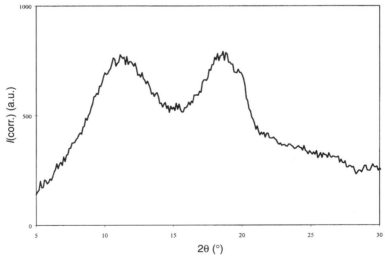

Figure 14   Typical WAXS of an amorphous polynorbornene (Cp_2ZrCl_2/MAO catalyzed polymerization) featuring two amorphous halos

In contrast, the oligomer distribution obtained by using $[Ph_2C(Flu)(Cp)]ZrCl_2/$ MAO as the catalyst indicates more stereoerrors, which correlates with the amorphic nature of the polymer and its solubility in organic solvents.

By using $Cp_2ZrCl_2/$MAO, as a consequence of chain end control no *rac,rac*-trimer is produced. Calculation of $\Delta G^{\ddagger}_{meso,meso} - \Delta G^{\ddagger}_{meso,rac}$ the difference in the free energies of activation between a *meso* linkage following a *meso* and racemic sequence and a racemic linkage following a *meso* sequence shows that alternating sequences are favored over diisotactic linkages ($30°C$, $-1.96\,kJ/mol$; $-30°C$, $-1.43\,kJ/mol$). This behavior results from the influence of the penultimate monomeric unic on the insertion reaction.

The homopolymers of cycloolefins such as norbornene and tetracyclododecene are not processable owing to their high melting-points and their insolubility in common organic solvents. By copolymerization of these cyclic olefins with ethene or α-olefins, cycloolefin copolymers (COC) are produced, representing a new class of thermoplastic amorphous materials [21]. Cycloolefin copolymers are characterized by excellent transparency and high, long-life service temperatures. They are solvent and chemical resistant and can be melt-processed. Owing to their high carbon/hydrogen ratio, these polymers have a high refractive index (1.53 for an ethene/norbornene copolymer at 50 mol% incorporation). Their stability against hydrolysis and chemical degradation, in combination with their stiffness, makes them interesting materials for optical applications, for example in compact discs, lenses or films.

Early attempts to produce such copolymers were made using heterogeneous $TiCl_4/AlEt_2Cl$ or vanadium catalysts, but real progress was made utilizing metallocene catalysts for this purpose. They are about 10 times more active than vanadium systems and by choosing the metallocene, the comonomer distribution may be varied from statistical to alternating.

Ethene–norbornene copolymers may divided into three subtypes [22] according to their ^{13}C NMR spectra. Although a full interpretation of the ^{13}C NMR-spectra of ethene–norbornene copolymers has not been achieved so far, the major differences in the subtypes are reflections of the different abilities of metallocenes to form blocks of cyclic monomer units and different stereoregularities within the norbornene blocks [23]. Figure 15 shows representative examples of the three types of copolymers at an incorporation rate of 49 %.

$[Me_2C(3-MeCp)(Flu)]ZrCl_2/$MAO produces a copolymer consisting mainly of alternating NENEN sequences (sharp resonance at 48 ppm) and only a few (E)NN(E) blocks (small resonances at 51–49 ppm and at 47.5–47 and 29.5–28 ppm). In contrast, at the same incorporation rate of norbornene the copolymer produced by $[Me_2C(Cp)(Flu)]ZrCl_2/$MAO features a significantly higher amount of (E)NN(E) blocks. The stereochemistry of the NN linkage most probably is predominantly racemic in accordance with the stereoselectivity of this catalyst in the hydrooligomerization of norbornene. In contrast to this, the polymer produced by *rac*-$[Me_2Si(Ind)_2]ZrCl_2/$MAO features *meso* (E)NN(E) blocks and additionally

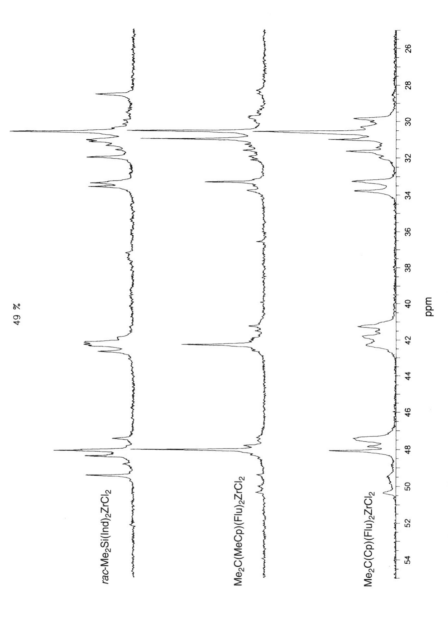

Figure 15 High-resolution ^{13}C NMR spectra of ethene–norbornene copolymers containing 49 mol% of norbornene, representing the three basic types of copolymer microstructures observed

already longer N(N)N sequences are indicated by the appearance of resonances between 35 and 40 ppm which grow significantly as the incorporation rate increases.

Whereas in the case of the [Me$_2$C(3-MeCp)(Flu)]ZrCl$_2$/MAO-catalyzed copolymerization the alternating structure is 'disturbed' by a small amount of blocks [Me$_2$C(3-^{t}BuCp)(Flu)]ZrCl$_2$/MAO gives copolymers containing no blocks of norbornene [24]. These copolymers are crystalline if the norbornene content is above 37 mol%. They feature glass transition temperatures of 100–130°C and melting-points of 270–320°C (Figure 16). They have an even higher resistance to solvents than statistical copolymers although they are still transparent owing to the small size of the crystalline regions (5 µm).

In contrast to other alternating polymers, the alternating structure in this case is not due to a chain end control mechanism (after insertion of the bulky monomer, no other sterically demanding monomer can insert) but to the structure of the catalyst in combination with the mechanism of polymerization, involving chain migratory insertion. While both monomers may insert only from position A, ethene can insert from position B, which results in copolymers containing only odd-numbered ethene sequence lengths, and in the case of high norbornene/ethene ratios an alternating structure results (Figure 17). This is the first example of a metallocene tailored to form a regular (crystalline) microstructure in copolymerization.

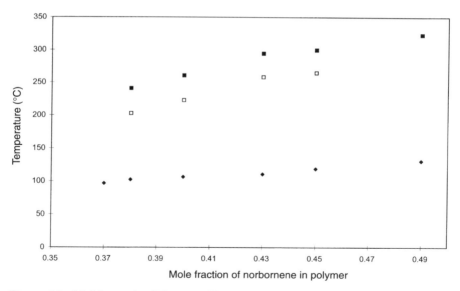

Figure 16 Melting point (■), crystallization temperature (□) and glass transition temperature (◆) of crystalline ethene–norbornene copolymers produced by [Me$_2$C(3-^{t}BuCp)(Flu)]ZrCl$_2$/MAO

Active species:

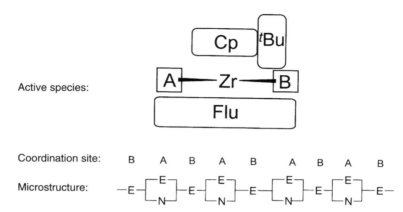

Coordination site:

Microstructure:

Figure 17 Mechanism of the formation of alternating copolymers from [Me$_2$C(3-^{t}BuCp)(Flu)]ZrCl$_2$/MAO-catalyzed polymerization

The glass transition temperature of ethene–norbornene copolymers (both statistical and alternating types) are correlated with the incorporation (weight%) by the FOX equation (Figure 18). The glass transition temperature of polynorbornene extrapolated from this correlation is about 220°C.

By variation of the cyclic monomer, the glass transition temperature at a given mol% incorporation rate in may be varied (30 mol% cyclic monomer: norbornene

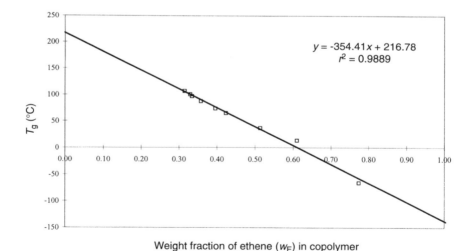

$$y = -354.41x + 216.78$$
$$r^2 = 0.9889$$

Weight fraction of ethene (w_E) in copolymer

Figure 18 Dependence of the glass transition temperature of ethene–norbornene copolymers on the weight% incorporation of ethene according to the Fox equation

$T_g = 72°C$, DMON $T_g = 105°C$), but a comparison of the FOX equation for ethene–norbornene copolymers and ethene–DMON copolymers shows similar glass transition temperatures at the same weight% content of the two cyclic monomers. With increasing bulk of the monomer, the incorporation of the cycloolefin becomes more and more difficult. Whereas the r_1 parameter for norbornene is close to that of propene, those of DMON and TMDA are comparable to those of butene and hexene (Table 5).

Because polypropene has a much higher glass transition temperature than polyethene, the copolymerization of propene and norbornene yields copolymers featuring a higher glass transition temperature at the same mol% incorporation. The rate of copolymerization is significantly lower than that observed for ethene–norbornene copolymerization, most reasonably explained by the difficulty of inserting a propene after the cycloolefin owing to steric interactions of the methyl group and the cyclic monomer unit. Therefore, also the rate of insertion is lowered relative to the rate of chain termination and the molecular weight of these copolymers decreases dramatically with increasing cycloolefin/propene ratio. Attempts to copolymerize hexene with norbornene yielded low molecular weight oligomer containing only a very few monomer units (<10). Contrary to the effects observed in cyclopentene–propene copolymerization, the chain ends are vinylic groups formed after α-olefin insertion, most probably formed by an H-transfer

Table 5 Copolymerization parameters for the copolymerization of ethene with norbornene, DMON and TMDA as calculated from the composition of the copolymers by non-linear regression[a]

Metallocene	Cycloolefin	Temperature (°C)	r_{ethene}	$r_{cyclolefin}$
rac-[En(IndH$_4$)$_2$ZrCl$_2$	Norbornene	−25	1.7	0.022
		0	2.2	0.026
		25	2.4	0.026
		50	3.0	—
	DMON	25	5.4	—
rac-[En(Ind)$_2$]ZrCl$_2$	Norbornene	25	7.1	—
	DMON	25	1.4	—
rac-[Me$_2$Si(Ind)$_2$]ZrCl$_2$	Norbornene	30	3.2	0.025
rac-[Ph$_2$Si(Ind)$_2$]ZrCl$_2$	Norbornene	30	4.0	0.035
Cp$_2$ZrCl$_2$	Norbornene	25	1.7	—
[Me$_2$C(Cp)(Flu)]ZrCl$_2$	Norbornene	0	1.6	0.024
		30	4.2	0.034
[Ph$_2$C(Cp)(Flu)]ZrCl$_2$	Norbornene	0	1.3	0.029
		30	4.1	0.34
	DMON	50	7.9	0.033
	TMDA	50	14.2	0.028
[Ph$_2$C(Cp)(Ind)]ZrCl$_2$	DMON	50	6.5	0.097
[Me$_2$C(Cp)(Flu)]HfCl$_2$	DMON	50	11.0	0.105

[a] It is not possible to analyze the sequence distribution of these copolymers on a level permitting the test of the first-order Markovian model on which the Lewis–Mayo equation is based.

reaction of the growing chain and an α-olefin monomer. ^{13}C NMR investigations of propene–norbornene copolymers and propene–DMON copolymers show that there is a strong tendency to avoid the formation of blocks of cyclic monomer units. Thus even at incorporation rates close to 50 mol% of norbornene and DMON no blocks are observed.

4   HOMO- AND COPOLYMERIZATION OF CYCLOBUTENE

Homopolymerization of cyclobutene by double bond opening was first achieved by Natta and co-workers [25] using $Cr(acac)_3/Et_2AlCl$ as the catalyst. Using *rac*-[En(IndH$_4$)$_2$]ZrCl$_2$/MAO, a semicrystalline polymer is produced (crystalline reflexes $2\theta = 16$ and $22°$, crystallinity about 55%) (Figure 19) [9e,f,26]. The polymer shows no softening or melting before decomposition at 400°C is observed.

Like polycyclopentene and polynorbornene produced by this catalyst, the polymer is hardly soluble in organic solvents. Comparison of the ^{13}C CP/MAS NMR spectrum with those of model compounds such as *cis*- and *trans*-1,2- and -1,3-dimethylcyclobutene shows that a *cis*-1,3-enchainment of the monomer units is reasonable (Figure 20).

Cyclobutene has been copolymerized with ethene ($r_{ethene} = 13$, $r_{cyclobutene} = 0.03$ at 20°C) and propene ($r_{propene} = 0.66$, $r_{cyclobutene} = 0.70$ at 20°C). In cyclobutene–ethene copolymerization the molecular weight does not decrease significantly with incorporation of the cyclobutene and the decrease in the melting-points with increasing comonomer content is small. For cyclobutene–propene copolymerization the copolymers are semicrystalline over the whole range of composition [9f]. The melting-point decreases from 130°C for the polypropene to about 55°C at 20%

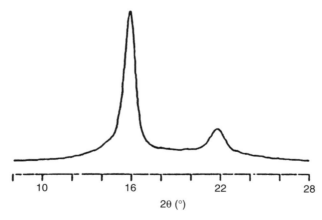

Figure 19   Wide-angle X-ray scattering of a semicrystalline poly(cyclobutene) produced by using *rac*-[En(IndH$_4$)$_2$]ZrCl$_2$/MAO at 20 °C

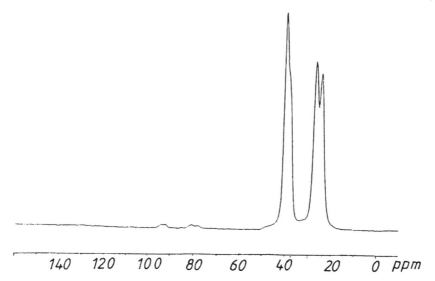

140   120   100   80   60   40   20   0 ppm

Figure 20 CP/MAS NMR spectrum of a semicrystalline poly(cyclobutene) produced by using rac-[En(IndH$_4$)$_2$]ZrCl$_2$/MAO at 20 °C

incorporation before it starts to increase even above the melting-point of the polypropene (40 mol% cyclobutene, $T_m = 155$°C).

5   REACTIVITY OF CYCLOOLEFINS IN CATALYTIC POLYMERIZATION [11a]

The reactivity of cycloolefins in metallocene-catalyzed polymerization is a function of the metallocene structure, the bulkiness and the 'intrinsic' reactivity of the cycloolefin. The low copolymerization parameters of norbornene and its derivatives reflect the high reactivity, which is able to compensate for steric stress at the active site. Based on concepts which have been developed in organic chemistry, the 'intrinsic' reactivity of cycloolefins is determined by two factors: (i) the olefin strain, which may be quantified by the difference in energy of the olefin and the corresponding alkane as calculated by force field methods, and (ii) the non-planarity of the reacting double bond, which can be classified into symmetric and asymmetric components (Figure 21).

Force field calculations of the olefin strain and the magnitude and type of non-planarity of the double bond yield the following interpretation of the cycloolefin reactivity (see Table 6):

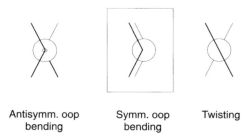

| Antisymm. oop | Symm. oop | Twisting |
| bending | bending | |

Figure 21 Schematic diagram of the possible non-planar deformation of double bonds. Any twist may be described as a linear combination of symmetric and antisymmetric oop bendings

Table 6 'Intrinsic reactivity' of cycloolefins based on force field calculations of the olefin strain (difference of the steric energy of the olefin and the corresponding alkane) and the symmetric and antisymmetric oop bending components which describe the non-planarity of the double bonds

Cycloolefin	Olefin strain (kcal/mol)	Symm. oop bending (°)	Antisymm. oop bending (°)
Cyclobutene	4.0	0	0
Cyclopentene	−1.3	1.56	0
Cyclohexene	−0.5	0	0.7
Cycloheptene	−2.5	0.14	0
Cyclooctene	2.2	0	1.61
Norbornene	4.4	1.13	0
Bicyclo[2.2.2]octene	0.5	0	0
DMON	4.6	1.13	0

(i) Cycloolefins with symmetric out-of-plane (oop) bending of the double bond protons can be polymerized. Symmetric oop bending makes the coordination of the monomer to the transition metal and the insertion easier owing to the lower barrier for the geometric changes occurring when the sp^3-hybridization develops. Typical examples are norbornene, DMON (high olefin strain, symmetrical oop bending; high reactivity for insertion) and cyclopentene (no olefin strain compensated by symmetrical oop bending; active in homo- and copolymerization).

(ii) Cycloolefins with antisymmetric oop bending cannot be polymerized. Typical examples are cyclohexene (low olefin strain, antisymmetric oop bending; no polymerization observed) and cyclooctene (presumably the olefin strain can compensate the antisymmetric bending; only active for copolymerization with ethene).

(iii) The reactivity of cycloolefins with planar double bonds is determined by the olefin strain only. Typical examples are cyclobutene (high olefin strain; can be polymerized) and bicyclo[2.2.2]octene (low olefin strain; no polymerization observed).

6 CONCLUSIONS

Metallocene catalysts have been shown to be able to generate new semicrystalline polymers from homopolymerization of cycloolefins and by alternating copolymerization with ethene. Another typical feature is their excellent performance in the copolymerization of ethene with norbornene type monomers, which is now commercialized.

The polymerization mechanisms reflect the sometimes difficult incorporation of cyclic monomers into the transition metal–carbon bond and also the fact that after an insertion of a cycloolefin the next insertion should preferably be that of a small monomer such as ethene. The development of new catalysts which will take into account the features observed in cycloolefin homo- and copolymerization promises a whole series of new, interesting materials from hydrocarbon monomers.

7 REFERENCES

1. For a comprehensive review on ring opening polymerization of cycloolefins, see Ivin, K. J., *Olefin Metathesis*, Academic Press, New York, 1983.
2. Demming, T. J. and Novak, B. M., *Macromolecules*, **26**, 7089 (1993).
3. Goodall, B. L., Barnes, D. A., Benedict, G. M., McIntosh, L. H. and Rhodes, L. F., *Polym. Mater. Sci. Eng.*, **76**, 56 (1997).
4. Sen, A., Lai, T. W. and Thomas, R. R., *J. Organomet. Chem.*, **358**, 567 (1988).
5. Heitz, W. and Haselwander, T. F. A., *Macromol. Rapid Commun.*, **18**, 689 (1997).
6. (a) Mehler, C. and Risse, W., *Makromol. Chem. Rapid Commun.*, **12**, 255 (1991); (b) Melia, J., Rush, S., Connor, E., Breunig, S., Mehler, C. and Risse, W., *Macromol. Symp.*, **89**, 433 (1992); (c) Mehler, C. and Risse, W., *Makromol. Chem. Rapid Commun.*, **13**, 455 (1992); (d) Mehler, C. and Risse, W., *Macromolecules*, **25**, 4226 (1992); (e) Breunig, S. and Risse, W., *Makromol. Chem.*, **193**, 2915 (1992); (f) Reinmuth, A., Mathew, J. P., Melia, J. and Risse, W., *Macromol. Rapid Commun.*, **17**, 173 (1996).
7. Kaminsky, W. and Arndt, M., *Adv. Polym. Sci.*, **127**, 143 (1997), and references cited therein.
8. (a) Sargusa, T., Tsujino, T. and Furukawa, J., *Makromol. Chem.*, **78**, 231 (1964); (b) Tsujino, T., Saegusa, T. and Furukawa, J., *Makromol. Chem.*, **85**, 71 (1965).
9. (a) Kaminsky, W., Bark, A., Spiehl, R., Möller-Lindenhof, N. and Niedoba, S., in Sinn, H. and Kaminsky, W. (Eds), *Transition Metals and Organometallics as Catalysts for Olefin Polymerisation*, Springer, Berlin, 1988, p. 291; (b) Kaminsky, W. and Steiger, R., *Polyhedron*, **7**, 2375 (1988); (c) Kaminsky, W. and Spiehl, R., *Makromol. Chem.*, **190**, 515 (1989); (d) Spiehl, R., PhD Thesis, University of Hamburg (1987); (e) Kaminsky, W.,

Bark, A., and Däke, I., in Keii, T. and Soga, K. (Eds), *Catalytic Olefin Polymerization*, Kodansha, Tokyo, 1990, p. 425; (f) Däke, I., PhD Thesis, University of Hamburg (1989).

10. (a) Collins, S. and Kelly, W. M., *Macromolecules*, **25**, 233 (1992); (b) Kelly, W. M., Taylor, N. J. and Collins, S., *Macromolecules*, **27**, 4477 (1994); (c) Kelly, W. M., Whang, S. and Collins, S., *Macromolecules*, **30**, 3151 (1997).

11. (a) Arndt, M., PhD Thesis, University of Hamburg (1994); (b) Arndt, M. and Kaminsky, W., *Macromol. Symp.*, **95**, 225 (1995); (c) Arndt, M. and Kaminsky, W., *Macromol. Symp.*, **97**, 225 (1995).

12. (a) Land, H. T. in *Proceedings of the International Congress on Metallocene Polymers, Metallocenes '95*, Brussels, Belgium, Schotland Business Research, Skillman, NY, USA, 1995, p. 217; (b) Land, H. T., Osan, F. and Wehrmeister, T., *Polym. Mater. Sci. Eng.*, **76**, 22 (1997); (c) Toyata, A. and Yamaguchi, M., *Polym. Mater. Sci. Eng.*, **76**, 24 (1997).

13. (a) Benedikt, G. M., Goodall, B. L., Marchant, N. S. and Rhodes, L. F., in *Proceedings of the Worldwide Metallocene Conference, MetCon '94*, Houston, TX, Catalyst Consultants, Spring House, PA, USA, 1994; (b) Benedikt, G. M., Goodall, B. L., Marchant, N. S. and Rhodes, L. F., *New J. Chem.*, **18**, 105 (1994).

14. (a) Arndt, M., Diploma Thesis, University of Hamburg (1990); (b) Arnold, M., Henschke, O. and Köller, F., *J. Macromol. Sci., Macromol. Rep.*, **A33** (Suppl. 3–4), 219 (1996).

15. (a) Kaminsky, W., Bark, A. and Arndt, M., *Makromol. Chem., Macromol. Symp.*, **47**, 83 (1991); (b) Kaminsky, W., Arndt, M. and Bark, A., *Polym. Prepr. Am. Chem. Soc.*, **32**, 467 (1991); (c) Bark, A., PhD Thesis, University of Hamburg (1990); (d) Kaminsky, W., in *Proceedings of the Worldwide Metallocene Conference, MetCon '93*, Houston, TX, Catalyst Consultants, Spring House, PA, USA, 1993, p. 325; (e) Kaminsky, W. and Noll, A., *Polym. Bull.*, **31**, 175 (1993); (f) Noll, A., PhD Thesis, University of Hamburg (1993); (g) Arndt, M., Kaminsky, W. and Schupfner, G. U., in *Proceedings of the International Congress on Metallocene Polymers, Metallocenes '95*, Brussels, Belgium, Schotland Business Research, Skillman, NY, USA, 1995, p. 403; (h) Engehausen, R., PhD Thesis, University of Hamburg (1994).

16. (a) Arndt, M., Engehausen, R., Kaminsky, W. and Zoumis, K., *J. Mol. Catal. A: Chem.*, **101**, 171 (1995); (b) Kaminsky, W. and Arndt, M., *Polym. Prepr. Am. Chem. Soc.*, **35**, 520 (1994).

17. Arndt, M. and Gosmann, M., *Polym. Bull.*, submitted for publication.

18. Carr, N., Dunne, B. J., Orpen, A. G. and Spencer, J. C., *J. Chem. Soc., Chem. Commun.*, 826 (1988).

19. (a) Zocchi, M., Tieghi, G. and Albinati, A., *J. Organomet. Chem.*, **33**, C47 (1971); (b) Zocchi, M. and Tieghi, G., *J. Chem. Soc., Dalton Trans.*, 1740 (1975); (c) Galazzi, M. C., Hanlon, T. L., Vitalli, G. and Porri, L., *J. Organomet. Chem.*, **33**, C45 (1971); (d) Galazzi, M. C., Porri, L. and Vitalli, G., *J. Organomet. Chem.*, **97**, 131 (1975); (e) Hughes, R. P. and Powell, J., *J. Organomet. Chem.*, **60**, 387 (1973); (f) Larock, R. C., Takagi, K., Herohberger, S. S. and Mitchell, M. A., *Tetrahedron Lett.*, **22**, 5231 (1981); (g) Larock, R. C., Hirohberger, S. S., Takagi, K. and Mitchell, M. A., *J. Org. Chem.*, **51**, 2450 (1986); (h) Lin, Z. and Marks, T. J., *J. Am. Chem. Soc.*, **112**, 5515 (1990).

20. Houk, K. N., in Watson, W. H. (Ed.), *Stereochemistry and Reactivity of Systems Containing p-Electrons*, Verlag Chemie, Deerfield Beach, FL, 1982, p. 1, and references cited therein.

21. (a) Kaminsky, W., Bark, A. and Arndt, M., *Makromol. Chem. Macromol. Symp.*, **47**, 83 (1991); (b) Kaminsky, W., Arndt, M. and Bark, A., *Polym. Prepr. Am. Chem. Soc.*, **32**, 467 (1991); (c) Bark, A., PhD Thesis, University of Hamburg (1990); (d) Kaminsky, W., in *Proceedings of the Worldwide Metallocene Conference, MetCon '93*, Houston, TX, Catalyst Consultants, Spring House, PA, USA, 1993, p. 325; (e) Kaminsky, W. and Noll, A., *Polym. Bull.*, **31**, 175 (1993); (f) Arndt, M., Kaminsky, W. and Schupfner, G. U.,

in *Proceedings of the International Congress on Metallocene Polymers, Metallocenes '95*, Brussels, Belgium, Schotland Business Research, Skillman, NY, USA, 1995, p. 403.

22. Chedron, H., Brekner, M. J. and Osan, F., *Angew. Makromol. Chem.*, **223**, 121 (1994).
23. Arndt, M. and Beulich, I., unpublished results.
24. (a) Arndt, M., Kaminsky, W. and Beulich, I., in *Proceedings of the Worldwide Metallocene Conference, MetCon '96*, Houston, TX, 1996, Catalyst Consultants, Spring House, PA, USA, (b) Kaminsky, W., Arndt, M. and Beulich, I., *Polym. Mater. Sci. Eng.*, **76**, 19 (1997); (c) Arndt, M. and Beulich, I., *Macromol. Chem.*, in press.
25. Dall'Asta, G., Natta, G. and Motroni, G., *Makromol. Chem.*, **69**, 163 (1963).
26. Zoumis, K., PhD Thesis, University of Hamburg (1994).

6

Metallocene Catalysts for 1,3-Diene Polymerization

LIDO PORRI AND ANTONINO GIARRUSSO
Politecnico di Milano, Milan, Italy

GIOVANNI RICCI
Istituto di Chimica delle Macromolecole CNR, Milan, Italy

1 INTRODUCTION

Metallocene-based catalysts have initiated a new era in polyolefin synthesis. The most important advantage of these systems over those previously known is that they have a well characterized molecular structure and consist of well defined single catalytic sites. In polyethylene polymerization, they allow one to control molecular weight, molecular weight distribution and comonomer distribution and content. In 1-alkene polymerization, they allow the tacticity of a polymer to be controlled by suitable modification of the cyclopentadienyl ligands. In addition, metallocene catalysts have allowed entirely new polymers to be obtained, such as cycloolefin polymers and cycloolefin–olefin copolymers, all of practical interest [1].

Some metallocene-based systems are also active for 1,3-diene polymerization. Although the polymerization mechanism of dienes is different from that of mono-alkenes, metallocene catalysts have proved to be of interest also for the polymerization of this class of monomers. They have permitted new polymers to be obtained, some of which are highly stereoregular, and have provided an opportunity to gain a deeper insight into the mechanism of polymerization. This chapter reports a critical examination of the work carried out on the polymerization of dienes with metallocene catalysts.

Metallocene-based Polyolefins Edited by J. Scheirs and W. Kaminsky
© 2000 John Wiley & Sons Ltd

2 PECULIAR FEATURES OF 1,3-DIENE POLYMERIZATION

There are some differences between the mechanism of polymerization of a 1,3-diene and that of propylene [2]. Consideration of these differences can give an idea of the possibilities and limitations of metallocene catalysts for diene polymerization.

The first difference relates to the type of bond between the transition metal of the active species, Mt, and the growing polymer chain: an η^3-allyl bond in diene polymerization, a σ-type bond in propylene polymerization (Figure 1). Owing to this difference, the factors determining or affecting activity and stereospecificity are different in diene and 1-alkene polymerization, respectively. In propylene polymerization, the last-inserted unit contains an asymmetric carbon atom, but it is now well established that stereospecificity is not controlled by the asymmetry of this unit, at least in polymerization at room temperature or above. Theoretical calculations and experimental data have led to the conclusion that enantioselectivity in propylene polymerization is mainly due to steric interactions of the methyl group of the coordinated monomer with the chirally oriented growing chain. Steric interactions of the monomer with the cyclopentadienyl ligand are of very minor importance [3]. In other words, the monomer reacts with the enantioface that minimizes the steric interaction between the growing polymer chain and the monomer itself. If during the

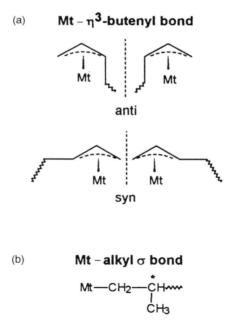

(a) **Mt – η^3-butenyl bond**

anti

syn

(b) **Mt – alkyl σ bond**

$$Mt—CH_2—\overset{*}{C}H\text{ww}$$
$$|$$
$$CH_3$$

Figure 1 Bonds between the growing polymer chain and the transition metal of the catalyst in (a) 1,3-diene and (b) propylene polymerization

polymerization there is freedom of rotation around the Mt—C bond, the monomer reacts with one or the other enantioface depending on the instantaneous orientation of the polymer chain. The result will be formation of an atactic polymer. The function of the cyclopentadienyl ligand, in catalysts for propylene polymerization, is to arrange the last-inserted unit in one chiral orientation.

If this is the case, the presence of a cyclopentadienyl ligand does not appear essential to stereospecificity in catalysts for diene polymerization. In the polymerization of these monomers, the growing polymer chain is bonded to Mt by an η^3-allyl bond, and this type of bond gives rise to a conformationally rigid chiral group.

These considerations explain why diene polymers with an isotactic structure could be obtained, since the early days of stereospecific polymerization, using soluble catalysts prepared from non-chiral catalyst precursors. Isotactic 1,2-poly(1,2-butadiene) was first prepared in 1955, using catalysts based on $AlEt_3$ and $Cr(CO)_6$ or its derivatives [4]; poly(1,3-pentadiene) with a *cis*-1,4 isotactic structure was obtained in 1963 with $Ti(OBu)_4$-based catalysts [5].

Another important difference between the polymerization of 1-alkenes and 1,3-dienes concerns the different mode of coordination of the two classes of monomers. In principle, a 1,3-diene can coordinate *cis*-η^4, *trans*-η^4 or *trans*-η^2, but for most of the dienes the mode of coordination by far energetically favored is *cis*-η^4. Only a few monomers [e.g. (*Z*)-1,3-pentadiene, 4-methyl-1,3-pentadiene] are exceptions, in the sense that for them, owing to steric factors, as will be shown later, the *cis*-η^4 coordination is not energetically much more favored than the *trans*-η^2 coordination.

A diene coordinated *cis*-η^4 can take two different orientations with respect to the Mt–polymer bond, as shown in Figure 2. Orientation as in Figure 2(a) takes place when no ligand is bonded to Mt. This is the case for some ionic Ni catalysts, such as [(allyl)Ni(COD)]$^+$[PF$_6$]$^-$ and analogous complexes. Orientation as in Figure 2(b) occurs when one or more anionic ligands, L, are bonded to Mt. L can be an alkoxy group, a chlorine atom or a cyclopentadienyl group. The validity of this rationalization has been verified by using deuterated monomers, such as CHD=CH—CH=CHD and CHD=CH—CH=CHMe. These have afforded stereoregular polymers having the structures expected from the schemes in Figure 2(a) and (b) [2a,c,6–15].

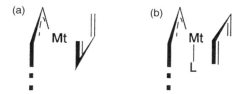

Figure 2 Possible orientations of the new monomer with respect to the last-inserted unit. L is a generic ligand (Cl, OR, Cp, etc.)

The two factors indicated above, i.e. a growing polymer chain η^3-bonded to the catalyst and a monomer coordinated η^4, make diene polymerization with metallocene catalysts different from that of 1-alkenes.

In the following, we shall examine first the polymerization of dienes with catalysts based on monocyclopentadienyl compounds, which are by far the most active, then the polymerization with catalysts based on bis-cyclopentadienyl compounds.

3 POLYMERIZATION WITH CATALYSTS BASED ON MONOCYCLOPENTADIENYL COMPOUNDS

3.1 TITANIUM-BASED CATALYSTS

3.1.1 Diene homopolymerization

The systems based on CpTiCl$_3$, CpTi(O^{n}Bu)$_3$ and CpTiCl$_2$, in combination with methylaluminoxane (MAO), are highly active and stereospecific for diene polymerization [13,15–20]. Most of the work has been carried out with CpTiCl$_3$–MAO. This system, which is soluble in toluene, polymerizes various types of monomers, including (Z)-1,3-pentadiene (ZP) and 4-methyl-1,3-pentadiene (4MP), which are not polymerized by the most common soluble catalysts used for the polymerization of dienes, e.g. Co-, Ni- and Nd-based catalysts (Table 1).

The polymer structure depends on the type of monomer. 2,3-Dimethyl-1,3-butadiene (DMB) and (E)-2-methyl-1,3-pentadiene (2MP) give polymers consisting almost only of cis-1,4 units, whereas 4-methyl-1,3-pentadiene (4MP) gives polymers consisting of 1,2 units only, with a syndiotactic structure [13,17,21]. The other monomers give polymers with a mixed cis/1,2 structure, with the exception of (E,E)-2,4-hexadiene (2,4-H), which gives a trans/1,2 polymer [13,17].

The results obtained with CpTiCl$_3$–MAO give some indications about the relationship between monomer structure and stereospecificity, and also allow an insight into the factors that determine the chemoselectivity, that is, formation of 1,2 versus 1,4 units, in diene polymerization.

The formation of cis or cis/1,2 polymers from butadiene (B), EP, DMB, 2MP and 3-methyl-1,3-pentadiene (3MP) probably derives from a situation as that represented in Figure 3, in which the monomer is cis-η^4 coordinated and the growing polymer chain is anti-η^3 bonded to the transition metal [2a,c,13,15]. With an arrangement of this type, cis-1,4 or 1,2 units can be obtained, depending on whether the new monomer reacts at C-1 or C-3 of the η^3-butenyl group. Symmetric monomers such as B and DMB react at C-1, which is less substituted and therefore more reactive than C-3, to give cis units. However, B also gives some 1,2 units (15–20%), whereas DMB gives cis units only. This is due to the fact that in the butenyl group derived from DMB (Figure 4), because of the presence of two substituents at C-3, the difference in reactivity between C-1 and C-3 is so great that the new monomer reacts only at C-1, with exclusive formation of cis units.

Table 1 Polymerization of some 1,3-dienes with CpTiCl$_3$–MAO[a]

Run	Monomer[b]	Temperature (°C)	Time (h)	Conversion (%)	[η][c] (dL/g)	M.p.[d] (°C)	Polymer microstructure[e]	Ref.
1	B	+20	0.15	81.7	4.1		81.4 % *cis*-1,4; 18.6 % 1,2	13,15,17,19
2	B	−30	5	6.8			80.9 % *cis*-1,4; 19.1 % 1,2	13,15,19
3	I	+20	72	5			≥ 95 % *cis*-1,4	13,17
4	EB	+20	100	Traces			≥ 95 % *cis*-1,4	13
5	DMB	+20	70	85.3		120	≥ 98 % *cis*-1,4	13,15
6	2MP	+20	18	14.7		136	≥ 98 % *cis*-1,4	13,15,17
7	2MP	−20	71	89.6	0.4	147	≥ 98 % *cis*-1,4	13,15
8	EP	+20	1	24.2	1.9		56.3 % *cis*-1,4; 43.7 % 1,2 *trans*	13,15,17,19
9	EP	−30	20	Traces			54.4 % *cis*-1,4; 45.6 % 1,2 *trans*	13,15,19
10	3MP	−20	6	55.9	2.5		50 % *cis*-1,4; 50 % 1,2	13,15,17
11	H	−20	20	47.2	1.9		55 % *cis*-1,4; 45 % 1,2	13,15
12	ZP	+20	2	4.1	Waxy		85 % *cis*-1,4; 15 % 1,2	13,15,17–19
13	ZP	−30	2	53	0.2	102.7	100 % *cis*-1,2 syndiotactic	18,19
14	4MP	+20	1/60	100	0.4		100 % 1,2 syndiotactic	13,15,17,19–21
15	4MP	−30	1/60	100	1.3	96.4	100 % 1,2 syndiotactic	13,15,17,19–21
16	2,4-H	+20	17	100	0.4		70 % 1,2; 30 % *trans*-1,4	13,15

[a] Polymerization conditions: monomer, 2 ml; toluene, 16 ml; CpTiCl$_3$, 1×10^{-5} mol; Al/Ti molar ratio, 1000; in runs 12 and 13, 5×10^{-5} mol of Ti and Al/Ti = 500 were used.
[b] B, butadiene; I, isoprene; EB, 2-ethyl-1,3-butadiene; DMB, 2,3-dimethyl-1,3-butadiene; 2MP, (E)-2-methyl-1,3-pentadiene; EP, (E)-1,3-pentadiene; 3MP, 3-methyl-1,3-pentadiene; H, 1,3-hexadiene; ZP, (Z)-1,3-pentadiene; 4MP, 4-methyl-1,3-pentadiene; 2,4-H, (E,E)-2,4-hexadiene.
[c] intrinsic viscosity, determined in toluene at 25 °C.
[d] Determined by DSC analysis.
[e] Determined by IR and NMR analysis.

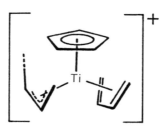

Figure 3 Possible structure of the cation formed in the reaction between CpTiCl$_3$ and MAO

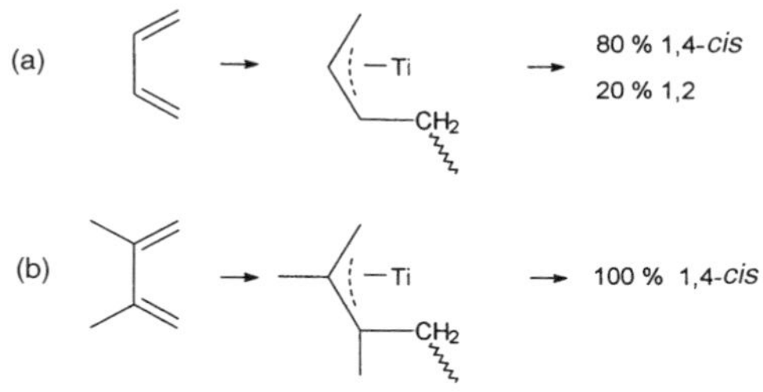

(a) [structure] $\longrightarrow$ [Ti structure with —CH$_2$] $\longrightarrow$ 80 % 1,4-*cis*
20 % 1,2

(b) [structure] $\longrightarrow$ [Ti structure with —CH$_2$] $\longrightarrow$ 100 % 1,4-*cis*

Figure 4 η^3-Butenyl groups derived from (a) butadiene and (b) 2,3-dimethylbutadiene. Only the *anti* form is shown

For non-symmetric monomers (EP, 2MP, 3MP, 4MP), the problem of chemoselectivity is more complex because each of these monomers (a) can give, at least in principle, two different η^3-butenyl groups on insertion into the growing polymer chain, and (b) can react with one or the other enantioface.

With regard to point (a), it has been found that, using CpTiCl$_3$–MAO and other transition metal catalysts, the monomers EP, 2MP, 3MP and 4MP give only one type of η^3-butenyl group, as indicated in Figure 5. This is shown by the fact that vinyl groups are completely absent in the polymers of EP, 3MP and 4MP, and that CH$_2$=C(Me)− groups are absent in the polymer of 2MP. These groups should be present, at least in a very small amount, if butenyl groups **b** of Figure 5 are formed.

With regard to point (b), information on whether EP, 2MP and 3MP react with both enantiofaces can be obtained from some structural features of the polymers. The formation of a *cis*-1,4/1,2 polymer of EP probably occurs according to the scheme in Figure 6 [13]. Reaction of the monomer as in (a) gives rise to a *cis*-1,4 unit, whereas reaction with the other enantioface, as in (b), gives a 1,2 unit. The *cis*-1,4/1,2 ratio depends on the relative concentrations of the species (a) and (b) and on the relative rates of insertion of the new monomer into the two species. A mechanism of this type should result in a polymer without inversions such as

$$-CH_2CH=CHCH(Me)CH(CH=CHMe)CH_2-$$

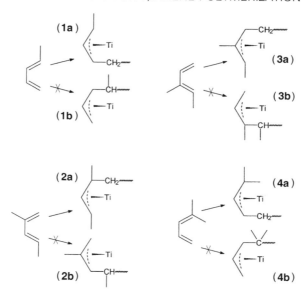

Figure 5 3-Butenyl groups that can be obtained from (*E*)-pentadiene (**1a, 1b**), (*E*)-2-methylpentadiene (**2a, 2b**), 3-methylpentadiene (**3a, 3b**) and 4-methylpentadiene (**4a, 4b**) on insertion into the growing polymer chain. Only the *anti* form is shown

and

$$-CH_2CH=CHCH(Me)CH(Me)CH=CHCH_2-$$

Inversions of this type have not been observed by NMR analysis, whereas they should be present if EP reacts with one enantioface only. This supports the validity of the schemes in Figure 6.

2MP gives a high-*cis* polymer with $CpTiCl_3$–MAO. Also this monomer, as EP, can likely coordinate with both enantiofaces, as shown in Figure 7. However, in this case the monomer coordinated as in (b), which should react at C-3 of the butenyl group, is practically unreactive, owing to the presence of two substituents on this carbon atom. The result is the formation of *cis* units only.

The case of ZP is less clear, because in the allyl groups derived from this monomer, C-1 and C-3 are equally substituted. It is possible that in this case steric effects due to the growing polymer chain favor reaction of incoming monomer at C-1.

With regard to the stereospecificity of the polymerization, the formation of a *cis*-1,4 isotactic polymer of 2MP is interpretable according to the schemes proposed previously for the formation of *cis*-1,4 isotactic polymers from terminally substituted dienes (Figure 8) [2a,c,8,9,11–14]. It is interesting that in this case the formation of an isotactic polymer does not derive from an enantioselectivity on the part of the

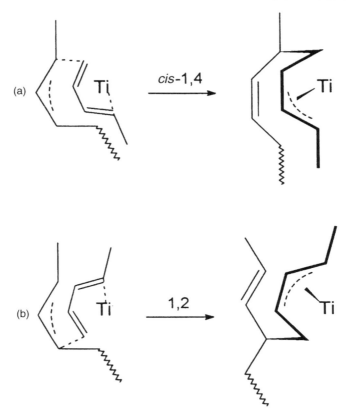

Figure 6 Scheme for the formation of a poly[(E)-1,3-pentadiene] with a mixed
cis-1,4/1,2 structure

catalyst but from the fact that coordination with one enantioface is practically
unreactive.

2,4-H differs from the other monomers because it gives polymers not containing
cis-1,4 units. It is well known that 2,4-H gives predominantly *trans* polymers, even
with catalysts based on Ti, Co and Nd, which give high-*cis* polymers from common
monomers such as butadiene, isoprene and (*E*)-pentadiene. It has been shown that
the formation of *trans* units is due to the fact that this monomer polymerizes much
more slowly than butadiene or isoprene, owing to the steric hindrance of the methyl
groups. The last-inserted unit, which is anti-η^3-bonded to Ti, has time to isomerize to
the more stable *syn* form prior to monomer incorporation (Figure 9) [22–25].
Reaction of the new monomer at C-1 of the *syn* butenyl group gives a *trans*-1,4 unit,
while reaction at C-3 gives a 1,2 unit. Probably C-1 and C-3 of the *syn* butenyl group

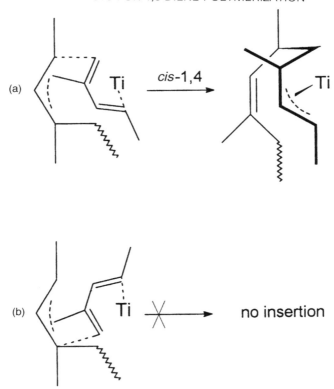

(a)

cis-1,4

(b)   no insertion

Figure 7 Scheme for the formation of a polymer of (*E*)-2-methylpentadiene with a *cis* structure

have a comparable reactivity and this accounts for the formation of a *trans*-1,4/1,2 polymer.

4MP gives, either with CpTiCl$_3$–MAO [13,15,17] or with other systems, homogeneous [15,26,27] and heterogeneous [28], homopolymers consisting practically of 1,2 units only. The polymer obtained with CpTiCl$_3$–MAO is crystalline by X-ray analysis [13,21] (Figure 10) and has a syndiotactic structure [17,21] (Figure 11). Formation of a 1,2 syndiotactic polymer of 4MP is, in principle, interpretable according to the scheme in Figure 8. In the butenyl group derived from 4MP (Figure 5), C-1 is much less reactive than C-3, owing to the presence of the methyl groups. Reaction of the incoming monomer at C-3 gives rise to a 1,2 unit and the new butenyl group has an opposite chirality with respect to the preceding one. A syndiotactic diad will result. This interpretation, apparently plausible, was proposed some years ago [13]. Recent data on the polymerization of ZP and 4MP with

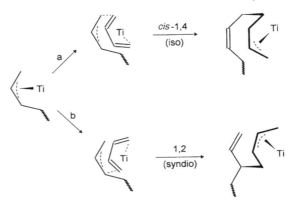

Figure 8   Scheme for the formation of a *cis*-1,4 isotactic and a 1,2 syndiotactic polymer from a generic 1,3-diene. In routes (**a**) and (**b**) the last-inserted unit is below the plane of the figure, the incoming monomer is above and Ti is on the plane

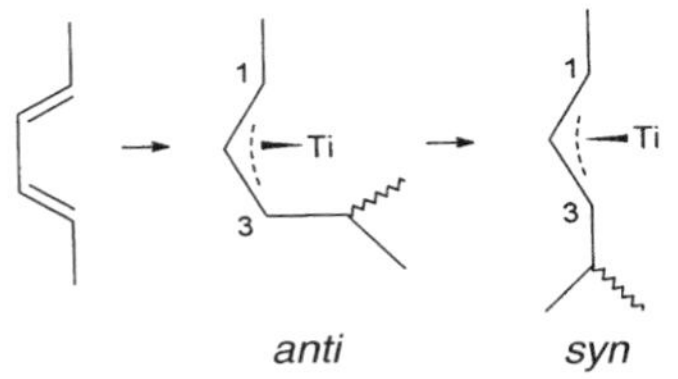

Figure 9   *Anti* → *syn* isomerization of the η^3-butenyl group derived from (*E,E*)-2,4-hexadiene

CpTiCl$_3$–MAO at different temperatures [15,18] have led to a revision of this interpretation [29].

The experimental data are as follows.

(i) ZP, which gives a predominantly *cis* isotactic polymer at room temperature [17], as mentioned above, gives instead a crystalline polymer having a 1,2 syndio-tactic structure at −20 °C [18]. In this polymer the double bonds of the side groups are all *cis*. This phenomenon has not been observed for the polymeriza-tion of the other monomers in Table 1, which give polymers with practically the same structure either at +20 or at −20 °C.

(ii) The polymerization of ZP is faster at −20 °C than at +20 °C (Figure 12).

The change in polymer structure and the enhancement of the polymerization rate on decreasing the polymerization temperature are indicative of a different mechan-ism of monomer insertion [18]. The phenomenon cannot be attributed to the existence of different catalytic centers at +20 and −20 °C, for the following reasons:

(a) if different catalytic centers are formed at different temperatures the phenom-enon should be observed also with monomers other than ZP;

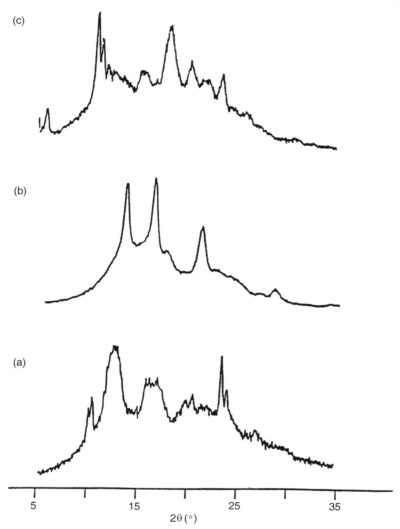

Figure 10 X-ray spectra of (a) 1,2-*cis*-syndiotactic polypentadiene, (b) *cis*-1,4 isotactic poly(2-methylpentadiene) and (c) 1,2 syndiotactic poly(4-methylpenta-diene) obtained with CpTiCl$_3$–MAO [13,18,21,30]

(b) in runs carried out by varying the temperature between +20 and −20 °C during the course of the polymerization it has been observed that the polymer obtained at +20 °C is predominantly *cis* whereas that obtained at −20 °C is 1,2 [30].

Factors (a) and (b) makes plausible the hypothesis that the temperature is the only factor responsible for the phenomenon observed in the polymerization of ZP. The

	C1	C2	C3	C4	C5	C6
(a)	42.54	32.39	137.06	122.98	13.33	
(b)	39.92	133.08	131.97	30.77	23.92	21.10
(c)	42.69	33.58	131.59	129.56	17.97	25.88

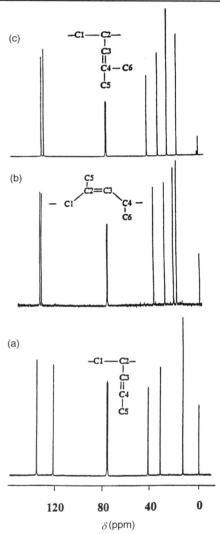

Figure 11 ^{13}C NMR spectra of (a) 1,2-*cis*-syndiotactic polypentadiene, (b) *cis*-1,4 isotactic poly(2-methylpentadiene) and (c) 1,2 syndiotactic poly(4-methylpenta-diene) obtained with CpTiCl$_3$–MAO (CDCl$_3$, 20 °C, TMS as internal standard) [13,18,21,30]

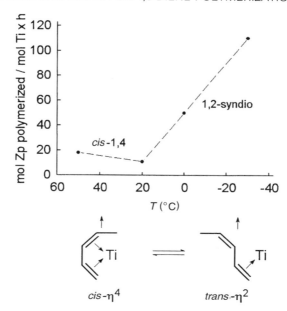

Figure 12 Temperature effect on the rate of (Z)-pentadiene polymerization with CpTiCl$_3$–MAO and a tentative interpretation

interpretation reported is shown in Figure 12. Whereas for dienes such as B, EP and 2MP the *cis-η⁴* coordination is by far the energetically most stable, in the case of ZP there is probably an equilibrium between *cis-η⁴* and *trans-η²* coordination. The equilibrium is probably shifted toward the η^2 form at low temperature. In our interpretation, 1,2 syndiotactic poly(ZP) derives from a *trans-η²* monomer coordination. The fact that poly(ZP) obtained at −20 °C consists exclusively of 1,2 units does not necessarily mean that at this temperature all ZP is coordinated η^2. The monomer so coordinated is probably incorporated much faster than that coordinated *cis-η⁴*.

4MP gives a 1,2 syndiotactic polymer either at room temperature or below [15]. For this monomer too there is probably an equilibrium between *cis-η⁴* and *trans-η²* coordination:

However, it is possible that in this case the η^4 coordination is virtually inactive, in the sense that insertion of the η^4-coordinated monomer is much slower than insertion of the monomer coordinated η^2 [15,29].

Also in the case of 4MP the temperature has an effect on polymerization rate, analogous to that observed for ZP, although much more modest: the polymerization has been found to have almost the same rate at −20 and +20 °C. A possible interpretation is that the concentration of the η^2 form, probably the only one active in the polymerization, increases on decreasing the temperature. However, this interpretation needs more experimental support.

On the basis of the above results, the scheme shown in Figure 13 has been proposed for the formation of 1,2 syndiotactic poly(4MP) [15]. With the orientation shown, the new monomer will give, after insertion, an allyl group of opposite chirality with respect to the previous one, and hence a syndiotactic polymer. An analogous scheme is probably valid for the formation of 1,2 syndiotactic poly(ZP).

It is worth noting that the rate of polymerization of 4MP is much higher than that of B, either at room temperature or, in a more pronounced way, at lower temperature (Table 2) [29]. This result is consistent with the hypothesis that poly(4MP) derives from an η^2 monomer coordination. In conclusion, ZP and 4MP exhibit the following differences with respect to the other monomers in Table 1: (a) ZP gives a predominantly *cis* isotactic polymer at +20 °C, whereas it gives a 1,2 syndiotactic polymer at −20 °C; (b) 4MP gives a 1,2 syndiotactic polymer independently of the polymerization temperature, in the interval from about +60 to about −70 °C; (c) the rate of polymerization increases (in particular for ZP) with decreasing polymerization temperature, in the range ca +20 to −20 °C.

The hypothesis that the different behavior of ZP and 4MP is due to the fact that for these compounds the *cis*-η^4 coordination is not so favored as for the other monomers seems plausible on the basis of the above data.

3.1.2 The role of the cyclopentadienyl ligand

It was mentioned in Section 2 that in catalysts for diene polymerization the cyclopentadienyl ligand cannot have the same function, with regard to the stereospecificity, as it has in catalysts for 1-alkene polymerization. An indication of the function of the cyclopentadienyl ligand can be gained by comparing the results obtained in the polymerization of some dienes with two soluble systems, CpTi(OBu)$_3$–MAO and Ti(OBu)$_4$–MAO (Table 3).

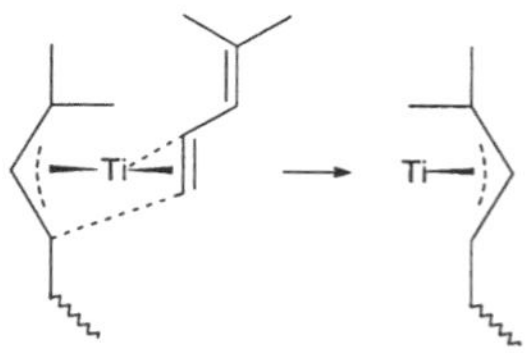

Figure 13 Scheme for the formation of 1,2 syndiotactic poly(4-methylpentadiene) with CpTiCl$_3$–MAO [15]

Table 2 Comparison between the polymerization of butadiene and 4-methylpentadiene with CpTiCl$_3$–MAO[a] [29,30]

Run	Monomer	Mol of Ti (×10⁵)	Temperature (°C)	Time (min)	Conversion (%)	N[b] (min⁻¹)	Polymer microstructure[c] cis-1,4	1,2
1	1,3-Butadiene	1	+20	9	81	233	81.4	18.6
2	1,3-Butadiene	1	-30	298	6.8	6	80.9	19.1
3	4-Methyl-1,3-pentadiene	0.44	+20	3	100	2238		100
4	4-Methyl-1,3-pentadiene	0.44	-30	1.5	86.5			100
5	4-Methyl-1,3-pentadiene	0.44	-78	55	21.6	15		100

[a] Polymerization conditions: monomer, 2 ml; toluene, 16 ml; Al/Ti = 1000 (molar ratio).
[b] Moles of monomer polymerized per mole of Ti per minute.
[c] Determined by IR and NMR analysis.

Table 3 Comparison between activity and stereospecificity of CpTi(OⁿBu)$_3$–MAO and Ti(OⁿBu)$_4$–MAO[a] [15,27,30]

| | CpTi(OⁿBu)₃–MAO | | | | | | Ti(OⁿBu)₄–MAO | | | | | |
| | | | | Polymer Microstructure[c] | | | | | | Polymer microstructure[c] | | |
Monomer	Time	Conversion (%)	N[b] (h⁻¹)	cis-1,4	trans-1,4	1,2	Time	Conversion (%)	N[b] (h⁻¹)	cis-1,4	trans-1,4	1,2
Butadiene	5 min	45	14000	82		18	1 h	23.1	600	81		19
(E)-1,3-Pentadiene	120 min	40	400	50		50	4 h	22	110	60	13	27
(Z)-1,3-Pentadiene	20 h	10	10	84		16	20 h	3.3	3	83		17
2,3-Dimethyl-1,3-butadiene	48 h	44	16	≥ 95			96 h	3.5	<1	≥ 95		
(E)-2-methyl-1,3-pentadiene	22 h	76.8	60	≥ 99			113 h	36	~5	40	60	
4-Methyl-1,3-pentadiene	1 min	75.3	~160 000			100	26 min	92.8	7300			100
(E,E)-2,4-Hexadiene	48 h	56	20		30	70	454 h	38.4	<1		≥ 99	

[a] Polymerization conditions; monomer, 2 ml; toluene, 16 ml; Al/Ti = 1000; 1 × 10⁻⁵ mol of Ti (5 × 10⁻⁶ mol of Ti for the polymerization of 4-methyl-1,3-pentadiene); +20 °C.
[b] Moles of monomer polymerized per mole of Ti per hour.
[c] Determined by IR and NMR analysis.

The significant points are as follows [15,27,30]:

(a) B, EP, ZP, DMB and 4MP give polymers with the same structure with both systems. 2MP gives a *cis* polymer with CpTi(OBu)$_3$–MAO, whereas it gives a polymer with a mixed *cis/trans* structure with Ti(OBu)$_4$–MAO. Some differences have been observed also in the polymerization of 2,4-H, which gives a practically 100 % *trans*-1,4 polymer with Ti(OBu)$_4$–MAO and a ca 30 % *trans*-1,4–70 % 1,2 polymer with CpTi(OBu)$_3$–MAO.

(b) The activity of CpTi(OBu)$_3$–MAO is much higher than that of Ti(OBu)$_4$–MAO. This may derive from a higher number of active centers and/or from a higher value of the kinetic constant of the propagation reaction, K_p. In the specific case of 2MP, the formation of *trans* units along with *cis* units with Ti(OBu)$_4$–MAO seems to indicate that K_p for this system is lower than that for CpTi(OBu)$_3$–MAO. In fact, it is known that in slow polymerizations the allyl group may undergo an *anti→syn* isomerization prior to monomer insertion, with consequent formation of *trans*-1,4 units [2a,c,25].

In conclusion, the data in Table 3 indicate that in CpTi(OBu)$_3$–MAO the Cp ligand has little influence on stereospecificity, whereas it has a substantial effect on activity. At present, it has not been clarified to what extent the higher activity depends on a higher number of active centers and/or on a higher value of K_p.

3.1.3 Diene copolymerization

The copolymerization of various dienes has been examined with CpTiCl$_3$–MAO, mainly for mechanistic studies.

In Section 3.1.1, the different behavior of ZP and 4MP with respect to the other monomers was tentatively attributed to the different mode of coordination of these monomers. According to the hypothesis proposed, the *cis* polymers of B, EP, 2MP, DMB and ZP derive from a *cis*-η^4 coordination of the monomer, whereas the 1,2 polymers of ZP and 4MP derive from a *trans*-η^2 coordination. With the aim of gaining more information on this problem, the following copolymerizations were investigated: 4MP–ZP, 4MP–B and B–EP. The idea was that pairs of monomers which, according to the hypothesis proposed, coordinate in the same way (e.g. B and EP, ZP and 4MP) would give different results to pairs that coordinate in a different way (e.g. B and 4MP).

The copolymerization results (Table 4) may be summarized as follows [29].

(a) The copolymerization of ZP with 4MP gives random copolymers, in which both ZP and 4MP have a 1,2 structure. The fact that ZP units have a 1,2 structure is by itself indicative that the product is a true copolymer and not a mixture of homopolymers, since ZP gives, at room temperature, homopolymers consisting predominantly of *cis* structure. Even with a proportion of 4MP in the copolymer of only 20 %, all the ZP units have a 1,2 structure. This finding seems to confirm

Table 4 Copolymerization of (Z)-1,3-pentadiene (ZP) with 4-methyl-1,3-pentadiene (4MP), butadiene (B) with 4MP and (E)-1,3-pentadiene (EP) with B, using $CpTiCl_3$–MAO as catalyst[a] [29]

				Feed composition				Copolymer composition		
Run	M1	mol (%)	M2	mol (%)	Temperature (°C)	Time (min)	Conversion (%)	M1 (mol %)	M2 (mol %)	$[\eta]$ (dl/g)
1	ZP	16.2	4MP	83.8	+20	0.75	36.8	9.6	90.4	0.16
2	ZP	53.2	4MP	46.8	+20	15	16.3	37.9	62.1	0.15
3	ZP	70.8	4MP	29.2	+20	14	8	50.7	49.3	0.12
4	ZP	83.7	4MP	16.3	+20	14	9.5	58.3	41.7	
5	ZP	95.2	4MP	4.8	+20	240	13	74.2	25.8	
6	ZP	83.3	4MP	16.7	+20	1440	48	73	27	
7	ZP	50	B	50	+20	18	20.3	7	93	
8	B	50	4MP	50	+20	27	21.6	99		
9	B	50	4MP	50	−30	1440	14.7	99		
10	B	20	4MP	80	+20	164	16.3	93	7	
11	B	20	EP	80	+20	12	16.9	48	52	2.1
12	B	50	EP	50	+20	9	17.7	68.5	31.5	2.8

[a] Polymerization conditions: $CpTiCl_3$, 1×10^{-5} mol; Al/Ti = 1000; toluene, 32 ml.

that the active coordination for 4MP is η^2. The coordination geometry around Ti of the active species is different when the diene is coordinated η^2 rather than η^4. If 4MP insertion derives from an η^2 coordination and the following monomer is ZP, this latter too will have the tendency to coordinate η^2. In other words, the catalytic complex is slow in modifying its coordination geometry if the monomer following 4MP is ZP, for which the η^2 and η^4 coordinations are not energetically much different.

In the ^{13}C NMR spectra of the copolymers having a ca 50 : 50 comonomer composition, the resonances typical of 1,2 syndiotactic poly(ZP) and 1,2 syndiotactic poly(4MP) are almost absent, which is indicative of a homogeneous distribution of the two comonomers along the polymer chains.

(b) B and 4MP give block copolymers. The difference between ZP–4MP and B–4MP copolymerizations is evident from the results of the runs in which the feed is ca 50 : 50 (Table 4, runs 2 and 8). ZP–4MP give random copolymers, whereas B–4MP give a product consisting practically of PB only. Copolymers containing ca 7 % of 4MP can be obtained only when the proportion of 4MP in the feed is ca 80 %. The ^{13}C NMR spectrum of these copolymers shows the resonances typical of 1,2 syndiotactic poly(4MP), which indicates that blocks of 4MP are present. It has been mentioned above that the rate of homopolymerization of 4MP is much higher than that of B. Notwithstanding this, B is incorporated much faster than 4MP, probably because the cis-η^4 coordination of B is so strong that the coordinated B cannot be easily displaced by 4MP.

(c) The copolymerization of B with EP gives copolymers with a random structure. This result is consistent with the fact that both comonomers coordinate cis-η^4.

In conclusion, the copolymerization results support the interpretation that the particular behavior of ZP and 4MP is due to the fact that they behave differently from other monomers with regard to their mode of coordination.

Butadiene–isoprene copolymerization has also been studied with $CpTiCl_3$–MAO [31]. The reactivity ratios have been found to be r_1 (butadiene) = 4.7, r_2 (isoprene) = 0.31. The comonomers copolymerize almost at random, although the $r_1 r_2$ product is >1. Isoprene homopolymerizes much more slowly than butadiene with $CpTiCl_3$–MAO and therefore it appears that monomer reactivities in homopolymerization do not parallel the reactivities in copolymerization.

3.1.4 Monoalkene–diene copolymerization

The copolymerization of styrene with butadiene and isoprene has been examined using the system $CpTiCl_3$–MAO. The reactivity ratios have been determined:

$$r_1 \text{ (styrene)} = 0.14 \quad r_2 \text{ (butadiene)} = 11.5 \quad r_1 r_2 = 1.6$$
$$r_1 \text{ (styrene)} = 0.35 \quad r_2 \text{ (isoprene)} = 6.3 \quad r_1 r_2 = 2.3$$

These values show that the diene is more reactive than styrene both when insertion occurs on growing chains ending with a styrene unit and when it occurs on chains ending with a diene unit [31,32].

The copolymerization C_2–4MP has also been examined [20]. It has been observed that, whereas 4MP in homopolymers consists of 1,2 units only, in the copolymers it has a mixed 1,2/1,4 structure. This finding has been put in relation with another result observed in the homopolymerization of 4MP: using a catalyst prepared from ^{13}C-enriched MAO, it has been possible to detect two different types of chain end-groups [16]:

These results have been tentatively interpreted assuming that chemoselectivity in 4MP polymerization (that is formation of 1,2 vs 1,4 units) may be influenced by a back-biting coordination of the penultimate monomer unit to the Ti of the active species. When the penultimate unit is ethylene or when the first monomer inserts into the Ti–$^{13}CH_3$ bond, the back-biting coordination cannot occur, and the absence of coordination would make possible the formation of 1,4 units from 4MP. This interpretation does not appear convincing in view of the fact that 1,2 syndiotactic poly(4MP) can be obtained also with Cp_2TiCl–MAO, as reported in Section 4. The back-biting coordination of the penultimate monomer unit appears improbable with this system, in which two Cp groups are bonded to the Ti of the active species. The results obtained with Cp_2TiCl–MAO seem to indicate that chemoselectivity in 4MP polymerization is not influenced by the back-biting coordination of the penultimate monomer unit. With regard to the two types of end-groups observed in 4MP homopolymers, a tentative alternative interpretation is possible. Insertion of 4MP into the $Ti-CH_3$ bond may give rise to both an *anti* and a *syn* allyl group, derived respectively from a *cis*-η^4 and a *trans*-η^2 monomer coordination. One of these two groups, probably the *anti* one, may give preferentially 1,4 units on reaction with the new monomer. However, more experimental data are needed to clarify this point.

3.2 ZIRCONIUM-BASED CATALYSTS

Whereas much work has been carried out, as shown in previous sections, on the polymerization of dienes with catalysts based on monocylopentadienyl-Ti compounds, only one paper has appeared on polymerization with analogous Zr catalysts. The system $Cp'ZrCl_3$–MAO ($Cp' = Me_5C_5$) was used [19]. This exhibits an activity much lower than the Ti-based systems. The polymers from butadiene (B) have a mixed *cis/trans* structure, whereas those from 4-methyl-1,3-pentadiene (4MP) have a 1,2 structure.

3.3 VANADIUM-BASED CATALYSTS

Catalysts based on monocyclopentadienyl compounds of vanadium have found to be highly active for the polymerization of dienes. Table 5 reports some data obtained with the system $(C_5H_4Me)VCl_2 \cdot 2PEt_3$–MAO [33]. The PEt_3 complex was used, since uncomplexed monocyclopentadienyl compounds of vanadium have a tendency to disproportionate [34]. The results of interest may be summarized as follows.

(a) The polymers obtained from butadiene, isoprene and (*E*)- and (*Z*)-1,3-pentadiene have a predominantly *cis* structure (about 80%), the other units being predominantly 1,2 (3,4 for polyisoprene). Soluble non-metallocene vanadium-based catalysts [e.g $AlEt_2Cl$–$V(acac)_3$; MAO–$V(acac)_3$] give from butadiene polymers with a *trans*-1,4 structure [2a,35]. The introduction of a cyclopentadienyl ligand causes a change in stereospecificity, a predominantly *cis* polymer being obtained. A plausible interpretation for this is as follows. Formation of a polybutadiene with a *trans* structure with soluble non-metallocene catalysts is probably determined by the fact that during the polymerization an *anti*→*syn* isomerization of the allyl group takes place, according to the schemes already reported in the literature [2a,c,35]. As a general rule, in diene polymerization, when *anti*→*syn* isomerization of the allyl group is more rapid than monomer insertion, a *trans* unit is formed instead of a *cis* unit. The formation of a *cis*-

Table 5 Polymerization of some diolefins with $(C_5H_4Me)VCl_2\cdot(PEt_3)_2$–MAO^a [33,35]

Run	Monomer	Catalyst		Polymerization				Polymer			
		mol of V (×10^5)	Al/V	Temperature (°C)	Time (h)	Conversion (%)	N^b (h^{-1})	cis-1,4 (%)	trans-1,4 (%)	1,2 (%)	$[\eta]^c$ (dl/g)
1	Butadiene	1	1000	+25	1/6	81.5	126778	80.4	2.4	17.2	3.3
2	Butadiene	1	100	+25	1	38.4	4978	82.4	2.1	15.5	1.4
3	Butadiene	1	1000	−30	1	26	3970	81.6		18.4	
4	Isoprene	1	1000	+25	5	75.6	302	78.2		21.8^d	
5	Isoprene	1	1000	0	1	16.9	338	83		17^d	
6	Isoprene	3	1000	−30	143	27.2		83		17^d	
7	(E)-1,3-Pentadiene	1	1000	0	4	22.7	114	71		29	
8	(Z)-1,3-Pentadiene	4	500	+20	116	11.4		85	15		
9	(E)-2-Methyl-1,3-pentadiene	4	500	−30	119	20		50	50		
10	4-Methyl-1,3-pentadiene	4	1000	−30	68	33.8		50	18	82	

^{a} Polymerization conditions: monomer, 2 ml; toluene, 16 ml (10 ml of monomer and 80 ml of solvent for butadiene).
^{b} Moles of monomer polymerized per mole of V per hour.
^{c} Intrinsic viscosity, determined in toluene at 25 °C.
^{d} 3,4 units.

polybutadiene with the vanadium metallocene system indicates that with this catalyst the insertion process is more rapid than the isomerization process. It is not clear if the cyclopentadienyl group makes the isomerization process slower or if it enhances the rate of the insertion process, or if it affects both these processes.

(b) The rate of polymerization of butadiene with $(C_5H_4Me)VCl_2 \cdot 2PEt_3$–MAO is particularly high, much higher than with other soluble vanadium-based systems [e.g. $V(acac)_3$–MAO] and also with $CpTiCl_3$–MAO.

4 CATALYSTS BASED ON BIS-CYCLOPENTADIENYL COMPOUNDS

4.1 TITANIUM- AND ZIRCONIUM-BASED CATALYSTS

For the reasons discussed in Section 2, catalysts based on bis-cyclopentadienyl compounds of Zr should not be active for diene polymerization. If the growing polymer chain is η^3-bonded to Mt of the active species and the monomer is η^4-coordinated, a 20-electron cation would result, which is improbable. We have examined the polymerization of butadiene and other dienes with Cp_2ZrCl_2–MAO under various conditions and have found that this system is actually inactive. However, this system can copolymerize butadiene with ethylene [36]. When the last polymerized unit is ethylene, butadiene can coordinate cis-η^4, and can be incorporated. On the other hand, when the last polymerized unit is butadiene, the polymer chain is bonded to Mt by an allyl bond and in this case only ethylene can coordinate.

The catalysts based on bis-cyclopentadienyl compounds of Zr are now of great interest for the polymerization of ethylene and propylene and also for the homo-polymerization of cycloalkenes and their copolymerization with monoalkenes. Various Zr metallocene compounds are now commercial products, owing to their interest as polymerization catalyst precursors. For the reasons mentioned above all these compounds are of no utility for diene polymerization.

In principle, the catalysts based on bis-cyclopentadienyl compounds of Ti should also be inactive, for the same reasons. However, Cp_2TiCl_2–MAO has been found to have some activity for the polymerization of butadiene and 4MP (Table 6) [37]. Other monomers give only oligomers or low molecular weight products, which have not been examined. The different behavior of the Zr and Ti catalyst has been attributed to the fact that Ti(III) is formed along with Ti(IV) on reaction of Cp_2TiCl_2 with MAO, and that Ti(III) is responsible for the catalytic activity. This hypothesis is supported by the fact that Cp_2TiCl–MAO is much more active than Cp_2TiCl_2–MAO for the polymerization of B and of 4MP [37]. One may ask which is the active

Table 6 Polymerization of butadiene (B) and 4-methyl-1,3-pentadiene (4MP): comparison of the results obtained with Cp_2TiCl_2–MAO, Cp_2TiCl–MAO and $CpTiCl_3$–MAO[a] [37]

Monomer	Temperature (°C)	Time (min)	Cp$_2$TiCl$_2$–MAO				Cp$_2$TiCl–MAO					CpTiCl$_3$–MAO				
			Conversions (%)	N^b (min^{-1})	$[\eta]$ (dl/g)	M.p. (°C)	Time (min)	Conversion (%)	N^b (min^{-1})	$[\eta]$ (dl/g)	M.p. (°C)	Time (min)	Conversion (%)	N^b (min^{-1})	$[\eta]$ (dl/g)	M.p. (°C)
B	+20	180	11.2	≈2			30	38.9	34			9	81.7	235		
B	0	1800	0.5				108	16.3	4							
B	−30	7000	traces				1350	6.8				300	6.8			
4MP	+20	17	42.9	44	1.1	89.3	2	66.4	581	0.7		1	100	1751	0.3	
4MP	0	12	24.6	36	1.7		2	59.1	517	1.6						
4MP	−30	14	9.7	12			2	18.7	164		92					
4MP	−78	195	5.8									1	100		1.1	96

[a] Polymerization conditions: monomer, 2 ml; toluene, 16 ml; Ti, 1×10^{-5} mol; Al/Ti = 1000.
[b] Moles of monomer polymerized per mole of Ti per minute.

species in Cp_2TiCl–MAO. When Cp_2TiCl reacts with MAO, the following reactions probably take place:

$$Cp_2TiCl + MAO \rightarrow Cp_2TiCH_3$$

$$Cp_2TiCH_3 + MAO \rightarrow [Cp_2Ti]^+ + [MAO-CH_3]^-$$

The cation $[Cp_2Ti]^+$ does not contain Ti–C bonds of σ type, which are essential for polymer chain growth. It has been hypothesized [37] that the cation $[Cp_2Ti]^+$ undergoes a rearrangement according to the following scheme:

The Ti–H bond would give the active Ti–C bond on reaction with the monomer. A rearrangement of this type has been observed [38] for the compound Cp_2Ti, in solution. With regard to this interpretation, what has been reported on the stability of $Cp_2Zr(CH_2Ph)_2$ and $Cp_2Ti(CH_2Ph)_2$ is also interesting [39]. The Zr compound can be recrystallized from boiling heptane, which means that it is stable at ca 100 °C. The Ti compound begins to decompose at ca 30 °C, according to the equation

$$nCp_2Ti(CH_2Ph)_2 \rightarrow [(C_5H_4)_2Ti]_n + 2nPhMe$$

The hydrogens of the Me group of toluene come from the Cp group.

There are other examples in the literature of hydrogen abstraction from Ti-bonded Cp groups [40,41]. All these data make plausible the interpretation proposed for the active species in Cp_2TiCl–MAO.

With regard to the polymers obtained with Cp_2TiCl–MAO, their structure is similar to that of the polymers obtained with $CpTiCl_3$–MAO. However, the activity of the $CpTiCl_3$ catalyst is higher, as shown by the data in Table 6.

4.2 VANADIUM-BASED CATALYSTS

Some data concerning the polymerization of butadiene and isoprene with the system Cp_2VCl–MAO have been reported (Table 7) [33]. This system is much less active than $(C_5H_4Me)VCl_2 \cdot PEt_3$–MAO. The polymers obtained have a predominantly *cis* structure (80–90 %). Nothing is known about the nature of the active species in Cp_2VCl–MAO. As a working hypothesis, it has been proposed that the active species derives from a rearrangement of the Cp ligands, with formation of a carbene-like structure, as in the case of Cp_2TiCl–MAO.

Table 7   Polymerization of butadiene and isoprene with $Cp_2VCl-MAO^a$ [33]

		Catalyst		Polymerization				Polymer		
Run	Monomer	Mol $(\times 10^5)$	Al/V	Temperature $(^{\circ}C)$	Time (min)	Conversion (%)	N^b (h^{-1})	cis-1,4 (%)	trans-1,4 (%)	1,2 (%)
1	Butadiene	2	1000	+15	56	40	2778	84.8	1.8	13.4
2	Butadiene	2	1000	-30	310	6	75	88.9		11.1
3	Butadiene	5	100	+15	305	8	41	87.2	1.1	11.7
4^c	Butadiene	5	100	+15	131	11.8	140	87.7	1.3	11
5^c	Butadiene	1.5	100	+15	26	22	4387	90	2	8
6	Isoprene	20	500	+25	24 h	7		82.1		17.9^d

^{a} Polymerization conditions: toluene, 80 ml; monomer, 10 ml.
^{b} Moles of monomer polymerized per mole of V per hour.
^{c} In these runs catalyst preformed and aged at 15 °C for 120 min was used [in the presence of a small amount of butadiene $(C_4H_6/V = 10$ molar ratio) in run 5].
^{d} 3,4 units.

5   CONCLUSIONS

Owing to the different mode of coordination of dienes (η^4 instead of η^2) and the different type of bond between the growing chain and Mt (η^3 instead of η^1), the cyclopentadienyl ligands cannot have the same role in catalysts for 1-alkene and 1,3-diene polymerization, respectively. The systems based on bis-cyclopentadienyl-Zr compounds, which are among the most active and stereospecific for ethylene and propylene polymerization, are inactive for 1,3-diene polymerization. Catalysts based on some bis-cyclopentadienyl-Ti and -V compounds (Cp_2TiCl_2, Cp_2TiCl, Cp_2VCl) exhibit some activity. This has been attributed to disproportionation of a Cp group, with the formation of a carbene-like structure.

The systems based on monocyclopentadienyl compounds are generally very active for diene polymerization. The cyclopentadienyl ligand does not seem to have a determining role in the stereospecificity, although in some cases ($CpVCl_2$, Cp_2VCl) it affects the *cis–trans* isomerism. However, in the cases examined, the cyclopentadienyl ligand enhances the catalyst activity. It is not clear to what extent the higher activity depends on a higher number of active centers or on a higher value of K_p.

From the industrial point of view, metallocene catalysts have not had the high impact on diene polymerization they have had on monoalkene polymerization. There are various reasons for this.

(a) The diene polymers of industrial interest are principally *cis*-polybutadiene and *cis*-polyisoprene, which are now produced on industrial scale with various catalysts, based on Ti, Co, Ni and Nd. These catalysts give a high-*cis* (95–97 %) polybutadiene and polyisoprene, while metallocene catalysts so far known give polymers with a *cis* content of ca 85 %, too low for use in elastomers.

(b) One of the advantages of metallocene catalysts over the conventional catalysts is that they are of 'single-site' type. Before the introduction of metallocene catalysts, no single-site catalyst was known for the polymerization of 1-alkenes. The situation is different in the case of dienes: single-site catalysts for the polymerization of this class of monomers have been known since the end of the 1950s. The system $Co(OCOR)_2-Al_2Cl_3Et_3$, proposed in 1958, is a soluble single-site catalyst. The same holds for the ionic Ni systems $[allylNiL_2]^+[An]^-$, where L = 1,5-cyclooctadiene and An = PF_6, BPh_4, etc., which were well known before the discovery of metallocene catalysts [2a].

(c) In catalysts for monoalkene polymerization, it is possible to control the stereospecificity by the type of cyclopentadienyl ligand used. In catalysts for diene polymerization, the ligand has little influence on stereospecificity, although the ligand affects other properties such as the solubility, the number of active centers and, at least in some cases, K_p also.

While metallocene catalysts for diene polymerization have not yet found an industrial application, they have a relevant scientific interest. Using these catalysts, various new polymers and copolymers have been synthesized, some of which are highly stereoregular {1,2 syndiotactic poly(4-methyl-1,3-pentadiene); 1,2 syndio-tactic poly[(Z)-1,3-pentadiene]}. In addition, metallocene catalysts have contributed greatly to a better understanding of the mechanism of diene polymerization. For the first time it has been possible to obtain information on the polymerization of monomers such as ZP and 4MP, which had been previously polymerized only with heterogeneous catalysts.

So far, only metallocene catalysts based on compounds of Ti, Zr and V have been intensively studied. Work on metallocene catalysts based on other transition metals will probably give interesting new results.

6 REFERENCES

1. For information on olefin polymerization with metallocene catalysts, see other chapters in this book and the following reviews: (a) Brintzinger, H. H., Fisher, D., Mülhaupt, R., Rieger, B. and Waymouth, R. M., *Angew. Chem. Int. Ed. Engl.*, **34**, 1143 (1995); (b) Bochmann, M., *J. Chem. Soc., Dalton Trans.*, 255 (1996).
2. For reviews on 1,3-diene polymerization with transition metal catalysts, see: (a) Porri, L. and Giarrusso, A., in Eastmond, G. C., Ledwith, A., Russo, S. and Sigwalt, P. (Eds), *Comprehensive Polymer Science*, Pergamon Press, Oxford, 1989, Vol. 4, Part II, pp. 53–108; (b) Porri, L., Giarrusso, A. and Ricci, G., *Polym. Sci., Ser. A*, **36**, 1421 (1994); (c) Porri, L., Giarrusso, A. and Ricci, G., *Prog. Polym. Sci.*, **16**, 405 (1991).
3. (a) Zambelli, A., Sacchi, M. C., Locatelli, P. and Zannoni, G., *Macromolecules*, **15**, 211 (1982); (b) Longo, P., Grassi, A., Pellecchia, C. and Zambelli, A., *Macromolecules*, **20**, 1015 (1987); (c) Pino, P., Cioni, P. and Wei, J., *J. Am. Chem. Soc.*, **109**, 6189 (1987); (d) Corradini, P. and Guerra, G., *Prog. Polym. Sci.*, **16**, 239 (1991); (e) Guerra, G., Cavallo, L., Moscardi, L., Vacatello, M. and Corradini, P., *J. Am. Chem. Soc.*, **116**, 2988 (1994); (f)

Bierwagen, E. P., Bercaw, J. E. and Goddard, W. A., *J. Am. Chem. Soc.*, **116**, 1481 (1994); for a more complete literature survey on the problem of enantioselectivity in propylene polymerization, see Chapter 1.

4. Natta, G., Porri, L., Zanini, G. and Palvarini, A., *Chim. Ind. (Milan)*, **41**, 1163 (1958).
5. Natta, G., Porri, L., Stoppa, G., Allegra, G. and Ciampelli, F., *J. Polym. Sci., Part B*, **1**, 67 (1963).
6. Porri, L. and Aglietto, M., *Makromol. Chem.*, **177**, 1465 (1976).
7. Destri, S., Gallazzi, M. C., Giarrusso, A. and Porri, L., *Makromol. Chem., Rapid Commun.*, **1**, 293 (1980).
8. Gallazzi, M. C., Giarrusso, A. and Porri, L., *Makromol. Chem., Rapid Commun.*, **2**, 59 (1981).
9. Bolognesi, A., Destri, S., Porri, L. and Wang, F., *Makromol. Chem., Rapid Commun.*, **3**, 187 (1982).
10. Porri, L., Gallazzi, M. C., Destri, S. and Bolognesi, A., in Quirk, R. P. (Ed.), *Transition Metal Catalyzed Polymerizations*, Harwood Academic, New York, 1983, Part B, pp. 555–567.
11. Porri, L., Gallazzi, M. C., Destri, S. and Bolognesi, A., *Makromol. Chem., Rapid Commun.*, **4**, 485 (1983).
12. Porri, L., Giarrusso, A. and Ricci, G., *Makromol. Chem., Macromol. Symp.*, **48/49**, 239 (1991).
13. Ricci, G., Italia, S., Giarrusso, A. and Porri, L., *J. Organomet. Chem.*, **451**, 67 (1993).
14. Porri, L., Giarrusso, A. and Ricci, G., *Makromol. Chem., Macromol. Symp.*, **66**, 231 (1993).
15. Porri, L., Giarrusso, A. and Ricci, L., *Macromol. Symp.*, **89**, 383 (1995).
16. Longo, P., Proto, A., Oliva, P. and Zambelli, A., *Macromolecules*, **29**, 5500 (1996).
17. Oliva, L., Longo, P., Grassi, A., Ammendola, P. and Pellecchia, C., *Makromol. Chem., Rapid Commun.*, **11**, 519 (1990).
18. Ricci, G., Italia, S. and Porri, L., *Macromolecules*, **27**, 868 (1994).
19. Longo, P., Oliva, P., Proto, A. and Zambelli, A., *Gazz. Chim. Ital.*, **126**, 377 (1996).
20. Longo, P., Grisi, F., Proto, A. and Zambelli, A., *Macromol. Rapid Commun.*, **18**, 183 (1997).
21. Meille, S. V., Capelli, S. and Ricci, G., *Macromol. Rapid Commun.*, **16**, 891 (1995).
22. Murahashi, S., Kamachi, M. and Wakabayashi, N., *J. Polym. Sci., Part B*, **7**, 35 (1969).
23. Kamachi, M., Wakabayashi, N. and Murahashi, S., *Macromolecules*, **7**, 744 (1974).
24. Wang, F., Bolognesi, A., Immirzi, A. and Porri, L., *Makromol. Chem.*, **182**, 3617 (1981).
25. Bolognesi, A., Destri, S., Zi-nan, Z. and Porri, L., *Makromol. Chem., Rapid Commun.*, **5**, 679 (1984).
26. Zambelli, A., Ammendola, P. and Proto, A., *Macromolecules*, **22**, 2126 (1989).
27. Ricci, G. and Porri, L., *Polymer*, **38**, 4499 (1997).
28. Porri, L. and Gallazzi, M. C., *Eur. Polym. J.*, **2**, 189 (1966).
29. Ricci, G. and Porri, L., *Macromol. Chem. Phys.*, **198**, 3647 (1997).
30. Porri, L., Giarrusso, A. and Ricci, G., unpublished data.
31. Zambelli, A., Proto, A., Longo, P. and Oliva, P., *Macromol. Chem. Phys.*, **195**, 2623 (1994).
32. Pellecchia, C., Proto, A. and Zambelli, A., *Macromolecules*, **25**, 4450 (1992).
33. Ricci, G., Panagia, A. and Porri, L., *Polymer*, **37**, 363 (1996).
34. Nieman, J., Teuben, J. H., Huffman, J. C. and Caulton, K. G., *J. Organomet. Chem.*, **255**, 193 (1983).
35. Ricci, G., Italia, S. and Porri, L., *Macromol. Chem. Phys.*, **195**, 1389 (1994).
36. Kaminsky, W. and Schlobohm, M., *Makromol. Chem., Macromol. Symp.*, **4**, 103 (1986).
37. Ricci, G., Bosisio, C. and Porri, L., *Makromol. Rapid Commun.*, **17**, 781 (1996).

38. Brintzinger, H. H. and Bercaw, J. J., *J. Am. Chem. Soc.*, **92**, 6182 (1970).
39. Fachinetti, G. and Floriani, C., *Chem. Commun.*, 654 (1972).
40. Volpin, M. E., Belyi, A. A., Shur, V. B., Lyakhovestsky, Yu. I., Kudryavtes, R. V. and Buhnov, N. N., *J. Organomet. Chem.*, **27**, C5 (1970).
41. van Tamelen, E. E., Seeley, D., Schneller, S., Rudler, H. and Cretney, W., *J. Am. Chem. Soc.*, **92**, 5251 (1970).

7

New Functionalized Olefin Copolymers Synthesized by Metallocenes and Novel Organometallic Catalysts

BARBRO LÖFGREN AND JUKKA SEPPÄLÄ
Helsinki University of Technology, Espoo, Finland

1 INTRODUCTION

One major advantage of the homogeneous metallocene/methylaluminoxane (MAO) catalysts over conventional Ziegler–Natta catalysts is their ability to polymerize a wide variety of bulky monomers, such as higher linear α-olefins [1–4], polar monomers [5–9], phenolic compounds [10,11], dienes [12–15] and cycloolefins [16–21]. They offer a promising new way not only to tailor polymer properties but also to produce entirely new polymeric materials.

Copolymerizations of ethene and long-chain α-olefins offer a way to generate polymers with a controlled level of long-chain branching (LCB) along the polymer backbone, which allows for improved rheological properties and enhanced processability. Polyolefins containing functional groups, in turn, show improved adhesive, thermal, rheological, morphological and mechanical properties, improved affinity for dyes and printing agents and greater compatibility with other polymers [22]. They also offer sites for initiating graft copolymerizations.

This chapter describes copolymerizations of olefins with higher linear α-olefins and with functional monomers: polar monomers, phenolic compounds, dienes and cycloolefins. We also report some attempts to prepare novel catalysts with lower

Metallocene-based Polyolefins Edited by J. Scheirs and W. Kaminsky
© 2000 John Wiley & Sons Ltd

Lewis acidity than the traditional metallocenes, which would permit the attack of a functional comonomer without deactivation of the catalyst. Finally, methods of characterizing the comonomer distributions in the copolymers are reviewed.

2   PREPARATION OF POLY(ETHENE–CO-LONG CHAIN α-OLEFIN)S

Relationships between the structure of metallocene catalysts and their polymerization behavior are of growing interest [23]. Indeed, an understanding of these relationships is essential for the goal-directed development of improved catalysts and the controlled tailoring of polymers. Besides choice of the central metal, there are two principal ways to modify the structure of metallocenes: (i) variation of the ligands and their substitution patterns and (ii) variation of the bridge connecting the ligands [24].

The effect of ligands and their substitution pattern is considered to be mediated through a combination of steric and electronic effects [23], which can be utilized to influence the polymerization behavior. Ligands have therefore been of interest in several studies [25–29]. Steric effects of bulky ligands or bulky ligand substituents influence the coordination of monomers, whereas their electronic effects affect the stability of active sites and influence chain growth mechanisms [25,30]. The structure of the interannular bridge influences not only the electronic effects but also catalyst rigidity and the coordination gap aperture [25,31,32]. Studies on metallocene-catalyzed copolymerizations of ethene or propene with long-chain α-olefins are nevertheless rare [3,33–35], and have seldom been combined with study of the influence of minor modifications in catalyst structure on copolymerization behavior [28,29].

α-Olefin copolymerizations with coordination catalysts have been explored for some time now, with the goal of achieving branches of intermediate length. Heterogeneous Ziegler–Natta catalysts are deactivated by long-chain α-olefins [36], whereas the activity of catalysts such as Cp_2ZrCl_2, $Et(Ind)_2ZrCl_2$, $Me_2Si(Ind)_2ZrCl_2$ and $^iPr(FluCp)ZrCl_2$ is enhanced for comonomers as long as 1-octadecene (C_{18}) [3,34,37–39]. Recently we reported [40] the behavior of the commercial catalysts $Et(Ind)_2ZrCl_2$, $Et(H_4Ind)_2ZrCl_2$ and $Me_2Si(Ind)_2ZrCl_2$ and modifications of the last [41] (Figure 1) in the copolymerization of ethene with 1-hexene and 1-hexadecene. The enhancement in polymerization activity known as the comonomer effect was clearly evident.

As can be seen for 1-hexadecene in Figure 2, the copolymerization behavior of the complexes **1–5** is more strongly influenced by the hydrogenation of the indenyl ligands than by their different interannular bridges. The hydrogenation of indenyl ligands and the replacement of silylene with ethylene bridges result in lower comonomer incorporation in ethene–α-olefin copolymer. The poorer incorporation appears to be the result of both steric and electronic effects. We have proposed that steric effects originate from the greater mobility of the hydrogenated indenyl ligands

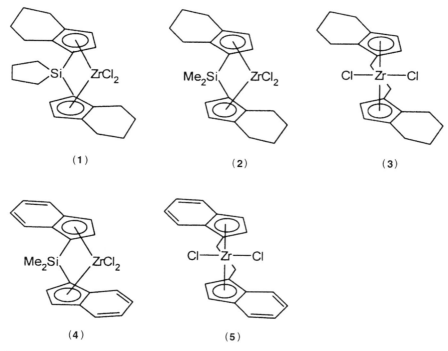

Figure 1. Structures of different *ansa*-metallocenes. Catalyst **1** = 1, 4-butanediylSi(H$_4$Ind)$_2$ZrCl$_2$; **2** = Me$_2$Si(H$_4$Ind)$_2$ZrCl$_2$; **3** = Et(H$_4$Ind)$_2$ZrCl$_2$; **4** = Me$_2$Si(Ind)$_2$ZrCl$_2$; **5** = Et(Ind)$_2$ZrCl$_2$

and the smaller coordination gap aperture resulting from the ethylene bridge, both of which factors may hinder the approach of the comonomer. Electronic effects may originate from the reduced electron density of the hydrogenated and ethylene-bonded ligands, increasing the Lewis acidity of the zirconium atom. This would lead to the formation of a stronger bond between the zirconium atom and the coordinated methylaluminoxane, which would make insertion of the α-olefin more difficult and increase the rate of chain termination by β-elimination.

3   PREPARATION OF FUNCTIONALIZED POLYOLEFINS

In general, the polymerization of monomers containing functional groups with Ziegler–Natta catalysts is limited by the intolerance of these catalysts to Lewis bases, which leads to catalyst deactivation, polymer degradation and comonomer homo-polymerization. Advances in metallocene catalyst chemistry offer routes for the preparation of a variety of polyolefinic materials, although even with metallocenes

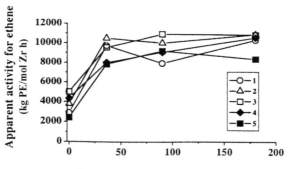

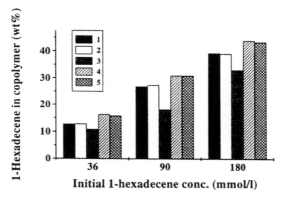

Figure 2. (a) Demonstration of the comonomer effect for different *ansa*-metallocenes. (b) Comonomer response of structurally modified metallocenes. Catalyst   **1** = 1,4-butanediylSi(H$_4$Ind)$_2$ZrCl$_2$;   **2** = Me$_2$Si(H$_4$Ind)$_2$ZrCl$_2$;   **3** = Et(H$_4$Ind)$_2$ZrCl$_2$; **4** = Me$_2$Si(Ind)$_2$ZrCl$_2$; **5** = Et (Ind)$_2$ZrCl$_2$

the Lewis acid components (Zr, Al) tend to form complexes with the non-bonded electron pairs of heteroatoms in preference to reacting with the π-electrons of the double bond of the incoming monomer. This undesirable interaction can be minimized in several ways: (i) by insulating the double bond from the heteroatom by a spacer group, (ii) by increasing the steric hindrance about the heteroatom, (iii) by reducing the electron-donating feature of the heteroatom (e.g. by attaching an electron-attracting group) and (iv) by choosing catalyst components that are inert to the functional groups.

In general, there are three main approaches to the functionalization of polymers: the direct polymerization or copolymerization of functional monomers [5–9], formation of precursors which are readily converted into various functional groups [22,42–44] and the post-modification of preformed polymers [14,15,45, 46].

3.1 POLAR COMONOMERS

The modification of polyolefins with functional monomers has seen good progress in recent years. Our investigations have shown that several metallocene catalysts favor the copolymerization of ethene and propene with comonomers that contain hydroxyl, acid or ester groups [5–8]. Some of our results are presented in Table 1. As can

Table 1 Copolymerizations of ethene and propene with polar comonomers

Run	Comonomer	In feed (mmol)	In polymer (mol%/wt%)	Activity (kg P/mol Zr h)	$M_w \times 10^{-3}$ (g/mol)	D	T_m (°C)
1	**ETHENE**[a]	–	–	7700	163	2.7	136.7
2	1,1-Dimethyl-2-propen-1-ol	1	None	1580	150	3.7	137.2
3		2	None	1640	183	3.1	136.0
4	5-Hexen-1-ol	1	0.09/0.3	1550	158	3.4	134.2
5		2	0.07/0.2	2900	130	3.1	134.8
6		3	0.1/0.35	2600	147	4.2	132.8
7	10-Undecen-1-ol	1	0.1/0.7	3450	203	3.3	134.7
8		2	0.2/1.1	2540	143	3.5	131.7
9		3	0.3/1.8	2370	121	5.1	132.4
10	10-Undecenoic acid	1	Traces	2500	125	3.2	132.6
11		3	0.3/2	2000	180	4.4	129.8
12	Methyl 9-decenoate	1	0.2/1.0	2900	98	3.1	131.5
13		2	0.2/1.1	2600	139	3.8	134.3
14		3		850	131	3.9	134.0
15	**PROPENE**[a]	–	–	16600	34	1.8	138.2
16	1,1-Dimethyl-2-propen-1-ol	1	None	8700	32	2.0	137.0
17		2	None	4300	31	2.0	136.4
18	5-Hexen-1-ol	1		3900	31	2.1	136.7
19		2		1800	27	2.0	136.9
20		5	0.3/0.6	180	21	2.0	133.9
21	10-Undecen-1-ol	1		2600	35	1.7	136.6
22		2		1500	31	1.8	134.9
23		5	0.7/2.7	100	28	1.7	130.6
24	12-Tridecen-2-ol	1		2600	35	1.7	137.3
25		3	0.3/1.3	500	28	1.8	132.5
26	1,1-Di-*tert*-butyl-9-decen-1-ol	2		3200	33	1.8	136.0
27	10-Undecenoic acid	1		3500	32	1.9	137.4
28		2	0.2/0.7	1100	30	1.9	135.7
29		3	0.6/2.4	600	28	1.9	134.6
30	Methyl 9-decenoate	0.5		1600	33	1.8	137.7
31	*tert*-Butyl 10-undecenoate	1	0.1/0.5	2400	32	1.9	137.1
32		2		1100	30	1.9	139.1

[a] Catalyst $= (^nBuCp)_2ZrCl_2$, $T_p = 60\,°C$, $p(C_2H_4) = 2.5$ bar, Al/Zr $= 4000$ mol/mol, $t_p = 40$ min.
[b] Catalyst $= Et(Ind)_2ZrCl_2$, $T_p = 30\,°C$, $p(C_3H_6) = 3$ bar, Al/Zr $= 4000$ mol/mol, $t_p = 60$ min.

be seen, a small addition of a functional comonomer containing oxygen as heteroatom nevertheless causes a significant deactivation of the metallocene catalyst. This is because the comonomer forms a stable complex with the catalyst, preventing the formation of a π-complex between the catalyst and double bond of the incoming comonomer, or it interacts with the Coulombic forces and hinders the formation of the π-complex, although the polymerization reaction may continue. The functional comonomer in runs 2–3 and 16–17 exemplifies the situation in which the double bond competes unsuccessfully with the hydroxyl group for the zirconium atom of the catalyst, with the consequence that no comonomer is incorporated.

Comonomers in runs 4–9 and 18–25 show the effect of a spacer. The length and flexibility of the spacer influence the population of the conformers in which the polar group comes into proximity with the double bond. There are four CH_2 groups in the spacer in the comonomer 5-hexen-1-ol and nine in 10-undecen-1-ol, which means that there are a larger number of conformers available for the second than the first comonomer. However, in most conformers the hydroxyl group is further than 3–4 Å from the double bond, and its energy is less than 3 kcal/mol above the global minimum. In most of the low-energy conformers, the distance between the groups exceeds 7 Å [47]. One would expect, therefore, that, under similar polymerization conditions, a larger amount of 10-undecen-1-ol than 5-hexen-1-ol would be incorporated in the polymer chain. This was also found to be the case. According to the ^{13}C NMR results, the hydroxyl group in 10-undecen-1-ol interacts both with the cocatalyst MAO and with the aluminium alkyl species present [48].

The comonomer in run 26 has approximately the same spacer length as 10-undecen-1-ol, but the OH group is now protected, and since this reduces the probability of conformers where the groups are close to each other, the deactivation of the catalyst is less pronounced. Similar effects can be seen for comonomers containing acid or ester groups, runs 10–14 and 27–32.

The surface properties, such as adhesion, are improved by the incorporation of functionality in the polyolefin matrix, as shown for peel strength in Table 2.

Table 2 Average peel strength of selected functionalized polyolefins

Sample	Average peel strength towards Al (N/mm)
Polyethenes	
PE–HD	0
PE–LD	0.02
—OH groups	0.2
—COOH groups	0.3–1.3
—COOCH$_3$ groups	0.4–1.4
Polypropenes	
Homopolymer	0
—OH groups	0.4–0.5

3.2 PHENOLIC COMONOMERS

Of particular interest has been the *in situ* copolymerization of hindered phenols, which act as stabilizers for both ethene and propene [10,11] (see Scheme 1). Relative to homopolymerization, the initial polymerization rate increases almost threefold when a hindered phenolic stabilizer is added during polymerization over a chiral bridged metallocene catalyst [$Me_2Si(IndH_4)_2ZrCl_2$ or $Et(IndH_4)_2ZrCl_2$]. The activity enhancement is attributed to the ability of the phenolic antioxidant to act as a large, weakly coordinating anion, which stabilizes the cationic center. The products are random copolymers containing isolated phenolic long-chain branches and possessing high thermo-oxidative stability.

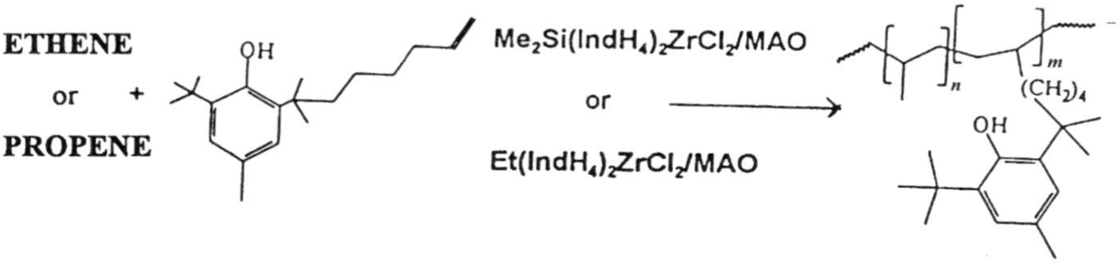

Scheme 1 Copolymerization of hindered phenols.

Chemically bound stabilizers offer several advantages over conventional admixed stabilizers. Chemically bound stabilizers will not be lost due to migration, volatilization, extraction or diffusion. Polypropene stabilized *in situ* through covalent bonds also exhibits reasonably good stability upon exposure to high-energy radiation [49] and, accordingly, is well suited for use in medical applications.

3.3 DIENES AS COMONOMERS

One way to avoid complex formation between the active catalyst site and the functional group of the comonomer is to copolymerize the olefins with non-conjugated dienes that do not interfere with the polymerization. As reported by Hackmann *et al.* [50], 7,5-dimethyl-1,6-octadiene can be used for this purpose, and its remaining double bonds can be used to introduce useful functionalities, such as epoxy groups, into the polymer (see Scheme 2). Work along the same lines for cyclic dienes has been reported by Marathe and Sivarann [14] and Kaminsky *et al.* [15].

Over the years, considerable attention has been paid to the molecular architecture of polyolefins, which is related to polymer chain length, the type and tacticity of side-chain branches and the distribution of side-chain branches along the main polymer chains. A method of producing long-chain branches is to copolymerize

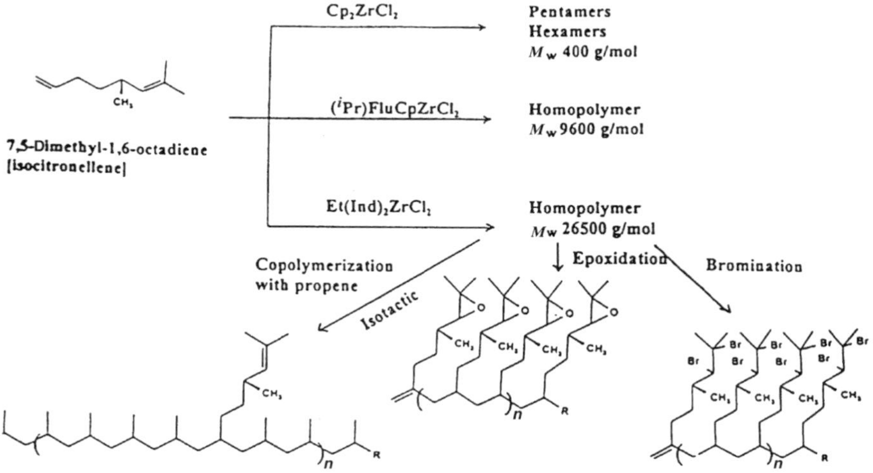

Scheme 2 Polymerizations of a diene [50]

ethene with α,ω-dienes [51–54], so obtaining cyclopentane structures interspaced with CH_2 groups.

3.4 CYCLIC COMONOMERS

Less crystalline and more elastomeric polymers can be produced through the use of cyclic monomers in place of the traditional alkenes [55]. Homopolymers of cyclic alkenes are ordinarily not processable owing to their high melting-points and insolubility. However, in the copolymerization of a norbornene with ethene, for example, the glass transition temperature (T_g) can be tailored by selecting a suitable norbornene and the appropriate incorporation level [16–21]. Norbornene–ethene copolymers are transparent and amorphous and are useful materials for compact discs and other optical applications. Although, at present, efforts are being directed towards high-T_g polymers, the copolymerization of norbornenes with α-olefins, the basis of EPDM technology, has gained a footing owing to the new processability-related properties originating from single-site catalyst technology [56].

4 NEW ORGANOMETALLIC CATALYSTS

Promising studies have recently been reported on the use of Ni(II) and Pd(II) complexes containing nitrogen donor ligands, activated by MAO, in α-olefin polymerizations [57,58] and functionalized olefin homopolymerizations [59,60]. These late transition metal complexes would be expected to tolerate a variety of

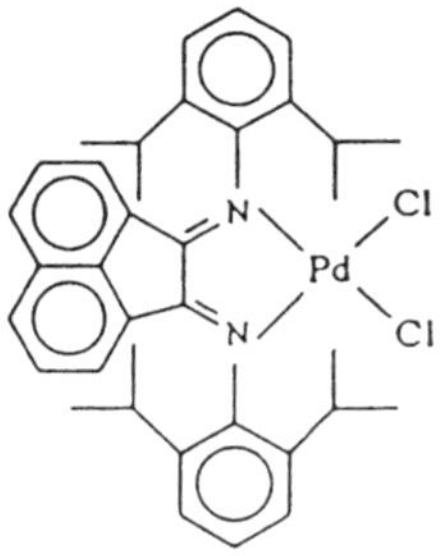

(6)   (7)   (8)

$PdCl_2[2,6-^iPr_rC_6H_3\text{-BIAN}]$ $Ta(BzNPy)_2Cl_3$ $Ta(PhNPyNHPh)_2Cl_3$

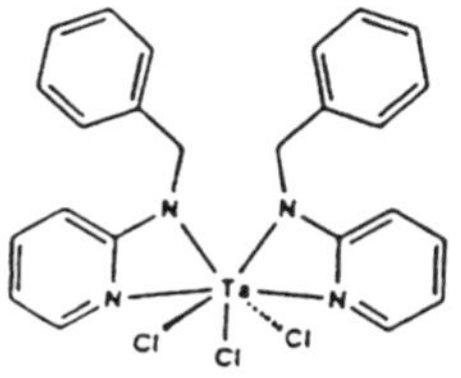

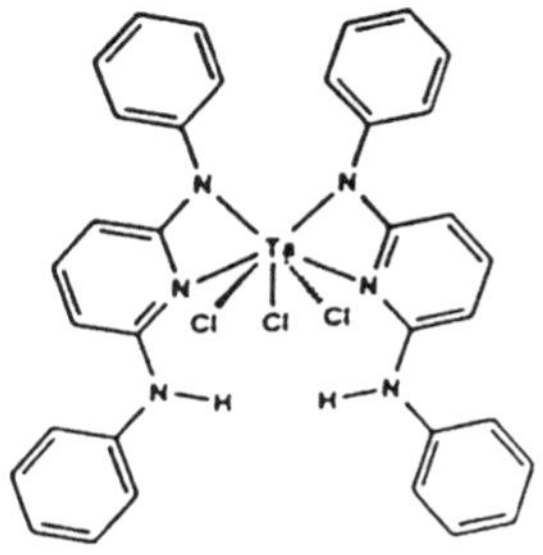

Figure 3 Schematic presentation of the complexes.

donors that deactivate early transition metal-based systems, and even to be useful in copolymerizations of carbon monoxide and ketones [61]. The activity of $PdCl_2[2,6-^iPrC_6H_3$-bis (imino)asenaphthene] (**6**, Figure 3) nevertheless dropped to one tenth when 3 mmol of comonomer was introduced in ethene copolymerizations with 10-undecen-1-ol (see Table 3) [60]. Moreover, in general, these late transition metal complexes are at present capable of producing high molar mass polymers only at impracticably low polymerization temperatures.

There is also rapidly growing interest in non-cyclopentadienyl derivatives of early transition metal complexes. In this area, Group 4 and 5 metal complexes containing nitrogen and oxygen donors have been introduced. For example, titanium dimethyl complexes with either ethene [62] or propene [63,64] bridged diamido ligands have been reported to be highly active in polyolefin polymerizations, and they have been shown to perform living polymerizations when tris(perfluorophenyl)boron is used as activator instead of MAO [64]. Another new contribution to the single-site catalyst family is the six-sided catalyst based on borabenzene rings used with a zirconium

Table 3 Results of ethene polymerization promoted by tantalum(V) aminopyridinato complexes and $PdCl_2[2, 6-^iPrC_6H_3$-bis (imino)asenaphthene] activated by MAO

Catalyst	[Catalyst] (μ mol)	T_p (°C)	Time (min)	Activity (kg P/mol Me h)	M_w (g/mol)	D
6	7.4	80	60	163	193 000	2.4
6[a]	7.4	80	120	10	136 000	2.3
7	6.7	60	7	5000	94 000	2.0
8	1.1	60	30	5800	85 000	1.9
8	1.1	80	7	23900	66 000	1.8

[a] 3 mmol of 10-undecen-1-ol was fed in this polymerization.

dichloride center [65,66]. These borabenzene catalysts produce a series of unusual olefinic polymers and copolymers.

Niobium(V) and tantalum(V) form mononuclear bis(aminopyridinato) complexes possessing pentagonal bipyramidal coordination spheres [67]. As shown in Table 3 [68], both tantalum(V) complexes (see Figure 3) are moderately active in ethene polymerization when activated with MAO. The activity of complex **8** at 80°C is close to the activities obtained with metallocenes.

Alternative systems based on various alkoxotitanium and alkoxozirconium complexes have been shown to catalyze α-olefin polymerization [69]. Chelating di-[70], tri-[71] and tetradentate [72] nitrogen–oxygen donor ligands have been used in various complexes.

5 COMONOMER DISTRIBUTION IN POLYMERS

Single-site catalysts yield polymers with a very narrow molar mass distribution (MWD) and narrow chemical composition distribution (CCD), in contrast to the broad MWD and CCD produced with Ziegler–Natta catalysts. Size-exclusion chromatography (SEC) is the most common method for the determination of molar masses, and in combination with additional on-line detectors such as viscometers or Fourier transform infrared detectors can also be applied to branched copolymers [73]. The narrow CCD means that the comonomer is uniformly distributed along the polymer chain regardless of chain length; no highly branched polymer is formed and the melting-point range measured by differential scanning calorimetry (DSC) is narrow.

The ^{13}C NMR technique has been demonstrated by Uozumi and Soga [74] to be extremely useful for determinations of the microstructure and reactivity ratios of copolymers [34,75]. The absolute comonomer contents of the copolymers and the amount of branches ($< C_6$) can today be determined by NMR in a standardized way, but methods other than NMR are required for determinations of the comonomer distribution.

So far, the temperature-rising elution fractionation (TREF) technique [76,77] continues to be the main method for the measurement of the short-chain branching distribution (SCBD) in branched polyethene, especially in PE–LLD. The drawback of the method is that it is time consuming, and a quicker and easier method, capable of giving information about the molecular structure and the homogeneity of the chains, is of interest.

A new technique to measure the SCBD in ethene copolymers has been developed and applied to metallocene copolymers [34,78,79]. The technique, which is referred to as crystallization analysis fractionation (CRYSTAF) [78], is based on a stepwise precipitation. A CRYSTAF analysis can be performed more quickly than a TREF analysis, but is not as convenient as a DSC measurement.

Adisson *et al.* [80] introduced a stepwise crystallization method based on successive annealing of the polymer melt and subsequent DSC analysis of the melting behavior of the treated sample. A similar method, based on stepwise isothermal segregation technique (SIST), was presented by Kamiya *et al.* [81] and Chiu *et al.* [82].

The segregation fractionation technique (SF) has been applied to both commercial and test grades of polyethenes produced by Ziegler–Natta and by single-site catalysts [34,39,83]. With SF, it is possible to achieve a separation efficiency similar to that of the TREF method. Moreover, the curves obtained after segregation fractionation provide information about the lamellar thickness distribution and the amount of short-chain branching, although complete quantitative calculations of branching distribution require the availability of reference comonomers. Figure 4 shows the differences in DSC endotherms after segregation fractionation of three functionalized polyethenes, with equal amounts of comonomer incorporated but prepared by different catalysts. The fractionations show a more homogeneous comonomer distribution when silyl-bridged catalysts (curves 2 and 3) are used instead of a non-bridged catalyst (curve 4).

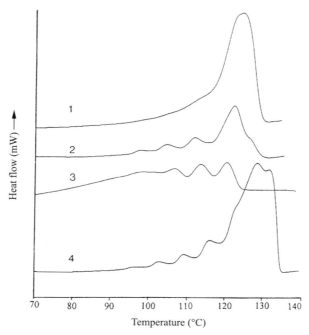

Figure 4   DSC endotherms after segregation fractionation for homopolyethene (1) and (ethene-10-undecen-1-ol) copolymers (2–4) with different catalysts: (1) and (2) $Me_2Si(Ind)_2ZrCl_2$; (3) $Me_2Si(MeInd)_2ZrCl_2$; and (4) $Et(Ind)_2ZrCl_2$

Recently, many studies have been carried out to obtain a better understanding of the dynamic mechanical properties of polyolefins with the purpose of using the results to define structure–property relationships and relate them to the performance of the polymer. Woo and co-workers [75,84], among others, have added fundamental information of assistance in the study of subambient relaxations of homo- and copolyethenes. The results [85,86] we have obtained from DMA measurements confirm that particularly the β-relaxations obtained from damping factor tan $\delta(= E''/E')$ and loss modulus curves, E'', can provide valuable information about the amounts of branching in different polyethene grades. The intensity of the tan δ maximum peak increases and the position of the corresponding peak temperature decreases with increase in the incorporation of comonomer, and these changes vary with the comonomer. Moreover, the width of the β-relaxation correlates with the SCBD.

DMA studies have shown that the mechanical properties of our functionalized polymers are at least as good as those of non-functionalized polymers [87]. The storage modulus (E'), which correlates with the Young's modulus, the flexural modulus and stiffness [88], is even slightly better for the functionalized polymers.

6   PERSPECTIVES

The most remarkable contribution of the new generations of highly active transition metal catalysts will be their ability to produce an unprecedented variety of polyolefins containing very bulky and even polar comonomers. As the development of advanced technologies continues there will be an increasing demand for higher value-in-use polyolefinic materials for speciality applications. Copolymers of ethene and higher α-olefins will be important commercial products and great efforts will be made to find new and more efficient catalysts for comonomer incorporation.

Other advances, such as the production of functionalized polyolefins, must await a more thorough understanding of the reaction mechanisms involved. In the production of functionalized polyolefins, the main emphasis will be on the design of catalysts with lower Lewis acidity than the traditional single-site catalysts, allowing the attack of a functional comonomer without deactivation of the catalyst active site. Nevertheless, polyolefins containing functional groups already are challenging engineering plastics. Particularly worth mentioning is the living polymerization of polar monomers, such as alkylacrylates and alkylmethacrylates, by organolanthanide complexes [89–91].

7   REFERENCES

1. Koivumäki, J. and Seppälä, J., *Polymer*, **34**, 1958 (1993).
2. Koivumäki, J., *Polym. Bull.*, **34**, 413 (1995).

3. Koivumäki, J., Fink, G. and Seppälä, J., *Macromolecules*, **27**, 6254 (1994).
4. Soga, K., Uozumi, T., Arai, T. and Nakamura, S., *Macromol. Chem. Rapid Commun.*, **16**, 379 (1995).
5. Aaltonen, P. and Löfgren, B., *Macromolecules*, **28**, 5353 (1995).
6. Aaltonen, P., Fink, G., Löfgren, B. and Seppälä, J., *Macromolecules*, **29**, 5255 (1996).
7. Aaltonen, P. and Löfgren, B., *Eur. Polym. J.*, **33**, 1187 (1997).
8. Hakala, K., Löfgren, B. and Helaja, T., *Eur. Polym. J.*, **34**, 1093 (1998).
9. Tsuchida, A., Bolln, C., Sernetz, F. G., Frey, H. and Mülhaupt, R., *Macromolecules*, **30**, 2818 (1997).
10. Wilèn, C.-E., Luttikhedde, H., Hjertberg, T. and Näsman, J. H., *Macromolecules*, **29**, 8569 (1996).
11. Wilèn, C.-E. and Näsman, J. H., *Macromolecules*, **27**, 4051 (1994).
12. Brintzinger, H. H., Fischer, D., Mülhaupt, R., Rieger, B. and Waymouth, M., *Angew. Chem., Int. Ed. Engl.*, **34**, 1143 (1995).
13. Ricci, G. and Italia, S., *Macromolecules*, **27**, 868 (1994).
14. Marathe, S. and Sivarann, S., *Macromolecules*, **27**, 1083 (1994).
15. Kaminsky, W., Arrowsmith, D. and Winkelbach, H. R., *Polym. Bull.*, **36**, 577 (1996).
16. Bergström, C. H. and Seppälä, J. V., *J. Appl. Polym. Sci.*, **63**, 1063 (1997).
17. Bergström, C. H., Väänänen, T. L. J. and Seppälä, J. V., *J. Appl. Polym. Sci.*, **63**, 1071 (1997).
18. Bergström, C. H., Starck, P. and Seppälä, J. V., *J. Appl. Polym. Sci.*, **67**, 385 (1998).
19. Benedikt, G. M., Goodall, B. L., Marchant, N. S. and Rhodes, L. F., *New J. Chem.*, **18**, 105 (1994).
20. Kaminsky, W. and Noll, A., in Fink, G., Mülhaupt, R. and Brintzinger, H. H. (Eds), *Ziegler Catalysts*, Springer, Berlin, 1995, p. 149.
21. Kaminsky, W., *Macromol. Chem. Phys.*, **197**, 3907 (1996).
22. Mülhaupt, R., Duschek, T. and Rieger, B., *Makromol. Chem., Macromol. Symp.*, **48/49**, 317 (1991).
23. Möhring, P. C. and Coville, N. J., *J. Organomet. Chem.*, **479**, 1 (1994).
24. Antberg, M., Dolle, V., Klein, R., Rohrmann, J., Spaleck, W. and Winter, A., in Keii, T. and Soga, K. (Eds), *Catalytic Olefin Polymerization, Proceedings of the International Symposium on Recent Developments in Olefin Polymerization Catalysts*, 23–25 October, 1989, Tokyo, Kodansha, Tokyo, 1990, p. 501.
25. Kaminsky, W., Engelhausen, R., Zoumis, K., Spaleck, W. and Rohrmann, J., *Makromol. Chem.*, **193**, 1643 (1992).
26. Burger, P., Hortmann, K. and Brintzinger, H. H., *Makromol. Chem., Macromol. Symp.*, **66**, 127 (1993).
27. Tian, J. and Huang, B., *Macromol. Rapid Commun.*, **15**, 923 (1994).
28. Schneider, M. J., Suhm, J., Mülhaupt, R., Prosenc. M.-H. and Brintzinger, H. H., *Macromolecules*, **30**, 3164 (1997).
29. Schneider, M. J. and Mülhaupt, R., *Macromol. Chem. Phys.*, **198**, 1121 (1997).
30. Ewen, J. A., in Keii, T. and Soga, K. (Eds), *Catalytic Olefin Polymerization, Proceedings of the International Symposium on Future Aspects of Olefin Polymerization Catalysts*, 4–6 July 1985, Tokyo, Kodansha, Tokyo, 1986, p. 271.
31. Quijada, R., Dupont, J., Silveira, D. C., Miranda, L. and Scipioni, R. B., *Macromol. Rapid Commun.*, **16**, 357 (1995).
32. Herfert, N., Montag, P. and Fink, G., *Makromol. Chem.*, **194**, 3167 (1993).
33. Koivumäki, J. and Seppälä, J. V., *Macromolecules*, **27**, 2008 (1994).
34. Lehtinen, C., Starck, P. and Löfgren, B., *J. Polym. Sci., Part A: Polym. Chem.*, **35**, 307 (1997).
35. Arnold, M., Henschke, O. and Knorr, J., *Macromol. Chem. Phys.*, **197**, 563 (1996).
36. Pasquet, V. and Spitz, R., *Makromol. Chem.*, **194**, 451 (1993).

37. Koivumäki, J. and Seppälä, J. V., *Macromolecules*, **26**, 5535 (1993).
38. Koivumäki, J., Liu, X. and Seppälä, J. V., *J. Polym. Sci., Part A: Polym. Chem.*, **31**, 3447 (1993).
39. Starck, P., Lehtinen, C. and Löfgren, B., *Angew. Makromol. Chem.*, **249**, 115 (1997).
40. Lehmus, P., Härkki, O., Leino, R., Luttikhedde, H. J. G., Näsman, J. and Seppälä, J. V., *Macromol. Chem. Phys.*, **199**, 1965 (1998).
41. Luttikhedde, H. J. G., Leino, R. P., Näsman, J. H., Ahlgren, M. and Pakkanen, T., *J. Organomet. Chem.*, **486**, 193 (1995).
42. Chung, T. C., *Macromol. Symp.*, **89**, 151 (1995).
43. Shiono, T. and Soga, K., *Macromolecules*, **25**, 3356 (1992).
44. Shiono, T., Kurosawa, H., Ishida, O. and Soga, K., *Macromolecules*, **26**, 2085 (1993).
45. Chung, T. C., Lu, H. and Li, C., *Macromolecules*, **27**, 7533 (1994).
46. Bruzaud, S., Cramarl, H., Duvig, L. and Deffieux, A., *Macromol. Chem. Phys.*, **198**, 291 (1997).
47. Ahjopalo, L. and Pietilä, L-O., *Research Report KET 1313/96*, VTT, Technical Research Centre of Finland, Chemical Technology, Polymer and Fibre Technology, 1996.
48. Turunen, J., Pakkanen, T. T. and Löfgren, B., *J. Mol. Catal. A: Chem.*, **123**, 35 (1997).
49. Ekman, K. B., Wilèn, C-E., Näsman, J. H. and Starck, P., *Polymer*, **34**, 3757 (1993).
50. Hackmann, M., Repo, T., Jany, G. and Rieger, G., *Macromol. Chem. Phys.*, submitted for publication.
51. *Eur. Pat. Appl.* 0 275 676 (1987), to Exxon.
52. *Eur. Pat. Appl.* 0 273 655 (1987), to Exxon.
53. *Eur. Pat. Appl.* 0 273 654 (1987), to Exxon.
54. Coates, G. W. and Waymouth, R. M., *J. Am. Chem. Soc.*, **113**, 6270 (1991).
55. Matsumoto, J. A., in *Proceedings of Fourth International Business Forum on Specialty Polyolefins, SPO '94*, Houston, September 21–23, Schotland Business Reseach, Inc., NY. 1994, p. 287.
56. Malmberg, A. and Löfgren, B., *J. Appl. Polym. Sci.*, **66**, 35 (1997).
57. Johnson, L. K., Killian, C. M. and Brookhart, M., *J. Am. Chem. Soc.*, **117**, 6414 (1995).
58. Pellechia, C. and Zambelli, A., *Macromol. Rapid Commun.*, **17**, 333 (1996).
59. Johnson, L. K., Mecking, S. and Brookhart, M., *J. Am. Chem. Soc.*, **11**, 267 (1996).
60. Löfgren, B. and Seppälä, J., in *Proceedings MetCon '97, 'Polymers in Transition'*, June 4–5, Houston, TX.
61. Drent, E., van Broekhoeven, J. A. M., Doyle, M. J. and Wong, P. K., in Fink, G., Mülhaupt, R. and Brintzinger, H. H. (Eds), *Ziegler Catalysts*, Springer Berlin, Heidelberg, 1995, p. 481.
62. Tinkler, S., Deeth, R. J., Duncalf, D. J. and McCamley, A., *J. Chem. Soc., Chem. Commun.*, 2623 (1996).
63. Scollard, J. D., McConville, D. H., Payne, N. C. and Vittal, J. J., *Macromolecules*, **29**, 5241 (1996).
64. Scollard, J.D. and McConville, D.H., *J. Am. Chem. Soc.*, **118**, 10008 (1996).
65. US Pat., 5 554 775 (1996), to Lyondell Petrochemical Co.
66. Bazan, G. C., Rogers, J. S. and Sperry, C. K., *J. Am. Chem. Soc.*, **119**, 9305 (1997).
67. (a) Polamo, M. and Leskelä, M., *Acta Chem. Scand.*, **51**, 449 (1997); (b) Polamo, M. and Leskelä, M., *J. Chem. Soc., Dalton Trans.*, 4345 (1996); (c) Polamo, M., *Acta Crystallogr., Sect. C*, **52**, 2977 (1996).
68. Hakala, K., Löfgren, B., Polamo, M. and Leskelä, M., *Macromol. Rapid. Commun.*, **18**, 635 (1997).
69. Miyatake, T., Mizunuma, K., Seki, Y. and Kakuago, M., *Macromol. Chem. Rapid Commun.*, **10**, 349 (1989).

70. Cozzi, P. G., Gallo, E., Floriani, C., Chiesi-Villa, A., and Rizzoli, C., *Organometallics*, **14**, 4994 (1995).
71. Baumann, R., Davis, W. M. and Schrock, R. R., *J. Am. Chem. Soc.*, **119**, 3830 (1997).
72. Repo, T., Klinga, M., Pietikäinen, P., Leskelä, M., Uusitalo, A.-M., Pakkanen, T., Hakala, K., Aaltonen, P. and Löfgren, B., *Macromolecules*, **30**, 171 (1997).
73. Foster, G. N. and Wasserman, S. H., in *Proceedings MetCon '97, 'Polymers in Transition'*, June 4–5, Houston, TX.
74. Uozumi, T. and Soga, K., *Makromol. Chem.*, **193**, 823 (1992).
75. Woo, L., Ling, M. T. K. and Westphal, S. P., Society of Plastics Engineers, Antec Conference Proceedings, New Orleans, May 9–13, 1993, Volume II, p. 1193.
76. Wild, L., *Adv. Polym. Sci.*, **98**, 1 (1991).
77. Glöckner, G. J., *J. Appl. Polym. Sci., Appl. Polym. Symp.*, **45**, 1 (1990).
78. Monrabal, B., *Macromol. Symp.*, **110**, 81 (1996).
79. Kim, J. D., Soares, J. B. P., Rempel, G. L. and Monrabal, B., in *Proceedings MetCon '97, 'Polymers in Transition'*, June 4–5, Houston, TX.
80. Adisson, E., Ribeiro, M., Deffieux, A. and Fontanille, M., *Polymer*, **33**, 4339 (1992).
81. Kamiya, T., Ishikawa, N., Kambe, S., Ikeyami, N., Nischibu, H. and Hattori, T., Society of Plastics Engineers, Antec Conference Proceedings, Dallas, May 7–11, 1990, p. 871.
82. Chiu, F. C., Keating, M. Y. and Cheng, S. Z. D., Society of Plastics Engineers, Antec Conference Proceedings, Boston, May 8–12, 1995, Volume II, p. 1503.
83. Starck, P., *Polym. Inter.*, **40**, 111 (1996).
84. Westphal, Y. S. P., Ling, M. T. K. and Woo, L., Society of Plastics Engineers, Antec Conference Proceedings, Boston, May 8–12, 1995, Volume II, p. 2293.
85. Starck, P., *Eur. Polym. J.*, **33**, 339 (1997).
86. Djupfors, R., Starck, P. and Löfgren, B., *Eur. Polym. J.*, **34**, 941 (1998)
87. Starck, P. and Löfgren, B., *J. Mater. Sci.*, submitted for publication.
88. Khanna, Y. P., Turi, E. A., Taylor, T. J., Vickroy, V. V. and Abbot, R. F., *Macromolecules*, **18**, 1302 (1985).
89. (a) Yasuda, H., Yamata, H., Yokota, K., Miyake, S. and Nakamura, A., *J. Am. Chem. Soc.*, **114**, 4908 (1992); (b) Yasuda, H., Yamata, H., Yamashita, M., Yokota, K., Nakamura, A., Miyake, S., Kai, Y. and Kanehisa, N., *Macromolecules*, **26**, 7134 (1993).
90. Giardello, M. A., Yamamoto, Y., Brad, L. and Marks, T. J., *J. Am. Chem. Soc.*, **117**, 3276 (1995).
91. Boffa, L. S. and Novac, B. M., *Macromolecules*, **27**, 6993 (1994).

PART III

Plastomers and Elastomers

8

Constrained Geometry Single-site Catalyst Technology Elastomers and Plastomers for Impact Modifications and Automotive Applications

K. SEHANOBISH AND S. WU
Dow Chemical Co., Freeport, TX, USA

J. A. DIBBERN AND M. K. LAUGHNER
DuPont Dow Elastomers L.L.C. Freeport, TX, USA

1 INTRODUCTION

The usage of thermoplastic polyolefins (TPOs) in interior and exterior automotive applications has increased rapidly over the past 5 years. TPOs are replacing traditional engineering thermoplastics and thermosets in automotive parts such as bumper fascia, claddings, air dams and instrument panels. The replacement is due mainly to the performance properties, formulation flexibility and low cost offered by TPOs. Amorphous EPDM and high molecular weight semicrystalline EPR have been widely used as impact modifiers for TPOs. However, a high percentage of an impact modifier will decrease the rigidity and heat properties of blends. Copolymer polypropylene has been introduced to enhance the toughness and rigidity with the trade-off of increasing of price.

Metallocene-based Polyolefins Edited by J. Scheirs and W. Kaminsky
© 2000 John Wiley & Sons Ltd

ENGAGE[†] polyolefin elastomer has been introduced recently as an impact modifier for toughening polypropylene. The low-temperature dart impact test results show that low density ENGAGE elastomer has achieved toughening efficiency similar to EPDM and high molecular weight EPR [1]. At comparable molecular weight and density, ENGAGE polyolefin elastomers have better low-temperature Izod impact properties than existing EPDM and EPR impact modifiers [2]. However, some of the very high molecular weight commercial EPDMs and amorphous EPRs can also lead to good impact properties. In this chapter we will explore polypropylene modified with ethylene–octene polyolefin elastomers (POE) and will cover the effect of dispersion, rheology and morphology along with discussions on the influence of elastomer molecular weight, crystallinity and comonomer choice.

A three-component blend system exhibiting core–shell-type morphology results in improved toughening efficiency and balance of physical properties. Optimization of the core–shell geometry and properties of materials will be discussed. The notched Izod impact and tensile tests show that the core–shell impact modified systems can achieve high low-temperature Izod impact strength and a balance of toughness and rigidity. Hence, the results suggest that we can achieve a similar or potentially better physical property balance than two-component blends of copolymer PP or in-reactor TPO and EPR or EPDM blends.

2 BACKGROUND

Toughening mechanisms of polymers with a dispersed rubbery phase have been discussed in the past two decades. Various explanations have been proposed based on the studies of amorphous plastics [3–6]. Rubber-toughening mechanisms in semicrystalline polymers have been studied to only a limited extent, because of their complex semicrystalline nature [7–9]. It is suggested that by acting as a nucleus for the crystallization of the matrix polymer, the dispersed rubber particles may decrease the average spherulite size, therefore increasing the impact strength of the matrix polymer [10]. Polypropylene toughening has been a subject of research of many authors, because PP has poor low-temperature fracture resistance [11–14]. Most of the work has been focused on failure mechanisms, particle size effect and morphological studies of the blends based on homopolymer polypropylene.

It is always a compromise between modulus, heat distortion temperature and toughness in designing practical TPO systems. It has now been demonstrated that a rigid-core–soft-shell type structure of impact modifiers can be used to balance the physical properties of TPO systems.

The stress analysis around the particles dispersed inside a matrix indicates that stress distributions are very similar for particles with or without a core, as long as the ratio of the core to the shell is maintained at a certain ratio [15]. If the toughening mechanisms mainly involve shear yielding or crazing due to stress concentration, a similar

---

† *Trademark of DuPont Dow Elastomers L.L.C.*

distribution of shear yielding or crazing for the matrix will be expected for homogeneous particles, as well as core–shell particles. This framework allowed the design of the morphology of an impact modifier to have a balance of properties and increased toughening efficiency. Another critical factor in improving toughening efficiency is by enhancing the PP-elastomer interfacial strength through comonomer choice. It will be shown that the strength of the interface is affected by the type of α-olefins used during copolymerization of the ethylene–α-olefin elastomers. Differences in interfacial strength will affect the impact performance and weld-line strength of TPOs.

3 EXPERIMENTAL

Most TPO compounds discussed in this paper were blended using a Farrell Banbury BR type mixer having a 1575 cm^3 capacity. A total of 1100 g was used for each formulation. The entire formulation amount was added to a warm Banbury mixer with the rotor speed at 200 rpm until the material began to flux (approximately 1 min), at which time the rotor speed was subsequently slowed to 175 rpm (or whatever rotor speed was required to maintain a melt temperature below 180 °C). Mixing continued for 3 min past flux. The mixed formulation then was discharged from the mixer and passed through a cold roll-mill to make a sheet. The sheet was ground into flake and the flake was subsequently injection molded into test specimens using an 'ASTM family mold'. A 70 ton Arburg injection molding unit was used with the following basic settings: a 190/210/210/210 °C temperature profile, 74 °F mold temperature, 300 psi injection pressure for 1.8 s, 250 psi hold pressure for 15 s and 30 s cooling time. Weld-lines produced using double-end gated tensile bars were tested via room temperature Izod impact (un-notched) and uni-axial tensile tests. Any other physical testing reported was done via the ASTM methods as noted. Some specified blends were melt mixed at 250 °C on a 30 mm, Werner Pfleiderer, co-rotating, twin screw extruder at a speed of 250 rpm (30–35 1b/h rate). The screw configuration was basic kneading blocks followed by gear mixer flights to produce a medium-shear, high-mixing configuration. Each extruded formulation was passed through a chilled water-bath, chopped into granules and collected for injection molding. ASTM samples were then prepared by injection molding on a 70 ton Arburg molding machine. Molding temperatures for the barrel were set at 90 °C (feed), 210 °C, 250 °C, 250 °C (barrel through nozzle), while the mold temperature was 38 °C. Injection cycles were maintained at 1.8 s injection, 15 s hold and 20 s cool. The injection/hold pressures were approximately 20–25 bar, which was adjusted to fill the mold cavities completely. Care was taken to compare materials prepared using the same processing conditions. The blend formulations of copolymer polypropylene (co-PP) with different impact modifiers are given in Table 1.

The notched Izod impact test on single-end gated bars (0.5 × 5.0 × 0.125 in) used a milled notch and conformed to ASTM D-256. The samples were notched in the center of the bar by a notcher with a notch depth of 0.400 ± 0.002 in. Five samples were tested for each case at −30 °C. The Izod impact testing used a standard unit

Table 1 Co-PP blend formulations

Sample	Matrix (59.5 wt%)	Rubber phase (25.5 wt%)	Filler (15 wt%)
1	Mitsui-Toatsu co-PP 30 MI	EPDM (0.86 g/cm^3, 35 Mooney)	VANTALC[a] 6H
2	Mitsui-Toatsu co-PP 30 MI	EPR (0.86 g/cm^3, 45 Mooney)	VANTALC 6H
3	Mitsui-Toatsu co-PP 30 MI	EPR (0.86 g/cm^{3}cc, 25 Mooney)	VANTALC 6H
4	Mitsui-Toatsu co-PP 30 MI	EG8180 (0.863 g/cm^3, 0.5 12)	VANTALC 6H
5	Mitsui-Toatsu co-PP 30 MI	EG8150 (0.868 g/cm^3, 0.5 12)	VANTALC 6H
6	Mitsui-Toatsu co-PP 30 MI	EG8100 (0.863 g/cm^3, 1.0 12)	VANTALC 6H

[a] Trademark of Vanderbilt, R.T. Company Inc.

equipped with a cold temperature chamber and a 2 ft-lb, free-falling hammer. The geometry and testing conditions for the falling dart penetration test are described in ASTM D3763-86. Five specimens were tested for each sample at $-30\,^\circ$C, using a model 5280 Dynatup drop tower equipped with a cold temperature chamber, a 100 lb drop weight and a 0.5 in Dynatup having an impact velocity of 5 mph.

4 RESULTS AND DISCUSSION

The morphologies of copolymer polypropylene and predominantly isotactic homopolymer polypropylene blends were analyzed by transmission electron microscopy (TEM). Figure 1(a) shows the morphology of i-PP–EPDM–talc (60:25:15) blend. The EPDM particles are well dispersed in the i-PP matrix. Figure 1(b) shows the morphology of co-PP, where three phases can be observed as a core–shell structure embedded in the matrix. The shell is formed by the amorphous (soft) phase of ethylene–propylene, while the core is formed by relatively high-density semicrystalline ethylene–propylene polymer. Another type of morphology has been reported in which the soft elastomer phase is continuous, with the rigid phase enhancing processability–modulus–toughness balance [16].

A review of the TPO performance requirements on a global basis reveals that the trend in automotive and other durable applications is moving towards higher melt flow, higher modulus, higher heat resistance and higher impact TPO formulations. The use of high melt flow i-PP or high melt flow co-PP and ENGAGE elastomers in TPO formulations offers a way to balance material performance. These elastomers, owing to their narrow molecular weight distribution, show better performance than broad molecular weight distribution EPDM and EPR systems. Owing to the severity of the $-30\,^\circ$C notched Izod test, it can serve as an effective tool for differentiating the toughness of various TPO systems. In Figure 2 dart and Izod impact results on the same set of materials are shown in order to demonstrate this fact clearly. On this premise, the notched Izod test will be mostly used for describing the performance attributes of the TPO systems.

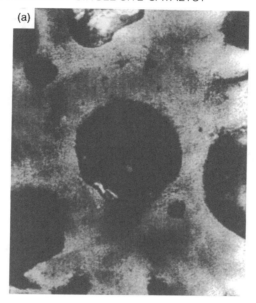

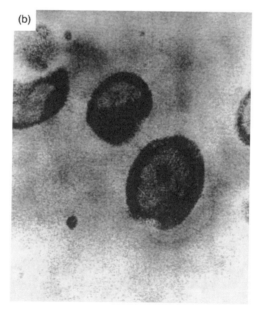

Figure 1 Morphology of (a) i-PP–EPDM–talc blend and (b) co-PP (1 mm = 0.026 μm

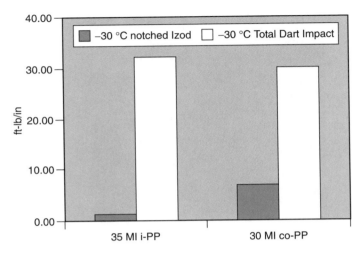

Figure 2 Dart and Izod impact comparison for the same set of materials: 35 MI i-PP or 30 MI co-PP-ENGAGE-talc (60:25:15)

4.1 ADVANCED ELASTOMER TOUGHENING CONCEPTS

Recently, a more advanced approach to toughening TPOs has been introduced in which a three-phase polymer structure is obtained by standard compounding techniques or an in-reactor approach [1]. The dispersed elastomer domain in the PP matrix can be extended if the soft elastomer encapsulates a rigid polymer domain. In simple terms this should lead to a better balance of rigidity and toughness. However, success of this concept lies in deciding the amount of elastomer shell and rigid core. Simply having a core–shell structure is not adequate. For example, impact co-PP inherently has a core–shell-type heterophasic morphology as shown in Figure 1(b), but it is extremely brittle in the low-temperature notched Izod test. The ratio of core size to shell size, the physical properties of core and shell material and how the core material is distributed inside the shell material are also very important in order to achieve optimal toughening.

4.1.1 Material Properties of the Core–Shell

According to simplified criteria for crazing, any critical parameter such as the maximum principle stress, principle strain, dilation and strain energy density must reach a certain threshold for crazing to occur [17]. The maximum stress usually occurs at the interface of matrix and particle. Stresses at the interface are the most significant of all and are controlled by the inhomogeniety, i.e. the ratio of the modulus of the matrix to the modulus of the particle. The larger the ratio is, the higher the stress concentration. Hence the shell material should be very soft at a

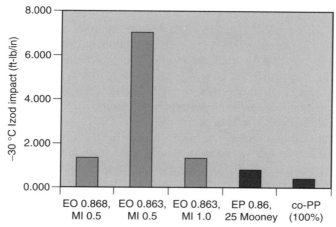

Figure 3 Notch Izod impact resistance of co-PP blends toughened by various EO copolymers, EPR and co-PP impact modifiers

given dispersion. The core material, on the other hand, should be stiff enough to balance the rigidity and thermal properties of blends.

The stress concentration at the interface is proportional to the ratio of the modulus of the matrix to the modulus of the particle, E_m/E_p. It is expected that a higher E_m/E_p ratio will result in higher stress concentrations. Hence this ratio as a function of temperature should be a key consideration for designing the shell material for the TPO systems. Many authors have discussed the effect of the molecular weight of the impact modifier on the toughening efficiency. An example of the effect of molecular weight is shown in Figure 3, where two impact modifiers, both semicrystalline ethylene–octene copolymers (EO) having similar density but with different molecular weights (MI = 0.5 vs 1), are compared. The blend with the higher molecular weight shell polymer shows higher impact resistance than that of lower molecular weight. A blend with a high molecular weight shell polymer had a relatively large crazing zone during impact testing and had an Izod impact resistance of 7.5 ft-lb/in, whereas a blend with a low molecular weight shell polymer fractured in a brittle fashion and had an impact resistance of 1.4 ft-lb/in for low molecular weight EO. This effect may be explained by the deformation of the shell and the interfacial strength. Higher molecular weight polymers have longer chains that may interdiffuse and form effective bonding with the PP matrix. Higher molecular weight shell polymers will also dissipate more energy during deformation.

4.1.2 Geometry of the Core–Shell

The main purpose of introducing a core is to increase toughening efficiency while maintaining the stiffness in terms of the amount of rubber and cost. The optimization of the geometry ratio of the core to the shell depends on the effect of the core on the stress-field at the interface. The stress-field near the interface for various core–shell geometry ratios has been analyzed [18]. The results show that the stress-field near the interface is very similar until the ratio of the core radius to the shell radius reaches 0.9. In other words the core material has very little influence on the stress-field at the interface if the radius ratio is < 0.9. For an ideal case of a sphere-shaped particle, the optimum ratio of volume fraction will be about 25 % core to 75 % shell. The notched Izod impact results of co-PP and co-PP–POE confirmed this prediction. The shell layer is too thin for a core–shell structure of co-PP, hence the sample fails in a brittle fashion at low temperature. The volume fraction of the shell reaching 25 % in a core–shell co-PP–POE blend results in ductile failure at low temperature. A similar effect of geometry ratio on impact resistance has been observed for a polycarbonate–latex blend [18].

4.1.3 Effective Modulus of Three-component System

The effective modulus of two-phase systems has been investigated extensively. The effective elastic modulus of a composite material containing spherical inclusions at dilute concentrations has already been derived [19]. For the three-phase system, the effective modulus of the blends will depend on the effective modulus of the core–shell particles on a macroscopic scale. The effective modulus of the core–shell particles, on the other hand, will depend on the modulus of the core and the shell materials as well as their organization, e.g., a single core–shell structure (SCS) vs a multiple core–shell structure (MCS). The schematic in Figure 4 considers two possible scenarios where a single hard segment or phase (SCS) or several hard segments or phases (MCS) are encapsulated by a soft matrix. On a macroscopic scale, the effective modulus of blends can be estimated based on the modulus of the matrix and particles. We will use an empirical method based on previous work [19, 20] to estimate the effective modulus of a multiple core and single core–shell particle problem. Thus, the effective modulus of a multi-core–shell three-phase system can be expressed as

$$\frac{E}{E_m} = 1 + \frac{v_{cs}(E_{cs}/E_m - 1)}{1 + (1 - v_{cs})[(E_{cs} - E_m)/(E_m + \tfrac{4}{3}\mu_m)]} \tag{1}$$

Using the self-consistency method again on the microscale, E_{cs} can be expressed as

$$\frac{E_{cs}}{E_{shell}} = 1 + \frac{v_{core}(E_{core}/E_{shell} - 1)}{1 + (1 - v_{core})[(E_{core} - E_{shell})/(E_{core} + \tfrac{4}{3}\mu_{shell})]} \tag{2}$$

where E is the overall modulus, E_m is the modulus of the matrix, E_{cs} is the modulus of the (multiple) core–shell rubber, E_{core} is the modulus of the core, E_{shell} is the

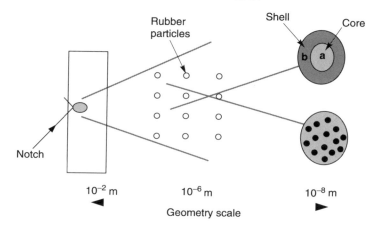

Figure 4 A schematic explanation of single- and multiple-core–shell structures

elastic modulus of the shell materials, v_{core} is the volume fraction of the core, v_{CS} is the volume fraction of the core–shell rubber and μ_{shell} and μ_m are the shear modulus of the shell material and the matrix material. Combining equations (2) and (1), the effective modulus of a three-phase system with a multiple core uniformly distributed inside the shell can be obtained. Based on this analysis, a multiple-core–shell morphology may provide an approximately 70 % increase in effective modulus of the impact modifier over a single-core–shell.

4.1.4 Demonstration of the Core–Shell Rubber Concept

Several three-component systems were selected to examine the core-shell structure concept. The materials are listed in Table 2. For the convenience of analysis, the matrix and shell materials were fixed. The core materials were selected by varying the density and molecular weight distribution. The densities of core materials were selected as 0.913 and 0.952 g/cm^3. Both AFFINITY † POP (with narrow molecular weight distribution and narrow comonomer distribution) and conventional linear low-density polyethylene (LLDPE) can be used for this purpose. Talc was used as a filler to enhance the rigidity and heat properties, with an additional benefit in the formulation cost.

The Izod impact test results of the several examples are listed in Table 3 and tensile test results are shown in Table 4. The effect of morphology on the notched Izod impact properties is shown in Figure 5. The effect of molecular weight and comonomer distribution on physical properties of the blends can be found in Tables 3 and 4. The notched Izod impact resistances are very similar for the third phase with

---

† *Trademark of Dow Chemical Company.*

Table 2 Formulation of three-phase systems

	Matrix	Shell phase	Filler	Core phase
Example 1	HIMONT PD-701 i-PP 35 MI	NORDEL 2522 EPDM	VANTALC 6H	AFFINITY plastomer 0.913 g/cm^3, 0.5 MI
wt %	50	25	15	10
Example 2	HIMONT PD-701 i-PP 35 MI	NORDEL 2522 EPDM	VANTALC 6H	AFFINITY plastomer 0.952 g/cm^3, 0.5 MI
wt %	50	25	15	10
Example 3	HIMONT PD-701 i-PP 35 MI	NORDEL 2522 EPDM	VANTALC 6H	LLDPE 0.913 g/cm^3, 0.5 MI
wt %	50	25	15	10
Example 4	HIMONT PD-701 i-PP 35 MI	NORDEL 2522 EPDM	VANTALC 6H	LLDPE 0.952 g/cm^3, 0.35 MI
wt %	60	25	15	10
Control 1	Mitsui-Toatsu co-PP 30 MI	NORDEL 2522 EPDM	VANTALC 6H	
wt %	60	25	15	0
Control 2	HIMONT PD-701 i-PP 35 MI	NORDEL 2522 EPDM	VANTALC 6H	
wt %	60	25	15	0

Table 3 Izod impact strength at $-30\,^\circ$C

	Example 1	Example 2	Example 3	Example 4	Control 1	Control 2
Average Izod impact strength (ft-lb/in) at $-30\,^\circ$C	11.24	10.22	10.0	8.35	5.57	11.64
Standard deviation	0.41	0.22	0.9	1.04	3.37	0.08

Table 4 Tensile properties of controls at room temperature

	Yield stress (kpsi)	Strain at break (%)
Example 1	2.30 ± 0.04	84.7 ± 16
Example 2	2.50 ± 0.01	51.3 ± 3.1
Example 3	2.28 ± 0.02	55.9 ± 8.6
Example 4	2.47 ± 0.01	47.95 ± 14.11
Control 1	2.689 ± 0.012	82.8 ± 29.8
Control 2	2.4 ± 0.01	25.9 ± 6.3

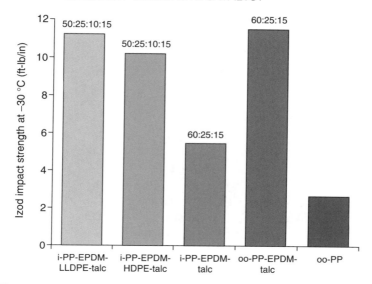

Figure 5 Effect of morphology on the notched Izod impact properties

narrow or broad molecular weight and comonomer distribution. The sample with a narrow molecular weight and comonomer distribution third phase, however, has a significantly higher elongation to break than that of samples with conventional LLDPE as the third phase. The effect of the modulus of the third phase material on the effective elastic modulus of blends can be found in Table 4. The elastic modulus increases as the density of the third component material increases according to equation (1). The elongation to break of a blend, however, decreases as the density of the third component material increases. It is clear from Table 5 that core–shell morphology can provide a better balance of modulus and toughness than co-continuous systems such as Toyota Super Olefin (TSOP) and conventional TPOs.

4.2 ROLE OF COMONOMER IN THE INTERFACE STRENGTH

As mentioned earlier, the strength of the interface between the rubber phase and PP plays a critical role in the final toughness of TPOs. The role of comonomer type in the interfacial strength was examined using the following approaches. The tensile

Table 5 Comparison of modulus–toughness balance of morphology-controlled structures

Material type	Core–shell	TSOP	Conventional TPO
Flex modulus (kpsi)	~1100	~1500	~1000
Notched Izod at $-30\,°C$ (ft-lb/in)	~11.2	~1.2	~3.74

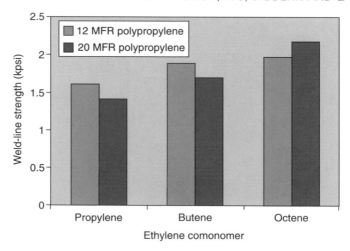

Figure 6 Weld-line strength (yield stress) as a function of comonomer type for a 70:20:10 polypropylene–elastomer–talc blend

strength of a double-gated tensile bar (Figure 6) shows that with a change to this simpler comparison and use of the right data for Figure 6, the weld-line strength of the EO rubber based TPOs will be higher relative to ethylene–butene copolymer (EB) rubber- and EPR-based TPOs. In an earlier publication it was shown that an EB rubber would provide significantly better weld-line strengths than an EO rubber [21]. Our present data (Figure 6) do not confirm that. In order to verify this claim further, weld-line strength was measured using an Izod impact test at room temperature. The results are shown in Figure 7. Once again it was found that the EO rubber-based

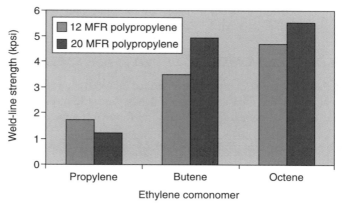

Figure 7 Weld-line Izod impact as a function of comonomer type for a 70:20:10 polypropylene–elastomer–talc blend

TPOs have the best weld-line strength measured using both the tensile strength measurement (Figure 6) and Izod impact measurement (Figure 7).

5 CONCLUSIONS

Rigid-core–soft-shell design of impact modifiers offers a unique opportunity to balance modulus and toughness. Theoretical optimization of the geometry and properties of core–shell materials recommends low-rigidity and -crystallinity material as the shell phase material and a ratio of volume fraction for core–shell structure with ~75% shell and ~25% core material. The effective modulus of a three-component system was analyzed and the results showed that multi-core–shell systems will have a higher effective elastic modulus than single-core–shell structure systems.

Several three-component systems (i-PP–EPDM–LLDPE) were investigated to examine the core–shell structure concept. The notched Izod impact and tensile tests showed that the core–shell-type impact modifier can achieve high low-temperature Izod impact strength ($-30\,^\circ$C) with a balance of toughness and modulus. LLDPE with densities varying from 0.91 to 0.95 can be used as the rigid core phase.

6 ACKNOWLEDGMENTS

Thanks are due to Clive Bosnyak and C.-I. Kao for valuable discussions and suggestions. We also express our gratitude to Larry Meiske for his help in fabrication and providing valuable inputs in conventional TPO designs.

7 REFERENCES

1. Chum, S., Sehanobish, K. and Wu, S., New TPOs based on single-site constrained geometry catalyst technology impact modifiers, presented at SAE Annual Meeting, Chicago, 1997.
2. Dibbern, J. A., Meiske, L. A., Wu, S. and Sehanobish, K., Impact performance evaluation of polypropylene modified with polyolefin elastomers, presented at *TPOs in Automotive 95*, Novi, MI, October 30–31, 1995.
3. Silvis, C. H., Murray, D. J., Fiske, T. R., Betso, S. R. and Turley, R. R., *PCT Int. Appl.*, WO94/06859 (1994).
4. Riew, C. K. and Gillham, J. K. (Eds), *Rubber-modified Thermoset Resins*, Advances in Chemistry Series, No. 208, American Chemical Society, Washington, DC (1984).
5. Riew, C. K. (Ed.), *Rubber-toughened Plastics*, Advances in Chemistry Series No. 222, American Chemical Society, Washington, DC (1989).
6. Riew, C. K. and Kinloch A. J. (Eds), *Toughened Plastics I, Science and Engineering*, Advances in Chemistry Series No. 233, American Chemical Society, Washington, DC (1993).

7. Bucknall, C. B. *Toughened Plastics*, Applied Science, London, 1977.
8. Speri, W. M. and Patrick, G. R., *Polym. Eng. Sci.*, **15**, 668 (1975).
9. Wu, S., *J. Polym. Sci., Polym. Phys. Ed.*, **21**, 699 (1983).
10. Bucknall, C. B. and Page, C. J., *J. Mater. Sci.*, **17**, 808 (1982).
11. Flexman, E. A. Jr, *Polym. Eng. Sci.*, **19**, 564 (1979).
12. Jang, B. Z., Uhlmann, D. R. and Vander Sande, J. B., *Polym. Eng. Sci.*, **25**, (1985).
13. Chou, J., Vijayan, K., Kibby, D., Hiltner A. and Baer, E., *J. Mater. Sci.*, **23**, 2521 (1988).
14. Jang, B. Z., Uhlmann D. R. and Vander Sande, J. B., *Polym. Eng. Sci.*, **25**, 643 (1985).
15. Jang, B. Z., Uhlmann, D. R. and Vander Sande, J. B. *J. Appl. Polym. Sci.*, **30**, 2485 (1985).
16. D'orazio, L., Mancarella, C., Martuscelli, E., Sticotti, G. and Ghisellini, R., *J. App. Polym. Sci.*, **53**, 387 (1994).
17. Nishio, T., Nomura, T., Yokoi, T., Iwai, H. and Kawamura, N., Toyota Motor Corp., Personal communication (1990).
18. Ricco, T., Pavan, A. and Danusso, F., *Polym. Eng. Sci.*, **18**, 774 (1978).
19. Segall, I., Dimonie, Y. L., El-Aasser, M. S., Soskey, P. R. and Mylonakis, S. G. *J. App. Polym. Sci.*, **58**, 419 (1995).
20. Halpin, J. C. and Tsai, S. W., *Air Force Scientific Research Report AFML-TR67-423*, June (1969).
21. Yu, T. C., *Proceedings of SPE ANTEC 53rd Annual Conference*, 1995, p. 2374.

9

Structure, Properties and Applications of Polyolefin Elastomers Produced by Constrained Geometry Catalysts

T. HO AND J. M. MARTIN
Dow Chemical Company, Freeport, TX, USA

1   INTRODUCTION

The polyolefin industry has made significant advances in technology in the past 5 years. Specifically the development of recent homogeneous catalysis systems permits the production of very low density copolymers. The Dow Chemical Company has developed the constrained geometry catalyst (CGC) as its single-site catalyst (SSC) technology. This technology allows the production of elastomers and plastomers with narrow molecular weight and branching distributions and the addition of long-chain branches along the main chain. The general structure–property relationships of CGC copolymers are covered elsewhere (Chapter 12, Vol. 1). In this chapter we will focus on the relationships of polyolefin elastomers (POEs) and ethylene–propylene–diene terpolymers (CGC-EPDMs). The density range for POEs is 0.86–0.89 g/cm^3. In this chapter, CGC-EPDM and Z-N EPDM refer to EPDMs made with CGC and Ziegler–Natta catalysts, respectively. Commercially available SSC POEs and EPDMs include AFFINITY[†] polyolefins from The Dow Chemical Company, NORDEL[‡] IP (EPDM) and ENGAGE[‡] (POE) from

---

[†] *Trademark of Dow Chemical Company.*
[‡] *Trademark of DuPont Dow Elastomers L.L.C.*

Metallocene-based Polyolefins Edited by J. Scheirs and W. Kaminsky
© 2000 John Wiley & Sons Ltd

DuPont Dow Elastomers L.L.C. The mechanical and thermal properties will be discussed in the light of the structure of both uncross-linked and cross-linked polymers. The relevance of these properties to specific applications will also be discussed.

2 STRUCTURE AND PROPERTIES OF POLYOLEFIN ELASTOMERS

This section will contain discussions on the morphology of POEs and its effect on the key properties of POEs, such as tensile and thermal properties, elastic recovery after cyclic loading and oxidative stability.

2.1 MORPHOLOGY

Semicrystalline homopolymers such as polyethylene and polypropylene crystallize from the melt into a lamellar morphology, a regular packing of the polymer chains [1, 2]. Ethylene copolymers with densities greater than $0.91 \, g/cm^3$ crystallize into spherulites with diameters of the order of $10–20 \, \mu m$. The lamellae that make up these spherulites range in thickness from 100 to 300 Å, with much greater lateral dimensions. The lamellar thickness distribution is broad as a result of the intramolecular heterogeneity of the short-chain branch and molecular weight distribution. Single-site catalysts allow for the production of very low density copolymers ($<0.900 \, g/cm^3$). Copolymers of propylene, 1-butene, 1-hexene and 1-octene contain

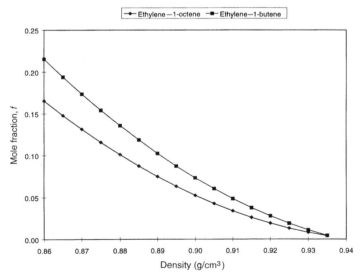

Figure 1 Mole percentage of α-olefin comonomer as a function of density for ethylene–1-octene and ethylene–1-butene copolymers [3]

methyl, ethyl, butyl and hexyl branches, respectively. Figure 1 shows the density of two CGC-catalyzed copolymers as a function of mole percentage of comonomer.

For a given density, the mole percentage of comonomer increases as the length of the comonomer decreases. The high level of incorporation into the crystals of the shorter, ethyl branches associated with the 1-butene comonomer necessitates a higher comonomer concentration to achieve the same density. The crystalline order is disrupted and smaller crystallites begin to appear [4–6]. These crystallites are referred to as fringed micelles. Previous work [7,8] has shown that ethylene–1-octene copolymers change gradually as a function of density from a well defined lamellar microstructure associated with high-density homopolymers to shorter and thinner lamellae. The microstructure of a 0.870 g/cm^3 ethylene–1-octene copolymer is shown in Chapter 12, Vol. 1 (Figure 14). The granular appearance is punctuated irregularly by the lamellae seen in higher density materials. Figure 2 shows the DSC endotherms for POEs with densities between 0.863 and 0.885 g/cm^3 at similar molecular weights. The lower melting-point shoulders in all three copolymers are associated with the melting of fringed micelles. The shoulder intensity is increased as a result of room temperature annealing, although the location does not change after reheating. The higher melting, most prominent in the 0.885 g/cm^3 copolymer, is associated with lamellar crystals. As the density increases the melting temperature also increases. Additionally, the relative fraction of fringed micelles to lamellar crystals decreases [9]. The melting temperature and crystal structure have the most profound effect on upper service temperature and mechanical properties.

The crystallinity of CGC polymers ranges from approximately 3 % for ethylene–1-propylene copolymers to as high as 60 % for homopolymers. The crystallinity

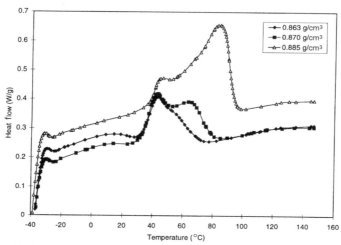

Figure 2 Endotherms of POEs at densities of 0.863, 0.870 and 0.885 g/cm^3 ethylene–1-octene copolymers. First heats, unannealed samples scanned at 10 °C/ min

consists exclusively of lamellae at the higher crystallinity levels. The proximity of the lamellae to each other in higher crystallinity resins leads to the formation of tie-molecules. The portion of the chain that cannot be incorporated into the lamellae, due to a defect such as a long- or short-chain branch, traverses the interlamellar distance and can be reincorporated upon crystallization into another lamellae. The subsequent interconnectivity has a profound effect on the mechanical properties. Copolymers made using single-site catalysts and classified as elastomeric have crystallinities less than 25 %. The density of these copolymers is less than $0.885 \, g/cm^3$. The population of fringed micelles increases as the density decreases. The fringed micelles are chains organized into loose bundles. The number of chains per bundle is fairly low in comparison with the number of chains associated with a single lamella. As a result of the small number of chains connecting the crystals and the lack of incorporation of the short-chain branches into the crystal, the tie-molecule concentration is fairly low [10]. If interconnectivity does exist, the fringed micellar junctions are weaker than the lamellar counterparts. The yielding of crystals during uniaxial extension depends primarily on the size of the crystal [11,12]. The low density copolymers have smaller crystals that cause the yield stress to decrease. Since the overall crystallinity and crystal size are considerably lower, plastic deformation is not as prevalent. These copolymers therefore act more like elastomers with highly recoverable deformation.

2.2   TENSILE PROPERTIES

The tensile properties as a function of density have been discussed previously, but here the focus remains on the elastomeric range. Figure 3 shows the stress–strain curves of CGC copolymers at densities between 0.863 and $0.885 \, g/cm^3$ at a similar molecular weight. For contrast, a copolymer with a density of $0.918 \, g/cm^3$ is also shown. This curve has a sharp yield point followed by drawing at a lower draw stress. Strain-hardening occurs as the chains begin to orient along the main axis. Higher stresses are necessary to increase orientation. After failure, the surfaces of the tensile bar were observed to have fibrillar characteristics. In contrast, the lower density copolymers exhibit homogeneous deformation without a sharp yield point. Localized yielding does not occur above the glass transition temperature (ca $-50\,^{\circ}C$) for copolymers at this density, owing to the fringed micellar morphology. The level of strain-hardening is much less than that seen in the higher density copolymer. The shapes of the curves are similar to each other in that a distinctive yield point is not observed. Observance of the tensile bars during uniaxial extension showed uniform elongation without any necking.

The modulus of each copolymer can be gauged by the slope of the initial portion of the stress–strain curve. The magnitude depends primarily upon the crystallinity. Therefore, the slope of the initial portion of the curve increases with increase in density. Additionally, the slope of the curve associated with strain-hardening is steeper as the density increases. The slope is associated with the strain-hardening

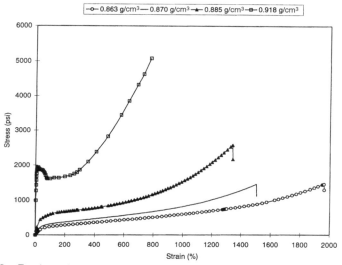

Figure 3 Engineering stress–strain curves of ethylene–1-octene copolymers as a function of density

modulus in conventional rubber elasticity theory [13]. The number of tie-chains dictates the degree of strain-hardening as well as the modulus. The increase with density is then consistent with calculations of the tie-chain concentration [7]. Following fracture, the sample at lower density shows a high degree of instantaneous recovery followed by a slow recovery with less than 20 % permanent set. Table 1 lists the tensile properties as a function of density. The elongation at break decreases with increasing density. The strain-hardened crystals are less extensible than the fringed micelles. The tensile stress at break increases with density, since the fibrillar crystals are much stronger than the elastomeric fringed micelles.

Molecular weight also plays a role in the tensile properties. The decrease in entanglements at lower molecular weights leads to fewer tie-junctions and, hence, less strain-hardening as the resistance to orientation is fairly low. Figure 4 shows a stress–strain curve for ethylene–1-octene copolymers at a $0.870\,g/cm^3$ density and three different molecular weights. Since the crystallinity is the same, the moduli are equivalent.

Table 1 Tensile properties of ethylene–1-octene POEs as a function of density

Density (g/cm^3)	100 % Modulus (psi)	Elongation at break (%)	Ultimate tensile stress (psi)
0.863	270	1000	1240
0.870	357	850	1550
0.885	680	750	2700

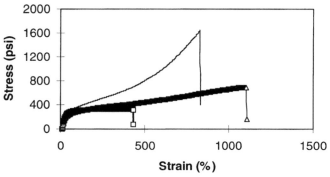

Figure 4 Engineering stress–strain curves of 0.870 g/cm^3 ethylene–1-octene copolymers made using CGC as a function of molecular weight (no symbol, 1 I_2; triangle, 30 I_2; square, 200 I_2)

Additionally, no change in the yield stress is observed, since the fringed micelle size remains the same. The tensile stress at failure is lower owing to the lack of strain-hardening, although the elongation at break is slightly higher in the lower molecular weight material.

Figure 5 shows the stress–strain curves of four ethylene–α-olefin copolymers at a density of ca 0.870 g/cm^3 at approximately the same melt index. Table 2 lists the

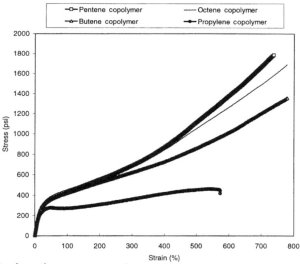

Figure 5 Engineering stress–strain curves of ethylene–α-olefin copolymers at a density of 0.870 g/cm^3. All samples pulled at 20 in/min using ASTM D412 sample geometry

Table 2 Tensile properties of ethylene–α-olefin copolymers

Comonomer	Density (g/cm^3)	Tensile at break (psi)	Elongation at break (%)
Propylene	0.8743	460	550
Butene	0.8753	1360	780
Pentene	0.8731	1790	740
Octene	0.8755	1690	780

tensile properties of these copolymers. The results in Table 2 indicate that the tensile strength of POE increases with the increase in the length of α-olefin comonomer. It is reported that the short-chain branching introduced by 1-propylene or 1-butene comonomer can be incorporated to some degree into the crystalline unit cell, thereby weakening the fringed micelle and reducing the tensile strength [14].

2.3 ELASTIC RECOVERY OF THERMOPLASTIC ELASTOMERS

Elastic recovery is defined as the ability of a material to return to its initial configuration after deformation. It can be quantified in terms of the compression set, instantaneous recovery after uniaxial extension and hysteresis. Measurement of these properties is necessary for the use of these materials in applications such as foams in shoe soles, timing belts and gaskets. In high-density polyethylene, recovery after elongation beyond the yield point is fairly low because permanent, plastic deformation of the lamellae occurs. The recovery increases as the polymer density decreases.

For most elastomeric applications, recovery after dynamic displacement, in which the sample is subjected to stress or strain cycles for a period of time, is the most commonly measured property. The loading–unloading loops are an indication of the amount of hysteresis or energy loss associated with the deformation process. The energy losses consist primarily of irrecoverable heat loss to the system, which usually results in melting on a microscale. Hysteresis can also be attributed to the changes in internal energy when the polymer network undergoes configurational changes during rubber elastic deformation. Materials with zero hysteresis are ideal rubbers for which the energy that is put into deforming the sample is exactly recovered as the sample is unloaded. Table 3 shows the hysteresis values of ethylene–1-octene copolymers as a function of density. The hysteresis is lower for materials that recover the imposed deformation.

These samples were subjected to 20 % compressive strain with 5 % amplitude at a frequency of 20 Hz for 12 000 cycles. The key structural differences between these copolymers are the crystallinity and crystal morphology. The crystallinity or density and population of fringed micelles are inversely related. It has been discussed earlier that a 0.870 g/cm^3 copolymer will form predominantly a fringed micellar structure, while a 0.885 g/cm^3 copolymer still has a combination of both lamellar and fringed micellar microstructures. It is well known that irreversible deformation under either

Table 3 Dynamic data for ethylene–1-octene copolymers as a function of density

Density	Dynamic compression set (%)	Hysteresis (%)
0.863	4.5	3
0.870	4.4	3.3
0.885	5.8	6.1
0.895	6.4	12.2

tensile or compressive stress occurs due to shear deformation of the lamellar structure [15]. The higher population of lamellae in the 0.885 and 0.895 g/cm^3 copolymers is clearly responsible for the higher hysteresis.

The dynamic compression set reflects the recovery of the material after it is subjected to cyclic loading. It is calculated in the same way as static compression set [16]. Dynamic compression set data, as a function of density at approximately the same molecular weight, are presented in Table 3. Clearly, the dynamic compression set reflects the same behavior as the hysteresis in that POEs with high crystallinity and large lamellae exhibit greater irrecoverable deformation.

As the molecular weight increases, the number of physical entanglements that can also act like cross-links increases. Lower molecular weight materials exhibit the opposite behavior. Shorter chains slip past one another more readily than longer chains; therefore, the lower molecular weight chains are not restricted and deformation leads to permanent set. A comparison of the dynamic properties of a 0.885 g/cm^3 ethylene–1-octene copolymer at a narrow and broad molecular weight distribution is shown in Table 4.

Compression set is a result of lamellar deformation and chain slippage at the entanglements. It is speculated that the low molecular weight fractions in the broad MWD copolymer lead to relatively easy slippage during compression that leads to a higher compression set and hysteresis.

Elastomers are required to perform in both tensile and compressive modes depending on the application. The tensile set is measured by extending the sample

Table 4 Comparison of dynamic properties as a function of molecular weight and molecular weight distribution (MWD) for three 0.885 g/cm^3 ethylene–1-octene copolymers

MWD	I_2	Dynamic compression set (%)	Hysteresis (%)
Narrow	0.5	4.9	4.1
Narrow	1.0	5.8	6.1
Broad	0.5	7.1	6.4

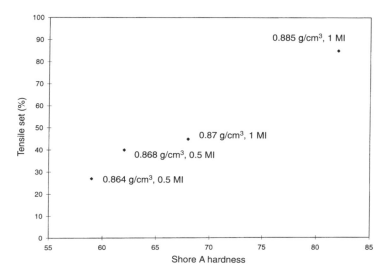

Figure 6 Tensile set as a function of Shore A hardness for CGC copolymers as a function of density and molecular weight

to 100 % displacement, holding for a specified time and then returning the sample to zero strain. After a pre-determined recovery period, the sample is again extended and the extension at which the onset of loading is observed is recorded. This value is compared with the initial extension and recorded as a percentage change over the initial deformation. The tensile set as a function of Shore A hardness is shown in Figure 6. The Shore A hardness correlated directly with the density of the copolymer.

Similar to the compression measurements, the tensile set increases with increase in density. The reasoning is the same as that discussed for the dynamic measurements.

2.4 THERMAL PROPERTIES

The melting temperature dictates the upper service temperature of the resin while the glass transition temperature (T_g) is responsible for low-temperature mechanical properties. At temperatures close to the melting temperature the polymer is too soft to maintain any mechanical integrity. Thermal behavior can be characterized using the melting-point from differential scanning calorimetry (DSC), modulus as a function of temperature from dynamic–mechanical analysis (DMA), and softening temperature under load via thermal–mechanical analysis (TMA). The DMA traces of two CGC copolymers are shown in Figure 7. A homopolymer prepared using CGC

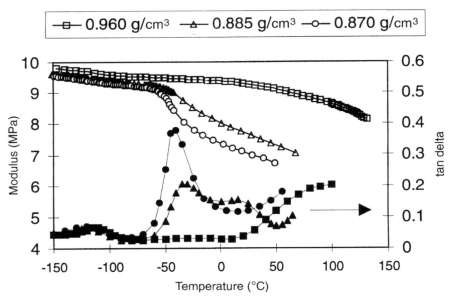

Figure 7 Dynamic mechanical spectra of CGC copolymers and a homopolymer (0.960 g/cm³) as a function of density. Filled symbols correspond to the tan delta data

is shown for comparison. Both the storage modulus, G', and tan delta are shown. Tan delta is the ratio of the storage to the loss modulus. The storage modulus reflects both the stiffness and the elasticity of the polymer. The glass transition temperature (T_g) is the temperature below which the molecular mobility of the amorphous domains is highly restricted. Semicrystalline polymers have both a T_g and a melting temperature (T_m). The latter indicates the onset of melting in the crystalline phase. The crystalline phase can consist of either fringed micelles or lamellae. From the tan delta peak, the glass transition temperature of the 0.870 and 0.885 g/cm³ copolymers are -50 and $-40\,°C$, respectively. The homopolymer does not exhibit any well defined transition in the same temperature range.

The range of operating temperatures for any of the copolymers can be gauged from the modulus–temperature relationship. For example, the 0.870 g/cm³ CGC copolymer has a room temperature modulus of 10^6 dyn/cm² and softening temperature is $65\,°C$. The range of operating temperatures for this copolymer is therefore between $-50\,°C$, the glass transition temperature, and $65\,°C$, the crystalline melting temperature. The relationship between density and G', T_g and T_m is shown in Table 5. All three of these properties increase with increase in the density of the elastomer.

For comparison, Table 6 shows the same properties for styrene block copolymers (SBCs), a flexible PVC (f-PVC) and an ethylene–vinyl acetate (EVA) copolymer

Table 5 Effect of density on dynamic mechanical properties and under load service temperature (ULST) of POEs

Density (g/cm^3)	T_m (°C)^a	G × 10 (dyn/cm^2)	T_g (°C)	Softening temperature by DMA (°C)	ULST (°C)	Hardness (Shore A)
0.863	43	0.7	−55	55	–	–
0.870	56	2	−52	65	65	73
0.877	70	4	−50	75	75	78
0.885	82	6	−49	85	85	85
0.895	95	9	−42	95	95	90

^{a} Melting temperatures taken from the second heats after the polymer has been heated above T_m and cooled to remove the effects of thermal history.

containing 28 wt% vinyl acetate. SBS and SIS are block copolymers with a midblock of butadiene and isoprene, respectively. SEBS is styrene–ethylene–butene–styrene, which contains a saturated midblock. All three copolymers have about 30 wt% of styrene comonomer.

Since parts made from POEs and EPDMs are often under load at elevated temperatures, TMA can be used to determine the upper service temperature. This technique uses a probe to apply a 100 g load, which is maintained as the temperature of the sample is raised. As the material softens, the probe indents the sample. The temperature at which the probe penetrates 1 mm into the sample is called the under-load softening temperature (ULST). This technique mimics the behavior of the

Table 6 Thermal properties of competitive materials

Polymer	Density (g/cm^3)	T_m (°C)	G × 10^7 (dyn/cm^2)	T_g (°C)	Softening temperature by DMA (°C)	ULST (°C)	Hardness (Shore A)
POE	0.870	56	2	−52	65	65	73
SBS (KRATON G1650)^a	0.94	–	3	−60	100	100	75
SIS (VECTOR 4211)^b	0.94	–	0.5	−90	90	80	60
SEBS (VECTOR 8508)	0.91	–	5	−60	90	106	80
f-PVC		–	4	−10	–	100	70
EVA (28% VA)	0.93	72	–	−30	–	70	75

^{a} KRATON is a registered trademark of Shell Chemical Company.
^{b} VECTOR is a registered trademark of Dexco Polymers.

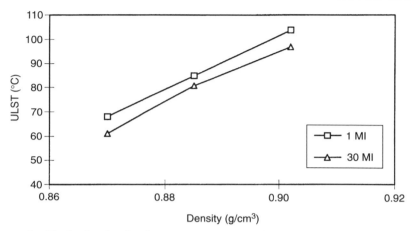

Figure 8   Under-load softening temperature (ULST) as a function of density and molecular weight for CGC polymers

material under load at elevated temperatures. The softening temperature is dependent on both molecular weight and density (Figure 8).

The results in Table 5 suggest that materials offering a wide range of properties can be tailored by the molecular architecture using comonomer and molecular weight control. By correct selection of the POE some of the thermal mechanical properties of block copolymers, EVA and f-PVC can be achieved (Table 6). For a given hardness, the POE has a lower T_g than EVA and f-PVC. This suggests that POEs would have better low-temperature performance than EVA and f-PVC.

Generally, for a given hardness, POE has a lower under-load softening temperature than a styrene block copolymer and f-PVC. However, the softening temperature can be improved by blending two different density POEs [17]. A blend containing POEs with a density difference of 0.04 g/cm^3 shows an increase in the ULST of 10 °C over the copolymer with the same melt index and density. A second technique used to increase the ULST is through grafting of a high-density polyethylene (HDPE) on to the copolymer backbone using peroxide [18]. Table 7 shows that the increase in ULST via grafting is 35 °C.

Table 7   Under-load service temperature of POE and POE–POE blends and grafted material

Polymer	Density (g/cm^3)	ULST (°C)
POE	0.87	65
Bimodal blend (0.86 + 0.90)	0.87	75
HDPE–POE blend	0.885	75
HDPE–g-POE	0.885	110

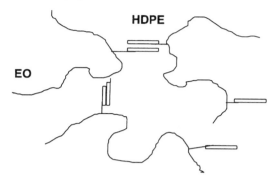

Figure 9 Schematic diagram illustrating the appearance of the grafted HDPE chains on the POE backbone. The POE backbone contains no unsaturation

Figure 9 illustrates the grafting mechanism of HDPE on to the POE backbone (HDPE–g-POE). The high-density chains form physical cross-links or crystallites, which serve to tie the elastomer chains into a three-dimensional network. The elastomer acts as a soft segment to provide flexibility from room temperature down to temperatures approaching the glass transition temperature of the elastomer. An additional contribution to the increase in upper service temperature comes from the higher melting temperature of the high-density phase. The lamellae of the high-density phase are segregated from the POE matrix if the materials are blended in the solution phase (Figure 10). However, by grafting the HDPE onto the backbone using peroxide, an even distribution of HDPE lamellae can be accomplished. This leads to a blend with properties of both components (Figure 11).

2.5 OXIDATIVE STABILITY

Oxidative stability can be discussed in terms of processing stability and weatherability. Processing stability relates to the high temperatures encountered during fabrication or extrusion. For example, Figure 12 compares the torque–time relationships of a POE to SIS and SBS during blending in a Haake rheometer. The POE shows a stable torque reading at 210 °C over time, indicating that no change to the backbone of the polymer has occurred. In comparison, the SBS, which was also processed at 210 °C, begins to cross-link after approximately 20 min. At a lower temperature (190 °C) the SIS begins to degrade rapidly via chain scission, as evidenced by the decrease in torque.

Weatherability is the degradation of the polymer during exposure to ultraviolet radiation [19]. In order to evaluate the weatherability of POEs compared with SBCs and a Z-N EPDM, plaques of each sample were placed in a xenon arc Weather-O-Meter against a black panel. The panel temperature was maintained at 63 °C with water spray for 18 min every 102 h. The humidity was kept constant at 50 %. The

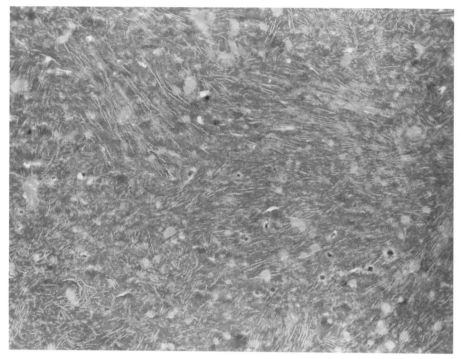

Figure 10 Transmission electron micrograph of a solution blend of a 0.885 g/cm^3 CGC ethylene–l-octene copolymer with a high-density polyethylene

inner and outer filters were borosilicate with a radiance of 0.35 W/m^3 at 340 nm. The extent of degradation was measured by measuring the change in tensile properties, specifically toughness. The toughness of the polymer was calculated from the area under a tensile stress–strain curve. The half-life values shown in Table 8 indicate the time necessary for the toughness to be reduced by 50 %. A POE at a density of 0.87 g/cm^3 is compared with styrene block copolymers and a Z-N EPDM. The weatherability decreases in the order POE > SEBS > Z-N EPDM > SBS > SIS. As shown in Table 8, the half-life of the POE is approximately twice that of the SEBS. SEBS does not contain unsaturation but does contain benzylic hydrogens that are susceptible to oxidation. Z-N EPDM has better oxidative stability than SIS and SBS, since it has less unsaturation overall with the saturated groups (4 wt%) situated on the side-chains. The SIS and SBS have considerable amounts (~30 mol%) of unsaturation along the main chain.

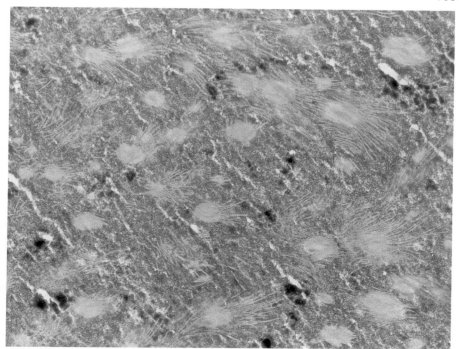

Figure 11 Transmission electron micrograph of the solution blend pictured in Figure 10 after grafting using 0.3 wt% peroxide.

3 APPLICATIONS OF POLYOLEFIN THERMOPLASTIC ELASTOMERS VIA CONSTRAINED GEOMETRY CATALYSTS

POEs prepared via CGC are relatively new in the market. Many applications are in the developmental stage. Consequently, the application information summarized in this chapter is mainly based on an understanding of structure–property relationships of CGC copolymers and the proposed utility found in patents and the open literature. The physical properties of CGC polymers suggest that they can be used to compete with f-PVC, styrene block copolymers (SIS, SBS and SEBS) and EVA. The applications of POEs include impact modification, blends with SBCs, injection- and blow-molded parts, profile extrusion articles, calendered sheets and adhesives.

POEs can be used as impact modifiers for polypropylene and polyethylene in thermoplastic polyolefins (TPOs) for instrument panels and bumper fascias. Amorphous POEs with densities less than $0.88 \, \text{g/cm}^3$ (preferably lower than $0.865 \, \text{g/cm}^3$)

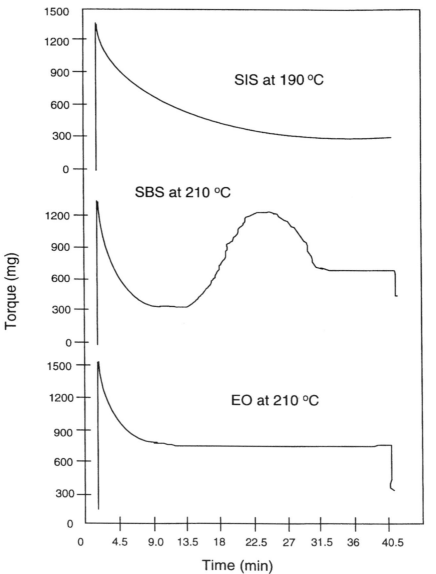

Figure 12 Torque–time diagrams from a Haake rheometer for SIS, SBS and POE at different blending temperatures

Table 8   Half-life of POE compared with block copolymers
and a Z-N EPDM

Polymer[a]	Half-life (h)[b]
POE ($0.87\,g/cm^3$, 1 MI)	800
EPDM (NORDEL 2722)	120
SEBS (KRATON G 1650, 29% styrene)	470
SBS (VECTOR 8508, 28% styrene)	50
SIS (VECTOR 4211, 30% styrene)	<50

[a] NORDEL is a trademark of DuPont Dow Elastomers LLC and KRATON is a
trademark of Shell Chemical Company.
[b] Irradiation time for the polymer to lose 50% of its tensile toughness (ASTM
D-2565).

are employed in this application. The low-temperature impact properties and good
miscibility with both matrices make POEs alternatives to the broad molecular weight
distribution Z-N EPDM. TPOs are discussed further in Chapter 8, Vol. 2. Blends
with SBCs, specifically SEBS, may also provide cost savings without substantial
loss in properties [20,21].

POEs with densities $\geq 0.870\,g/cm^3$ find applicability in injection-and blow-
molded parts, owing to their good processability and color dispersion. The selection
of the density of POE for an application is based on the required upper service
temperature. Molded goods include medical devices, toys, automotive parts, appliances and footwear. The combination of transparency (for POE with density of
$0.870\,g/cm^3$), flexibility, elasticity and cleanliness (low catalyst residue and low
oligomer content) makes POEs attractive for certain uses. In the flexible toy and
automotive markets, POEs can be used to replace f-PVC, offering potential cost
savings in terms of tooling and fabrication owing to the reduction in part weight.
Furthermore, POE has an added advantage of not requiring the addition of
plasticizer to achieve flexibility.

In the footwear market SBCs, EVA and f-PVC are commonly used. As discussed
in the previous section, POEs offer a good combination of flexibility, UV stability
and colorability. POEs would therefore be a good fit for this market also. Rubber
boots can be made with POEs [22]. The POE rubber boot has greater flexibility and
lower weight than f-PVC rubber boots.

Gaskets in large appliances such as refrigerators require good flexibility and low
compression set so that the part does not deform under repeated loading. The
excellent elasticity of these low density materials provides the desirable combination
of properties for this application.

It has been reported that POEs can be used to make hot melt adhesives [23]. The
hot melt adhesive is comprised of polymer, tackifier and wax. The adhesives are

useful in a variety of applications including book binding, case and carton seals and packaging. EVA currently is used in hot melt adhesive applications. POE has better heat stability and lower T_g than EVA. This suggests that the adhesive from POE should have better pot-life stability and better performance at low temperature than an EVA-based adhesive.

4   STRUCTURE AND PROPERTIES OF THERMOSET ELASTOMERS FROM POE AND CGC EPDM

The previous section discussed the fact that the melting-point and crystallinity of POEs limit the upper service temperature. Some applications also require better recovery at higher temperatures. In order to overcome this limitation, the material is often cross-linked, creating chemical junctions that are still present after the physical cross-links or crystals have melted. This section discusses the methods of cross-linking of POE and CGC EPDM and the mechanical properties of the resulting systems.

4.1   CHEMISTRY OF CROSS-LINKING

POEs can be cross-linked by conventional methods for polyolefins such as peroxide [24] and E-beam radiation [25]. Co-agents such as triallyl cyanurate (TAC) can be used in combination with peroxide to improve the cross-linking efficiency [26]. A dual curing system using a combination of peroxide and silane chemistry can also be employed [27]. Vinylsilanes such as vinyltrimethylsiloxane (VTMS) are grafted onto POE using peroxide. The latter system uses moisture to cross-link and heat to create the free radical using peroxide [28]. Another form of functionalized cross-linking is achieved by grafting POEs with maleic anhydride (MAH) using peroxide initiator and then reacting the MAH-grafted POE with curing agents such as multifunctional epoxies, amines and alcohols [29].

The narrow molecular weight and comonomer distribution lead to a high cross-linking efficiency in POEs compared with Ziegler–Natta catalyzed copolymers (Chapter 23, Vol. 2). Tertiary hydrogens from the α-olefin comonomer cause chain scission. Therefore, for a given density, POEs with higher α-olefin comonomer contents have higher cross-linking efficiency, since the backbone contains fewer tertiary hydrogens [30]. In addition to the cross-linking chemistry for POEs, CGC-EPDMs can be cured by conventional curing agents such as sulfur and phenolic resins.

4.2 TENSILE PROPERTIES

Figure 13 compares the tensile deformation behavior as a function of density for three CGC POEs reacted with 1 wt% peroxide. The same statements that were made about the unmodified materials can be made about the effect of density on cross-linked POEs. At room temperature the elastic recovery will still be better for the 0.863 g/cm^3 material than the 0.885 g/cm^3 material. However, the recovery compared with the uncross-linked material across the entire elastomer density range is improved, since the network causes less non-recoverable deformation. At temperatures above the crystalline melting-point the recovery is purely dependent upon the efficiency of the cross-linking process. Table 9 shows the tensile properties as a function of density and cross-linking level. The distance between junction points decreases as the level of reacted peroxide increases. The ultimate tensile strength decreases slightly since the material can no longer strain-harden. A decrease in elongation-to-break is also observed, which is a result of the restriction of the chains due to cross-linking. However, the recovery, as observed after testing, does improve significantly as the level of cross-linking increases.

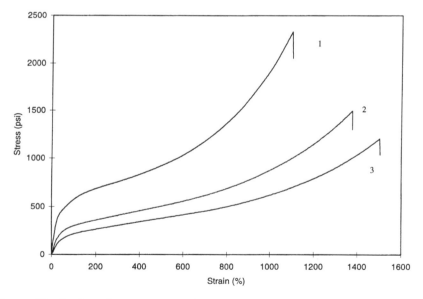

Figure 13 Engineering stress–strain curves of ethylene–1-octene POEs with 1 wt% dicumyl peroxide as a cross linking agent. Curves: 1, 0.885 g/cm^3 POE; 2, 0.870 g/cm^{-3} POE; 3, 0.863 g/cm^3 POE

Table 9 Tensile properties of ethylene–1-octene copolymers as a function of cross-linking level

Density	Peroxide (wt%)	Tensile strength (psi)	Elongation at failure (%)
0.863	0	1240	1000
	1	1175	800
	4	470	450
0.870	0	830	1550
	1	740	1540
	4	610	760
0.885	0	750	2700
	1	600	2245
	4	680	1450

4.3 ELASTIC RECOVERY: DYNAMIC PROPERTIES

The dynamic performance of cross-linked semicrystalline materials is dictated by the combination of crystals and the elastic network created by chemical cross-links. The size of the lamellae and entanglement density will determine the relaxation behavior or degree of hysteresis. Table 10 lists the dynamic properties of cross-linked POEs as a function of density, cross-linking level and temperature. At room temperature, the hysteresis (irrecoverable energy loss) decreases over the unmodified material upon reacting with 1 wt% peroxide (see Table 3). The proximity of the chemical junctions increases with the addition of peroxide, thereby reducing the molecular weight between entanglements. This serves also to reduce the crystal size as crystallization becomes diffusion-limited (see discussion of thermal properties). At smaller crystal sizes POEs exhibit lower hysteresis and compression set, regardless of the density.

Table 10 Dynamic properties as a function of cross-linking level, temperature and density for ethylene–1-octene copolymers

Density (g/cm^3)	Peroxide (wt %)	Temperature (°C)	Hysteresis (%)	Dynamic compression set (%)
0.863	1	23	2.1	3
	4	23	1.9	2.5
	1	120	5.4	2.5
	4	120	0.8	0.5
0.870	1	23	2.4	3.3
	4	23	2.0	3.2
	1	120	6.9	6.3
	4	120	0.8	0.5
0.885	1	23	3.3	5.7
	4	23	2.7	4.7
	1	120	5.5	5.8
	4	120	1.2	2.5

At 120 °C the hysteresis is higher than at room temperature for all densities at the same peroxide concentration. Although peroxide cross-linking permits the use of these materials at temperatures above the melting-point, the effective number of junction points must reach a certain level for the material to achieve the minimum hysteresis. Addition of 4 wt% percent peroxide yields a hysteresis loss number of approximately 1 % for all densities. The dynamic compression set decreases concomitantly.

The effects of narrow molecular weight and short-chain branching distributions have been discussed in the methods of cross-linking POEs. The role of molecular weight is to increase the tightness of the physically cross-linked network at similar crystallinity levels. A narrow molecular weight POE exhibits slightly lower hysteresis and compression set than the broader molecular weight POE at room temperature. The difference is negligible at elevated temperatures (Table 11). However, the higher molecular weight (lower melt index) resin does exhibit better hysteresis and compression set comparatively at room temperature. At 120 °C, the compression set is similar for both the high and low melt index materials with a narrow molecular weight distribution.

With the addition of peroxide the dynamic properties improve, but further addition of peroxide offers enhancement only for the broad molecular weight distribution polymers. At temperatures above the melting-point the network junction point density appears to be the same. This infers that the cross-linked higher molecular weight material in both materials dominates the dynamic performance. The following statements can be made about the effects of density, molecular weight and molecular weight distribution on the dynamic performance of POEs [31]:

1. At room temperature, lower density gives better compression set and hysteresis.
2. At temperatures above the melting temperature, the quality of the network dictates the performance; this seems to be the same for resins with densities of 0.87 and 0.885 g/cm³.
3. High molecular weight materials form better networks for a given density owing to the rapid build-up of a network.

Table 11 Comparison of CGC ethylene–1-octene copolymers at a density of 0.885 g/cm³ as a function of molecular weight and molecular weight distribution (1 wt% peroxide)

MWD	Temperature (°C)	I_2	Dynamic compression set (%)	Hysteresis (%)
Narrow	23	0.5	3.6	2.5
	120		5.8	4.2
Narrow	23	1.0	5.7	3.3
	120		5.8	5.5
Broad	23	0.5	5.4	3.2
	120		5.2	4.3

Deformation of these elastomers consisting of both chemical and physical cross-links (at room temperature) can be broken down into two processes based on the nature of each type of cross-link. The physical crystalline cross-links rearrange under compressive loading by a shear mechanism. Even at these low strains, plastic, irreversible deformation of lamellae is plausible. It is proposed that the fringed micelles, owing to looser spatial constraints, may rapidly re-form. The destruction and regeneration of fringed micelles are therefore closer to an equilibrium process leading to better recovery.

Entanglements, more numerous in high molecular weight copolymers, play a role at temperatures above the crystalline melting-point. Longer molecules in the coiled melt state have a greater end-to-end distance than a lower molecular weight polymer. Under any loading conditions the entropy of these molecules will be greater than that of lower molecular weight molecules. The resilience or elasticity of these molecules ought to lead to better recovery based on this thermodynamic consideration.

Elastic recovery is, however, dominated by the network of chemical cross-links. A combination of a good network and fewer crystals gives the best performance at room temperature. At high temperatures only the chemical cross-links are present. Here, the number of cross-link sites decreases the molecular weight between entanglements, leading to a more homogeneous network. The network homogeneity is also enhanced by the narrow molecular weight distribution. Cross-linking between long and short chains in a broad molecular weight distribution resin does not promote recovery, since the molecular weight between cross-link points does not decrease uniformly across the molecular weight distribution.

4.4 THERMAL PROPERTIES OF CROSS-LINKED POEs

Figure 14 compares the DMA curve of an uncross-linked $0.870 \, g/cm^3$ POE with a $0.870 \, g/cm^3$ ethylene–1-octene copolymer cross-linked with 4 wt% dicumyl peroxide. The uncross-linked POE shows a decrease in storage modulus, E', at the melt temperature of 55 °C. The same POE cross-linked with 4 wt% dicumyl peroxide shows a decrease in modulus at 55 °C, but then plateaus and extends up to temperatures significantly above the crystalline melting-point. This can be attributed to the introduction of chemical cross-links that create a tight network structure. Table 12 shows that the overall crystallinity of POEs at a density less than $0.885 \, g/cm^3$ does not decrease significantly. However, cross-linking reduces the melting temperature, T_m, of the $0.870 \, g/cm^3$ POE from 65 to 56 °C (Table 12). Therefore, it is suggested that the cross-linking hinders the crystallization of the lamellae to a greater extent than in the fringed micellar microstructure.

4.5 EPDM TERPOLYMERS PREPARED USING CGC TECHNOLOGY

CGC-EPDM can be a terpolymer of ethylene, propylene and ethylidenenorbornene (ENB). ENB is introduced so that the EPDM can be cross-linked with sulfur or

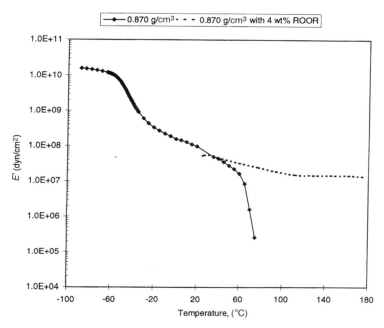

Figure 14 DMA curves of 0.870 g/cm^3 ethylene–1-octene POE and a 0.870 g/cm^3 ethylene–1-octene POE cross-linked with 4 wt% dicumyl peroxide

Table 12 Melting temperatures and heat of fusion (ΔH_m) of ethylene–1-octene copolymers cross-linked with dicumyl peroxide

Density (g/cm^3)	Dicumyl peroxide (wt%)	T_m(°C)	ΔH_m (J/g)
0.863	0	43	28.4
	1	44	28.1
	4	43	27.7
0.870	0	65	38.9
	1	61	37
	4	56	37.7
0.885	0	82	69.9
	1	80	59.1
	4	74	56.4
0.896	0	95	82.0
	1	92	75.7

phenolic cross-linking agents. The majority of Z-N EPDM products are made using a vanadium catalyst with an organoaluminum co-catalyst [32]. Z-N EPDM products are one of the major forces in the rubber industry (about 1.5 billion lb/year), since they have a good balance of properties such as resistance to aging, retention of mechanical properties at high temperatures, high extensibility and relative processing ease [33]. The structure and properties of Z-N EPDM have been well documented [34]. This section will focus on the properties of CGC-EPDM. The CGC provides substantial improvements in the production and products of EPDM. The high-efficiency CGC catalyst eliminates the costly residual catalyst washing step, which is necessary in the production of Z-N EPDM [35]. The CGC EPDM products therefore have much lower catalyst residue and a low yellowness index and lack the characteristic odor associated with commercial Z-N EPDM. Typical total metal content in CGC EPDM and Z-N EPDM are 20 and 400 ppm, respectively. CGC catalyst is homogeneous with a single active site, which produces CGC EPDM with a narrow molecular weight distribution and homogeneous distribution of comonomer across different molecular weight fractions [35]. The homogeneous structure should lead to a better cross-linking efficiency and better mechanical properties of the cross-linked products for reasons similar to those given for ethylene–1-octene copolymers. Most importantly, the well defined structure of the CGC are reported to allow the manipulation of molecular structure to deliver the desired performance, and also to produce products with high lot-to-lot consistency [36].

The detailed structure–property relationship of CGC-EPDM has been reported by Parikh and co-workers [36,37]. The process of fabrication using cross-linked CGC-EPDM includes three steps: melt mixing the ingredients, fabrication into shaped articles and curing. Consequently, the rheology of the uncured polymer, curing rate and efficiency and mechanical properties of the cross-linked polymer are considered as the main criteria for evaluating the performance of CGC-EPDM. Generally, higher shear thinning behavior (non-Newtonian) is preferred for most fabrication processes. Narrow molecular weight distribution CGC-EPDM is less shear thinning than broad molecular weight distribution CGC-EPDM. However, the combination of catalyst design and manufacturing process technology can lead to terpolymers with a broad range of shear thinning behavior [33].

The mechanical properties of the cross-linked CGC-EPDM depends on thermal properties (T_g and T_m) and cross-linking efficiency. As mentioned before, the homogenous distribution of ENB comonomer in CGC-EPDM and the narrow molecular weight distribution should increase the cross-linking efficiency. The crystallinity of CGC-EPDM increases with increased ethylene content. For example, the crystallinities of CGC-EPDM with 62, 70, and 75 wt% ethylene are approximately 7, 15 and 20 %, respectively. Crystallinity increases both the tensile strength and modulus, but has a detrimental effect on the elastic recovery.

For a constant ethylene content (72 wt%), the effects of ENB content on thermal properties of CGC EPDM are shown in Table 13. An increase in ENB and decrease

Table 13 Effects of ENB content on thermal properties of EPDM

ENB (wt%)	Propylene (wt%)	Ethylene (wt%)[a]	$T_g(^\circ C)$	$T_m(^\circ C)$	Crystallinity (%)
3	25	72 (80.6)	−40	37	13.5
5	23	72 (81.4)	−36	47	14.7
9	19	72 (83)	−27	55	15.7

[a] The values in parentheses are the mole percentages of ethylene.

in propylene slightly increase the crystallinity but significantly increase T_g and T_m. The increase in crystallinity was attributed to the increase in mole percentage of ethylene. The increase in T_g may be due to the cyclic structure of ENB that reduces the local segment mobility.

Preliminary evaluation of CGC-EPDM compared with commercial Z-N EPDM products using a sulfur curing system suggests that CGC-EPDM products meet the required mechanical properties for major applications such as hose, extruded goods, calendered sheets and molded goods that currently use Z-N EPDM products [31]. The results also show that CGC-EDPM systems have a better cross-linking efficiency, better modulus and tear strength and greater ease of processing than the corresponding Z-N EPDM. These data are in agreement with the structure–property relationships discussed above.

4.6 OXIDATIVE STABILITY OF CROSS-LINKED POE AND EPDM

Since products from cross-linked POE and EPDM may be exposed to high temperature and UV radiation, the long-term heat aging stability and weatherability are the main criteria for evaluation of oxidative stability of cross-linked POE and EPDM. Polymer structure, additives and cross-linking agents can affect the oxidative stability of cross-linked systems. A peroxide-cured system is more oxidatively stable than a sulfur-cured system, since the sulfur-cured system introduces weak sulfur–sulfur linkages into the polymer network. The effects of polymer structure on the oxidative stability of cross-linked POE and CGC-EPDM are similar to those for the corresponding uncross-linked system, i.e. the polymer that contains unsaturation or tertiary hydrogens is more sensitive to oxidative stability. Consequently, it is expected that POE is more stable than CGC-EPDM, since CGC-EPDM contains unsaturation. For a given density an ethylene–1-octene elastomer (EO) is more stable than an ethylene–1-propylene elastomer (EP), since the ethylene–1-octene copolymer contains fewer tertiary hydrogens. Heat aging studies of peroxide cross-linked EO, EP and Z-N EPDM in wire-cable formulations [38] and automotive formulations [39] indicate that the oxidative stability increases in the order EO > EP > Z-N EPDM. POEs and CGC-EPDMs are expected to have better oxidative stability than the corresponding materials made using Ziegler–Natta catalysts since the polymers produced using CGC have much less catalyst residue than polymers made with Ziegler–Natta catalysts.

5 APPLICATIONS OF CROSS-LINKED POES AND EPDMS

Thermoset elastomers from ethylene–1-propylene (EP) and EPDM made using Ziegler–Natta catalysts are known for their good ozone, oxidation, heat and chemical resistance and very good electrical insulation properties, as compared with the butadiene-based rubber [33]. The primary applications include hoses (automotive coolant hose, garden hose), extruded goods (gasket, channel along automobile windows, tubing, wire and cable), calendered sheets (roofing membrane) and molded goods (automotive ducts and boots) [31,37]. Z-N EPDM can be used as a rubber component in thermoplastic vulcanizate (TPV) [29]. Cross-linked POEs and CGC-EPDMs can also be used in these markets, providing the advantage of improved oxidative stability and better mechanical properties, due to the narrow molecular weight distribution.

5.1 MOISTURE-CURABLE APPLICATIONS

It has been reported that weather-stripping for cars and building windows, fibers and gaskets can be prepared from silane cross-linkable POE [24]. Vinylsilanes such as vinyltrimethylsiloxane (VTMS) are grafted on to POE using peroxide. The silane provides a means of cross-linking using moisture. The resulting silane-grafted POE can be extruded into parts such as fibers, weather stripping and gaskets. The part is exposed to moisture for curing. By cross-linking the POE, the elastic behavior of the material is retained even after exposure to elevated temperatures. The potential benefit of using POE rather than EPDM for weather-stripping is again related to the better weatherability.

5.2 FOAMS

Cross-linked foams made from POE exhibit a unique combination of resiliency, tensile and tear properties that make foams an excellent fit for cushioning applications where dynamic and static loading are experienced [40]. These foams have the resiliency of cross-linked LLDPE, but with greater strength and resiliency than a cross-linked LDPE. Additionally, the softness of these foams matches that of EVA and blends of PVC with nitrile rubber. Unlike EVA-based foams, POE foams are thermally stable and will not cause unpleasant odors during fabrication. These foams find application in sporting goods, medical devices and cushioning applications [41].

6 CONCLUSION

The structure–property relationships and applications of new elastomeric polyolefins made using CGC single-site catalysts have been discussed. Materials offering a wide range of properties can be tailored by molecular architecture using comonomer and molecular weight control.

POEs crystallize into a predominantly fringed micellar microstructure rather than lamellae. The primary consequence of this microstructure is the attainment of high elastic recoverability. This technology allows the production of new materials that are a cost-effective substitute for EVAs, SBCs and f-PVC. POEs have good processability and oxidative stability.

Owing to the narrow molecular weight distribution, cross-linked POE and CGC-EPDM products have better mechanical properties than the corresponding materials made using conventional Ziegler–Natta catalysts. The design flexibility of CGC polymers further establishes the viability of producing materials that better meet performance requirements in applications not attainable with traditional catalysts. Products made from POEs will continue to appear, specifically where the combination of processability and oxidative stability are important.

7 ACKNOWLEDGEMENTS

The authors acknowledge discussions with DuPont Dow Elastomers personnel, especially Deepak Parikh and Kim Walton. Seema Karande also provided insightful discussions on foam performance and Rajen Patel aided in the interpretation of the thermal behavior. Dynamic mechanical data for this chapter were provided by Teresa P. Karjala and Selim Bensason.

8 REFERENCES

1. Wunderlich, B., *Macromolecular Physics*, Academic Press, New York, 1973, Vol. 1, pp. 315–339.
2. Geil, P., *Polymer Single Crystals*, Robert C. Krieger, Huntington, 1963, pp. 235–304.
3. Kale, L. T., Plumley, T. A., Patel, R. M. and Jain, P., Structure–property relationships of ethylene/1-octene and ethylene/1-butene copolymers made using INSITE™ technology, in *SPE ANTEC '95*, Boston, 1995, p. 2249; available from the Society of Plastics Engineers, 14 Fairfield Drive, Brookfield, CT, USA.
4. Hwang, Y., Chum, S., Guerra, R. and Sehanobish, K., Morphology and deformation behavior of homogeneous polyolefin copolymers made with INSITE™ technology, in *SPE ANTEC '94*, San Francisco, 1994, p. 3414.
5. Flory, P. J., *J. Am. Chem. Soc.*, **84**, 2857 (1962).
6. Turek, D. E., Landes, B. G., Winter, J. M., Sehanobish, K. and Chum, S., Comparison of the solid-state properties of ethylene–octene elastomers with ethylene–propylene elastomers, in *SPE ANTEC '95*, Boston, 1995, p. 2270.

7. Bensason, S., Minick, J., Moet, A., Chum, S., Hiltner, A. and Baer, E, *J. Polym. Sci., Part B: Polym. Phys.*, **34**, 1301 (1996).
8. Bensason, S., Minick, J., Moet, A., Hiltner, A., Baer, E., Chum, S. and Sehanobish, K., The tensile deformation behavior of homogeneous ethylene–octene copolymers produced by the constrained geometry catalyst technology, in *SPE ANTEC '95*, Boston, 1995, p. 2256.
9. Minick, J., Moet, A., Hiltner, A., Baer, E. and Chum, S. P., *J. Appl. Polym. Sci.*, **58**, 1371 (1995).
10. Patel, R. M., Sehanobish, K., Jain, P., Chum, S. P. and Knight, G. W., *J. Appl. Polym. Sci.*, **60**, 749 (1996).
11. Graham, J. T., Alamo, R. G. and Mandelkern, L., *J. Polym. Sci. Part B: Polym. Phys.*, **35**, 213 (1997).
12. Darras, O. and Séguéla, R., *J. Polym. Sci. Part B: Polym. Phys.*, **31**, 759 (1993).
13. Haward, R. N., *Macromolecules*, **26**, 5860 (1993).
14. MacFaddin, D. C., Russell, K. E. and Kelusky, E. C., *Polym. Commun.*, **29**, 258 (1988).
15. Fatou, J. G., *et al.*, *J. Mater. Sci.*, **31**, 3095 (1996).
16. ASTM D-395, Standard Test Methods for Rubber Property–Compression Set, *Annual Book of ASTM Standards*, American Society for Testing and Materials, Philadelphia, PA, 1989.
17. Parikh, D, Chum, S., Jain, P., Patel, R., Clayfield, T., McKeand, T., Kummer, K., Khan, W. and Markovich, R., to Dow Chemical, *PCT Int. Appl.*, WO 97/26297 (1997).
18. Johnston, R., Morrison, E., Mangold, D. and Ho, T., patent pending.
19. ASTM D-2565, Standard Practice for D2565-92a Operating Xenon Arc-type Light-exposure Apparatus With and Without Water for Exposure of Plastics, *Annual Book of ASTM Standards*, American Society for Testing and Materials, Philadelphia, PA, 1997.
20. Da Silva, A. L. N., Tavares, M. I. B., Politano, D. P., Coutinho, F. M. B. and Rocha, M. C. G., *J. Appl. Polym. Sci.*, **66**, 2005 (1997).
21. Parikh, D., Guest, M., Patel, R., Ahmed, W., Betso, S., Ho, T., Guerra, R. and Allen, J. D., Structure and properties of metallocene olefin elastomer blends with styrene block copolymers, in *SPE ANTEC 96*, Indianapolis, 1996, p. 1638.
22. Lancaster, G. M., and Graves, P., Molecular architecture: designing elastomers for automotive and footwear applications, in *Metallocenes Asia*, Singapore, 1997, p. 301. Available from Schotland Business Research, 16 Ducan Lane, Skillman, NJ08558, USA.
23. Simmons, E. R., Bunnelle, W. G., Malcolm, D. B., Knutson, K. C., Kauffman, T. F., Kroll, M. S., Keehr, M. S., Parikh, D. R., Allen, J. D., Yalvac, S., Finlayson, M. F. and Rickey, C. L., to Dow Chemical and H. B. Fuller Licensing and Financing, *PCT Int. Appl.*, WO 97/33921 (1997).
24. McKay, K. W., Blanchard, R. R., Feig, E. R. and Kummer, K. G., to Dow Chemical *US Pat.*, 5 414 044 (1995).
25. Reed, J. F., to Minnesota Mining & Manufacturing, *US Pat.*, 5 324 576 (1994).
26. Walton, K. L., and Karande, S. V., Selected properties of crosslinked bun foams produced from substantially linear homogeneous polyolefin elastomers, in *SPE ANTEC '96*, Indianapolis, 1996, p. 1926.
27. Walton, K. L., in *SPE ANTEC '97*, Toronto, 1997, p. 3250.
28. Penfold, J., Hughes, M. and Brann, J., to Dow Chemical, *PCT Int. Appl.*, WO 95/29197 (1995).
29. Ho, T., Johnston, R. T., Hughes, M. M. and Allen, J. D. *PCT Int. Appl.*, WO 98/02489 (1998).
30. Capta, M. and Borsig, E., *Eur. Polym. J.*, **16**, 611 (1980).
31. Minick, J., and Sehanobish, K., Structural model of crosslinked thermoplastic elastomers produced via INSITE™ technology, in *SPE ANTEC '96*, Indianapolis, 1996, p. 1883.
32. Makino, K., in *Elastomer Technology Handbook*, Cheremisinoff, N. P., (Ed.), CRC Press, Boca Raton, FL, 1993, p. 445.

33. Laird, J. L. and Riedel, J., Evaluation of EPDM materials as produced by constrained geometry catalyst chemistry against current commercial EPDM products and performance requirements, presented at the 148th Technical Meeting of the *Rubber Division, American Chemical Society*, Cleveland, OH, October 1995.

34. Brydson, J. A., *Rubbery Materials and Their Compounds*, Elsevier Applied Science, London, 1988, p. 147.

35. Sylvest, R. T. and Pillow J. G., NORDEL^{TM} IP—the first commercial EPDM based on metallocene technology, presented at a meeting of the *Danish Society of Rubber Technology*, Helsingoer, Denmark, May 1997.

36. Parkih, D. R, Edmondson, M. S., Smith, B. W., Castille, M. J., Mangold, D. and Winter, J. M. Composition and structure–property relationships of single site constrained geometry EPDM elastomer, presented at a meeting of the *Rubber Division, American Chemical Society*, May 1997.

37. Parikh, D. R, Edmondson, M. S., McGirk, R. H., Castille, M. J., Meiske, L. A., Vara, R. and Winter, J. M., Influence of molecular structure on viscoelastic and solid state behavior of EPDM, presented at a meeting of the *Rubber Division, American Chemical Society*, Cleveland, OH, October 1997.

38. Hemphill, J., Processing and performance of new polyolefin elastomers, presented at a meeting of the *SPE*, Houston, TX, February 1995.

39. Brann, J., Comparison of ENGAGE[*] polyolefin elastomers and EPM/EPDM elastomers, presented at a meeting of the *Rubber Division, American Chemical Society*, Pittsburgh, PA, October 1994.

40. Karande, S. V. and Walton, K. L., Selected properties of crosslink bun foams produced from substantially linear homogeneous polyolefin elastomers, in *SPE ANTEC '96*, Indianapolis, 1996, p. 1926.

41. Park, C., Chum, P. S. and Knight, G. W., to Dow Chemical, *US Pat.*, 5 387 620 (1995).

10

Packaging Applications Using Enhanced Polyethylenes and Polyolefin Plastomers Produced by Constrained Geometry Single-site Catalysts

KAELYN KOCH, JACKIE deGROOT AND JOHN LASTOVICA
Dow Chemical Company, Freeport, TX, USA

1 INTRODUCTION

In previous chapters, three key distinctions were noted between ethylene–α-olefin copolymers produced with single-site catalysts (SSC) and ethylene–α-olefin copolymers produced with Ziegler–Natta (Z-N) catalysts: an ability to incorporate higher levels of the α-olefin to achieve low polymer density or crystallinity, a uniform comonomer distribution and a narrow molecular weight distribution (MWD) with M_w/M_n of about 2. The ability to incorporate higher levels of comonomer has made possible new product families, e.g. polyolefin plastomers (POP) and polyolefin elastomers (POE). For example, elastomers with densities of less than 0.87 g/cm^3 are available from SSC, whereas traditional Z-N ethylene–α-olefin copolymers typically are not available below about 0.905 g/cm^3. The narrow molecular weight (MW) and comonomer distributions contribute to several unique properties, including controllable melting-points, reduced extractables, reduced blocking, excellent optics and excellent mechanical properties. The narrow MWD also results in less shear

Metallocene-based Polyolefins Edited by J. Scheirs and W. Kaminsky
© 2000 John Wiley & Sons Ltd

Table 1   Benefits of plastomers in flexible packaging

Polymer properties	Packaging benefits
Toughness of a linear polyethylene	Protection of goods, gauge reduction
Low sealing temperatures	Faster packaging line speeds
Low extractables	Better taste and odor characteristics
Excellent optics	Package appeal
High oxygen transmission	Breathable films
Elasticity	High recovery and flexibility
Hydrocarbon composition	Polyolefin compatibility, recycle friendly
Thermal stability	Processing flexibility
Moisture insensitivity	Bulk handling, good moisture barrier

thinning, which can translate to reduced bubble stability, increased melt fracture and higher amps and back-pressure.

The presence of long-chain branching (LCB) is one feature that differentiates plastomers and elastomers made with constrained geometry catalyst (CGC) technology from other narrow MWD polymers (e.g. Tafmer polymers made by Mitsui with a soluble vanadium catalyst or homogeneous, linear SSC polymers such as Exact made by Exxon). As discussed in previous chapters, increased shear sensitivity provided by the LCB results in improved melt strength, reduced horsepower requirements during extrusion and greater resistance to melt fracture. The increased melt strength and greater ease of extrusion can translate into higher output on an extrusion line.

The unique characteristics of homogeneous ethylene–α-olefin copolymers described above make them useful for packaging applications. For example, the ability to control melting and crystallization behavior, the excellent optics and excellent mechanical strength make plastomers an ideal candidate to compete with the traditional high-performance sealants such as ethylene–vinyl acetate (EVA) copolymers containing 12–18 % vinyl acetate and ionomers. Indeed, one important commercial application for plastomers is use as a high-performance sealant in multilayer coextruded or laminated films. Table 1 shows several examples of key properties of plastomers produced with SSC and the resulting packaging benefits.

2   MOLECULAR DESIGN

In addition to delivering new polymer families such as plastomers and elastomers, SSC have also changed product development methods and capabilities. The availability of homogeneous materials and SSC has significantly improved the ability to develop structure–property models, in addition to kinetic and thermo-dynamic models. Increased reliance on models, versus more labor-intensive laboratory experimentation, can reduce the time and expense required to bring a new product to commercialization.

Molecular design is a product development approach that leverages this enhanced modeling capability, along with material science understanding and a flexible polymer manufacturing process, to deliver polymers which meet specific performance requirements. Molecular design results in polymers which have new performance combinations that were not previously available in a single resin.

2.1 THE TRADITIONAL APPROACH [1,2]

Traditionally, many of the molecular parameters used to design polyethylenes (PE) were dictated by the choice of process, comonomer and catalyst. The conventional approach to PE product design was to specify the melt index and density (or comonomer level), and occasionally, the MWD to vary performance. Once the melt index and density were selected, many other key molecular parameters such as MWD, tie-chain concentration and comonomer distribution were also predetermined with little ability to manipulate independently. Since the key molecular parameters for controlling polymer performance were interdependent, the polymer designer and customer were often required to make performance compromises or trade-offs.

2.2 THE MOLECULAR DESIGN APPROACH

However, with CGC technology and a flexible manufacturing process, polymers with controllable molecular architecture may be produced, because parameters such as density, tie-chain concentration, MW and MWD can be controlled more independently than before [3]. For example, Figure 1 shows the MWD of two molecularly designed ethylene–octene copolymers made in the solution process with INSITE technology; both resins have very similar melt indices, densities and I_{10}/I_2s. Clearly, gel permeation chromatography (GPC) shows differences in MWD in spite of the similarities in the traditional specification parameters. Also, in fact, the performance properties of these two PEs also differ measurably. Kale *et al.* described several performance differences between these two resins in wire and cable jacketing applications, including notch sensitivity, surface quality and cable jacket shrinkback [4]. The performance is controlled by the balance between MW, MWD and LCB.

Design flexibility means that new performance combinations are possible because some of the traditional limitations and rules have changed, making it possible to begin the resin development process with a greater emphasis on performance requirements [5]. Performance requirements may include fabrication performance, esthetic considerations, mechanical requirements, economic limitations and other parameters. If the performance requirements are clearly understood and accurate models are available, development speed and cost are favorably impacted by reducing resource-intensive, iterative validation and qualifications steps.

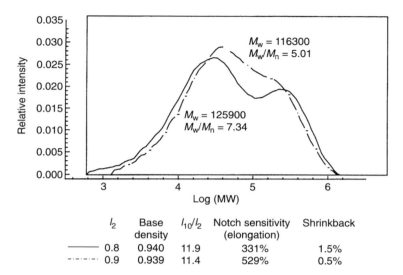

I_2	Base density	I_{10}/I_2	Notch sensitivity (elongation)	Shrinkback
—— 0.8	0.940	11.9	331%	1.5%
---- 0.9	0.939	11.4	529%	0.5%

Figure 1 Polymers with similar traditional specifications yield different performances

2.3 EXAMPLES OF NEW PERFORMANCE COMBINATIONS

The molecular design principles discussed above can be leveraged to deliver new performance combinations, e.g., high impact and high modulus. To increase impact strength of a traditional Ziegler–Natta PE, one needed to decrease the density or increase the MW; however, these options had the undesirable side-effects of reducing stiffness or processability, respectively. Through more independent control of resin design parameters, molecular design can be used to increase tie-chain

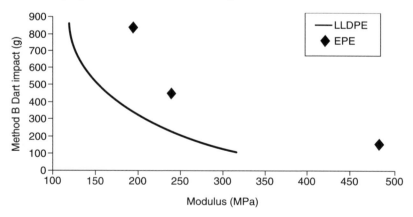

Figure 2 New performance combination: high dart impact and high modulus

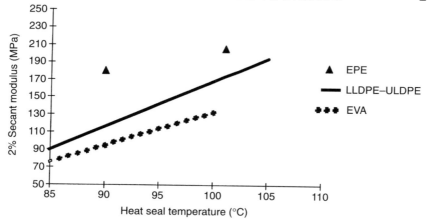

Figure 3 New performance combination: low heat seal initiation and high modulus

concentration through comonomer distribution and MWD without significantly increasing the melt index or decreasing the density. Figure 2 shows the relationship between dart impact and modulus for three enhanced polyethylenes (EPE), relative to Z-N ethylene–octene copolymers [3].

Two other examples of new performance combinations are demonstrated in Figures 3 and 4. All of the solid lines represent ethylene–octene copolymers produced with conventional Z-N technology, including high-density PE (HDPE), linear low-density PE (LLDPE) and ultra-low-density PE (ULDPE), which has a density range of 0.905–0.915. The EPEs are molecularly designed ethylene–octene copolymers [3].

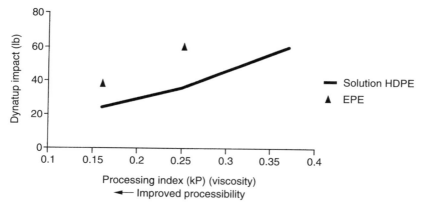

Figure 4 New performance combination: processability and impact strength

In the sections that follow, we will describe how both the homogenous plastomers and elastomers, as well as molecularly designed ethylene–α-olefin copolymers, are useful in packaging applications. Examples of films, coatings and molded containers will be discussed.

3   FOOD PACKAGING FILM APPLICATIONS

The primary distinguishing feature of food packaging is the need to prevent contamination of the food and to maintain or extend the shelf-life. The package acts as a barrier to the environment. Barrier properties will vary depending upon the specific type of food being packaged. The barrier of a flexible package can be compromised when (i) hermetic sealing of the package is not obtained during packaging, (ii) the seal of the package is not maintained during the packaging, distribution and sales of the product or (iii) the package itself fails owing to inadequate puncture, tear, impact or abrasion resistance. Seal performance and abuse resistance are two of the critical performance requirements for many, if not all, of the food packaging applications described below.

3.1   HIGH PERFORMANCE SEALANTS

3.1.1   Processed Meat and Cheese Applications [6]

Examples of processed meats include such items as bacon, hot dogs, ground meat chubs, link sausage and 'cook-in' hams and turkeys. Cheese applications include packaging for chunk cheese, shredded cheese and sliced cheese. These foods have relatively short shelf-life spans. As a result, these products require a high level of oxygen barrier and are refrigerated or frozen to maintain freshness throughout the packaging, distribution and sales chain. Typical barrier polymers include ethylene–vinyl alcohol (EVOH) and poly(vinylidine chloride) (PVDC).

Both meat and cheese applications utilize coextruded or laminated film structures for the package. An example of a package comprised of both a coextrusion and a lamination is hot dog packaging. These packages are typically made using horizontal, thermoform, fill and seal equipment where the bottom web is formed into a cavity, the cavity is filled with product, and the product is covered with the printed non-forming top web. A representative forming structure is approximately. 3–4 mil (1 mil = 0.001 in) in total thickness and is comprised of the following layers:

   10 % nylon/15 % tie-layer/15 % barrier/15 % tie-layer/25 % sealant

The non-forming top web is often a lamination with a thickness ranging from 2.5 to 3.5 mil with the following representative structure:

   PVDC-coated PET [poly(ethyleneterephthalate)] film/adhesive/sealant

Table 2   High-performance sealants for processed meat packaging

Property	POP 1	ULDPE 1	Na ionomer
Comonomer	octene	octene	—
Melt index (dg/min)	1.0	1.0	1.3
Density (g/cm^3)	0.909	0.912	—
2 mil monolayer blown film properties			
Puncture (ft-lb/in^3)	330	310	140
Dart impact, B(g)	$\geqslant$ 850	710	750
MD Elmendorf tear A(g)	480	925	22
Haze (%)	3.0	7.2	3.2

Prior to the introduction of SSC technology, ionomers were the standard sealants for many processed meat and cheese applications. Table 2 describes three resins that could be considered for sealants in a structure similar to the non-forming web described above: a polyolefin plastomer (POP 1), an ultra-low-density linear polyethylene (ULDPE 1) and a sodium-based ionomer (Na ionomer). ULDPE is generally not classified as a high-performance sealant but is included to show the evolution in performance from linear polyethylenes to POPs. The critical properties of a high-performance sealant are briefly described below. A more in-depth description of the sealing technique and sealant parameters can be found elsewhere [7,8].

Seal Plateau Strength

Since strong hermetic seals are required to prevent product exposure to the environment, the highest level at which the seal strength plateaus is considered a critical property. As illustrated in Figure 5, both POP 1 and ULDPE 1 provide approximately 100 % higher seal plateau strengths than the Na ionomer. The improved seal strength also translates to a practical seal application in which the seal area is not clean, but is contaminated with a residue that simulates hot dog juices.

Heat Seal Initiation Temperature

The heat seal initiation temperature is often referred to as the temperature where a 1 lb/in seal strength is achieved. The temperature at which the seal plateau strength is achieved is referred to as the seal plateau temperature. Low seal initiation and plateau temperatures are desirable in that they allow for faster packaging line speeds because less heat must be added and removed to achieve a strong, hermetic seal. Second, when less heat is added to the package, there is less chance that the food will be discolored due to heat exposure. Finally, a sealant that provides lower seal temperatures also tends to have a broader heat-seal temperature range and, in turn, is more forgiving of temperature variability in packaging equipment.

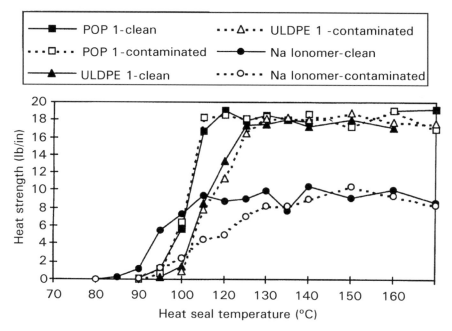

Figure 5 Comparison of high-performance sealants POP, ULDPE and Na ionomer

The seal initiation and plateau temperatures of the three sealants are illustrated in Figure 5. The seal initiation of the clean Na ionomer is the lowest (89 °C), followed by POP 1 (94 °C), and finally the ULDPE (98 °C). The Na ionomer and the POP 1 with clean seals provide the same seal plateau temperature, approximately 105 °C, while the ULDPE provides a significantly higher seal plateau temperature, 115 °C. The lower seal initiation and seal plateau temperature of the POP versus the ULDPE is achieved via the ability to manipulate the melting characteristics of this homogeneous polymer via SSC technology. A final note is that the seal plateau temperature of the Na ionomer increased significantly (from 105 to 120 °C) when the seal area was contaminated; POP 1 and ULDPE 1 were unaffected by the contamination.

In addition to sealability, the sealant can also contribute to the other functional attributes of the package. High abuse resistance is important in preventing damage to the package, and ultimately failure of the barrier, during packaging and distribution. As illustrated by the data in Table 2, the tear and puncture properties are higher for ULDPE 1 and POP 1 than the Na ionomer. Furthermore, the POP provides at least a 15–20 % improvement in impact strength vs the Na ionomer and ULDPE. Another critical feature for retail applications is the package esthetics or consumer appeal.

POP 1 and the Na ionomer provide very low haze, allowing the customer to view the contents of the package easily. The combination of sealability, abuse resistance and optics of the POP makes it a very good option for retail processed meat and cheese applications.

3.1.2   Cereal and Cake Mix Inner Liners [9]

Another segment of high-performance sealants is the box inner liners for dry foods such as cake mix, cereal and crackers. While a high oxygen barrier is critical to preserve the freshness of meat and cheese, a high moisture barrier is essential for preserving the freshness of dry products. A typical plastic film structure for cereal and cake-mix inner liners is a coextrusion of HDPE with a sealant. The sealants typically used for these applications are ionomers or EVAs containing high levels ($\geqslant 18\%$) of vinyl acetate (VA).

Again, since a strong, hermetic seal is critical for maintaining the barrier of the package, POPs are another option for replacing ionomers or EVA copolymers (18 % VA). A comparison of the other critical sealant performance requirements for a polyolefin plastomer (POP 2), an ionomer (Zn ionomer), and an EVA copolymer (18 % VA) is discussed below and summarized in Table 3.

Heat-seal Initiation Temperature

POP 2 and EVA copolymers (18 % VA) have comparable seal initiation temperatures, 69 and 72 °C, respectively. They are followed by the Zn ionomer with a seal initiation temperature of 84 °C.

Table 3   High-performance sealants for box inner liners

Property	POP 2	EVA (18 % VA)	Zn ionomer 1
Melt index (dg/min)	1.6	0.8	5.5
Density (g/cm^3)	0.8965	Not available	Not available
2.0 mil monolayer blown film properties			
Puncture (ft-1b/in^3)	270	222	104
Tear resistance (g)	300	118	190
Haze (%)	1.3	1.5	2.9
Odor intensity rating	1.2	Not available	2.7
Water vapor transmission rate (g mil/100 in^2 day atm)	2.0	6.2	Not available
1.5 mil HDPE–0.5 mil sealant, blown coextruded			
Heat-seal initiation (°C)	69	72	84
Ultimate hot-tack strength (N/in)	7.5	3.4	4.4

High Ultimate Hot-tack Strengths

Hot-tack strength is a measure of the force required to separate a semi-molten, just-made seal. This test is designed to predict the performance on vertical, form, fill and seal (VFFS) equipment, in which the strength of the seal after it has just been made may be the rate-limiting step in this automated process. High hot-tack strength reduces the likelihood that the food product will break through the seal when being dropped into the forming package. The ultimate hot-tack strength is the highest hot-tack strength achieved over a full range of sealing temperatures. Table 3 shows that POP 2 provides the highest ultimate hot-tack strength of the three products, followed by the Zn ionomer and the EVA copolymer. Note that the 18 % VA copolymer provides very low hot-tack levels and is generally not suitable for the packaging of heavy products, such as cake mixes.

'Easy Open' Seal

Since cereal is intended primarily for retail consumption, these packages are designed to open easily. To facilitate the 'easy open' feature, a peelable seal is often used. A peelable seal is usually obtained by one of two failure mechanisms: delamination of the sealant layer from the rest of the structure or face-to-face seal failure at a relatively low separation force (approximately 1–3 lb/in) via the blending of small amounts of incompatible polymers with the sealant. Formulations for POP peelable seals are not provided here, but have been given in other publications [9].

Since the inner liner is typically protected by a box, the importance of optical and abuse resistance properties is reduced, but not insignificant. Irregularly shaped cereals tend to require the greatest puncture and tear resistance. As illustrated in Table 3, the puncture and tear resistances of POP 2 are significantly higher than those of the EVA copolymer and Zn ionomer.

The organoleptic performance of the sealant, or the amount of off-taste or odor imparted to the food by the sealant, is also critical when packaging bland products such as cereals. As shown in Table 3, an odor intensity rating for POP 2 and the Zn ionomer shows significantly lower odor (on a scale from 0 to 3) for the POP. In a separate paired comparison test of POP 2 with the EVA (18 % VA) copolymer, 24 out of 24 panelists identified the POP as having the lower off-odor.

Finally, the primary function of the inner liner package is to provide a moisture barrier to preserve freshness. The purpose of the HDPE is to provide low moisture transmission; however, since the sealant comprises about 25 % of the total structure, it too can have an effect on the moisture transmission properties. The moisture barrier of POP 2 was significantly higher than that of EVA copolymer.

3.2 PRODUCE PACKAGING

A common example of fresh-cut produce packages is bags of washed and pre-cut lettuce mixes for salads. The role of the package in this application is to extend the shelf-life by creating an atmosphere which is capable of slowing the respiration rate of the produce. Details of how this atmosphere can be created via package design are provided elsewhere [10,11].

The fresh-cut produce application takes advantage of the relatively high gas transmission properties of POPs to achieve an appropriate oxygen atmosphere for the various produce items. The performance requirements for produce packaging include (i) specified gas transmission rates, as dictated by the respiration rates of the various types of produce, (ii) low water vapor transmission to prevent the produce from drying out, (iii) strong, hermetic seals to maintain the desired modified atmosphere within the package, (iv) low heat-seal initiation temperatures and high hot-tack strengths to facilitate faster packaging line speeds, (v) excellent optics, particularly for retail applications, (vi) stiffness for automatic packaging machinability and for consumer appeal and (vii) good resin extrudability on existing blown and cast film equipment.

Figure 6 [12] shows the relationship between film modulus and oxygen transmission rate (OTR) for POP, ULDPE and EVA films. The lower end of the produce packaging market, e.g. packaging salad mixes for food service, can often be satisfied by monolayer films containing LLDPE alone or in blends with other materials, since optics are not critical and only medium OTR is required.

The higher end of the fresh-cut produce market is in the packaging of vegetables, such as broccoli or spinach, for retail consumption. These applications require significantly higher OTRs and excellent optical properties. Table 4 provides a

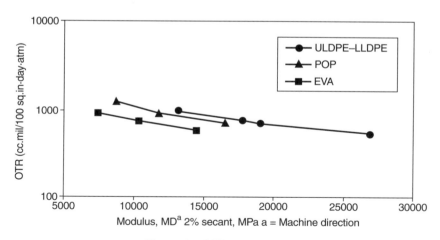

Figure 6 OTR vs modulus

Table 4 Produce packaging[a]

Property	POP 2	POP 3	EVA (12 % VA)	EVA (18 % VA)	ULDPE 1	ULDPE 2
Comonomer	Octene	Octene	VA	VA	Octene	Octene
Melt Index (dg/min)	1.6	1.0	0.32	0.8	1.0	0.8
Density (g/cm^3)	0.8965	0.902	–	–	0.912	0.905
OTR (cm^2 mil/100 in^3 day atm)	1459	1110	948	1050	857	1186
WVTR (g mil/100 in^2 day atm)	2	1.5	4.7	6.2	1.2	1.4
Puncture (ft-lb/in^3)	270	348	180	222	310	304
Haze (%)	1.3	1.3	3	1.5	7.2	6
Heat-seal Initiation (°C)	73	81	86	78	96	86
Ultimate heat-seal strength, (lb/in)	11.3	11.4	9.9	8.1	12.1	10.4
Ultimate hot-tack strength (N/in)	11.2	14.6	1.8	2.6	11.1	8.6

[a] Heat-seal tests were conducted on 1.0 mil nylon/1.0 mil tie-layer / 1.5 mil sealant blown coextrusions, unless noted otherwise. Physical property tests were conducted on 2 mil monolayer blown films.

comparative summary of the key performance properties of several polymer options for produce packaging. The most likely candidates for the high-end applications are the POPs and EVAs, since they offer the highest OTRs. The LLDPE–ULDPE family is limited in its maximum OTR level due to Food and Drug Administration (FDA) regulations on extractables. Since LLDPE–ULDPE resins have a heterogeneous comonomer distribution, the extractables at a given density are higher than those for a homogeneous POP. EVA copolymers can provide high OTR but, as shown in Figure 6, they have a lower modulus than that of the POP at the same OTR and a higher water vapor transmission rate (WVTR, Table 4), which can result in dried, shriveled produce. EVA copolymers provide excellent seal initiation temperatures, but they have relatively low hot-tack strengths, making them unsuitable for use in large VFFS packages. Also, as shown in Section 3.1, they can contribute more off-taste and odor to the package than the other comparative materials [12].

POPs provide a combination of high OTRs and low WVTRs, excellent optics, high puncture resistance and outstanding heat-seal and hot-tack performance (Table 4); however, they typically do not have the modulus necessary for use in VFFS packaging. As a result, laminations of oriented polypropylene (OPP) with a sealant such as a POP or EVA copolymer have been used, but they are limited by the maximum OTR that can be achieved. Coextrusions of a thin layer of a higher modulus, more heat-resistant resin with a POP can provide higher OTRs than the aforementioned laminations and are commonly used for the packaging of higher respiration rate produce applications [12].

3.3 ORIENTED SHRINK-FILMS

Oriented shrink-films can be differentiated from conventional hot blown shrink-films by the process in which they are fabricated. The conventional blown shrink-film method involves extrusion of the polymer through an annular die at a certain draw-down ratio and a certain blow-up ratio and then cooling the bubble via ambient or chilled air. Orientation of the film takes place in a completely molten state and is primarily in the machine direction, owing to the draw-down ratio, and secondarily in the cross direction, as a result of the blow-up ratio. In contrast, the oriented shrink-film process can be described generically by the following steps:

extrusion–quenching–reheating–biaxial stretching–cooling

The reheating step involves temperatures just above the glass transition temperature in the case of amorphous polymers or below the peak melting temperature in the case of semicrystalline polymers.

There are two basic methods to produce biaxially oriented films. The first method is the tenter frame process and the second method is often referred to as the 'double-bubble' or 'trapped-bubble' process. More in-depth descriptions of these processes can be found elsewhere [13]. Orientation by these processes takes place in a semi-molten state, i.e. at a temperature below the melting-point of the polymer. As a result, a significantly higher degree of orientation is obtained compared with hot blown shrink-films. These orientation processes produce high-tensile, high-modulus films with high shrink and shrink tension, in addition to excellent clarity [13]. Biaxially oriented films made via the 'double-bubble' or 'trapped-bubble' process are used to produce packages for very large, sub-primal cuts of meat after slaughter. The packages are often referred to as barrier bags. The sub-primal cuts of meat are inserted into a barrier bag that is generally sealed at one end, the air is evacuated and then the bag is heat sealed. The package is then heated, usually via hot water, to obtain shrinkage of the film around the meat [14]. Typical barrier bag structures range from three to five layers and minimally contain an abuse/shrink layer, an oxygen barrier layer such as PVDC or EVOH and a sealant layer. In addition, adhesive layers may be used to tie the structure together.

The use of LLDPEs and ULDPEs alone or as blends with EVA copolymers in both the abuse and sealant layers has been reported [14]. The broad comonomer distributions and melting ranges of LLDPEs and ULDPEs create a sufficiently broad orientation window for the double-bubble process. Furthermore, the excellent toughness of these linear polymers provides the high level of abuse resistance needed to prevent punctures from the bones in subprimal meats and to sustain impact abuse during the distribution process. ULDPE resins offer improvements in shrinkage over higher density LLDPE resins, owing to their reduced crystallinity. Further improvements in shrinkage can be obtained using POPs. POPs can offer low temperature shrinkage improvements over ULDPEs at the same density because of their lower melting-points. Figure 7 illustrates the improved shrinkage at 90 °C for

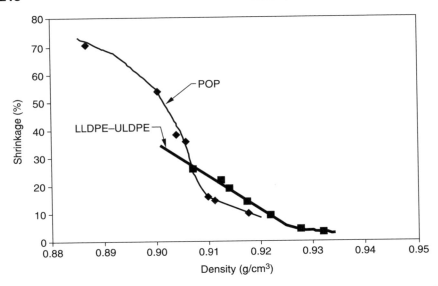

Figure 7 Hot water shrinkage of films oriented on T.M. Long stretcher under isothermal conditions at a 4.5 × 4.5 draw ratio at 5 in/s and 90 °C

POPs at a density below about 0.908 g/cm^3 [15]. In addition it is assumed that the improved sealability and abuse resistance properties for POPs in conventional blown films also translate to improvements in oriented shrink-films.

As mentioned earlier, blends of EVA with ULDPE or LLDPE have been used in barrier bag structures. A particular disadvantage of these blends in retail applications is the poor optics that can result from these blends. As shown in Table 5, the POP–LLDPE blend provides significantly better optics in an oriented shrink-film than the EVA–LLDPE blend. Both the POP and EVA have comparable melting-points and crystallinity. The poor optics of the EVA–LLDPE blend are believed to be due to the immiscibility of these two polymers [16]. The improvement in optics of POP–LLDPE blends versus EVA–LLDPE blends is another benefit that POPs can bring to oriented shrink-film applications.

Table 5 Optics of oriented films[a]

	EVA (12 % VA)–LLDPE 1[b]	POP 3–LLDPE 1
Blend ratio	30:70	30:70
Haze (%)	7.6	0.5

[a] Films oriented on T.M. Long stretcher under isothermal conditions at a 4.5 × 4.5 draw ratio at 5 in/s.
[b] LLDPE 1 is an ethylene–octene copolymer having a melt index of 1.0 °C/min and a density of 0.920 g/cm^3.

3.4 HIGH-PERFORMANCE FILMS [17]

This broad category includes bag-in-box packaging, ice bags, frozen poultry, frozen produce, stand-up coffee and dry soup pouches, and also vertical form, fill and seal (VFFS) packaging of flowable liquids. The films in this category are generally simple monolayer or coextruded polyolefin structures, where the polyolefin provides all or most of the key functional attributes necessary for the package. These films can also be laminated to other structures to obtain additional functionality, such as increased barrier or very high modulus. LLDPE, ULDPE and EVA copolymers (4–12 % VA) and blend combinations thereof are typical materials used in these applications.

3.4.1 Lower Modulus Blown Films

For applications requiring the stiffness necessary for VFFS-type applications, e.g. ice bags, individually quick frozen (IQF) poultry and frozen produce, LLDPE can satisfy the modulus requirements. However, packaging line speeds are then limited by the high sealing temperature. ULDPE and EVA copolymers provide low-temperature sealability, but have low modulus and higher blocking tendencies that can result in poor convertibility in bag-and pouch-making processes. Furthermore, EVA copolymers typically lack the mechanical strength of a linear PE. As a result, the performance requirements for these applications were often satisfied by blends or coextrusions of the aforementioned materials.

Prior to the introduction of CGC technology and molecular design, a polyolefin resin that provided low-temperature sealability, high abuse resistance and good convertibility, while maintaining the processability of traditional LLDPE resins, was not available.

As illustrated in Figure 3, the relationship between modulus and sealability has been redefined with the introduction of molecularly designed EPEs. At a seal initiation temperature comparable to those for ULDPE or EVA (9 % VA), significantly higher stiffness can be obtained. The data in Table 6 illustrate the overall physical performance of an enhanced polyethylene (EPE 1) relative to a traditional LLDPE, ULDPE and a gas-phase, homogeneous, linear SSC polymer. EPE 1 not only provides an excellent combination of abuse resistance (low temperature and room temperature), convertibility and sealability, but also provides the excellent processability of LLDPEs. One feature that distinguishes enhanced polyethylene, produced via INSITE technology, from homogeneous, linear SSC polymers is the ability to provide equal to or better processability than LLDPE, via controlled LCB and MWD. This unique combination of properties in one polymer can allow film converters to consolidate resins by reducing the number of layers in a structure or replacing a blend with a single resin that provides all of the functional attributes of their film.

Table 6 High-performance, lower modulus blown films

Property	EPE 1	LLDPE 1	ULDPE 1	EVA (9 % VA)	Homogeneous, linear SSC polymer
Comonomer	Octene	Octene	Octene	Vinyl acetate	Hexene
Resin properties					
Melt index, (dg/min)	1.2	1.0	1.0	2.0	1.0
Density (g/cm^3)	0.917	0.920	0.912	N/A	0.918
Film abuse resistance[a]					
Dart impact (type B) (g)	>850	250	760	250	>850
Impact (−20°C) (g)	574	<332	<332	<332	470
Elmendorf tear MD (type A) (g)	768	656	972	102	467
Film convertibility[a]					
2 % secant modulus MD (MPa)	175	193	124	99	182
Block force[b] (g)	Low	Low	Medium/high	Medium/high	Low
Film sealability[c]					
Heat-seal initiation at 1 lb/in, (°C)	90	107	98	91	96
Film esthetics					
Haze (%)	9.5	11	7.5	4.1	12–15
Processability[a]					
Output (kg/h)	54	54	54	N/A	49[d]
Melt temperature (°C)	232	232	232	N/A	238
Amps	80	79	79	N/A	92
Head pressure (bar)	399	347	414	N/A	432

[a] 50.8 μm films were processed on a Gloucester blown film line, 64 mm extruder, 152 mm die, 1.8 mm die gap, 2.5:1 blow-up ratio (BUR). Except where noted, film properties were performed according to standard ASTM procedures.

[b] Low block = 0–50 g, medium block = 50–100 g, high block => 100 g.

[c] Heat-seal testing was performed on 89 μm coextrusions consisting of nylon-6–6, 6-tie-layer–sealant (28.5:28.5:43 layer ratio). Heat-seal initiation is defined as the temperature at which 4.4 N/25 mm force is achieved.

[d] The homogeneous, linear SSC polymers could not be fabricated at 120 lb/h owing to excessive temperatures, amps and head pressures on the equipment.

3.4.2 Higher Modulus Blown Films

The limiting factors for downgauging high-performance films tend to be the loss of convertibility and toughness at thinner gauges. Tough films of $0.920 \, \text{g/cm}^3$ density LLDPE tend to become 'soft and stretchy' at thinner gauges, while higher modulus HDPE films lack the tear and puncture resistance needed in many thin-film applications. Poorer convertibility of downgauged structures is a result of the reduced dimensional stability, which comes from a tendency to deform, stretch or split in VFFS and bag making equipment. A typical structure for high modulus applications such as stand-up pouches includes a coextrusion of a sealant and a medium density polyethylene (MDPE) or HDPE layer, often laminated to a reverse printed film. Efforts to toughen the structure by adding an abuse-resistant blend resin or an additional layer can reduce the stiffness of the film, requiring thicker overall films to maintain the same level of dimensional stability.

As illustrated in Figure 2, the traditional LLDPE structure–property relationship between modulus and impact strength has been broken with these new EPEs made by INSITE technology. The impact strength of EPE 2 (Table 7) at a $0.939 \, \text{g/cm}^3$ density is remarkable, considering that only high molecular weight HDPE and high-stalk blown-film extrusion have been able to deliver this kind of performance combination in the past. The molecular design of EPE 2 is well suited to the stand-up pouch application, allowing the resin converter potentially to combine the stiffness- and abuse-resistant layers into one, with a potential reduction in overall film thickness. In addition, EPE 2 uses molecular design to provide excellent shear

Table 7 High-performance, higher modulus blown films

Property	EPE 2	MDPE
Comonomer	Octene	Octene
Resin properties		
Melt index, (dg/min)	0.84	1.0
Density (g/cm^3)	0.939	0.935
I_{10}/I_2	9.2	7.7
Film abuse resistance[a]		
Dart impact, B (g)	148	<100
Elmendorf tear MD (g)	148	86
Film convertibility[a]		
2 % secant modulus MD (MPa)	484	344
Film processability[b]		
Output (kg/h)	54.4	54.4
Melt temperature (°C)	232	232
Amps	76	85
Head pressure (bar)	390	448

[a] Films were tested according to standard ASTM procedures.
[b] 50.8 µm films were processed on a Gloucester blown film line, 64 mm extruder, 152 mm die, 1.8 mm die gap, 2.5:1 BUR.

thinning behavior, good bubble stability and reduced amps and pressures to meet the rigors of high output blown film extrusion (Table 7).

4   INDUSTRIAL FILM PACKAGING APPLICATIONS

4.1   BATCH INCLUSION BAGS [18]

4.1.1   Description and Performance Requirements

Batch inclusion packages are used to contain materials such as carbon black, titanium dioxide powder, elastomers (rubbers) and polystyrene pellets. Other materials packaged in batch inclusion bags include clay, silicates, calcium carbonate, organic dyes, color pigments and zinc oxide. A major function of the batch inclusion package is to minimize dust health hazards and possible fire hazards, in addition to waste, since the entire package and its contents are processed during use of the contents.

Performance requirements for batch inclusion bags include compatibility with the packaged material, low melting temperature, a narrow softening range and extrudability. Unvulcanized rubber, for example, can be packaged in a film made from an EVA copolymer containing an antiblock agent. However, polyethylene film is not suitable for overwrapping bales of rubber, because when the bale (and film wrapper) is charged into a suitable mixer, the polyethylene film is not sufficiently dispersed and causes defects in the end product.

4.1.2   Plastomer Performance

Polyolefin elastomers and plastomers are useful in packaging various articles for use in batch inclusion applications. Traditional Ziegler–Natta PEs have a high-density fraction with high-temperature melting peaks, even for lower density products. However, POPs and POEs have lower peak melting-points and narrower melting ranges, which enable the polymer to melt quickly upon reaching the appropriate temperature. The melting range of a polymer may be measured by the difference between the peak melting point and the softening point, shown in Table 8.

Table 8   Melting and softening point temperatures for POP and Z-N polymers

Polymer type	Melt index (g/10 min)	Density (g/cm³)	Peak m.p. (°C)	Vicat softening (°C)	Melting range (°C)
POP 3	1.0	0.902	96.3	88.3	8
POP 1	1.0	0.909	103	96.8	6.2
ULDPE 1	1.0	0.912	121	94	27
LLDPE 1	1.0	0.920	121	101	20

4.2 BLOWN AND CAST STRETCH FILMS

4.2.1 General Description

Blown and cast stretch films are polyolefin films that are used to protect and unitize manufactured goods or items during transportation or storage. Typically the unitization of goods takes place on a pallet. For load palletization operations, film is stretched and wrapped tightly around articles on a pallet. The film adheres to itself to create a secure, unitized package. The stretch film market can be segmented into several categories: (i) hand-held stretch films, having the lowest stretch requirements (<100 %), (ii) conventional or roller stretch films which are stretched about 150 %, (iii) power pre-stretch films, requiring the highest stretch, typically 200–250 % but as high as 300 %, and (iv) specialty stretch films for silage, bundling and paper reel wraps. Polymers produced via SSC are primarily targeted at the higher performing stretch films in the power pre-stretch and specialty segments. For stretch films, the key performance requirements are the ability to stretch to high levels without film breakage, adequate load holding forces to maintain a secure load, on-pallet puncture resistance and sometimes good optical properties. Sufficient cling to self-adhere the film to itself after wrapping a pallet is also important.

4.2.2 Plastomers as Cling Layers [19]

Typically, one-sided stretch cling films are multilayer stretch films, with one outer layer providing cling, another outer layer with lower tack or cling and a core layer (or layers) for stretching, load holding and puncture performance. Materials used for traditional one-sided cling layers have included EVA, ethylene–methacrylic acid copolymers (EMA) and LLDPE–ULDPE resins that are blended with a 'tackifier' compound such as polybutylene.

Low-crystallinity POEs have shown utility as non-migratory cling layers in one-sided cling films. Table 9 compares the cling performance of one-sided cling stretch films formulated with a typical EMA copolymer with a POP as the cling layer. The cast film is tested under two conditions: stretched at 200 % elongation and unstretched. The narrow MWD of POPs also reduces the build-up of low-MW or

Table 9 One-sided cling layer comparison[a]

Film	Cling layer	Center layer	Obverse layer	Stretched cling[b]	Unstretched cling
A	EMA	LLDPE	PP	47 g	123 g
B	POP 4[c]	LLDPE	PP	122 g	220 g

[a] Three-layer coextruded 0.8 mil film with polypropylene as the obverse non-cling layer and LLDPE as the core layer.
[b] 200 % film stretching.
[c] POP with 0.870 g/cm^3 density.

migratory material on cast film dies. The similar monomer chemistry of POPs to LLDPE provides good polymer compatibility for recycling purposes.

4.2.3 Enhanced PE (EPE) for High-performance Stretch Films

Bullard [20] and Cook [21] reported that one of the key benefits provided by 'metallocene-catalyzed' [20] or 'metallocene-based' [21] polyethylenes for stretch film is improved puncture resistance or dart impact strength. As illustrated in Tables 10 and 11, the enhanced polyethylenes made with the INSITE technology for blown and cast stretch films exhibit significantly higher on-pallet puncture strength at moderate pre-stretch levels and higher dart impact strengths than comparable Z-N catalyzed LLDPEs. This benefit can be directed toward the production of a higher performing film for applications requiring outstanding puncture resistance, or it can be used to facilitate the downgauging of films to reduce raw material costs. Another benefit of EPE 4 over LLDPE 2 is the ability to provide higher levels of on-pallet stretch without sacrificing impact strength (Table 11). With increased extensibility, the end-users could realize more wrapped pallets per roll of film and hence lower their costs.

One particular drawback of 'metallocene-catalyzed' [20] or 'metallocene-based' [21] polyethylenes is the poorer processability vs comparable melt index (MI) LLDPEs, as measured by higher amps and pressures. This is attributed to the reduced shear sensitivity of narrow MWD polymers [20, 21]. Owing to molecular design capabilities, the enhanced polyethylene resins exhibit comparable processability to their LLDPE counterparts (Tables 10 and 11).

Table 10 Blown stretch films

Property	LLDPE 1	EPE 3
Comonomer	Octene	Octene
Melt index (dg/min)	1.0	0.85
Density (g/cm^3)	0.920	0.920
Film properties		
Film gauge (mil)	0.8	0.8
MD modulus (MPa)	200	220
Dart impact (g)	145	>850
MD Elmendorf tear (g)	323	360
Haze (%)	10	13
Extrusion parameters		
Output Rate (lb/h)	80	80
Temperature ($^\circ$F)	445	445
Amps	65	70

Table 11 Cast stretch films

Property	LLDPE 2	EPE 4	EPE 5
Comonomer	Octene	Octene	Octene
Melt index (dg/min)	2.3	4.0	3.5
Density (g/cm^3)	0.917	0.917	0.915
Film properties			
Film gauge (mil)	0.8	0.8	0.8
Dart impact, method B (g)	110	300	630
On-pallet puncture^a (lb)	8	10	11
On-pallet ultimate stretch (%)	300	320	300
MD/CD Elmendorf tear (g)	300/500	320/460	450/650
Extrusion			
Output (lb/h)	299	290	292
Melt temperature (°F)	525	525	525
Amps	99	80	87

^{a} On-pallet puncture was measured at 250 % pre-stretch with a rectangular, angle-iron probe mounted on a frame. The frame is stretched wrapped at a specified dancer bar tension, F2. The F2 is increased until the film fails. The highest tension obtained without film failure is the on-pallet puncture.

4.3 HEAVY DUTY SHIPPING SACKS (HDSS)

4.3.1 Description and Performance Requirements

An HDSS is typically a large bag, designed to carry approximately 50 lb of product. Some examples of packaged goods include chemicals, plastics, fertilizers, lawn and garden products, insulation, pet food, salt and feed seed. Bags can be 2.5–7 mil thick, with 2.5–4 mil being most common.

Performance requirements of specific HDSSs vary, but must address items such as fabrication ease, often on narrow die gaps, bag performance on the packaging line, bag performance during shipping and storage and economics. The bag must open easily, fill easily, contain product, re-open and empty effectively. Typically, tear strength, dart impact, film stiffness and extrusion output rate are critical properties for all HDSS market segments. Depending on the market segment, other properties may also be important, e.g. hot tack and heat seal, resistance to creep, printability and acceptable COF.

4.3.2 Enhanced PE (EPE) Performance

Molecularly designed polymers provide unique combinations of film properties unmatched by traditional Z-N catalyzed LLDPE resins. These properties provide the means to meet the performance requirements of HDSS.

As discussed in previous chapters, an increased concentration of tie-chain molecules provides improved dart impact and puncture strength at equal stiffness; this allows converters either to downgauge their bags or to supply an improved

Table 12   Resin and blown film properties for HDSS

Property	LLDPE 3	EPE 6
Melt index (dg/min)	1.0	0.85
Density (g/cm^3)	0.926	0.925
Comonomer	Octene	Octene
Blown film, 3 mil		
Maximum output (lb/h/in die circumference)	13.3	13.3
Horsepower at maximum output	37.2	36.5
Dart impact (g)	410	610
MD Elmendorf tear (g)	740	725
MD 1 % modulus (psi)	52000	53400
Ultimate hot tack strength, (N/in)	3.9	4.6

performing product to the market. The presence of controlled amounts of LCB play a role in reducing or eliminating melt fracture of films made from narrow (40 mil, for example) blown film die gaps. The improved toughness properties provided by EPE resins allow a converter the opportunity to produce either a stiffer bag at equal strength, or a stronger bag at equal stiffness.

Blown film properties of a solution process, Z-N LLDPE are compared with those of a molecularly designed EPE in Table 12.

4.4   LOW-DENSITY LOOK-ALIKE RESINS WITH EXCELLENT PROCESSIBILITY [22]

4.4.1   General Description

The term 'low-density look-alikes' or 'high-processability polymers' (HPP) refers to those polymers that exhibit the exceptional processability performance of low-density polyethylene (LDPE) resins, but are made using either solution-, slurry- or gas-phase manufacturing processes. The advantages of producing a high-processability polymer using the solution-, slurry- or gas-phase processes instead of building a new, high-pressure LDPE plant are numerous. A greater range of product capabilities and performance, plus lower capital costs per pound of polymer, are the main advantages over building high-pressure PE plants.

The unique processability of LDPE has made this a difficult polymer to substitute. Large amounts of LCB and a broad MWD provide shear thinning and melt strength properties unmatched by traditional LLDPE resins. Non-Newtonian shear thinning provides the high-shear, low-melt viscosity for good extruder processability and the low-shear, high-melt viscosity for superior melt strength and blown film bubble stability.

4.4.2 Performance Requirements

The performance requirements vary depending upon the application, but include elements of (i) the polymer 'extrudability' (high-shear rheology) and melt strength (low-shear rheology), (ii) mechanical properties of the fabricated article and (iii) optical properties of the fabricated article. Processing performance is often measured in terms of the bubble stability, polymer output rate (lb/h) and extruder performance (pressure, melt temperature and motor amps). Mechanical strength of the fabricated article may be measured by tensile strength, resistance to tear, resistance to puncture, etc., while optics of fabricated articles are measured by clarity, haze and gloss.

4.4.3 Product Attributes

Adjusting the molecular architecture to have a controlled amount of LCB and a designed MWD will improve the resin melt strength and, therefore, its processability. These elements are critical to the successful design of an easy processability resin that exhibits both extrusion performance and bubble stability similar to those of LDPE, and retains the key mechanical properties required by the applications. Using CGC technology, a controlled amount of LCB can be incorporated in the resin, giving improved extruder processability and melt elasticity without sacrificing toughness.

4.4.4 Performance Data and Comparisons

HPP With LDPE–LLDPE Blend Performance (Table 13)

Many applications derive advantages by blending small amounts (< 25 %) of LDPE into an LLDPE resin to improve the polymer processability and film optics of the 100 % LLDPE resins. The performance depends upon the LDPE concentration in the blend, but can be described as (i) mechanical properties slightly less than the LLDPE resin, (ii) extruder processability and bubble stability better than LLDPE resins and (iii) optics similar to LLDPE resins.

Although the Elmendorf tear values are lower with the HPP resin than the LLDPE, an increase in dart impact is obtained, owing to the presence of tie-chains. The combination of LCB and a broader MWD enables the HPP resin to process with a lower extrusion melt temperature, pressure and motor load than the LLDPE resin. This combination also gives a higher low-shear viscosity that enables the HPP resin to provide a stable bubble configuration, compared with the LLDPE resin. The broader MWD is also responsible for slightly higher film haze.

HPP With Optical-grade LDPE Performance (Table 14)

Performance requirements for optical-grade LDPE film applications, such as clarity liner and bakery film, include (i) extruder processability and bubble stability similar

Table 13 HPP with LDPE–LLDPE blend performance[a]

Property	LDPE 1	HPP 1	LLDPE 4
Melt index (g/10 min)	2.0	1.9	2.3
I_{10}/I_2	—	8.3	7.7
Density (g/cm^3)	0.922	0.922	0.917
Comonomer	Homopolymer	Octene	Octene
Mechanical properties			
MD Elmendorf tear (g)	250	609	846
CD Elmendorf tear (g)	295	881	925
Dart impact (g)	110	284	246
Extrudability			
Output (lb/h)	120	120	120
Melt temperature, (°C)	450	450	455
Extruder pressure (psi)	1700	2810	3130
Motor amps	40	54	61
Bubble stability	Stable	Stable	Unstable
Optics			
Haze (%)	6.4	25.6	20.7

[a] 2.0 mil films produced from a 2.5 in extruder with a single-flight screw at a 2.5 blow-up ratio (BUR) and 70 mil die gap.

Table 14 HPP with optical-grade LDPE performance[a]

Property	LDPE 2	HPP 2	LLDPE 1
Melt index (g/10 min)	1.9	1.6	1.00
I_{10}/I_2	—	13	8.0
Density (g/cm^3)	0.922	0.923	0.920
Comonomer	Homopolymer	Octene	Octene
Mechanical properties			
MD Elmendorf tear (g)	414	287	691
CD Elmendorf tear (g)	310	729	819
Dart impact (g)	103	172	236
Extrudability			
Melt temperature (°C)	379	390	462
Extruder pressure (psi)	3390	3950	5040
Motor amps	47	59	69
Optics			
Haze (%)	5.6	4.6	12

[a] 2.0 mil films produced from a 2.5 in extruder with a single-flight screw at a 2.5 BUR and 70 mil die gap.

to those of LDPE resins, (ii) optics similar to those of clarity-grade LDPE and (iii) mechanical properties better than those of LDPE polymers.

The HPP resin has similar processability and optics to the LDPE resin, with a different balance in Elmendorf tear values and higher dart impact strength. Fabrication conditions play a critical role in determining the desired film orientation and, therefore, the balance of properties obtained from a given resin.

Table 15 HPP with non-optical-grade LDPE performance[a]

Property	LDPE 2	HPP 3	LLDPE 1
Melt index (g/10 min)	1.9	2.2	1.00
I_{10}/I_2	–	13	8.0
Density (g/cm^3)	0.922	0.917	0.920
Comonomer	Homopolymer	Octene	Octene
Mechanical properties			
MD Elmendorf tear (g)	438	685	705
CD Elmendorf tear (g)	440	1136	835
Dart impact (g)	110	188	245
Extrudability			
Melt temperature (°C)	445	451	450
Extruder pressure (psi)	1430	2040	4710
Motor amp	33	43	76
Optics			
Haze (%)	8.5	26	14

[a] 2 mil films produced from a 2.5 in extruder with a single-flight screw at a 2.5 BUR and 70 mil die gap.

HPP with non-optical-grade LDPE performance (Table 15)

Some LDPE film applications, such as mulch films and construction films, do not require good optical properties. Performance requirements for such applications include (i) extrudability and bubble stability similar to those of LDPE, (ii) improved mechanical properties and (iii) optics similar to those of LLDPE.

5 COATING APPLICATIONS

5.1 DESCRIPTION AND MARKETS

Extrusion coating is the application of a thin polymer film on to a flat substrate of paper, paperboard, foil, fabric or film [e.g. OPP, poly(ethylene terephthalate), (PET)] It is an economical method of adding performance properties to the substrate. For example, the coating may provide moisture barrier, mechanical strength, heat sealability, product resistance, esthetics or adhesion to another substrate. LDPE, LLDPE, EVA, EAA (ethylene–acrylic acid copolymers) and ionomers are among the traditional polyolefin resins commonly used in extrusion coating processes.

LDPE and/or LLDPE are often coated on to paper and paperboard to provide sealing, abuse resistance and liquid resistance for drink cups, drink boxes, milk cartons and ice cream containers. LLDPE provides mechanical and heat-seal strength performance advantages over LDPE, but its relatively lower melt strength increases neck-in, draw resonance and horsepower requirements during the extrusion coating process. Dry soups, drink boxes, coffee brick packs and metallized balloons are examples of extrusion coating applications where the mechanical properties of LLDPE are useful.

Ionomers and EVA copolymers (9–18 % VA) are used in flexible applications to package such things as cheese, dry mixes, flavored teas and snack foods. These high-performance sealants are used when seal performance is a critical performance requirement. Hammond and Potts described specific examples of flexible structures that are useful for aseptic packaging, laminate tubes, pet and moist food packaging, condiment packaging, snack and dry food packaging, skin packaging and meat packaging [23]. They also described the role played by various polymers in meeting the functional requirements of different packaging applications.

5.2 PLASTOMER PERFORMANCE [24]

Kelley *et al.* [24] compared the performances of a POP, an EVA (15 % VA) copolymer, an LDPE, an LLDPE and an ionomer in PET structures, foil structures and paperboard structures. The monolayer coating data in Table 16 show that neck-in is much higher for POP 5 than the LDPE or ionomer and marginally higher than the LLDPE and the EVA copolymer. Less energy is required for extrusion of the POP than LLDPE, as demonstrated by 8 % lower amps; however, the POP processed with higher amps than the LDPE, EVA copolymer and ionomer resins. The drawdown characteristics of the POP match the excellent drawdown of the LLDPE and are superior to those of the other resins. As an example of mechanical properties, the Elmendorf tear of the POP-coated paper is superior to those of all materials compared.

The POP has excellent adhesion (>400 g) to the treated high-energy OPP and to the primed PET films, but only moderate adhesion to foil. If higher foil adhesion is

Table 16 Extrusion coating performance comparison

Property	POP 5	LDPE 4	LLDPE 5	EVA (15 % VA)	Zn ionomer 2
Melt index (g/10 min)	7.5	8.0	5.5	8.0	5.0
Density	0.902	0.916	0.923	0.940	0.940
Melt temperature (°F)	600	600	600	450	550
Amps	138	94	150	121	126
Neck-in (in)[a]	5.3	1.6	4.5	4.6	1.8
Minimum thickness (mil)	<0.25	0.50	<0.25	0.35	0.63
MD Elmendorf tear (g)[b]	373	133	271	152	134
Adhesion to IG OPP (g)[c]	335	123	77	36	45
Adhesion to foil (g)[d]	241	45	159	54	368
Adhesion to primed PET (g)[e]	636	268	822	41	318

[a] 1 mil monolayer coating at 440 ft/min.
[b] 1 mil on 50 lb kraft paper.
[c] Mobil BICOR industrial grade OPP, one side high energy surface treat on PP core, 70 gauge.
[d] Paper/poly/foil structure.
[e] Hoechst Diafoil Hostaphan 2DEF, 48 gauge.

required, the POP can be coextruded with an ethylene–acrylic acid copolymer (EAA) tie-layer that will provide excellent foil adhesion. Kelley *et al.* showed that adhesion to primed PET and foil increases with increasing melt temperature for LDPE, LLDPE and POP [24].

On 50 lb kraft paper (Figure 8) the POP's heat-seal strength is twice that of the LDPE and even greater than that of the LLDPE. The POP heat-seal initiation temperature is 20 °C lower than those of the LDPE and LLDPE. The heat-seal strengths of the ionomer and the plastomer are similar. The EVA, which was coated at 450 °F, has the lowest heat-seal strength and does not show an advantage in heat-seal initiation.

Figure 9 shows the heat-seal initiation data for primed PET films. The POP has comparable seal initiation temperatures to the high-performance EVA (15 % VA) copolymer and ionomer, and heat-seal strength that is superior to those of all resins compared. Figure 10 shows that POPs ultimate hot-tack strength is much better than that of the LDPE and similar to that of the LLDPE, but at an approximately 20 °C lower temperature. The hot-tack window of the POP is narrower than that of the ionomer sealant.

Plastomers may be used alone or in blends to replace LDPE, LLDPE, EVA or ionomer in extrusion-coated structures. The most notable benefit of a plastomer over LDPE or LLDPE is lower heat-seal and hot-tack initiation temperatures. Key advantages over EVA copolymers include mechanical properties and hot-tack

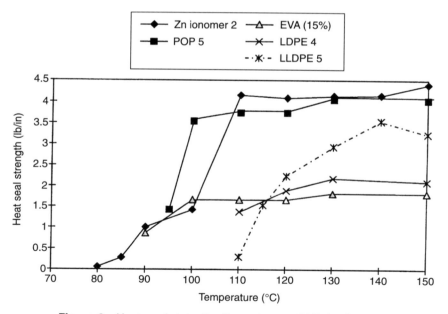

Figure 8 Heat-seal data (1 mil coating on 50 lb kraft paper)

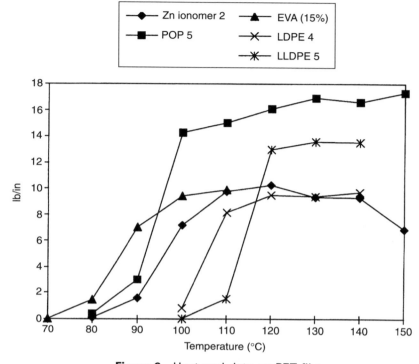

Figure 9   Heat-seal data on PET film

strength. The plastomer can provide comparable package performance to an ionomer at lower cost. Additionally, Potts and Pope have described how converters can use extrusion coating-grade plastomers to eliminate costly laminating steps [25].

6   MOLDED THIN-WALL CONTAINERS [26]

6.1   PERFORMANCE REQUIREMENTS AND TRADITIONAL HDPE

Three key performance requirements for HDPE resins for injection-molded, thin-wall containers are (a) adequate modulus for container stiffness and stacking, (b) ease of processing for thin-wall parts and multicavity molds and (c) impact resistance. Over the last several years, molding equipment advances, combined with market pressure to make lighter weight containers and reduce machine cycle times, have driven the thin-wall industry towards easier processing materials. As a result, converters have increasingly favored resins with higher melt indices and broader MWD. Unfortunately, increasing melt index and/or broadening MWD can result in an undesirable decrease in impact strength.

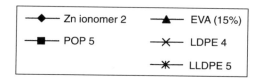

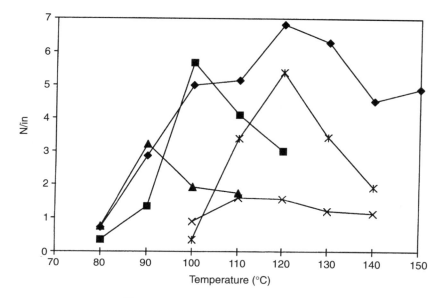

Figure 10 Hot-tack data on PET film

HDPEs 1–4 in Table 17 show the performance available with Z-N technology. The data in this table show the trade-offs between processing ease (and aggressive mold fill) and toughness as measured by a drop test. The broader MWD resins, HDPE 1 and HDPE 4, have the best processability, but they had the lowest drop testing results. Comparing HDPE 1 with HDPE 4, one can see the same trade-off between impact and processing as a result of changing melt index at a similar M_w/M_n. The slightly lower density of HDPE 4 reduces some of the sacrifice in impact resistance that would normally accompany higher melt index and broader MWD; while HDPE 4 has excellent processability and acceptable impact, it has lower stiffness than the other three HDPEs.

6.2   IMPACT MODIFICATION VIA POP BLENDING

Tie-chain concentration in conventional HDPEs is very low. One way to improve the mechanical performance would be to increase the tie-chain concentration via

Table 17   Properties of HDPE container resins

Resin	Melt index (g/10 min)	Density	M_w/M_n (GPC)	Drop[a] test (m)	Topload[b] (N)	Process index[c] (kP)	Aggressive mold fill?[d]
Ziegler–Natta HDPE							
HDPE 1	60	0.952	3.8	0.8	560	0.25	Yes
HDPE 2	65	0.952	3.2	1.1	565	0.37	No
HDPE 3	80	0.952	2.6	1.9	694	0.30	No
HDPE 4	90	0.950	3.7	0.9	534	0.19	Yes
15 % POP in Z-N HDPE 'A'							
Blend	55	0.945	N/A	1.5	467	0.29	Yes
Molecularly designed HDPE							
EPE 7	60	0.950	2.8	2.5	498	0.25	Yes

[a] A measure of impact strength. Sample containers were filled with ice–water and dropped at room temperature from a variable height platform using the Bruceton staircase method. Result is the height at which 50 % of the samples failed.
[b] A measure of stacking strength. An Instron 4500 tensiometer was used in compressive mode at a rate of 1.27 cm/min to determine the force required to buckle each container as load was applied to the top of the container.
[c] A high-shear viscosity measurement to indicate processability. The apparent viscosity measured at a constant stress of 2.15×10^6 dyn/cm^{2}X on a gas extrusion rheometer.
[d] Did the resin fill a thin 444 μm container mold using 95 % machine injection velocity (0.25 s fill time) at 288 °C melt temperature?

blending a plastomer produced with SSC. In fact, plastomers have proven to be excellent impact modifiers for injection-molded parts [27]. This is also illustrated by the blend resin in Table 17. The blend of 15 % POP with HDPE 1 shows significant improvements in impact resistance while maintaining excellent processability. However, the overall density was lowered to 0.945, which resulted in a lower top load strength.

6.3   MOLECULAR DESIGN

High-density EPE (EPE 7) is molecularly designed to increase tie-chain concentration while maintaining modulus. It also balances overall MW with a moderate MWD to provide good processing, but avoid excessive orientation, which can detract from the impact strength of the finished part. Referring back to Figure 4, the Dynatup impact of EPE 7 (at 60 lb) is compared with a family of Z-N resins designed for excellent processability; this shows the excellent combination of processability and impact strength that can be achieved with molecular design and CGC technology.

A resin with greater impact strength can be used to produce thinner containers for reduced cost, yet still provide the required impact strength. The increased impact strength could also be used in existing applications that require the impact of lower melt index resins, but which would benefit from improved processability.

7 SUMMARY

Enhanced polyethylenes and polyolefin plastomers provide excellent utility in numerous packaging applications, ranging from retail meat packages to industrial stretch films. These resins may be fabricated into packaging films or containers via blown and cast film methods, biaxial oriented film processes, extrusion coating, lamination and injection molding. In packaging applications the resins based on CGC technology have successfully replaced and/or upgraded EVA, ionomer, LDPE, LLDPE and ULDPE.

8 ACKNOWLEDGMENTS

The authors gratefully acknowledge the contributions of the following colleagues to this chapter: Sharon Baker, Staci DeKunder, Brian Faulkner, Dave Kelley, Laura Mergenhagen, Rajen Patel, Tim Pope, Mike Potts, David Ramsey, Deb Walker, Alan Whetten, and Jeff Wooster.

9 REFERENCES

1. Swogger, K. W. and Lancaster, G. M., Novel molecular structure opens up new applications for INSITE[TM] based polymers, in *Proceedings of SPO Conference*, Houston, TX, September 1993.
2. Whiteman, N. F., deGroot, J. A., Mergenhagen, L. K. and Stewart, K. B., Constrained Geometry Catalyst Technology Benefits High Performance Packaging Resins in *Proceedings of RETEC Conference*, Society of Plastics Engineers, South Texas Section, Houston, TX, February, 575–90 1995.
3. Koch, K. C. and Van Volkenburgh, W. R., Enhancing Polyethylene Performance with INSITE Technology and Molecular Design, in *Proceedings of ANTEC Conference*, The Society of Plastics Engineers, Indianapolis, IN, 1879–82, May 1996.
4. Kale, L. T., Iaccino, T. L. and Bow, K. E., Enhanced polyethylene resins in cable jacketing applications, in *Proceedings of ANTEC Conference*, The Society of Plastics Engineers Indianapolis, IN, 1647–51, May 1996.
5. Lancaster, G. M., Daen, J., Orozco, C. and Moody, J., Global product and application development utilizing INSITE[TM] technology in *Proceedings of METCON Conference*, Catalyst Consultants Inc. Houston, TX, May 1994.
6. DeKunder, S. A., Achieving higher sealant performance at lower overall cost using AFFINITY[TM] polyolefin plastomers, in *Proceedings of Future Pak '97 Conference*, George O. Schroeder Associates, Chicago, IL, 1647–51, October 1997.
7. Simpson, M. F. and Presa, J. L., Seal through contamination: a comparison of AFFINITY[TM] plastomers produced via INSITE[TM] technology with ethylene/acrylic acid copolymers, ionomers, and ULDPE, in *Proceedings of Future Pak '96 Conference*, George O. Schroeder Associates, Chicago, IL, 1–18 Session IV Processing, Converting and Testing, November 1996.

8. DeKunder, S. A. and Simpson, M. F., Sealant choices for barrier packaging applications, in *Proceedings of SME '97, Barrier Technology for The Food Packaging Industry*, Society of Manufacturing Engineers, Atlanta, GA, June 1997.
9. Mergenhagen, L. K. and Pope, T. J., Polyolefin plastomer-based peelable seal formulations, in *TAPPI Polymers, Laminations, and Coatings Conference*, Boston, MA, TAPPI Press, 63–71, September 1996.
10. Young, G. L., Designing packages for fresh cut produce, in *TAPPI Polymers, Laminations, and Coatings Conference*, Chicago, IL, TAPPI Press, 383–86, August 1995.
11. Wooster, J. J., Step-by-step procedures for extending the shelf-life of fresh cut produce, in *Proceedings of MAPack '95 Conference*, Anaheim, CA, Institute of Packaging Professionals, October 1995.
12. Wooster, J. J., New resins for fresh-cut produce, in *TAPPI Polymers, Laminations, and Coatings Conference*, Toronto, TAPPI Press, 567–75, August, 1997.
13. Benning, C. J., *Plastic Films for Packaging*, Technomic Publishing, Lancaster, PA, USA, 1983.
14. Newsome, D. L., to American National Can, *US Pat.*, 4 457 960 (1984).
15. Patel, R. M., Langohr, M. F., Walton, K. L. and McKinney, O. K., to Dow Chemical Company, *PCT Int. Appl.*, WO 97/30111 (1997).
16. Patel, R. M., Saavedra, P., deGroot, J. A., Hinton, C. and Guerra, R., Comparison of EVA and polyolefin plastomer as a blend component in various film applications, in *Proceedings of ANTEC '97 Conference*, Toronto, Society of Plastics Engineers, 1950–53, May 1997.
17. deGroot, J. A., New polyethylenes via INSITE™ technology, in *Proceedings of SPO '96 Conference*, Houston, TX, Shotland Business Research Inc., 269–80, September 1996.
18. Falla, D. J. and Walker, D., to Dow Chemical Company, *US Pat.*, 5 525 659 (1996).
19. Ramsey, D. B. and Stewart, K. B., to Dow Chemical Company, *US Pat.*, 5,789,029 & 5,840,430.
20. Bullard, E., Tenneco Packaging's experience with metallocene catalyzed polyethylenes, in *Proceedings of SPO '97 Conference*, Houston, TX, Schotland Business Research, Inc., September 1997, pp. 273–287.
21. Cook, J., The role of metallocene based polyethylenes in stretch film, in *Proceedings of SPO '96 Conference*, Houston, TX, Schotland Business Reseach, Inc., September 1996.
22. Lastovica, J. E., A new family of LLDPEs with LDPE processability using INSITE™ technology, in *Proceedings of SPO '97 Conference*, Houston, TX, Schotland Business Reseach, Inc., September 1997.
23. Hammond, F. M. and Potts, M. W., *TAPPI J.*, TAPPI Press, 81–88, Vol. 72 (1989).
24. Kelley, D. C., Baker, S. L. and Potts, M. W., DPT-1450: a polyolefin plastomer for extrusion coating applications, in *TAPPI Polymers, Laminations, and Coatings Conference*, Chicago, IL, TAPPI Press, 103–22. September 1994.
25. Potts, M. W. and Pope, T. J., Extrusion coating vs. film lamination: how AFFINITY™ polyolefin plastomers will offer more, in *Proceedings of Future Pak '96 Conference*, Chicago, IL, George O. Schroeder Associates, Inc., Session 2, Resin and Film Technologies, November 1996.
26. Faulkner, B. J. and Whetten, A. R., New HDPE resins for thin wall injection molding: designed for the future, in *Proceedings of ANTEC Conference*, Toronto, The Society of Plastics Engineers, 2107–11, May 1997.
27. Silvis, C. H., Murray, D. J., Fiske, T. R., Betso, S. R. and Turley, R. R., to Dow Chemical Company, *US Pat.*, 5 688 866 (1997).

11

Measurement, Mathematical Modelling and Control of Distribution of Molecular Weight, Chemical Composition and Long-chain Branching of Polyolefins Made with Metallocene Catalysts

JOÃO B.P. SOARES AND ALEXANDER PENLIDIS
University of Waterloo, Waterloo, Ontario, Canada

1   INTRODUCTION

The major impact of metallocene catalysts on the polyolefin industry is related to the fact that these catalysts can produce polymers with controlled microstructure, as measured by their distributions of molecular weight (MWD), chemical composition (CCD) and long-chain branching (LCBD). The heterogeneous Ziegler–Natta catalysts currently used in several industrial processes for olefin polymerization typically produce polyolefins with non-uniform microstructure. This is generally attributed to the presence of several active site types on heterogenous Ziegler–Natta catalysts, while most metallocene catalysts show single-site-type behavior [1–3].

Several polyolefin applications benefit from microstructural uniformity. Polymers with narrow MWD have greater dimensional stability, higher impact resistance,

Metallocene-based Polyolefins Edited by J. Scheirs and W. Kaminsky
© 2000 John Wiley & Sons Ltd

greater toughness at low temperatures and higher resistance to environmental stress cracking. However, several other applications such as extrusion molding and thermoforming require broad and sometimes multimodal microstructural distributions that can not be obtained with single-site metallocene catalysts. Polyolefins made with metallocene catalysts would not, in principle, be adequate for these types of applications. In order to produce such polyolefins with metallocene catalysts one has three different options: (i) variation of operation conditions in a single reactor during polymerization, such as temperature and concentrations of comonomer and hydrogen, (ii) use of reactors in series where each reactor is maintained at different polymerization conditions, commonly referred to as tandem reactor technology, and (iii) simultaneous use of more than one type of metallocene catalyst during polymerization.

Variation of polymerization conditions in a single reactor has been tried in laboratory-scale reactors [4] but it is unlikely that this technique can be effectively extended to industrial reactors. Non-steady-state operation of polymerization reactors is subject to quality control problems and appreciable production of off-specification material. It can hardly be considered an option for the manufacture of commodity polyolefins.

Tandem reactor technologies are already commonly used with heterogeneous Ziegler–Natta catalysts for the production of impact copolymers and a variety of reactor blends [5–7].

Although tandem reactor technologies do permit fairly versatile control of polymer properties, they are subject to higher costs, since at least two reactors in series are necessary.

Finally, the selective combination of metallocene catalysts that produce polymers with different average molecular weights, chemical composition and long-chain branching can also be used to control the microstructure of polyolefins. If appropriate metallocene types and polymerization conditions can be found, only one reactor is necessary to produce a polymer with well controlled non-uniform microstructure.

At first sight, these polymerization procedures seem to contradict one of the main advantages of metallocene catalysis, i.e. the production of polymers with uniform molecular structure. For all three described polymerization conditions, the main objective is to produce polymers with non-uniform microstructure, apparently similar to the molecular structure obtained with conventional heterogeneous Ziegler–Natta catalysts. However, what must be realized is that the non-uniformity of polyolefins made with heterogeneous Ziegler–Natta catalysts arises from the intrinsic multiple-site-type nature of these catalysts, which is very difficult, if not impossible, to control with a high degree of confidence. On the other hand, the selective combination of different metallocene types under appropriate polymerization conditions opens the door to much improved control of the microstructure of these polyolefin chains.

2 MATHEMATICAL MODELS FOR THE MOLECULAR STRUCTURE OF POLYOLEFINS

Most instantaneous distributions of molecular structure of polymers made with single-site-type catalysts such as metallocenes can be described by simple mathematical expressions. The distributions that have been most commonly associated with polyolefins made with single-site-type metallocenes are Flory's most probable chain length distribution for homopolymers [8] and Stockmayer's bivariate distribution for copolymers [9]. Some other empirical distributions have also been used to describe the MWD of polymers made with metallocene catalysts [10]. In some cases, these empirical distributions might provide a more adequate fit of the data from a purely statistical point of view but they do not provide the fundamental insight on the polymerization mechanisms given by the theoretically sound distributions derived by Flory and Stockmayer.

2.1 MOLECULAR WEIGHT DISTRIBUTION

For linear chains, Flory's most probable distribution [8] can be used to describe the instantaneous MWD of polyolefins made with single-site-type catalysts:

$$w(r) = \tau^2 r \exp(-\tau r) \tag{1}$$

where $w(r)$ is the weight chain length distribution of chains of length r and τ is the ratio of overall transfer to propagation rates. Several soluble metallocene and Ziegler–Natta catalysts produce polyolefins with MWD that can be closely described by Flory's distribution [11,12]. For the case of homopolymers, this is generally considered to be solid evidence for the single-site-type nature of the catalyst.

Polymers that obey Flory's distribution have a number-average chain length, r_n, equal to $1/\tau$ and a polydispersity index (PDI) of 2. The ratio of transfer to propagation rates, τ, can be calculated using the following generic equation:

$$\tau = \frac{R_\beta}{R_p} + \frac{R_M}{R_p} + \frac{R_{CTA}}{R_p} + \frac{R_{Al}}{R_p} = \frac{k_\beta}{k_p[M]} + \frac{k_M}{k_p} + \frac{k_{CTA}[CTA]}{k_p[M]} + \frac{k_{Al}[Al]}{k_p[M]} \tag{2}$$

where $R_p, R_\beta, R_M, R_{CTA}$ and R_{Al} are the rates of propagation, β-hydride elimination, transfer to monomer, to chain transfer agent and to cocatalyst, respectively, and $k_p, k_\beta, k_M, k_{CTA}$ and k_{Al} are their equivalent rate constants, $[M]$ is monomer concentration, $[CTA]$ is the concentration of chain transfer agent and $[Al]$ is the concentration of cocatalyst. Methylaluminoxane (MAO) is the most frequent choice of cocatalyst with metallocene catalysts and hydrogen is commonly used as chain transfer agent.

For the case of multiple-site-type catalysts, it has been proposed that each site type produces polymer chains that instantaneously follows Flory's distribution [13–16]. A similar approach can be adopted if several metallocenes having single-site-type behavior are used simultaneously during polymerization. In this case, the instantaneous MWD for the whole polymer will be a weighted sum of individual Flory's distributions [17]:

$$\overline{w}(r) = \sum_i m_i w_i(r) = \sum_i m_i \tau_i^2 r \exp(-\tau_i r) \tag{3}$$

where i indicates metallocene type and m_i is the mass fraction of polymer produced with each metallocene type.

2.2 MOLECULAR WEIGHT AND CHEMICAL COMPOSITION DISTRIBUTION

Heterogeneous Ziegler–Natta catalysts produce copolymers of ethylene and α-olefins with very broad and generally multimodal CCD. Temperature-rising elution fractionation (TREF) has been a technique fundamental to the elucidation of this phenomenon [18,19]. A newly developed technique called crystallization analysis fractionation (CRYSTAF), which has a shorter analysis time than TREF, is also contributing to the elucidation of the CCD of these polyolefins [20,21].

It is now generally accepted that different site types produce copolymer chains with different average chemical compositions and comonomer sequence lengths [22, 23].

Analogously to the approach adopted for describing the MWD of linear homopolymers, the instantaneous MWD and CCD of linear binary copolymers made on single-site-type catalysts can be conveniently described by Stockmayer's distribution [9]:

$$w(r, y)\mathrm{d}r\,\mathrm{d}y = r\tau^2 \exp(-r\tau)\mathrm{d}r \frac{1}{\sqrt{2\pi\beta/r}} \exp\left(-\frac{y^2 r}{2\beta}\right)\mathrm{d}y \tag{4}$$

where $w(r, y)$ is the bivariate weight distribution of chains of length r and chemical composition y, y is the deviation from the average molar fraction of comonomer type 1 incorporated in the copolymer, $\overline{F}_1$, and

$$\beta = \overline{F}_1(1 - \overline{F}_1)K \tag{5}$$

$$K = [1 + 4\overline{F}_1(1 - \overline{F}_1)(r_1 r_2 - 1)]^{0.5} \tag{6}$$

where r_1 and r_2 are reactivity ratios for copolymerization.

For copolymers made with multiple-site-type catalysts or combinations of metallocenes, the overall instantaneous CCD and MWD have been proposed to be a weighted sum of the individual site type distributions [3,23]:

$$\overline{w}(r, y) = \sum_i m_i w_i(r, y) \tag{7}$$

This model has been used to describe qualitatively analytical TREF curves of linear low-density polyethylene (LLDPE) and isotactic polypropylene [23]. Recently, this model has been applied to describe CRYSTAF profiles of a family of poly(ethylene–co-1-octene) resins of varying molecular weight averages made with a single-site-type catalyst [24].

2.3.   MOLECULAR WEIGHT, CHEMICAL COMPOSITION AND LONG-CHAIN BRANCHING DISTRIBUTIONS

Constrained geometry metallocene catalysts (CGC) can be used to produce poly-ethylene and ethylene–α-olefin copolymers with long-chain branches (LCB) [25, 26]. The mechanism of LCB formation in these catalysts seems to be terminal branching. In terminal branching, polymer chains containing a terminal vinyl unsaturation (formed, for instance, via β-hydride elimination) are able to propagate again, adding LCBs to the living polymer chains.

These polyolefins have remarkable new properties since they combine the good mechanical properties of polyolefins with narrow MWD with the easy processability of branched polyolefins. It has been demonstrated [27,28] that the instantaneous molecular weight and chemical composition distribution for polymer populations of different number of long chain branches per chain, n, can be simply described with an extension of Stockmayer's distribution:

$$w(r, y, n) = \frac{1}{(2n + 1)!} r^{2n+1} \tau^{2n+2} \exp(-r\tau)\mathrm{d}r \times \frac{1}{\sqrt{2\pi\beta/r}} \exp\left(-\frac{y^2 r}{2\beta}\right)\mathrm{d}y \quad (8)$$

Equation (8) permits a detailed description of the polymer populations (i.e. linear chains, chains with 1 LCB/chain, 2 LCB/chain, etc.) produced with single-site-type catalysts that form LCBs via a terminal branching mechanism.

There is no complete analytical solution for these distributions when two or more metallocene catalysts are present simultaneously during polymerization. One of the main difficulties in this case is that the chain formed in one of the catalysts can, in principle, form LCBs with chains growing on the other catalyst, therefore altering the distributions of long-chain branching, molecular weight and composition.

A partial analytical solution for the MWD of polyolefins made with a binary metallocene system has been proposed by Zhu and Li [29]. In their model, one of the catalysts produces polymers with LCB while the other catalyst only produces linear polymer chains. However, their solution only applies to the MWD of the branched polymer chains. The MWD of the linear chains that were not incorporated as LCBs is not considered and therefore the MWD of the overall polymer cannot be obtained.

Beigzadeh et al. [30,31] used population balances and the method of moments to calculate the average molecular weight, chemical composition and long-chain branching for the case when more than one metallocene is present during poly-merization. Based on simulation results, recipes for maximizing the degree of long-chain branching for these binary systems have been proposed. Some of these results will be reviewed later.

3 CONTROL OF THE DISTRIBUTIONS OF MOLECULAR WEIGHT AND CHEMICAL COMPOSITION

3.1 SINGLE-SITE-TYPE CATALYSTS

The type of metallocene used for polymerization has a very large effect on the MWD of the polymer produced. Additionally, the MWD can also be influenced significantly by polymerization conditions such as temperature and concentrations of monomer, catalyst and cocatalyst. These effects have been studied in detail in the literature [32–38] and are the subject of several recent review articles [39–46]. In this section, only basic characteristics of these catalysts will be highlighted to illustrate different ways of controlling the microstructure of the polymer produced. For a thorough review of different catalyst systems, the reader is referred to the cited review articles.

3.1.1 Effect of Metallocene Type

Figure 1 and Table 1 compare the MWDs of polyethylene made in a slurry reactor with the different metallocene types: (**1**) Cp_2ZrCl_2/MAO; (**2**) Cp_2TiCl_2/MAO; (**3**) Cp_2HfCl_2/MAO; (**4**) $Et(Ind)_2ZrCl_2/MAO$; and (**5**) $Et(H_4Ind)_2ZrCl_2/MAO$ [38]. Although the MWDs differ significantly in average values, their PDIs range from 2.1 to 2.5, which is slightly higher than the value of 2.0 predicted with Flory's most probable distribution. Several researchers have reported PDIs of polyolefins made with metallocene catalysts in the range 2.0–4.0 [33,47,48]. Obtaining MWDs with PDIs slightly higher than the theoretical value of 2.0 is generally attributed to uncertain physical and chemical phenomena taking place during polymerization, e.g. mass transfer resistances and temperature oscillations [38]. This could also be related to peak broadening during MWD analysis via high-temperature size-exclusion chromatography (HT-SEC) [49], a phenomenon that is almost always neglected during HT-SEC analysis of polyolefins.

Direct quantitative comparisons with other published results have to be made carefully because differing polymerization conditions will necessarily lead to different catalyst behavior. Kaminsky [34] found that the molecular weights of polyethylene made with catalysts **1–5** at 30 °C and 2.5 bar ethylene pressure decrease in the same order as shown in Figure 1. Han et al. [4] observed that the Ti-based catalyst **2** produced polyethylene with higher molecular weight than that produced with the Zr-based catalyst **1**, as observed in Figure 1. However, unlike Figure 1, they found that catalysts **1** and **4** produced polyethylene of the same molecular weight. Similarly to Figure 1, Heiland and Kaminsky [50] observed that Hf-based catalysts

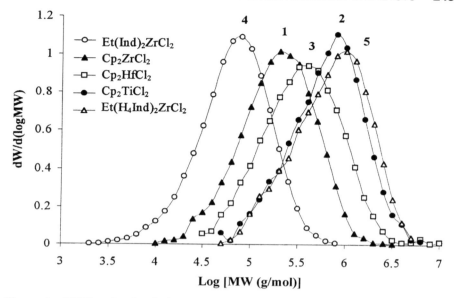

Figure 1 MWDs of polyethylene produced by metallocene/MAO catalysts **1–5**. See Table 1 for experimental conditions

produce polymers with higher molecular weight than do Zr-based catalysts, a phenomenon that has been attributed to the higher transition metal–carbon bond enthalpy for hafnium [51]. As expected, trends in the molecular weight of polypropylene may differ from those observed with polyethylene. Huang and Rempel [35] observed that the molecular weight of polypropylene decreases in the order **4** > **5** > **1**, where for polyethylene the molecular weight decreases in the order **5** > **1** > **4**, as shown in Figure 1.

Table 1 Results of polymerizations using catalysts **1–5**[a]

Catalyst	Concentration (μM)	$\overline{M}_w$ (g/mol)	PDI
Cp$_2$ZrCl$_2$ (**1**)	0.27	340 000	2.19
Cp$_2$TiCl$_2$ (**2**)	1.0	871 000	2.16
Cp$_2$HfCl$_2$ (**3**)	1.0	490 000	2.46
Et(Ind)$_2$ZrCl$_2$ (**4**)	1.0	104 400	2.19
Et(H$_4$Ind)$_2$ZrCl$_2$ (**5**)	1.0	966 000	2.24

[a] Other experimental conditions: $[Al]_{MAO} = 3.0$ mM, $T = 60\,°C$, $P_{ethylene} = 3.4$ bar, 400 rpm.

3.1.2   Effect of Polymerization Temperature

The effect of polymerization temperature on the MWD of polymer made with metallocene catalysts has been widely studied [36,43]. The average molecular weight decreases with increasing polymerization temperature, as illustrated in Figure 2. This has been attributed to higher activation energies for chain transfer reactions (notably β-hydride elimination) than for chain propagation. Note that, for single-site-type catalysts, PDI should not be influenced by polymerization temperature, contrary to what is observed with conventional heterogeneous Ziegler–Natta catalysts [1,52], where the activation energies for propagation and deactivation of the different site types might differ. This characteristic of metallocene catalysts permits good control of the average molecular weights by varying the polymerization temperature without significantly influencing the breadth of the MWD.

3.1.3   Effect of Catalyst and Cocatalyst Concentrations

Catalyst and cocatalyst concentrations have also been found to influence the average molecular weight of polymers made with metallocene catalysts. D'Agnillo *et al.* [38] applied a factorial experimental design to assess the influence of catalyst and cocatalyst concentration in the system Cp_2ZrCl_2/MAO for ethylene polymerization. They concluded that an increase in both catalyst or cocatalyst concentration would

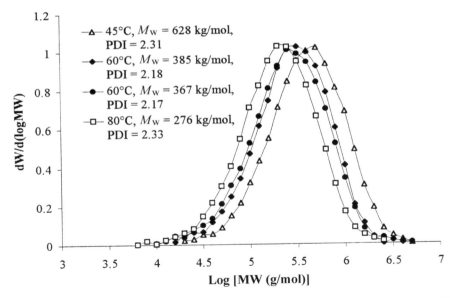

Figure 2   Effect of temperature on MWD in polymerizations with **1**/MAO. $T = 45$, 60 and 80 °C, with a replicate trial at $T = 60$ °C

lead to a decrease in weight-average molecular weight. Similar results for catalyst concentration were observed by other researchers [33,34,37]. Kaminsky [34] suggested that the decrease in molecular weight with increasing catalyst concentration is due to a bimetallic transfer mechanism and that the decrease in weight-average molecular weight with increasing cocatalyst concentration is caused by increased transfer to cocatalyst, which is consistent with the mechanism described in equation (2).

The substitution of MAO with trimethylaluminum (TMA) has been extensively investigated, since TMA is less expensive than MAO. Figure 3 illustrates the effect of changing the TMA/MAO ratio on the MWD of polyethylene made with Cp_2ZrCl_2/MAO/TMA [38]. It has been observed that molecular weight decreases with increasing TMA/MAO ratio, as also observed by other investigators [32,33]. These results also suggest that the MWD broadens with increasing [TMA]/[MAO], which has been tentatively explained as caused by the formation of two different active site types, i.e. Cp_2ZrCl_2/MAO and Cp_2ZrCl_2/TMA [38].

3.1.4 Effect of Hydrogen

The effect of hydrogen as a transfer agent during polymerization has been thoroughly investigated by several researchers. For most polymerization conditions with several metallocene catalysts, the molecular weight of the polymer decreases

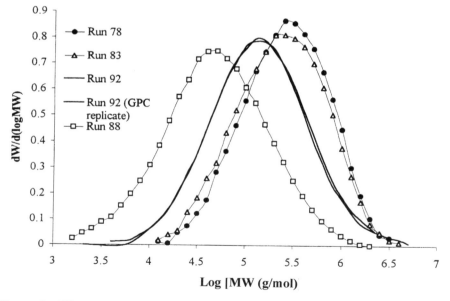

Figure 3   Effect of replacing MAO with TMA on MWD using **1**/MAO. See Table 2 for experimental conditions and molecular weights

Table 2 Effect of TMA in polymerizations with $1/MAO^a$

Run No.	[Zr] (μM)	$[Al]_{MAO}$ (mM)	$[Al]_{TMA}$ (mM)	Activity [kg PE/(mol Zr h)]	$\overline{M}_w$ (kg/mol)	$\overline{M}_w/\overline{M}_n$
78	0.30	3.0	0	77400	403	2.67
83	0.30	3.0	0.7	117700	371	2.96
92	0.30	0.6	2.40	9300	268	3.64
					263	3.80
88	0.30	0.3	2.70	2800	94	4.35

^{a} For all polymerizations, $Al_{(MAO+TMA)}/Zr = 10\,000$ (except run 83). $T = 60\,^\circ C$, $P_{ethylene} = 3.4$ bar.

significantly even with low hydrogen concentrations, as illustrated in Figure 4 [4,34, 38]. However, it has been shown [53,54] that under some polymerization conditions, certain metallocene catalysts can be insensitive to transfer to hydrogen. This has been successfully used to control the MWD of polyethylene made with a binary metallocene catalyst, as discussed in the next section.

3.1.5 Effect of Monomer Pressure

The effect of monomer pressure can be of great importance when controlling the MWD of polyolefins made with metallocene catalysts. Remembering equation (2), it

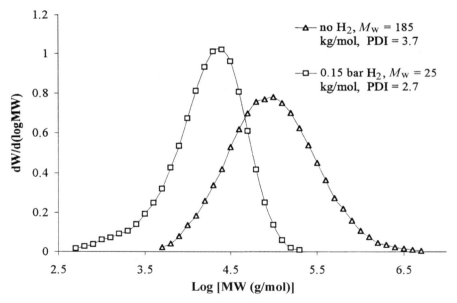

Figure 4 MWD of polyethylene produced with **1**/MAO in absence of hydrogen and with hydrogen partial pressure of 0.15 bar

is easy to note that if transfer to monomer is the controlling transfer mechanism, the MWD of the polymer produced will be unaffected by monomer pressure:

$$\tau = \frac{1}{r_n} = \frac{k_M}{k_p} \gg \frac{k_\beta}{k_p[M]} + \frac{k_{CTA}[CTA]}{k_p[M]} + \frac{k_{Al}[Al]}{k_p[M]} \qquad (9)$$

This behavior is clearly illustrated in Figure 5 for the polymerization of ethylene with Et(Ind)$_2$ZrCl$_2$/MAO [38]. Similar results have been observed by the same investigators for Cp$_2$ZrCl$_2$/MAO and by other researchers [55].

Conversely, when the chain termination mechanism is controlled by β-hydride elimination, transfer to cocatalyst or to chain transfer agent, the molecular weight should increase with increasing monomer pressure as predicted by

$$\tau = \frac{1}{r_n} = \frac{k_\beta}{k_p[M]} + \frac{k_{CTA}[CTA]}{k_p[M]} + \frac{k_{Al}[Al]}{k_p[M]} \gg \frac{k_M}{k_p} \qquad (10)$$

Figure 6 illustrates this behavior for ethylene polymerization catalyzed by Cp$_2$HfCl$_2$/MAO [38].

It is important to realize that, for single-site-type behavior, all these transfer mechanisms should not affect the PDI of the polymer produced.

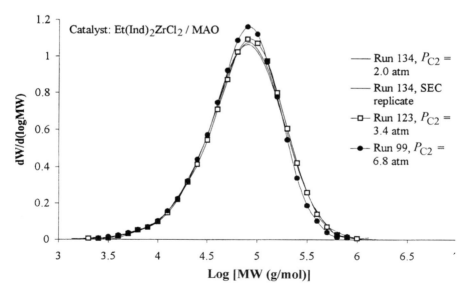

Figure 5   Effect of ethylene partial pressure on MWD of polyethylene made with 4/MAO (PC2: ethylene partial pressure)

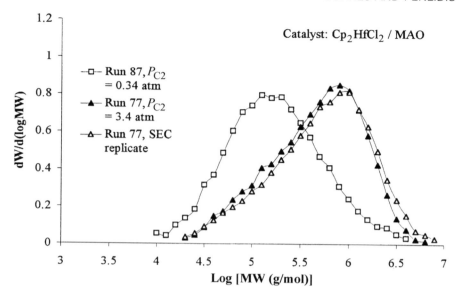

Figure 6   Effect of ethylene partial pressure on MWD of polyethylene made with 3/MAO (PC2: ethylene partial pressure)

3.1.6   Chemical Composition Distribution

The CCD of metallocene polyolefins has been less studied than the MWD. In contrast to the broad and multimodal CCD of olefin copolymers made with conventional heterogeneous Ziegler–Natta catalysts, the CCD of olefin copolymers made with metallocene catalysts is generally narrow and unimodal.

Soares *et al.* [24] analyzed a series of seven poly(ethylene–co-1-octene) samples made with a single-site-type catalyst by CRYSTAF. The samples had approximately the same average chemical compositions but different molecular weight averages. According to Stockmayer's bivariate distribution, owing to the statistical nature of polymerization, the CCD should become broader with decreasing molecular weight. This trend is clearly illustrated in Figure 7 for the seven copolymers analyzed.

3.2   MULTIPLE-SITE-TYPE CATALYSTS

3.2.1   Soluble Metallocenes

It has been claimed that through the careful combination of metallocene catalysts that individually produce polymer with known MWD and CCD, polyolefins with controlled MWD and CCD can be synthesized. This is the concept of tailor-made polyolefins.

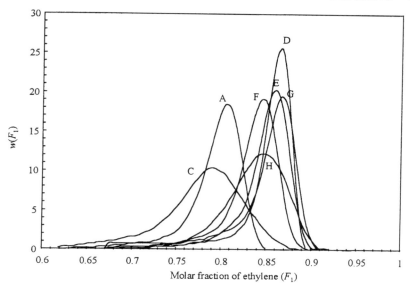

Figure 7   CCD distributions (measured by CRYSTAF) for poly(ethylene–co-1-octene) samples of varying number-average molecular weights made with a single-site-type catalyst. (A) 29 600; (C) 17 300; (D) 38 000; (E) 34 400; (F) 29 300; (G) 22 000; (H) 19 800

Heiland and Kaminsky [50] polymerized ethylene with combined soluble Et(Ind)$_2$ZrCl$_2$ and Et(Ind)$_2$HfCl$_2$. The polymer produced had a bimodal MWD. They concluded that each metallocene type produced polymer independently of each other. The relative activity of each metallocene over the range of polymerization temperatures (from -20 to $70\,°C$) changed significantly, which was reflected in the MWD of the polymer, as clearly indicated by equation (3). Lee *et al.* [56] observed similar trends for siloxane-bridged metallocenes.

Several patents have been issued for controlling the MWD and CCD of polyolefins made with combinations of different metallocenes or metallocenes and conventional Ziegler–Natta catalysts. These patents apply some of the principles described in the present chapter and have been reviewed by Soares and Hamielec [41]. For instance, Ewen [57] combined two soluble metallocene catalysts (Cp$_2$HfCl$_2$ and Cp$_2$ZrCl$_2$) and Welborn [58] developed a supported catalyst that incorporates one metallocene and one conventional Ziegler–Natta catalyst to produce polyethylene with a broad MWD.

When attempting to control the MWD via the combination of different metallocene types, attention should be paid to the possible effects of the catalyst components and polymerization conditions on the MWD. While catalyst type and polymerization temperature have a dominant effect on the MWD, control of MWD can still be exercised by manipulating other reactor operation conditions.

A detailed study of ethylene polymerization with soluble binary metallocene systems has been presented by D'Agnillo *et al.* [10]. Figure 8 illustrates how the MWD of polyethylene produced with a binary metallocene catalyst is equivalent to the superposition of their individual MWDs. This is strong evidence that the chemical nature of the active sites is not affected by the interaction with different site types during polymerization. In this way, the effect on MWD of combining different metallocene types is equivalent to blending polyethylene samples that were independently synthesized by each metallocene. The advantage of producing this *in situ* blend in the reactor using a binary catalyst system is simply that less energy is required for the process and a well dispersed polymer mixture is obtained. This is especially true if the polymers made by the two catalysts have a tendency to segregate in separate phases.

It is hardly surprising that the MWD of polyolefins made with a combination of metallocene catalysts would approximate the weighted average of the MWDs produced by the individual metallocenes, as postulated by equation (3). What is more interesting from an application perspective is how efficient this MWD control can be if one considers the experimental errors involved in the polymerization of

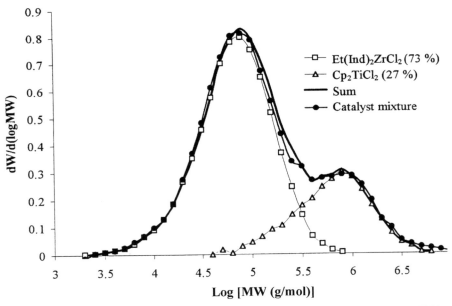

Figure 8 MWDs for polyethylene produced by Et(Ind)$_2$ZrCl$_2$, Cp$_2$TiCl$_2$, and a mixture of the two catalysts. Sum is the weighted sum of MWDs from individual catalysts. Experimental conditions: [MAO] = 3.0 mM, T = 60 °C, $P_{ethylene}$ = 3.4 bar, [Et(Ind)$_2$ZrCl$_2$] = 0.13μM, [Cp$_2$TiCl$_2$] = 0.50μM, [Et(Ind)$_2$ZrCl$_2$ + Cp$_2$TiCl$_2$] = 0.14μM +0.20μM

olefins with metallocene catalysts. When assessing the effectiveness of this MWD control, it is imperative to consider the influence of catalyst activity. Although for replicate polymerizations with individual metallocenes the MWD is very reproducible [59], the activity of soluble metallocene catalysts in semi-batch laboratory-scale slurry reactors can be highly variable [10]. It is therefore difficult to predict with a high degree of confidence how much polymer will be produced by each metallocene during polymerization. In other words, according to equation (3), although the τ_i values do not vary appreciably, the m_i values are subject to significant variation, which affects the MWD of the whole polymer. This is illustrated in Figure 9, where the MWD of polyethylene made in two replicate runs is shown. Although the underlying distributions for the two metallocenes are the same, it is clear that their relative amounts vary from trial 1 to trial 2 (i.e. the relative activity of the metallocenes varies between replicates), leading to significantly different MWDs. Consequently, when mixing soluble metallocene catalysts, the high variance in catalyst activity translates into a difficulty in predicting the resulting global MWD.

In their companion paper, D'Agnillo *et al.* [60] suggested a model to predict the MWD of polyethylene produced with a soluble binary metallocene system. Their model incorporates the variance in catalyst activity to obtain a level of confidence in the prediction of the final MWD as given by equation (3). Some of their main results are illustrated in Figures 10 to 12. In Figure 10, the experimentally measured MWD

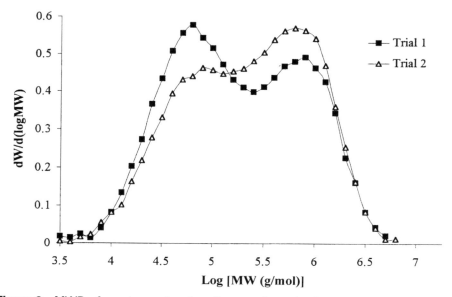

Figure 9 MWDs from two sets of replicate polymerizations with Et(Ind)$_2$ZrCl$_2$ and Cp$_2$TiCl$_2$. Experimental conditions: [MAO] = 3.0 mM, T = 60 °C, $P_{ethylene}$ = 3.4 bar, [Et(Ind)$_2$ZrCl$_2$ + Cp$_2$TiCl$_2$] = 0.13μM +1.01μM

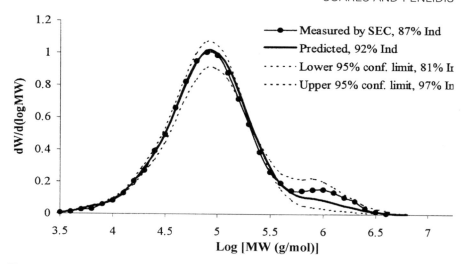

Figure 10 MWD and confidence limits predicted for run 13 (combining catalysts Ind = Et(Ind)$_2$ZrCl$_2$ and Tita = Cp$_2$TiCl$_2$) compared with that measured for run 13. Experimental conditions: [MAO] = 3.0 mM, T = 60 °C, $P_{ethylene}$ = 50 psig, [Et(Ind)$_2$ZrCl$_2$ + Cp$_2$TiCl$_2$] = 0.13 μM +0.25 μM

falls within the 95 % confidence bands, which indicates that satisfactory control was achieved during this polymerization run. In Figure 11, the experimentally measured MWD also falls within the 95 % confidence bands but in this case the bands exhibit a considerably wider variation. In Figure 12, however, the experimentally measured MWD falls outside the 95 % confidence bands, indicating that the MWD was not well controlled in this particular case.

Two main reasons were pointed out as main causes for improper MWD control: (i) the confidence bands are too wide, and (ii) polymerizations give results which lie outside the confidence bands more often than expected. The first case is a reflection of the variance in catalyst activity. The second case might indicate a shortcoming of the model or the occurrence of an unexpected event such as selective contamination of one metallocene type.

The methodology proposed is very powerful because it permits the prediction of the outcome of a given polymerization with a certain degree of confidence, reaching in this way the final objective of tailor-made polymerization. As mentioned, the breadth of the MWD confidence interval is a strong function of the variance in catalyst activity. Enhanced control of polymerization conditions could probably be used to obtain narrower confidence intervals. However, since the ultimate goal of tailor-made polymerization is to control end-use rheological and mechanical properties of the polymer produced, it should be interesting to evaluate the impact of this MWD variation on these macroscopic properties. If these end-use properties are not

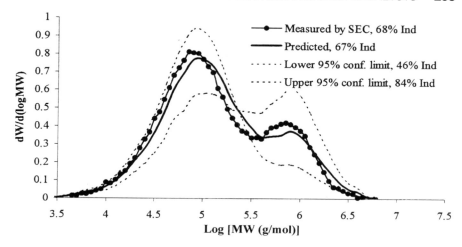

Figure 11 MWD and confidence limits predicted for run 9 (combining catalysts Ind = Et(Ind)$_2$ZrCl$_2$ and Tita = Cp$_2$TiCl$_2$) compared with that measured for run 9. Experimental conditions: [MAO] = 3.0 mM, $T = 60\,^{\circ}$C, $P_{ethylene} = 50$ psig, [Et(Ind)$_2$ZrCl$_2$ + Cp$_2$TiCl$_2$] = 0.06μM +0.60μM

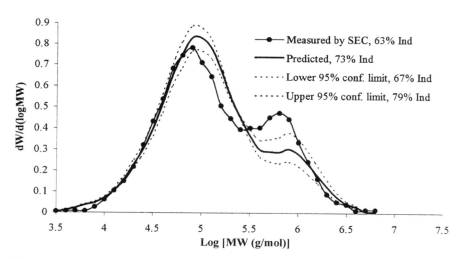

Figure 12 MWD and confidence limits predicted for run 10 (combining catalysts Ind = Et(Ind)$_2$ZrCl$_2$ and Tita = Cp$_2$TiCl$_2$) compared with that measured for run 10. Experimental conditions: [MAO] = 3.0 mM, $T = 60\,^{\circ}$C, $P_{ethylene} = 50$ psig, [Et(Ind)$_2$ZrCl$_2$ + Cp$_2$TiCl$_2$] = 0.07μM +0.50μM

severely affected by variations in MWD (i.e. by wide confidence bands), it might not be worth trying to tighten the control over MWD.

It is also likely that for the continuous polymerization systems usually employed in the commercial production of polyolefins, these difficulties can be easier to overcome. If MWD is regularly monitored, catalyst activity changes can be compensated for as part of the product quality control scheme of the reactor operation.

3.2.2 Supported Metallocenes

Supported metallocene catalysts are generally favored for the commercial production of polyolefins since most commercial processes were developed for heterogeneous Ziegler–Natta catalysts. It is likely that the same difficulties in tailoring MWD that occur with soluble catalysts will be present with supported systems. At the time of catalyst supporting, the relative amounts of each metallocene deposited on the support must be carefully adjusted to control the resulting MWD. A high variance in MWD can result if the relative activities of the different metallocenes are not properly controlled at this stage.

The control of MWD by combined metallocene catalysts in supported systems follows trends similar to those discussed earlier for soluble systems. However, for supported systems, it seems less practical to control MWD (and CCD, as required) by varying the relative amount of the different metallocenes on the support, since each MWD grade would require a new catalyst formulation. In this case, it seems more adequate to examine the effect of polymerization conditions on the MWD of the polymer produced using a given binary (or multicomponent) metallocene combination.

The effect of polymerization conditions on the MWD of polyethylene made with a binary metallocene-supported catalyst has been thoroughly investigated for several metallocene types [17,53,54].

Figure 13 shows that the MWD of polyethylene made with the combined catalyst $Et(Ind)_2ZrCl_2/Cp_2HfCl_2$ can be well described as a weighted superposition of two Flory most probable distributions. Under these operating conditions, the low and high molecular weight populations correspond to polymers made with $Et(Ind)_2ZrCl_2$ and Cp_2HfCl_2, respectively.

Figure 14 shows the effect of polymerization temperature on MWD. From the comparison of the peak areas, it is clear that the relative amount of polyethylene produced on Cp_2HfCl_2 sites decreases with increasing temperature. This is possibly related to a higher thermal deactivation rate of Cp_2HfCl_2 sites and provides a convenient way of varying the MWD of the polymer produced with this catalyst.

As already noted for soluble systems, the termination mechanism of $Et(Ind)_2ZrCl_2$ for ethylene partial pressures higher than 100 psi is apparently

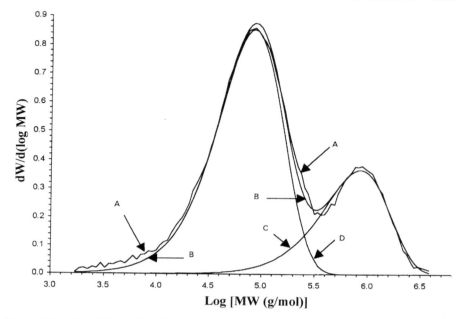

Figure 13 Fit of bimodal MWD with two Flory distributions: (A) experimental distribution measured by HT-SEC {40 °C, $P_{ethylene}$ = 120 psi, combined catalyst (Et[Ind]$_2$ZrCl$_2$/Cp$_2$HfCl$_2$ = 0.36 mol/mol)}; (B) superposition of C and D, where C and D = Flory's distributions for polyethylene produced by Cp$_2$HfCl$_2$ and Et[Ind]$_2$ZrCl$_2$, respectively (w_{Zr} = 0.70, W_{Hf} = 0.30, r^2 = 0.995)

dominated by transfer to monomer, whereas for Cp$_2$HfCl$_2$ the main transfer mechanism in absence of hydrogen is probably β-hydride elimination. This can be effectively used to control the MWD of polymer made with this combined catalyst, since with increasing ethylene pressure one should expect the molecular weight of polymer made with Et(Ind)$_2$ZrCl$_2$ to remain unaltered, whereas the molecular weight of polymer made with Cp$_2$HfCl$_2$ should increase. This is illustrated in Figure 15. Since the polymerization rate is first order with respect to monomer concentration, the relative areas under the peaks should not depend on polymerization pressure. For this particular case, it was possible to show that this variation was related to selective poisoning of the Cp$_2$HfCl$_2$ sites [54].

Surprisingly, it was found that the effects of ethylene and hydrogen partial pressures on the MWD of polyethylene made with Et(Ind)$_2$ZrCl$_2$ and Cp$_2$HfCl$_2$ are different. This can be used to one's benefit when controlling the MWD of the resulting polymer. When the ethylene pressure is $\geqslant$ 100 psi, hydrogen can be used to control the MWD of polyethylene produced with Et(Ind)$_2$ZrCl$_2$. It has been proposed that, under these polymerization conditions, the controlling transfer

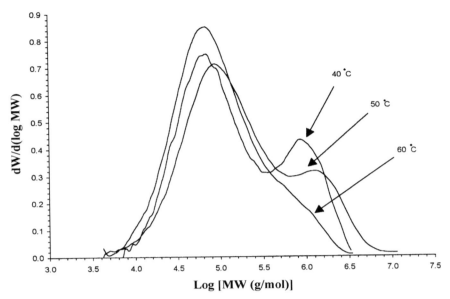

Figure 14 Effect of polymerization temperature on MWD: combined catalyst $(Et[Ind]_2ZrCl_2/Cp_2HfCl_2 = 0.36mol/mol)$, $P_{ethylene} = 50\,psi$

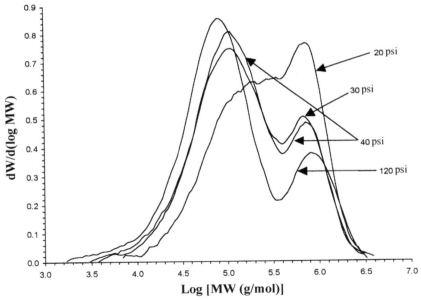

Figure 15 Effect of ethylene pressure during polymerization: combined catalyst $(Et[Ind]_2ZrCl_2/Cp_2HfCl_2 = 0.36\,mol/mol)$, $40\,°C$

mechanism for Et(Ind)$_2$ZrCl$_2$ is transfer to monomer whereas that for Cp$_2$HfCl$_2$ is transfer to hydrogen. According to equation (2),

$$\tau_{zr} = \frac{1}{r_n} = \frac{k_M}{k_p} \gg \frac{k_\beta}{k_p[M]} + \frac{k_{CTA}[CTA]}{k_p[M]} + \frac{k_{Al}[Al]}{k_p[M]} \tag{11}$$

and

$$\tau_{Hf} = \frac{1}{r_n} = \frac{k_{CTA}[CTA]}{k_p[M]} \gg \frac{k_M}{k_p} + \frac{k_\beta}{k_p[M]} + \frac{k_{Al}[Al]}{k_p[M]} \tag{12}$$

Figures 16 and 17 illustrate how the overall MWD can be controlled by simply varying hydrogen concentration during the polymerization. As shown in Figure 16, regardless of the presence of hydrogen, the MWD of polymer produced by the combined catalyst is the superposition of the MWDs of the individually produced polymers. In Figure 17 it is clearly illustrated how the MWD of the polymer population made on Cp$_2$HfCl$_2$ sites varies from high to low molecular weight, while the MWD of polymer made on Et(Ind)$_2$ZrCl$_2$ sites remains virtually unaltered. In fact, for high hydrogen concentrations, polyethylene made on Cp$_2$HfCl$_2$ sites has a lower molecular weight than that made on Et(Ind)$_2$ZrCl$_2$ sites. Table 3 summarizes the molecular weight averages for these polyethylenes. It is interesting to observe how the PDI initially decreases as the two molecular weight peaks overlap at increasing hydrogen concentration and then finally increases again as the peak corresponding to polyethylene made on Cp$_2$HfCl$_2$ sites moves towards the low molecular weight region. Interestingly, when lower hydrogen pressures are used, the main transfer mechanism for Et(Ind)$_2$ZrCl$_2$ sites becomes transfer to hydrogen [in good agreement with equation (2)], and increasing hydrogen pressures will cause the whole MWD to move to lower average values, as shown in Figure 18.

The chemical composition distribution for combined metallocene systems is largely unreported in the open literature. Kim et al. [61] presented some preliminary results on the control of CCD for poly(ethylene–co-hexene) made with different binary metallocene systems composed of Et(Ind)$_2$ZrCl$_2$, Cp$_2$HfCl$_2$ and a CGC catalyst. Some of the resins synthesized had a higher degree of comonomer incorporation on the high molecular weight end of the MWD (as opposed to the regular behaviour of heterogeneous Ziegler–Natta catalysts), a characteristic that is attractive for pipe applications [7].

The combination of metallocenes with different stereochemical control can also be used to produce new polymer products. In a very interesting study, Chien et al. [62] used binary mixtures of isospecific and aspecific metallocenes supported on SiO$_2$ to produce thermoplastic elastomers of polypropylene. The ratio of the two catalysts could be adjusted to permit the synthesis of polymers with properties varying from plastomers to elastomers.

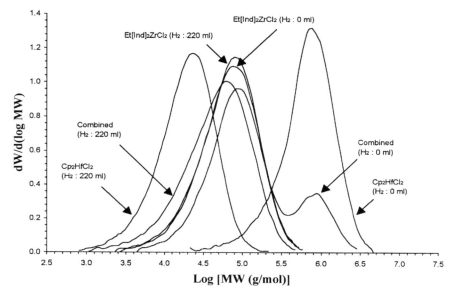

Figure 16 Effect of hydrogen concentration: 50 °C, $P_{ethylene}$ = 100 psi

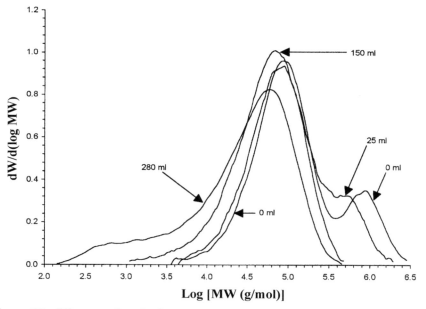

Figure 17 Effect of hydrogen concentration: combined catalyst (Et[Ind]$_2$ZrCl$_2$/Cp$_2$HfCl$_2$ = 0.36 mol/mol), 50 °C, $P_{ethylene}$ = 100 psi

Table 3 Effect of hydrogen concentration on average molecular weights of polymer produced by combined catalyst: $50\,°C$, $P_{ethylene} = 100\,psi$

Hydrogen (ml)	$\overline{M}_n$	$\overline{M}_p$	$\overline{M}_w$	PDI
0	64 000	84 900	264 400	4.13
25	54 900	87 800	183 700	3.35
150	27 600	65 300	78 900	2.86
220	5500	58 800	53 800	9.78

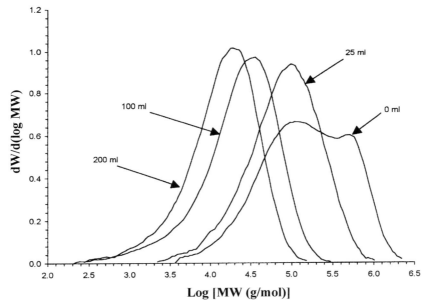

Figure 18 Effect of hydrogen concentration: combined catalyst $(Et[Ind]_2ZrCl_2/Cp_2HfCl_2 = 0.36\,mol/mol)$, $50\,°C$, $P_{ethylene} = 20\,psi$

3.3 TANDEM REACTOR TECHNOLOGIES

One way of controlling MWD and CCD of polyolefins is using tandem reactor technology [63,64]. In one application of this method, two polymerization reactors are used in series to produce polymers with bimodal MWD and/or CCD. In the first reactor the polymerization is conducted without hydrogen to produce high molecular weight polymer. The polymer is then transferred to the second reactor to be further polymerized in the presence of hydrogen, producing lower molecular weight chains. The polyethylene manufactured with this technology shows bimodal MWD, having the strength and stiffness of high-density polyethylene, while retaining the high-stress-crack resistance and processability of unimodal medium-density grades [7].

3.4   NON-STEADY-STATE POLYMERIZATION

Han *et al.* [4] attempted to control the MWD of polyethylene by combining soluble metallocenes, in addition to changing the temperature and hydrogen concentration during semi-batch polymerization, in a simulated tandem reactor operation. By selecting two catalysts in equimolar amounts from Cp_2ZrCl_2, Cp_2TiCl_2 and $Et(Ind)_2ZrCl_2$, they were able to achieve bimodality and overall broadening of the MWD. Similarly, Eskelinen and Seppälä [36] produced bimodal MWD polyethylene with Cp_2ZrCl_2/MAO by varying the temperature during polymerization.

4   CONTROL OF LONG-CHAIN BRANCHING

The most suitable single-site-type catalysts for LCB formation appear to be monocyclopentadienyl complexes of Hf, Ti and Zr. These catalysts are commonly called 'open site' or 'constrained geometry' catalysts [25,26,65]. The active center of these catalysts is based on Group IV transition metals that are covalently bonded to a monocyclopentadienyl ring and bridged with a heteroatom, forming a constrained cyclic structure with the transition metal center. This geometry allows the transition metal center to be more accessible (or 'open') to the addition of higher α-olefins. For instance, the reactivity ratio of 1-octene to ethylene with these catalysts can be 30 times higher than that of a typical $MgCl_2$-supported Ziegler–Natta catalyst [66]. Additionally, this active center structure seems to favor the addition of vinyl-terminated polymer macromolecules, allowing the production of polymer chains with LCBs [67].

The most likely LCB formation mechanism with these catalysts is terminal branching, a common mechanism in free-radical polymerization [12]. With metallocene catalysts, β-hydride elimination appears to be responsible for *in situ* formation of dead polymer chains with terminal vinyl unsaturation. These chains can be considered 'macromonomers', which upon coordination with the active center and insertion in the carbon–metal bond will generate an LCB [27]. Other reaction types, such as β-methyl elimination, also give the desired terminal vinyl group [68]. It is generally accepted that the most effective macromonomer for addition to the active center for the generation of a long trifunctional branch is the one with terminal vinyl unsaturation.

Shiono *et al.* [69] were also able to copolymerize ethylene with atactic polypropylene containing terminal double bonds using a CGC, $Et[IndH_4]_2ZrCl_2$, Cp_2ZrCl_2 and Cp_2TiCl_2, leading to the formation of polyethylene with atactic polypropylene LCBs. They found that the CGC was more active toward macromonomer incorporation, and that the other metallocene catalysts could also incorporate macromonomer in the order $Cp_2ZrCl_2 \ll Cp_2TiCl_2 \ll Et[IndH_4]_2ZrCl_2$.

Polymerization reactor engineering for the production of polyolefins with long-chain branches using metallocene catalysts has been discussed in a recent review [43]. Solution processes are favored for LCB synthesis for several reasons. Macro-monomers with terminal vinyl unsaturation are formed at higher rates at elevated temperatures owing to increased β-hydride elimination. These dissolved polymer chains are mobile and have relatively large self-diffusion coefficients. It can also be speculated that at the higher temperatures necessary for solution polymerization the steric barrier for the addition of a macromonomer to the active center is relatively less important than at lower temperatures. However, recent publications [70] seem to indicate that LCB can also be formed even in gas-phase reactors with supported catalysts. For supported metallocene catalysts, each polymer–catalyst particle can be considered a microreactor with high polymer and low monomer concentration. Terminal branching is maximized for these conditions [27], supposing that the unsaturated chain ends are mobile enough to approach close to the active sites and undergo polymerization.

For continuous operation at the steady state, the residence-time distribution plays a very significant role in LCB generation. Consider the two extremes of a plug flow tubular reactor (PFTR) with a narrow residence-time distribution, and of a continuous stirred-tank reactor (CSTR) with a very broad residence-time distribution. With the PFTR, the polymer concentration increases while the monomer concentration decreases as one moves along the tube, similarly to a batch reactor in time. High polymer concentrations are obtained near the exit of the PFTR, while much of the polymerizing mass in the reactor is at low polymer concentrations. With the CSTR, the entire reacting mass is at the same high polymer and low monomer concentration. Rates of LCB-forming reactions are higher at higher concentrations of polymer with terminal vinyl unsaturation, and rates of monomer consumption are lower at lower monomer concentrations, giving high levels of LCB per 1000 carbon atoms. It is perfectly clear, therefore, that the optimal reactor type and operation conditions for maximum LCB formation are the steady-state operation of a CSTR.

The degree of LCB of polyolefins can be controlled by variation of the polymerization conditions and reactor type, leading to polymers with a narrow MWD and a certain degree of LCB. According to the polymerization mechanisms proposed for these systems, the rate of LCB formation is proportional to the concentration of macromonomer in the polymerization reactor. Therefore, an increase in macromonomer concentration will accelerate the rate of LCB formation. One possible way to achieve this objective is by feeding macromonomer to the polymerization reactor. This approach is covered in some recent patents [71,72].

Some theoretical studies [30,31] have focused on the use of a combination of a CGC catalyst (called LCB catalyst) and a conventional metallocene catalyst that is assumed not to form LCBs to any appreciable degree (called linear catalyst). The linear catalyst is used as a source of macromonomer to form LCBs for the polymer made by the LCB catalyst, as illustrated in Figure 19. However, since not all polymer made with the linear catalyst will become incorporated as LCBs, this population must be considered when calculating the MWD and CCD of the whole polymer. An analytical solution to calculate these distributions has not yet been found. However, their averages can be calculated using population balances and the method of moments for both non-steady-state [30] and steady-state polymerizations [31].

Figure 20 shows the variation of LCB per 1000 carbon atoms (λ_n) as a function of LCB catalyst mole fraction at different reactor residence times for a simulated catalyst system. As expected, higher LCB levels are obtained at higher residence times, but by increasing the LCB catalyst mole fraction, λ_n passes through a maximum. This behavior can be related to the lower macromonomer concentration as the mole fraction of linear catalyst decreases. For these simulated results, the linear catalyst was assumed to have a higher β-hydride elimination rate constant than the LCB catalyst. Therefore, at higher mole fraction ratios of LCB catalyst, the rate of formation of macromonomer will be lower and consequently λ_n will decrease.

Figure 21 shows the variation of M_w as a function of LCB catalyst mole fraction ratio for different residence times. Since higher reactor residence times favor long-chain branching reactions (high polymer concentration) and since LCB formation increases molecular weight, higher M_w values will be obtained at higher residence times. This can explain the diverging behavior of the plots of M_w versus C_{LCB}/C_{tot}

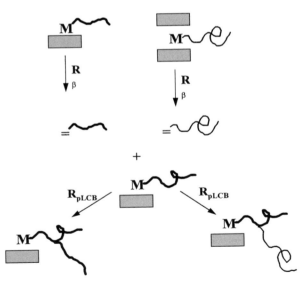

Figure 19 Schematic LCB formation for the binary system CGC (LCB catalyst) and conventional metallocene catalyst (linear catalyst). The rectangles represent the cyclopentadienyl rings and M indicates the transition metal

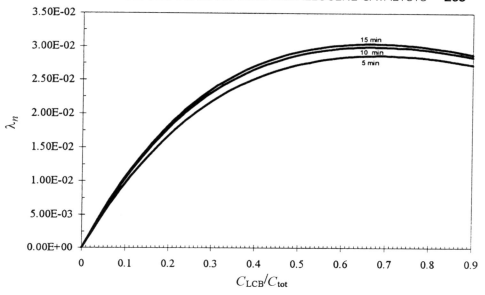

Figure 20 Variation in the number of branching points per 1000 carbon atoms (λ_n) versus mole fraction of LCB catalyst (in catalyst feed) for different reactor residence times of a copolymer made with a binary catalyst system comprised of a CGC and a linear metallocene catalyst (for simulation data, refer to Beigzadeh *et al.* [31])

for different reactor residence times, where C_{LCB} is the concentration of the LCB catalyst and C_{tot} is the total concentration of LCB and linear catalysts.

Figure 22 illustrates the variation of PDI as a function of LCB catalyst mole fraction for different reactor residence times. As shown, as the C_{LCB}/C_{tot} ratio increases, PDI passes through a maximum. This behavior can be attributed to two competing phenomena. First the β-hydride elimination rate constant assumed for the linear catalyst is higher than that for the LCB catalyst, while their propagation rate constants are assumed to be equal. Therefore, the molecular weights of chains made with the linear catalyst will be lower than those made used the LCB catalyst. This difference in molecular weights of polymer chains increases the PDI as the molar fraction of the LCB catalyst in the reactor increases. By increasing the molar fraction of the LCB catalyst after a certain C_{LCB}/C_{tot} ratio, this site type will be the major component in the combined catalyst system. As the fraction of the LCB catalyst increases, more uniform chains will be formed and PDI will decrease. Second, at different C_{LCB}/C_{tot} ratios, different LCB degrees will be obtained. By increasing the C_{LCB}/C_{tot} ratio, the concentration of LCB active sites increases but the rate of formation of macromonomers decreases (owing to the higher β-hydride elimination rate constant of the linear catalyst). Therefore, the plot of PDI as a function of

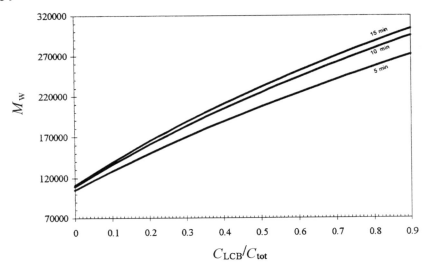

Figure 21 Variation in weight-average molecular weight (M_w) versus mole fraction of LCB catalyst (in catalyst feed) for different reactor residence times of a copolymer made with a binary catalyst system comprised of a CGC and a linear metallocene catalyst (for simulation data, refer to Beigzadeh *et al.* [31])

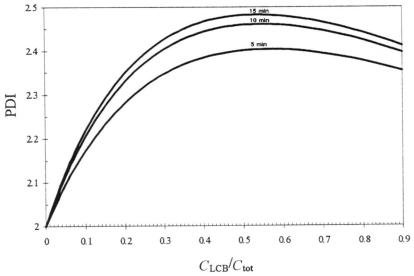

Figure 22 Variation in polydispersity index (PDI) versus mole fraction of LCB catalyst (in catalyst feed) for different reactor residence times of a copolymer made with a binary catalyst system comprised of a CGC and a linear metallocene catalyst (for simulation data, refer to Beigzadeh *et al.* [31])

C_{LCB}/C_{tot} passes through a maximum value which will be in accordance with the maximum LCB formation rate.

Although these results are based on simulations, they indicate the possibilities involved in the control of long-chain branching when these types of metallocene catalysts are used for polymerization.

5 ACKNOWLEDGMENTS

We acknowledge our graduate students involved in this long-term research project on metallocene polyolefins: Luigi D'Agnillo, Jung Dae Kim and Daryoosh Beigzadeh. Financial assistance from the Natural Sciences and Engineering Research Council (NSERC) of Canada is also gratefully acknowledged.

6 REFERENCES

1. Zucchini, U. and Cecchin, G., *Adv. Polym. Sci.*, **51**, 101 (1983).
2. Floyd, S., Heiskanen, R. and Ray, W. H., *Chem. Eng. Prog.*, **84**, 56 (1988).
3. Soares, J. B. P. and Hamielec, A. E., *Polym. React. Eng.*, **3**, 261 (1995).
4. Han, T. K., Choi, H. K., Jeung, D. W., Ko, Y. S. and Woo, S. I., *Macromol. Chem. Phys.*, **196**, 2637 (1995).
5. Whiteley, K. S., Heggs, T. G., Koch, H., Mawler, R. L. and Immel, W., In Elvers, B., Hawkins, S. and Schuttz, G. (Eds), *Ulmann's Encyclopedia of Industrial Chemistry*, VCH, Weinheim 1992, Vol. A21, p. 487.
6. Soares, J. B. P. and Hamielec, A. E., *Polym. React. Eng.*, **4**, 153 (1996).
7. Scheirs, J., Böhm, L. L., Booth, J. C. and Leevers, P. S., *Trends Polym. Sci.*, **4**, 408 (1996).
8. Flory, P. J., *Principles of Polymer Chemistry*, Cornell University Press, Ithaca, NY (1953).
9. Stockmayer, W. H., *J. Chem. Phys.*, **13**, 199 (1945).
10. D'Agnillo, L., Soares, J. B. P. and Penlidis, A., *J. Polym. Sci., Part A: Polym. Chem.*, **36**, 831 (1998).
11. Kissin, Y. V., *J. Polym. Sci., Part A: Polym. Chem.*, **33**, 227 (1995).
12. Hamielec, A. E., MacGregor, J. F. and Penlidis, A. *Macromol. Chem., Macromol. Symp.*, **10/11**, 521 (1987).
13. McLaughlin, K. W. and Hoeve, C. A. J., in Quirk, R. P. (Ed.), *Transition Metal Catalyzed Polymerization*, Harwood, New York, 1988, p. 337.
14. Vickroy, V. V., Schneider, H. and Abbott, R. F., *J. Appl. Polym. Sci.*, **50**, 551 (1993).
15. Kissin, Y. V., *Macromol. Chem., Macromol. Symp.*, **66**, 83 (1993).
16. Soares, J. B. P. and Hamielec, A. E., *Polymer*, **36**, 2257 (1995).
17. Soares, J. B. P., Kim, J. D. and Rempel, G. L., *Ind. Eng. Chem. Res.*, **36**, 1144 (1997).
18. Wild, L., *Adv. Polym. Sci.*, **98**, 1 (1990).
19. Soares, J. B. P. and Hamielec, A. E., in Pethrick, R. A. (Ed.), *Experimental Methods in Polymer Characterization*, Wiley, Chichester, p. 15, 1999.
20. Monrabal, B., *J. Appl. Polym. Sci.*, **52**, 491 (1994).
21. Monrabal, B., *Macromol. Symp.*, **110**, 81 (1996).
22. Usami, T., Gotoh, Y. and Takayama, S., *Macromolecules*, **19**, 2722 (1986).
23. Soares, J. B. P. and Hamielec, A. E., *Macromol. Theory Simul.*, **4**, 305 (1995).

24. Soares, J. B. P., Monrabal, B., Blanco, J. and Nieto, J., *Macromol. Phys. Chem.*, **199**, 1917 (1998).
25. Lai, S. W., Wilson, J. R., Knight, G. W., Stevens, J. C. and Chum, P. W. S., *US Pat.*, 5 272 236 (1993).
26. Montagna, A. A., *Chemtech*, October, 44 (1995).
27. Soares, J. B. P. and Hamielec, A. E., *Macromol. Theory Simul.*, **5**, 547 (1996).
28. Soares, J. B. P. and Hamielec, A. E., *Macromol. Theory Simul.*, **6**, 591 (1997).
29. Zhu, S. and Li, D., *Macromol. Theory Simul.*, **6**, 793 (1997).
30. Beigzadeh, D., Soares, J. B. P. and Hamielec, A. E., *Polym. React. Eng.*, **5**, 141 (1997).
31. Beigzadeh, D., Soares, J. B. P. and Hamielec, A. E., *J. Appl. Polym. Sci.*, **71**, 1753 (1999).
32. Chien, J. C. W. and Wang, B. P., *J. Polym. Sci., Part A: Polym. Chem.*, **26**, 3089 (1988).
33. Rieger, B. and Janiak, C., *Angew. Macromol. Chem.*, **215**, 35 (1996).
34. Kaminsky, W., *Macromol Chem Phys.*, **197**, 3907 (1996).
35. Huang, J. and Rempel, G. L., *Polym. React. Eng.*, **5**, 125 (1997).
36. Eskelinen, M. and Seppälä, J. V., *Eur. Polym. J.*, **32**, 331 (1996).
37. Vela-Estrada, J. M. and Hamielec, A. E., *Polymer*, **35**, 808 (1994).
38. D'Agnillo, L., Soares, J. B. P. and Penlidis, A., *Macromol. Chem. Phys.*, **199**, 955 (1998).
39. Huang, J. and Rempel, G. L., *Prog. Polym. Sci.*, **20**, 459 (1995).
40. Reddy, S. S. and Sivaram, S., *Prog. Polym. Sci.*, **20**, 309 (1995).
41. Soares, J. B. P. and Hamielec, A. E., *Polym. React. Eng.*, **3**, 131 (1995).
42. Gupta, V. K., Satish, S. and Bhardwaj, I. S., *J. Macromol. Sci., Rev. Macromol. Chem. Phys.*, **C34**, 439 (1994).
43. Hamielec, A. E. and Soares, J. B. P., *Prog. Polym. Sci.*, **20**, 651 (1996).
44. Po, R. and Cardi, N., *Prog. Polym. Sci.*, **21**, 47 (1996).
45. Kashiwa, N. and Imuta, J. I., *Catal. Surv. Jpn.*, **1**, 125 (1997).
46. Olabisi, O., Atiqullah, M. and Kaminsky, W., *J. Macromol. Sci., Rev. Macromol. Chem. Phys.*, **C37**, 519 (1997).
47. Quijada, R., Dupont, J., Silveria, D. C., Lacerda Miranda, M. S. and Scipioni, R. B., *Macromol. Rapid Commun.*, **16**, 357 (1995).
48. Läthi, M., Koivumäki, J. and Seppälä, J., *Angew. Macromol.*, **236**, 139 (1996).
49. Penlidis, A., Hamielec, A. E. and MacGregor, J. F., *J. Liq. Chromatogr.*, **6**, 179 (1983).
50. Heiland, K. and Kaminsky, W., *Macromol. Chem.*, **193**, 601 (1992).
51. Siedle, A. R., Lamanna, W. M., Newmark, R. A., Stevens, J., Richardson, D. E. and Ryan, M., *Makromol. Chem., Macromol. Symp.*, **66**, 215 (1993).
52. Jejelowo, M. O., Lynch, D. T. and Wanke, S. E., *Macromolecules*, **24**, 1755 (1991).
53. Kim, J. D., Soares, J. B. P., and Rempel, G. L., *Macromol. Rapid Commum.* **19**, 197 (1998).
54. Kim, J. D., Soares, J. B. P. and Rempel, G. L., *J. Polym. Sci., Part A: Polym. Chem.*, **37**, 331 (1999).
55. Thorshaug, K., Rytter, E. and Ystenes, M., *Macromol. Rapid Commun.*, **18**, 715 (1997).
56. Lee, D. K., Yoon, K., Noh, S., Kim, S. and Huh, W., *Macromol. Rapid Commun.*, **17**, 639 (1996).
57. Ewen, J. A., *US Pat.*, 4 975 403 (1990).
58. Welborn, H. C., Jr, *US Pat.*, 5 183 867 (1993).
59. D'Agnillo, L., Soares, J. B. P. and Penlidis, A., *Polym. React. Eng.*, **7**, 259 (1999).
60. D'Agnillo, L., Soares, J. B. P. and Penlidis, A., *Polym. Int.*, **47**, 351 (1998).
61. Kim, J. D., Soares, J. B. P., Rempel, G. L. and Monrabal, B., Control of molecular weight and chemical composition distributions of polyolefins made with supported metallocene catalysts, presented at MetCon'97 Polymers in Transition, Houston, TX, June 4–5 (1997).
62. Chien, J. C. W., Iwamoto, Y., Rausch, M. D., Wedler, W. and Winter, H. H., *Macromolecules*, **30**, 3447 (1997).
63. Galli, P. and Haylock, J. C., *Makromol. Chem., Macromol. Symp.*, **63**, 19 (1992).

64. Collina, G., Pelliconi, A., Sgarzi, P., Sartori, F. and Baruzzi, G., *Polym. Bull.*, **39**, 241 (1997).
65. Swogger, K. W. and Kao, C. I., presented at Polyolefins VIII, Tech. Pap., Reg. Tech. Conf.—Soc. Plast. Eng., 14, Houston, February 22–24 (1993).
66. Knight, G. W. and Lai, S., Dow constrained geometry catalyst technology: new rules for ethylene–α-olefin interpolymers—unique structure and property relationships, presented at Polyolefins VIII, Tech. Pap., Reg. Tech. Conf.—Soc. Plast. Eng., p. 28 (1993).
67. Woo, T. K., Fan, L. and Ziegler, T., *Organometallics*, **13**, 2252 (1994).
68. Resconi, L., Piemontesi, F., Fransiscono, G., Abis, L. and Fiorani, T., *J. Am. Chem. Soc.*, **114**, 1025 (1992).
69. Shiono, T., Moriki, Y. and Soga, K., *Macromol. Symp.*, **97**, 161 (1995).
70. Karol, F. J., Kao, S. C., Wasserman, E. P., Brady, R. C., Yu, Z., Use of copolymerization studies with metallocene catalysts to probe the nature of active sites, presented at Polymer Reaction Engineering III, March 16–21, Palm Coast, FL (1997).
71. Brant, P., Canich, J. A. M. and Merrill, N. A., *US Pat.*, 5 444 145 (1995).
72. Brant, P. and Canich, J. A. M., *US Pat.*, 5 475 075 (1995).

PART IV
Styrene Polymerization

12

Structure, Properties and Applications of Ethylene–Styrene Interpolymers

M. J. GUEST, Y. W. CHEUNG, C. F. DIEHL AND S. M. HOENIG
Dow Chemical Company, Freeport, TX, USA

1 INTRODUCTION TO ETHYLENE–STYRENE COPOLYMERS

Polyethylene and polystyrene are two of the most thoroughly investigated polymers, in part due to their commercial value. In addition, they are often used as model materials for elucidating the semicrystalline and amorphous nature of macromolecules, respectively. Consequently, there has been considerable interest to produce ethylene–styrene (ES) copolymers. Many molecular architectures are available. In addition to the monomer distributions (random, alternating or blocky nature), there are possibilities for chain branching and tacticity in the chain microstructure. These molecular architectures have a profound influence on the melt and solid-state morphology and hence on the processability and material properties of the copolymers.

ES copolymers containing low levels ($<4\,mol\%$) of styrene incorporation have been prepared by both conventional Ziegler–Natta catalyst [1–3] and free radical processes [4]. Many reported copolymerizations produce heterogeneous polymers consisting of mixtures of copolymers and homopolymers [5]. It has proved difficult to efficiently synthesize ES copolymers with high levels of styrene incorporation and having homogeneous structures with comonomer distributions of the polymers which are 'random' rather than 'blocky' in nature.

Recent developments using metallocene catalysis include the demonstrated ability to polymerize and copolymerize efficiently a wide range of hydrocarbon-based vinyl monomers, including ethylene (E), higher α-olefins, dienes and vinyl aromatics such

Metallocene-based Polyolefins Edited by J. Scheirs and W. Kaminsky
© 2000 John Wiley & Sons Ltd

as styrene (S). In particular, Dow INSITE[†] technology, including developments of single-site, constrained geometry, addition polymerization catalysts [6,7], has provided a route to produce ethylene–styrene interpolymers (ESI), offering the possibility for a novel class of materials. The term 'interpolymer' is used in this chapter to describe the specific ethylene–styrene copolymers produced via Dow INSITE Technology.

There is increasing interest, as evidenced by the available literature, in the preparation and characterization of such polymeric materials. Ethylene and styrene can be copolymerized using catalysts based on titanocene or zirconocene compounds with cocatalyst methylaluminoxane (MAO). Depending on the structure and symmetry of the catalysts, so-called 'pseudo-random' [7,8] and alternating [9, 10] ES copolymers with high levels of styrene incorporation have been produced. Copolymers have been described as having a highly isotactic (ES) alternating structure combined with further monomer unit chain insertions to give a 'so-called' random structure [11,12]. Heterogeneous ES copolymers containing up to about 85 mol% styrene with high isotacticity were synthesized using a coordination $(TiCl_4/NdCl_3/MgCl_2/AlEt_3)$ catalyst [13] designed for the stereospecific polymerization of styrene. The copolymers were separated from the polymer mixture and some mechanical properties were reported. Copolymerization of styrene with other α-olefins such as propylene [14,15] and 1-butene [15,16] have been described.

Sernetz, Mulhaupt and co-workers at the University of Freiburg have reported on the copolymerization of ethylene with styrene [17,18], the influence of polymerization conditions on copolymerization [19], characterization [20] and viscoelastic properties [21] of copolymers. Notable research from the University of Salerno includes the copolymerization of ethylene and styrene which was achieved with at least one catalyst more closely associated with the production of syndiotactic polystyrene [22,23]. Correlations between microstructure and physical properties of an ES copolymer (ca 50 wt% styrene) have been examined by d'Aniello et al. [24], who commented that the material showed good elastic properties such as strain recovery.

Patents also describe further catalyst systems [25] or manufacturing processes [26,27] for the preparation of ethylene–styrene copolymers.

2 INSITE TECHNOLOGY INTERPOLYMERS: STRUCTURE–PROPERTY RELATIONSHIPS

Table 1 presents the characterization data and tensile stress–strain data for a range of ESI synthesized via INSITE technology in a continuous miniplant process. Based on ^{13}C NMR analyses, these particular copolymers containing typically up to about 50 mol% (ca 80 wt%) styrene content have been described as 'pseudo-random', since successive head-to-tail styrene chain insertions were shown to be absent, even at high levels of styrene incorporation [7]. Table 1 shows that these polymers contain

---

[†]Trademark of Dow Chemical Company.

Table 1 Ethylene–styrene interpolymers: characterization and property data[a]

Styrene (%)	$10^{-3} M_w$	M_w/M_n	aPS (%)	T_c (°C)	T_m/% Xtyl	T_g (DSC) (°C)	Shore A	Modulus (psi)	E_f (%)	σ_f (psi)	SR (%)
20.7	270	2.3	0.3	72.8	84.0/24.9	−19.7	92/91	6150	472	3146	26.4
20.5	205.6	2.2	0.2	74.4	86.3/26.1	−19.7	93/93	6449	509	3246	24.7
21.4	156	2.2	0.2	74.4	86.9/25.9	−18.9	94/93	6273	592	3397	29.5
20.2	113	2.2	0.3	76.4	91.4/29.0	−19.7	92/92	7117	584	2947	28.3
30.9	210.2	2.3	0.3	48.5	65/15.1	−20.9	85/84	2425	514	2596	31.5
30.7	188.1	2.3	0.1	50.6	64.8/16.0	−20.2	86/85	2600	560	2920	31.4
30.7	115.1	2.1	0.3	55	69.2/17.0	−21	85/84	2785	742	3211	34.5
42.2	311.9	2.2	1	10	35.2/4.7	−18.9	70/67	825	519	1561	27.5
43.1	241.9	2.1	0.6	8.3	32.5/4.3	−19.6	69/66	786	593	1448	32.9
41.9	116.9	2.5	0.8	13.3	42.4/5.7	−19.7	67/63	912	699	1683	35.5
46	89.7	2.2	0.8	3.5	36.0/3.4	−19.2	64/59	482	918	640	67.7
60.8	272.6	2	1.2	n/a	n/a	0.3	66/64	574	619	626	45.3
59.8	209.5	1.9	1.9	n/a	n/a	−1.4	67/64	492	904	333	57.5
59.1	136.4	1.8	0.2	n/a	n/a	−3.7	64/59	370	2046	71	74
58.7	99.8	1.9	0.9	n/a	n/a	−4.5	63/56	397	1187	32	83.5
72.8	167.5	1.8	2.7	n/a	n/a	20.4	98/97	31022	239	1888	94.7
73.3	115.8	1.9	2.9	n/a	n/a	20.2	95/95	33333	259	1775	95.2

[a] DSC data: 10 °C/min; T_m (second heating) and T_c are the peak temperatures; % Xtly is the area under the melting endotherm; T_g is mid-point of the ΔC_p change associated with T; Shore A hardness, 23 °C, following ASTM D240. Tensile properties: 23 °C, ASTM D638 samples (compression molded), 5 min^{-1}, average of four test bars. E_f is elongation at fracture; σ_f is stress at fracture; SR is the stress relaxation following 10 min hold at 50 % tensile strain.

very small amounts of atactic polystyrene (aPS), primarily produced during unreacted monomer removal after copolymerization. Such concentrations of aPS have been shown independently not to affect significantly the mechanical properties described here. An introduction to the structure, thermal transitions and mechanical properties of selected ESI has been presented elsewhere, with the subsequent sections building on this information [28].

For convenience, all subsequent comonomer contents are expressed in weight percentages, unless stated otherwise. The code ES *xy* refers to an interpolymer having *xy* wt% styrene monomer content in the copolymer. All mechanical property data on ESI polymers and blends presented in this chapter are for compression molded test parts. Not all property data presented in the figures are specifically for materials represented in Table 1.

2.1 THERMAL TRANSITIONS, CRYSTALLINITY AND DENSITY

The effects of styrene incorporation into the polyethylene chain are profound. Crystallization and melting temperatures (T_c, T_m) and crystallinity are found to decrease with increasing styrene content. Crystallizability of ethylene chain sequences is suppressed, and ultimately inhibited, by the incorporation of increasing styrene (S) comonomer content into the crystallizing chains, somewhat analogous to other comonomers such as 1-octene [29]. Based on steric hindrance arguments, the styrene unit is believed to be excluded from the crystalline region of the copolymers.

Characterization of thermal transitions and phase behavior underlies the mapping of structure–property relationships for ESI. A general correlation of the differential scanning calorimetric (DSC) data for the glass transition temperature (T_g), T_m and T_c with the S content of the ESI is given in Figure 1. Figure 2 presents the crystallinity (determined from the area under the DSC melting endotherm) and density for ESI.

Transition from the semicrystalline state to an essentially amorphous solid-state structure occurs at about 50 % S incorporation. This is manifested in the DSC data as a change in the form of the melting endotherm. At low S contents (<40 %) the copolymers generally exhibit a relatively well defined melting process. This transition becomes a diffused melting process covering a much wider temperature range and shows no distinct peak melting temperature for 40–50 % S contents. At higher S content there is no discernible melting endotherm.

The ESI have glass transitions generally in the range −20 to +35 °C, as depicted in Figure 1. Below about 50 % S content, the single number T_g(ESI) appears to be relatively independent of styrene content, attributed in part to the crystalline domains present which restrict molecular mobility in the amorphous domains. In addition, the styrene units in the semicrystalline polymers are located in the amorphous phase, and thus give a higher T_g amorphous phase than would be found for a fully amorphous, low styrene content ES copolymer. Above 50 % S content, for essentially amorphous copolymers, T_g(ES) increases with increasing S content. The observed behavior for the amorphous ESI T_gs is anticipated, reflecting some

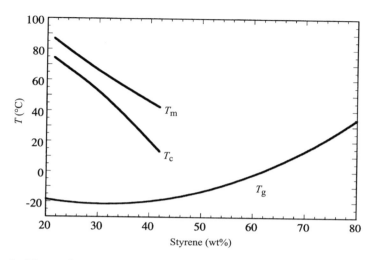

Figure 1 Thermal transitions of ethylene–styrene interpolymers (ESI) with T_m, T_c and T_g denoting the melting, crystallization and glass transition temperatures, respectively

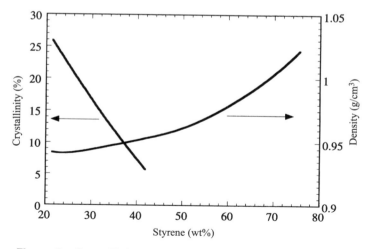

Figure 2 Crystallinity (measured by DSC) and density of ESI

form of weighted average of the homopolymer T_gs when taking account of the higher T_g(ca 105 °C) found for aPS relative to that of PE.

Numerous models have been proposed to predict the compositional variation of T_g for single-phase multicomponent systems [30]. The composition dependence of T_g for amorphous random copolymers (and also miscible amorphous polymer blends) is often correlated in some way with the weight fraction of the component species by using approaches such as the Fox [31] or Gordon–Taylor [32] equations, the latter being equation (1) below. Effects such as crystallinity and lack of additivity for component heat capacities are not accounted for in these models.

$$T_g = \frac{w_1 T_{g1} + k w_2 T_{g2}}{w_1 + k w_2} \tag{1}$$

where T_g is the glass transition temperature of copolymer, w_i the weight fraction and T_{gi} the glass transition temperature of component i and k ($= \Delta\alpha_2/\Delta\alpha_1$) is the ratio of the difference in thermal expansion coefficient between the liquid and glassy states at T_{gi}.

The composition dependence of T_g can be ascribed to three possible nearest-neighbor interactions including the two like (homopolymer) dyads and the unlike (alternating copolymer) dyad. Without going into detail on the relative merits of available models, satisfactory correlation between copolymer composition and T_g is found either by considering the ESI as random copolymers of ethylene and styrene or as random copolymers of ethylene and (ES) dyads, using an alternating atactic (ES) copolymer T_g of 44 °C [33].

In the low-styrene region (<40 % S), the density is almost independent of composition, as illustrated by Figure 2. Any decrease in polymer density anticipated due to the reduction of crystallinity is compensated by the increase in overall density imparted by the styrene monomer. In the high-styrene region (>40 % S), an almost linear increase in density with styrene content is found, within a range up to the density value [34] for aPS of 1.05 g/cm^3.

2.2 SOLID-STATE VISCOELASTIC BEHAVIOR

Dynamic mechanical spectroscopic (DMS) data provide a link between the phase transitions, microstructure, viscoelastic behavior and mechanical properties of polymeric materials. Both elastic and loss modulus data contribute to defining performance characteristics, and loss processes in particular can be correlated with toughness, vibration and sound damping and surface features such as frictional behavior, adhesion and abrasion. With T_g(ES) falling in the temperature range evident from Figures 1 and 3(b), it is clear that an understanding of the viscoelastic properties and time–temperature dependence of properties is a key element of the materials science of interpolymers, and ultimately in their utility. Stress relaxation is another important aspect of viscoelastic behavior; some information has been reported by Mudrich *et al.* [35].

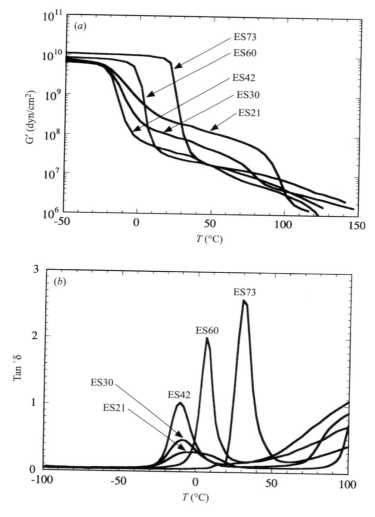

Figure 3 Dynamic mechanical spectroscopy (DMS) plots of (a) shear storage modulus versus temperature and (b) tan δ versus temperature, measured at 10 rad/s, for a series of ESI with styrene incorporation ranging from 21 wt% (ES21) to 73 wt% (ES73)

Figure 3(a) and (b) present elastic modulus and tan δ data for selected ESI having S contents between 20 and 75 % and generated over temperature ranges covering T_g and T_m. The smaller modulus drop at T_g and the modulus plateau evident between T_g and T_m for ES21 and ES30 is associated with load-bearing capabilities of the crystalline phase. The T_g loss peak for these polymers is broader relative to amorphous ESI, as a result of the mobility constraints imposed on the amorphous phase by the crystalline domains. ES42 is a polymer with a low level (<5 %) of crystallinity, but has a tan δ loss peak associated with T_g which is broader than those for fully amorphous polymers such as ES73.

The two amorphous polymers ES73 and ES60 have sharp moduli decrease on passing through T_g and intense tan δ loss processes associated with the amorphous phase T_g(ES). The loss peak width and amplitude of T_g for these polymers is broadly comparable to that of the aPS, a model amorphous polymer.

Further analysis of DMS data generated on ESI, and particularly the form of T_g master curves generated by classical superposition techniques of temperature–frequency data, has shown that the polymers can be considered as thermorheologically simple in the temperature range around the T_g region. Furthermore, from the data analysis, the apparent Arrhenius activation energy for the glass transition process is found to increase with increasing styrene content.

2.3 INTERPOLYMER MELT RHEOLOGY

Melt rheology studies extend the understanding of viscoelastic behavior to temperatures more associated with the processing of polymers. In addition to its contribution to understanding structure–property relationships, melt rheological behavior is a fundamental characteristic determining melt processability, and hence part fabrication. The latter is considered further in Section 5, because of its close association with the potential applicability of interpolymers. Some aspects of ethylene–styrene copolymer melt viscoelastic behavior have been reported by Lobbrecht et al. [21]. Data generation and analysis for ESI have been reported elsewhere [36] in more detail, where it was shown that rheology data showed excellent time–temperature superposition, hence allowing the estimation of key material parameters. Selected information is presented for the purpose of discussion.

Figure 4(a) presents complex viscosity–shear rate data for ES70 and ES30, which have similar weight-average molecular weights, the lower styrene containing material is seen to have significantly higher viscosity. This is rationalized as follows: many properties, including mechanical and rheological behavior, are determined by entanglement of polymer chains. The entanglement molecular weight (M_e) is a fundamental material characteristic. ES copolymers are a combination of two monomers where the M_e for the individual homopolymers differs significantly, PE being in the range 1400–4000 and PS 18 800–25 000 [37]. An estimation of M_e(ES) has been made from analysis of rheology data [36], confirming that M_e increases with increasing styrene content, with the net effect that, for equivalent molecular weight, an ESI with a higher styrene content will have a less chain entangled melt microstructure and lower viscosity.

Figure 4(b) presents complex viscosity–shear rate data for ES60 varying in weight-average molecular weight, M_W. As expected, increasing M_W results in

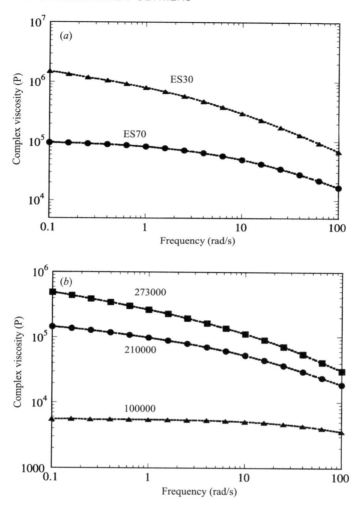

Figure 4 (a) Effect of wt% styrene on complex viscosities (at 190 °C) of ESI of approximately 250 000 weight-average MW (M_w) and (b) effect of M_w on complex viscosities (at 190 °C) of ESI with 60 wt% styrene incorporation

increased viscosity. The activation energy for flow, E_a, or temperature dependence of the viscosity is a strong function of styrene level. Base polymers of high-density polyethylene are typically quoted [38] as having an E_a of 6.5 kcal/mol and polystyrene as having an E_a of 23 kcal/mol. The E_as were found to be ca 15 kcal/mol at around 75 wt% copolymer S level, decreasing to 8.7 kcal/mol for ES30.

The generation and mapping of melt rheology data introduced above as a function of process parameters such as temperature and shear rate, and material parameters such as comonomer content and molecular weight/molecular weight distribution, are key elements in understanding processability, in matching molecular architecture with performance requirements and in the design of polymer blends.

2.4 INTERPOLYMER MECHANICAL STRESS–STRAIN, STRAIN RECOVERY BEHAVIOR

The tensile stress–strain behavior (23 °C, 5.7 min^{-1}) for selected ES copolymers is shown in Figure 5. The copolymers all show strain at rupture in excess of 200 %, and have been found to show uniform deformation behavior. ES20 and ES30 exhibit tensile properties broadly similar to those reported elsewhere for random ethylene–α-olefin copolymers [29]. The materials in the mid-styrene region, i.e. ES43 and ES60, are characterized by low modulus and show some additional elastomeric characteristics. Regarding the origin of this elastomeric-like behavior, it is believed that the residual crystallinity could contribute to some form of network structure. The effects of styrene become dominant in the high-styrene region where the modulus and yield stress are seen to increase. The ES73 copolymer shows some characteristics of a glassy material. There is a distinct change of slope point in the stress–strain curve, which has been shown to be due to the large strain rate dependence of the response when tested at temperatures close to T_g. Overall, a parabolic U-shaped composition dependence is apparent for yield stress and modulus when plotted as a function of copolymer S content, presented elsewhere [28].

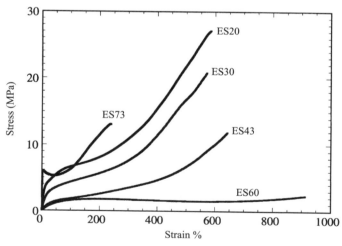

Figure 5 Engineering stress–strain behavior of ESI measured at 23 °C

Figure 6(a) and (b) present stress–strain data for ES20 and ES60 at a range of different molecular weights. The crystalline ES20 polymers show much less sensitivity to molecular weight, associated with the load-bearing capability of the crystalline regions networked together with tie-molecules. ES60 polymers show sensitivity to molecular weight due to the amorphous structure, and the link of properties to chain entanglements, introduced in the previous section. In addition to styrene content and molecular weight, the effects of temperature and strain-rate on

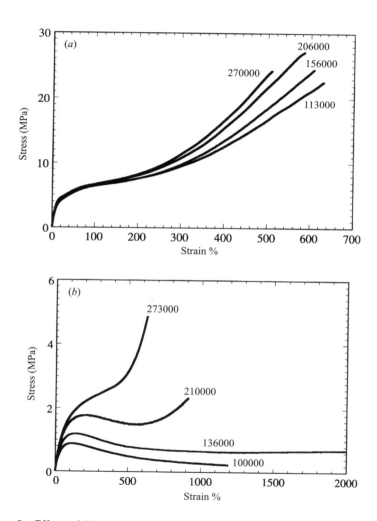

Figure 6 Effect of M_w on the engineering stress–strain behavior of (a) 21 and (b) 60 wt% styrene ESI at 23 °C

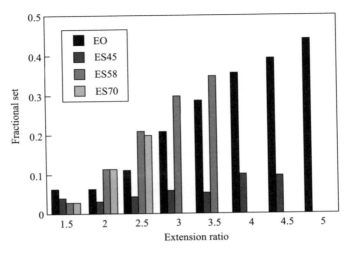

Figure 7 Fractional set versus extension, measured at $T_g + 40\,^{\circ}C$, following 5 min of recovery under no-load conditions (specimens loaded and unloaded at 100 %/min). EO is an ethylene–octene copolymer with a density of 0.87 g/cm^3 and ES45, ES58 and ES70 contain 45, 58 and 70 wt% styrene incorporation, respectively

mechanical behavior are important components of the materials science under-standing of ESI.

Correlations between microstructure and physical properties of an ES copolymer (ca50 wt% styrene) have been examined by d'Aniello *et al.* [24], who commented that the material showed good elastic properties such as strain recovery. Mudrich *et al.* [35] reported on the deformation and recovery behavior of ESI. The more extensive presentation of data in Figure 7 exemplifies the elastic behavior of selected ESI. For the experimental conditions summarized in the figure caption, all ESI show excellent strain recovery behavior, comparable to a commercially available thermo-plastic ethylene–1-octene elastomer for extension ratio of 1.5 (tensile strains of 50 %). As shown in Figure 7, ES45 is particularly noteworthy, showing excellent strain recovery characteristics for extension ratios in excess of 4 (tensile strains in excess of 300 %).

3 CLASSIFICATION OF INTERPOLYMERS

Interpolymers of styrene with ethylene span a broad range of material types, offering unique properties distinct from either polyethylene or polystyrene homopolymers. On the basis of their microstructure, thermal transitions and mechanical properties, three broad performance regimes can be assigned for ES polymers. These material regimes are defined according to the level of styrene (S) incorporation, being semi-crystalline (low-S), elastomeric (mid-S) and amorphous/glassy (high-S).

Structurally, the low-S materials are seen to have significant levels of crystallinity and have some similarity to random ethylene–α-olefin copolymers where the elastic properties (modulus, yield stress) are predominantly controlled by the crystallinity, and its connectivity through tie-molecules. It is believed that the styrene functionality is to be found in microstructural regions excluded from the crystalline region. Differences between α-olefins and styrene as comonomers for ethylene are related to inter- and intra-chain segment compatability (and consequently the nature of the crystalline/amorphous phase interfacial region), polarizability, chain conformation and packing, where these influence physical, rheological and mechanical properties.

The mid-styrene copolymers show many characteristics of thermoplastic elastomers at ambient temperatures (20–30 °C). These ESI have low to no measurable crystallinity and are characterized by very low modulus. Although the origin and nature of the elasticity is complex, it is believed that the residual crystallinity, together with the network resulting from chain entanglement or chain association, strongly influences the deformation behavior.

The effects of styrene become dominant for the high-styrene copolymers where the modulus and yield stress increase to a level comparable to those found for the low-styrene materials. The high-styrene materials are essentially amorphous, and their responses to mechanical deformation are governed by both the intra- and inter-chain interactions. Some of the unique viscoelastic properties observed in the high-styrene materials cannot be easily rationalized if the solid-state structure is that of a 'simple' amorphous material.

4 INTERPOLYMER MATERIAL ENGINEERING

Ethylene–styrene interpolymers have a wide range of interesting material engineering possibilities, with selected areas being introduced in this section.

4.1 CHEMISTRY/COMONOMER

The higher reactivity offered by the phenyl group of the styrene units leads to some additional chemistries not available for saturated olefin-based polymers. Examples include Friedel–Crafts modifications to achieve cross-linking, sulfonation, chlorination and hydrogenation of the aromatic ring. Although it is clear that comonomer content and molecular weight are primary controls for ESI structure–property relationships, INSITE technology also permits the copolymerization of ethylene and styrene with additional monomers [7], including dienes, higher α-olefins (e.g. 1-butene) and norbornene. These possibilities are further evident from the work of

Sernetz and Mulhaupt, who have reported terpolymerization with propylene, 1-octene and norbornene [39] and with 1,5-hexadiene [40]. By engineering the appropriate monomer contents into the polymer chain, materials can be produced covering a broader range of T_g; for example, the terpolymerization of ethylene–styrene–1-octene can result in T_g as low as $-45\,^\circ\text{C}$. The patent literature provides other references to terpolymers including dienes [41–43] (and thermoset compositions [41]) and α-olefins [44].

4.2 BLENDS AND COMPATIBILITY

Blending is recognized as a major materials engineering option for polymeric systems. Depending on the nature and extent of interactions, polymer blends may be classified as miscible, compatible and incompatible [45]. Blends are defined as miscible when the blend components are 'molecularly mixed', forming a homogeneous phase and having a single T_g process. Most polymer blends are immiscible, and have multiple T_gs reflecting the different phases. Immiscible blends may be further described as compatible or incompatible, and the properties which can be obtained are strongly dependent upon the morphology and interfacial regions between phases. The degree of compatibility may be assessed by the magnitude of shift and broadening in the glass transition region. Blends showing good mechanical integrity as reflected by the ultimate mechanical properties and diffused interfacial phase boundary are referred to as compatible. Severe phase debonding and significant losses in ultimate mechanical properties are typically found for incompatible blend systems. Blending offers an additional property control which could be used to expand the applicability of ESI. Some of the potential benefits of blends technology typically includes enhanced mechanical properties, improved processibility, modification in T_g, relaxation behavior and energy dissipation mechanisms. The styrene content and chain microstructure of ESI are potential contributors to the compatability between ESI and other materials including oils, low molecular weight materials such as plasticizers and tackifiers, processing aids and other polymers.

    Blending copolymers of similar comonomer types but differing in content is recognized as important, and is a good starting point for discussion of ESI blend systems. It is generally accepted that blends of two random copolymers will be immiscible when the composition difference in comonomer content is above a certain critical value. Polymer blend thermodynamics are often described using the classical Flory–Huggins free energy of mixing equation. By employing the solubility parameter equation to estimate the Flory χ interaction parameter, and making certain assumptions, it is possible to calculate the form of a miscibility map between copolymers. If the densities and molecular weights (MW) of the two ESI are assumed to be equal, the maximum volume fraction styrene difference (ΔS) between two ESI for which miscibility is achieved can be estimated from the equation

$$\Delta S = \left(\frac{2\rho RT}{(\delta_{PS} - \delta_{PE})^2 M} \right)^{1/2} \tag{2}$$

where ΔS is the maximum volume fraction styrene content between two copolymers for which miscibility is achieved, ρ is density, R is the gas constant, δ is the solubility parameter and M is molecular weight.

On the basis of equation (2), the critical composition difference for miscibility of ESI–ESI blends is inversely proportional to the difference in solubility parameter between the two monomer segments. Although the reported solubility parameters for polyethylene and polystyrene span a fairly wide range of values [46], two sets of values including the average and maximum difference were used to calculate the miscibility map in Figure 8.

The critical composition difference above which phase separation will occur for ESI–ESI blends as a function of molecular weight and solubility parameter difference is illustrated in Figure 8. The critical styrene content difference decreases with increasing molecular weight and mismatch in the solubility parameter. It can be seen that the critical comonomer difference in styrene content at which phase separation occurs is estimated as about 10 vol.% for copolymers with molecular weights around 10^5. For the purposes of the discussion, weight% can be considered as broadly equivalent to volume%. Figure 9 presents tan δ versus temperature data

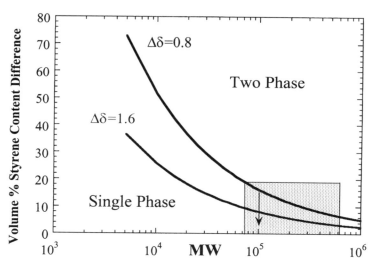

Figure 8 Miscibility map for ESI/ESI blends denoting the critical styrene content above which the blends are immiscible

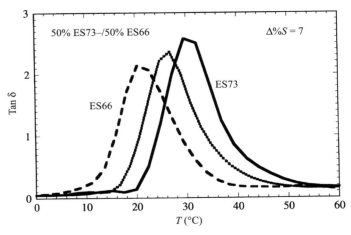

Figure 9 DMS of ES73–ES66 (50:50 weight ratio) blend. Note blend T_g (tan δ maximum) is intermediate between the pure component T_gs

for a blend of two interpolymers differing by 7 wt% styrene content, showing miscibility as evidenced by a single T_g for the blend intermediate between that of the two blend component interpolymers.

The general applicability of the above analysis has been reported [47] in more detail elsewhere. ESI–ESI blends also offer the opportunity to engineer materials with broadened T_g loss processes and enhancements in relaxation behavior, mechanical properties and melt rheology/processability.

Extending the above rationale regarding miscibility to blends of ESI with styrenic and olefinic polymers, it is evident that ESI having <80 % styrene content will be immiscible with PS, unless the molecular weights of the respective polymers are very low. Initial studies on ESI blends with PS show that significant toughening of brittle atactic PS can be achieved with selected ESI, in large part due to good compatibility between these polymers.

Similarly, the olefin nature of ESI helps with compatibility with olefin polymers and copolymers including polyethylenes and polypropylenes. Again, ESI with >10 % styrene content will, in general, be immiscible with unmodified olefin polymers. These polymers are immiscible but compatible, and blending gives the combined advantages of the unique viscoelasticity and functionality of ES copolymers and the elasticity of EO copolymers.

It is envisaged that interpolymers can be blended with many other polymers, including styrene block copolymers and olefin polymers such as ethylene–acrylic acid copolymer, ethylene–vinyl acetate copolymer and chlorinated polyethylene. Blends with PVC-based polymers may also be important.

As introduced above, most polymer blends are immiscible owing to the unfavorable free energy of mixing, and it is well known that polymer pairs such as polyethylene and polystyrene are highly incompatible. Generally, a blend compatibilizer is often used to lower the interfacial tension, improve phase adhesion and reduce coalescence between two immiscible polymers [48]. Diblock copolymers such as styrene–butadiene have been extensively used as blend compatibilizers. By weaving and forming small loops across the interface, random copolymers could be very effective in reducing interfacial tension and promoting interfacial adhesion [49]. Owing to their unique molecular structure, specific ESI have been demonstrated as effective blend compatibilizers for polystyrene–polyethylene blends [50].

5 COMMENTS ON POTENTIAL APPLICATIONS

Interpolymers offer a novel range of materials which are not widely available for investigation at the present time. Consequently, much of the application information to date is based either on an understanding of the structure–property relationships and materials engineering possibilities being reported, or on actual and proposed utility found in patent and open literature publications.

Processibility is an important attribute describing the ease with which a material can be transformed into a fabricated product. Melt rheology was considered in detail in Section 2.3. Increasing levels of styrene comonomer influence processability, contributing to shear thinning behavior and good molding characteristics more closely associated with polystyrene. Other aspects of note reported for ESI regarding processability and utility of ESI are good thermal stability and melt strength. As a general comment, melt processability of interpolymers is favorable towards most fabrication techniques. These characteristics have allowed the fabrication of ESI articles by a wide range of standard melt processing techniques including injection molding, blown film, cast film, blow molding operations, calendering and melt extrusion.

The structure–property relationships introduced above provide the basis for understanding what attributes and benefits interpolymer technology offers, and hence potential applications. The polymers themselves cover a broad range of material types, from semicrystalline to fully amorphous structures. Depending on styrene content, ESI are elastomeric and are tough, even at service temperatures below T_g, having high tensile elongations at rupture. It has been found [28] that high styrene copolymers exhibit relaxation behavior similar to some flexible PVC compositions. The latter materials are known for their unique relaxation behavior, which translates into application-related phenomena such as shape recovery, dead fold and rubbery behavior. The T_g process in polymers is associated with energy absorption, and the temperature–frequency ranges found for ESI, combined with other attributes such as excellent filler acceptance, indicate its utility in a range of applications including flooring, sound management and vibration damping systems. Potential utility in pressure-sensitive adhesives [44] and multilayer film structures [51], including layers using ES copolymers, has been identified.

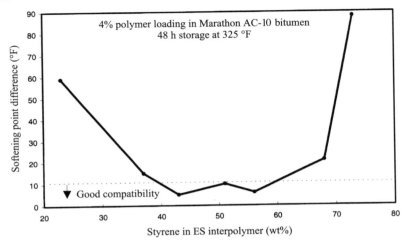

Figure 10   Compatibility of ESI in bitumen (Arab heavy crude)

Selected polymers also show high compatibility with bitumens, which is believed to be related to the chain microstructure and presence of aromatic functionality. The compatibility of ES interpolymers in bitumen is a function of the copolymer styrene content. Figure 10 shows the difference in softening point between the top and bottom layers of polymer-modified bitumens stored at elevated temperature for extended periods of time. A compatible system will be storage stable and show little difference in softening points. ESI with very low levels of styrene (high levels of crystallinity) have poor compatibility, as do interpolymers with too high a level of styrene incorporation.

Compatible ESI, in the intermediate styrene range, can be used to enhance the high-temperature rutting resistance of asphalt pavements, while not adversely

Table 2   Effect of addition of ESI43 on viscoelastic parameters

Property[a]	Neat Conoco AC-10 bitumen	4 % ESI43 in Conoco AC-10
Dynamic shear rheometer[b]		
$G^*/\sin \delta$ at 76 °C (initial unaged sample)	0.25	1.66
Bending beam rheometer[c]		
S (MPa) at -18 °C	222	244
m at -18 °C	0.303	0.304

[a] $G^* = $ complex modulus; $\sin \delta = $ sine of phase angle in dynamic shear; $S = $ creep stiffness; $m = $ log shear rate from bending–beam rheometer testing.
[b] Superpave specification is > 1.0 kPa.
[c] Superpave specification is $m \geqslant 0.30$ and $S \leqslant 300$ MPa.

impacting the low-temperature crack resistance. Table 2 demonstrates the impact of the addition of 4 wt% ES43 as measured by viscoelastic parameters defined in the Strategic Highway Research Program (SHRP) Superpave grading system [52].

Interpolymers offer attractive properties and characteristics including softness, esthetics and drape for a wide range of non-cross-linked and cross-linked foam applications. Other product technologies of interest are as injection-molded structural foams, as foamed layers in multilayer structures and as foamed blends of interpolymers with styrenic and olefinic polymers.

Interpolymers also have potential for film and sheet applications, as presented by Murphy [53]. Material attributes such as processability, mechanical properties and heat sealability are the basis for utility in elastic films, deadfold film and sheet, in laminated structures and in thermoformable film and sheet.

Ethylene–styrene copolymers have received attention for use in electrical power transmission systems, with one aspect being its utility as a new insulating material considered to show higher breakdown strength than low-density polyethylene [54]. Tanaka *et al.* [55] have reported a study of space charges in ethylene–styrene copolymers irradiated by low-dose γ-rays, using results from a range of electrical measurements. It was considered that the presence of benzene rings in the olefin structure contributed significantly to the electron energy absorption effect, which was seen as a benefit in d.c. conduction. The relationship between comonomer-induced morphological changes and the dielectric strength of copolymers was also described [56].

The current range of potential markets/applications for ESI which have been identified also include packaging, injection and blow molding articles, adhesives, toys, calendered films, wire and cable, foams, building and construction, automotive and extruded sheet and films [57]. The compatibility of ESI with many materials may find utility from a recycle/waste viewpoint. It is anticipated that a much broader applicability of interpolymers will be found in the future, as the materials become more available in the public domain.

6 CONCLUSION

The above overview of ethylene–styrene (ES) copolymer developments and technology highlights the utility and versatility of interpolymers from INSITE technology. Materials offering a wide range of properties can be tailored by the molecular architecture, via comonomer content and distribution, and via molecular weight control offering unique performance attributes. A range of other technologies for materials engineering have been introduced, including terpolymerization and blending. ES interpolymers have particularly good compatibility with many other materials, including styrenic and olefinic polymers, fillers and bitumens.

Inevitably, intermaterial substitutions in existing applications utilizing flexible PVC, styrenic block copolymers, ethylene–vinyl acetate copolymers and ethylene–propylene-based elastomers will be potential application areas. Interpolymers do, however, offer an exciting new generic class of materials, and some unique opportunities for innovative developments in basic polymer chemistry and materials science, materials engineering and application technology.

7 ACKNOWLEDGMENTS

The authors particularly thank Steve Chum, Scott Mudrich and Teresa Karjala for helpful comments and discussions. They further thank researchers at Case Western Reserve University, including Professor Eric Baer, Professor Anne Hiltner, Hong-Yu Chen and Andy Chang, for their contributions to our understanding of structure–property relationships and material classification. The authors also thank many others, especially Joe Huang and Ken Reichek for their help with providing the data presented in this chapter. Finally, they thank the Dow Chemical Company and the Interpolymer Business Management team for permission to publish this material.

8 REFERENCES

1. Kobayashi, S. and Nishioka, A., *J. Polym. Sci., Part A*, **2**, 3009 (1964).
2. Kawai, W. and Katsuta, S., *J. Polym. Sci., Part A*, **8**, 2421 (1970).
3. Soga, K. and Lee, D., *Polym. Bull.*, **20**, 237 (1988).
4. Orikasa, Y., Kojima, S., Inoue, T., Yamamoto, K., Sato, A. and Kawakami, S., to Nippon Petrochemicals, *US Pat.*, 4 748 209, 1988.
5. Gorham, W. F. and Farnham, A. G., to Union Carbide Corporation, *US Pat.*, 3 117 945, 1964.
6. Stevens, J. C., *Abstracts of the Metcon*, Houston, 1993, p. 157.
7. (a) Stevens, J. C., Timmers, F. J., Wilson, D. R., Schmidt, G. F., Nickias, P. N., Rosen, R. K., Knight, G. W. and Lai, S., to Dow Chemical Company, *Eur. Pat. Appl.*, EP 416 815, 1990; (b) Timmers, F. J., to Dow Chemical Company, *US Pat.*, 5 703 187, 1997.
8. Ren, J. and Hatfield, G. R., *Macromolecules*, **28**, 2588 (1995).
9. Miyatake, T., Mizunuma, K. and Kakugo, M., *Makromol. Chem., Macromol. Symp.*, **66**, 203 (1993).
10. Pellecchia, C., Pappalardo, D., D'Arco, M. and Zambelli, A., *Macromolecules*, **29**, 1158 (1996)
11. Arai, T., Ohtsu, T. and Suzuki, S., *Polym. Prepr., Am. Chem. Soc. Div. Polym. Chem.*, **38**, 349 (1997).
12. Okamoto, A., Nakamura, A., Suzuki, S., Otsu, T. and Arai, T., to Denki Kagaku Kogyo, *Ger. Pat. Appl.*, DE 1 9711 339, 1997.
13. Lu, Z., Liao, K. and Lin S., *J. Appl. Polym. Sci.*, **43**, 1453 (1994).
14. Xu, G. and Lin, S., *Polym. Prepr., Am. Chem. Soc. Div. Polym. Chem.*, **35**, 686 (1994)
15. Kawasaki, M., Kitani, H. and Yoshitake, J., to Mitsui Petrochemical Industries, *US Pat.*, 5 244 996, 1993.
16. Gao, F. and Lin, S., *Macromol. Chem. Phys.*, **197**, 4225 (1996).
17. Sernetz, F. G., Mulhaupt, R., Fokken, S. and Okuda, J., *Macromolecules*, **30**, 1562 (1997).
18. Sernetz, F. G., Mulhaupt, R., Amor, F., Eberle, T. and Okuda, J., *J. Polym. Sci., Part A: Polym. Chem.*, **35**, 1571 (1997).

19. Sernetz, F. G. and Mulhaupt, R., *Macromol. Chem. Phys.*, **197**, 1071 (1996).
20. Thomann, Y., Sernetz, F. G., Thomann, R., Kressler, J. and Mulhaupt, R., *Macromol. Chem. Phys.*, **198**, 739 (1997).
21. Lobbrecht, A., Friedrich, C., Sernetz, F. G. and Mulhaupt, R, *J. Appl. Polym. Sci.*, **65**, 209 (1997).
22. Longo, P., Grassi, A., and Oliva, L., *Makromol. Chem.*, **191**, 2387 (1990).
23. Oliva, L., Caporaso, L., Pellecchia, C. and Zambelli, A., *Macromolecules*, **28**, 4665 (1995).
24. D'Aniello, C., De Candia, R., Oliva, L. and Vittoria, V., *J Appl. Polym. Sci.*, **58**, 1701 (1995).
25. Mitsui Toatsu Chemicals, *US Pat.*, 5 652 315, 1997.
26. BASF, *Ger. Pat. Appl.*, DE 19 542 356, 1995.
27. DSM, *PCI Int. Appl.*, WO 9742240, 1997.
28. Cheung, Y. W. and Guest, M. J., in *Proceedings of SPE ANTEC 54th Annual Conference*, 1996, p. 1634.
29. Bensason, S., Minick, J., Moet, A., Chum, S., Hiltner, A. and Baer, E., *J. Polym. Sci., Part B: Polym. Phys.*, **34**, 1301 (1996).
30. Couchman, P. R., *Macromolecules*, **15**, 770 (1982).
31. Fox, T. G., *Bull. Am. Phys. Soc.*, **1**, 123 (1956).
32. Gordon, M. and Taylor, J. S., *J. Appl. Chem.*, **2**, 493 (1952).
33. Chen, H., Guest, M. J., Chum, P. S., Hiltner, A. and Baer, E., in *Proceedings of SPE ANTEC 56th Annual Conference*, 1998, p. 1808.
34. Brandrup, J. and Immergut, E. H. (Eds), *Polymer Handbook*, V/82 Wiley, New York, 3rd edn, 1989, p. V/82.
35. Mudrich, S. F., Cheung, Y. W. and Guest, M. J., in *Proceedings of SPE ANTEC 55th Annual Conference*, 1997, p. 1783.
36. Karjala, T. P., Cheung, Y. W. and Guest, M. J., in *Proceedings of SPE ANTEC 55th Annual Conference*, 1997, p.1086.
37. Wu, S., *J. Polym. Sci., Polym. Phys.*, **27**, 723 (1989).
38. Scott, R. L., *J. Polym. Sci.*, **9**, 423 (1952).
39. Sernetz, F. G. and Mulhaupt, R., *J. Polym. Sci.*, **35**, 2549 (1997).
40. Sernetz, F. G., Mulhaupt, R. and Waymouth, R. M., *Polym. Bull.*, **38**, 141 (1997).
41. Feig, E. R., McKay, K. W. and Timmers, F. J., to Dow Chemical Company, *PCT Int. Appl.*, WO 96/07681, 1995.
42. Sumitomo Chemical, *Eur. Pat. Appl.*, EP 718 323, 1995.
43. Tosoh, *Jpn. Pat.*, JP 08 085 740, 1994.
44. Tosoh, *Jpn. Pat.*, JP 07 278 230, 1994.
45. Krause, S. J., in Paul, D. R. and Newman, S. (Eds), *Polymer Blends*, Academic Press, New York, 1978, Vol. 1, p. 16.
46. Cheung, Y. W., Guest, M. J. and Chum, P. S., in *Proceedings of SPE ANTEC 56th Annual Conference*, 1998, p. 1798.
47. Coleman, M. M., Serman, C. J., Bhagwager, D. E. and Painter, C. P., *Macromolecules*, **18**, 2719 (1990).
48. Brown, H. R., Char, K., Deline, V. R. and Green, P. F., *Macromolecules*, **26**, 4155 (1993).
49. Char, K., Brown, H. R. and Deline, V. R., *Macromolecules*, **26**, 4164 (1993).
50. Park, C. P., Clingerman, G. P., Timmers, F. J., Stevens, J. C. and Henton, D. E., to Dow Chemical Company, *US Pat.*, 5 460 818, 1995.
51. Bradfute, J. G. *et al.*, to W. R. Grace, *PCT Int. Appl.* WO 95/32095, 1995.
52. Hussain U. B. and Anderson D. A., presented at the Transporation Research Board 74th Annual Meeting Washington, DC, January 1995.

53. Murphy, M. W., in *Proceedings of SP '97, 13th Annual World Congress*, Zurich, December 1–3, 1997.
54. Tanaka, Y. Ohki, Y. and Ikeda, M., *IEEE Trans. Electr. Insul.*, **27**, 432 (1992).
55. Tanaka, Y., Mita, Y., Ohki, Y., Hiroshi, Y., Ikeda, M. and Fumihiko, Y., *J. Phys. D, Appl. Phys.*, **23**, 1491, (1990).
56. Nippon Petrochem, *Jpn. Pat.*, JP 01 128 313, 1987.
57. *Mod. Plas.*, October, 33, (1997).

13

Metallocene-based Reactive Polyolefin Copolymers Containing p-Methylstyrene

T. C. CHUNG
Pennsylvania State University, University Park, PA, USA

1 INTRODUCTION

For many decades since the commercialization of PE and PP, the functionalization of polyolefins [1] has been of great scientific interest and a technologically important research subject with the aim of improving their adhesion to and compatibility with other materials. Unfortunately, the chemistry involved in the preparation of functional polyolefins is very limited both in direct and post-polymerization processes, owing to catalyst poisoning [2] and the inert nature [3] of polyolefins. Several recent new research approaches, including the use of late transition metals [4] and protecting groups [5] may provide some solutions.

Our functionalization approach has been focused on 'reactive' polyolefins, containing groups which can be easily incorporated in polyolefins and can be interconverted into various desirable functional groups, such as OH, NH_2, COOH, anhydrides and halides. Preferably, the reactive groups can be effectively transformed in to 'stable' initiators for 'living' graft-from polymerization reactions as illustrated in Scheme 1.

The resulting polyolefin graft copolymers, containing both polyolefin and functional polymer segments, not only offer high concentrations of functional groups but also preserve the desirable polyolefin properties (such as crystallinity, melting-point and glass transition temperature).

Our first discovery of 'reactive' polyolefin copolymers contained borane groups [6], which can be effectively incorporated into polyolefins by both direct [7] and

Metallocene-based Polyolefins Edited by J. Scheirs and W. Kaminsky
© 2000 John Wiley & Sons Ltd

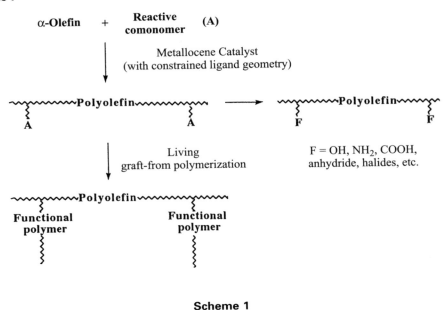

α-Olefin + Reactive comonomer (A)

Metallocene Catalyst
(with constrained ligand geometry)

Polyolefin
A   A

Polyolefin
F   F

Living
graft-from polymerization

F = OH, NH$_2$, COOH,
anhydride, halides, etc.

Polyolefin
Functional polymer   Functional polymer

Scheme 1

post-polymerization [8] processes. Borane groups in polyolefins not only can be easily interconverted in to various functional groups, as shown in borane chemistry [9], but can also be transformed in to 'stable' radical initiators [10] which then initiate 'living' radical graft-from polymerization of monomers containing functional groups. In other words, the reactive borane group serves as the intermediate for the transformation reaction from transition metal to 'living' radical polymerizations. Such a process produces graft [10] and block [11] copolymer structures, containing both polyolefin and functional polymer segments.

Recently, our 'reactive' polyolefin approach has been extended to another type of reactive comonomer, p-methylstyrene (p-MS) [12]. The major advantages of p-MS are its commercial availability, easy incorporation into polyolefins and versatility [13] in functionalization chemistry under various reaction mechanisms, including free radical, cationic and anionic processes. Benzylic protons are known to undergo many acile chemical reactions, such as halogenation [14], metallation [15] and oxidation [16], to form desirable functional groups at the benzylic position under mild reaction conditions. In addition. benzylic protons can also be interconverted to a stable anionic initiator for 'living' anionic graft-from polymerization [17].

In general, the preparation of 'reactive' polyolefin copolymers has been greatly enhanced by the exciting progress in metallocene technology [18]. Well defined metallocene catalysts with constrained ligand geometry, having spatially opened catalytic sites, allow the effective incorporation of large comonomers. Several important examples are LLDPE copolymers [19] with well defined composition and molecular weight distributions, Dow's long-chain branching polyethylene [20] and poly(ethylene–co-styrene) copolymer with pseudo-random [21] and alternating microstructures [22].

2 RESULTS AND DISCUSSION

In addition to the preparation of *p*-MS-containing polyolefins, it is very interesting to compare various metallocene catalysts for *p*-MS incorporation. Three metallocene catalysts were used, including a simple Cp_2ZrCl_2 catalyst and two bridged catalysts, $Et(Ind)_2ZrCl_2$ and $[C_5Me_4(SiMe_2N^tBu)]TiCl_2$, illustrated in Scheme 2, representing three generations of metallocene technology.

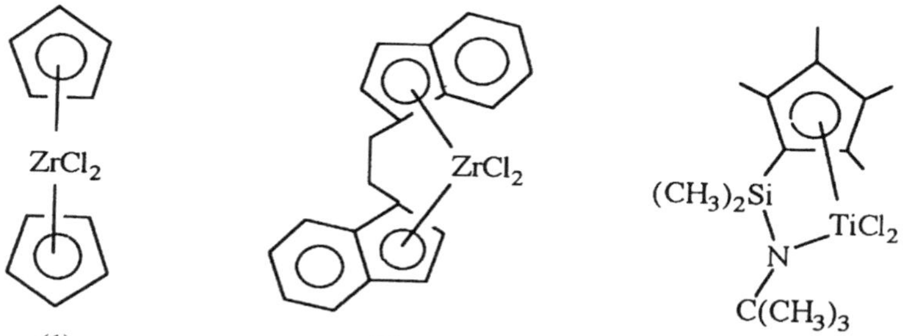

Scheme 2

Catalyst **1**, without a bridge between Cp ligands, has the zirconium active site sandwiched between two Cp rings, with a Cp—Zr—Cp angle of 180°. In catalyst **2**, the ethylene bridge induces constrained indenyl ligand geometry, with a Cp—Zr—Cp angle [23] of 125.8°. On the other hand, the silicon bridge in catalyst **3** pulls back both the Cp and amido ligands from their normal positions to form a highly constrained ligand geometry, with a Cp—Ti—N angle [24] of 107.6°. The spatial opening of the catalytic site decreases in the order **3** > **2** > **1**. Based on the structure–activity relationships of the metallocene catalysts, it is logical to predict that the incorporation of *p*-MS will follow the same trend of spatial opening. The bridged catalysts **2** and **3** with the spatially opened active sites will be more favorable than the non-bridged catalyst **1**, and the one-membered silicon bridged catalyst **3** will be preferable over the two-membered ethylene bridge in **2.** Each copolymerization reaction was systematically evaluated by studying the conversion of *p*-MS vs. reaction time, comonomer reactivity ratios at low monomer conversion and the molecular structure of the resulting copolymers. In addition, *p*-MS was compared with styrene and other methylstyrene derivatives (i.e. *o*-and *m*-methylstyrene).

2.1 POLY(ETHYLENE–CO-P-METHYLSTYRENE) COPOLYMERS

Equation (1) illustrates the copolymerization reaction of ethylene and p-methyl-styrene.

$$x\ CH_2 = CH_2 + y\ CH_2 = CH \quad \xrightarrow[\text{catalyst}]{\text{Metallocene}} \quad -(CH_2-CH_2)_x-(CH_2-CH)_y-$$

In a typical copolymerization, the reaction was started by the addition of the metallocene catalyst mixture to a solution of the two monomers in a solvent under an inert gas atmosphere. The appearance of the reacting polymer solution was very dependent on the quantity of p-MS used. In high p-MS cases, a homogeneous solution was observed through the whole copolymerization reaction. In low p-MS cases, a slurry solution with a white precipitate was observed at the beginning of the reaction. The precipitation is obviously due to the crystallinity of the copolymer which has long ethylene sequences. The copolymer was isolated by filtering, washed thoroughly with MeOH and dried under vacuum at 50°C for 8 h. Figure 1 shows typical GPC curves for the homo- and copolymers prepared with $[C_5Me_4(SiMe_2N^tBu)]TiCl_2$ catalyst.

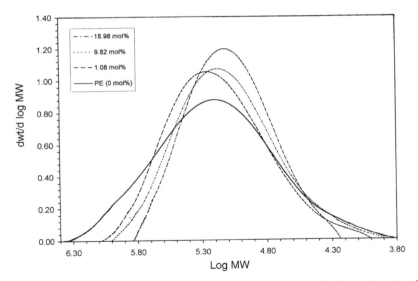

Figure 1 GPC curves of (a) polyethylene and three poly(ethylene–co-p-methyl-styrene) copolymers, containing (b) 1.08, (c) 9.82 and (d) 18.98 mol% of p-MS comonomers, prepared with $[C_5Me_4(SiMe_2N^tBu)]TiCl_2$ catalyst, and the molecular weights and molecular weight distributions (inset)

The uniform molecular weight distribution in all samples, with $\overline{M}_w/\overline{M}_n = 2\text{–}3$, implies a single-site polymerization mechanism. In fact, the GPC curves show a slight reduction of the molecular weight distribution in the copolymers, from $\overline{M}_w/\overline{M}_n = 2.86$ in PE to 1.68 in poly(ethylene–co-p-methylstyrene) containing 18.98 mol% of p-MS. Similar narrow molecular distribution results were also observed in the copolymers prepared with $Et(Ind)_2ZrCl_2$ catalyst. The better diffusibility of monomers in the copolymer structures (due to lower crystallinity) may help to provide the ideal polymerization conditions. It is very interesting that the average molecular weight of copolymers remains very high throughout the entire composition range, which may be attributed to the relatively high reactivity of p-MS.

Figure 2 shows a comparison of the DSC curves of PE homopolymer and poly-(ethylene–co-p-methylstyrene) copolymers prepared with $[C_5Me_4(SiMe_2N^tBu)]$ $TiCl_2$ catalyst. Even a small amount (ca 1 mol%) of p-MS comonomer incorporation has a significant effect on the crystallization of polyethylene. Overall, the melting-point (T_m) and crystallinity (χ_c) of the copolymer are strongly related to the density of the comonomer: the higher is the density, the lower are the T_m and χ_c. Only a single peak is observed throughout the whole composition range and the melting peak completely disappears at ca 10 mol% p-MS concentration. A similar general trend was also observed in the of DSC curves of poly(ethylene–co-p-methylstyrene) copolymers prepared with $Et(Ind)_2ZrCl_2$ catalyst. The systematic decrease in T_m and the uniform reduction of the crystalline curve imply homogeneous reduction of PE in consecutive sequences. It is interesting that the T_m peak [ranging from 80 to 45°C

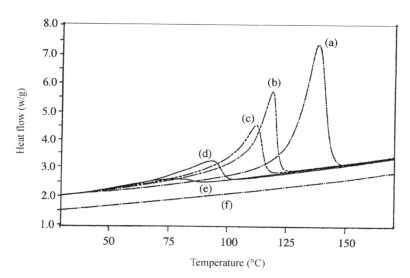

Figure 2   Comparison of DSC curves between (a) polyethylene homopolymer and poly(ethylene–co-p-methylstyrene) copolymers with (b) 1.08, (c) 2.11, (d) 5.40, (e) 9.82 and (f) 18.98 mol% of p-MS comonomer

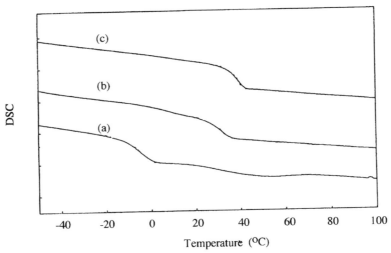

Figure 3 DSC curves of (a) poly(ethylene–co-*p*-methylstyrene) copolymers with (a) 18.98, (b) 32.8 and (c) 40 mol% of *p*-MS comonomer

as shown in Figure 2(e)] of the copolymer containing 9.82 mol% of *p*-MS covers a similar T_m range (45–55°C) to paraffin wax. This poly(ethylene–co-*p*-methylstyrene) copolymer consisted of on average 20–22 consecutive methylene units, which is in the molecular weight range of solid paraffin wax.

The removal of crystallinity provides the opportunity to obtain elastic properties in many thermoplastic polymers. Figure 3 shows the glass transition temperature (T_g) of poly(ethylene–co-*p*-methylstyrene) copolymers with 18.98, 32.8 and 40 mol% *p*-MS. The lowest observed T_g [Figure 3(a)] is −5.7 °C for the copolymer containing 18.98 mol% *p*-MS. With increase in *p*-MS concentration, the T_g of the copolymer systematically increases, as shown in Figure 3(b) and (c). The single T_g with a sharp thermal transition is indicative of a homogeneous copolymer structure. It is clear that poly(ethylene–co-*p*-methylstyrene) copolymer has difficulty in forming a good elastomer owing to the high T_g of poly(*p*-methylstyrene). To achieve an elastic polymer, the system requires a third monomer which can provide a low T_g. The detailed results for elastic poly(ethylene–ter-propylene–ter-*p*-MS) and poly-(ethylene–ter-1-octene–ter-*p*-MS) terpolymers will be discussed later.

Table 1 summarizes the copolymerization results, including all three catalyst systems. The copolymerization efficiency follows the sequence [C_5Me_4 ($SiMe_2N^tBu$)]$TiCl_2$ > Et(Ind)$_2ZrCl_2$ > Cp$_2ZrCl_2$, which is directly related to the spatial openings at the active sites. In comparable runs, p-363/p-358 and p-365/p-360, Cp$_2ZrCl_2$ (**1**) and Et(Ind)$_2ZrCl_2$ (**2**) catalysts produced copolymers with 2.2/5.16 and 1.84/4.76 mol% *p*-MS concentrations, respectively, and 26.2/51.1 and 7.72/32.8% overall *p*-MS conversions, respectively. On the other hand, the [$C_5Me_4(SiMe_2N^tBu$)]$TiCl_2$ catalyst gave the best results of the three catalysts in the

Table 1 Summary of the copolymerization reactions^a between ethylene and p-methylstyrene

Run No.	Catalyst^b (μ mol)	Ethylene (psi)	p-MS (mol/l)	Solvent	Temperature (°C)	Yield (g)	Catalyst efficiency [kg P/(mol M h)]	p-MS in copolymer (mol%)	Conversion of p-MS (%)
p-363	**1**, 17	45	0.678	Hexane	50	24.2	1423.5	2.20	26.2
p-365	**1**, 17	45	0.678	Toluene	50	8.44	496.5	1.84	7.72
p-356	**2**, 17	45	0.085	Hexane	50	6.5	382.4	1.83	47.2
p-372	**2**, 17	45	0.508	Hexane	50	19.8	1164.7	3.94	48.7
p-358	**2**, 17	45	0.678	Hexane	50	21.9	1288.2	5.16	51.1
p-361	**2**, 17	45	1.36	Hexane	50	18.9	1111.8	7.20	29.0
p-371	**2**, 17	45	2.03	Hexane	50	20.8	1223.5	8.94	25.4
p-357	**2**, 17	45	0.085	Toluene	50	5.70	335.3	1.30	29.9
p-392	**2**, 17	45	0.456	Toluene	50	8.63	507.6	3.85	23.2
p-360	**2**, 17	45	0.678	Toluene	50	15.1	888.2	4.76	32.8
p-362	**2**, 17	45	1.36	Toluene	50	19.5	1147.1	6.36	27.0
p-375	**2**, 17	45	2.03	Toluene	50	19.4	1141.2	8.49	22.8
p-270	**3**, 10	45	0	Toluene	30	4.4	440.0	–	–
p-377	**3**, 10	45	0.447	Hexane	30	12.0	1200.0	13.5	90.3
p-378	**3**, 10	45	0.912	Hexane	30	15.5	1550.0	22.6	81.3
p-267	**3**, 10	45	0.447	Toluene	30	13.0	1300.0	10.9	83.8
p-379	**3**, 10	45	0.912	Toluene	30	17.4	1740.0	21.6	86.9
p-380	**3**, 10	45	1.82	Toluene	30	24.2	2420.0	32.8	75.8
p-383	**3**, 10	10	1.82	Hexane	30	15.9	1590.0	40.0	54.6

^{a} 45 psi ethylene, ca 0.309 mol/l in toluene, 0.424 mol/l in hexane at 50 °C; ca 0.398 mol/l in toluene, 0.523 mol/l in hexane at 30 °C; 10 psi ethylene, ca 0.116 mol/l in hexane at 30 °C.
^{b} **1**, Cp$_2$ZrCl$_2$/MAO; **2**, Et(Ind)$_2$ZrCl$_2$/MAO; **3**, [(C$_5$Me$_4$)SiMe$_2$N(^{t}Bu)]TiCl$_2$/MAO.

preparation of poly(ethylene–co–p-methylstyrene) copolymers. In run p-377, about 90 % of p-MS was incorporated into copolymer in 1 h. In run p-383, the reaction produced copolymer containing 40 mol% of p-MS, which is close to the ideal 50 mol% (as will be discussed in relation to reactivity ratio and copolymer microstructure studies, the consecutive insertion of p-MS with all three catalyst systems is almost impossible). A high concentration of p-MS in the copolymer and a high p-MS conversion were achieved with catalyst **3**, which must provide excellent spatial freedom for p-MS to access the propagating chain end.

In general, the catalyst activity systematically increases with increase in p-MS content, which was also observed in 1,4-hexadiene copolymerization reaction [8] and could be a physical phenomenon relative to the improvement of monomer diffusion in the lower crystalline copolymer structures. With $[C_5Me_4(SiMe_2N^tBu)]$ $TiCl_2$, the catalyst activity attained a value of more than 2.4×10^6 g of copolymer per mole of Ti per hour in run p-380, which is about six times the value for the homopolymerization of ethylene in run p-270 under similar reaction conditions. It is very interesting that a very small solvent (hexane and toluene) effect on catalyst **3** activity was observed in comparative runs, p-377/p-267 and p-378/p-379, despite the significant difference at the beginning of the reaction conditions (heterogeneous in hexane and homogeneous in toluene). However, the solvent effect is very significant for both catalyst **1** and **2** systems. All comparative reaction pairs (p-363/p-365, p-356/p-357, p-358/p-360, p-361/p-362 and p-371/p-375), under similar reaction conditions, consistently show higher p-MS incorporation in hexane solution. The explanation of the solvent effect is not clear.

2.2 REACTIVITY RATIOS

The best way to investigate a copolymerization is to measure the reactivity ratio of the comonomers. To obtain meaningful results, a series of experiments were carried out by varying the monomer feed ratio and comparing the resulting copolymer composition at low monomer conversion ($<10\%$). The reactivity ratios between ethylene ($r_1 = k_{11}/k_{12}$) and p-MS ($r_2 = k_{22}/k_{21}$) were estimated by the Kelen–Tüdos method [25]. With $[C_5Me_4(SiMe_2N^tBu)]TiCl_2$, high r_1 ($r_1 = 19.6$ and 21.4 at 20 and 60°C, respectively) and low r_2 ($r_2 = 0.04$ and 0.08 at 20 and 60°C, respectively) indicate the strong tendency for ethylene consecutive insertion and the very low possibility of continuous p-MS insertion. The values of $r_1 r_2$ are close to unity at both temperatures, which suggests near ideal random copolymerization reactions and a very low probability of finding two adjacent p-MS units in the polymer chain. In other words, the p-MS units are homogeneously distributed in the polymer chain. With $Et(Ind)_2ZrCl_2$, the copolymerization reactions exhibit even higher r_1 ($r_1 > 60$), strongly favorable for ethylene incorporation, and almost no possibility of p-MS consecutive insertion ($r_2 \approx 0$). The less open active site in $Et(Ind)_2ZrCl_2$ catalyst may sterically prevent p-MS consecutive insertion.

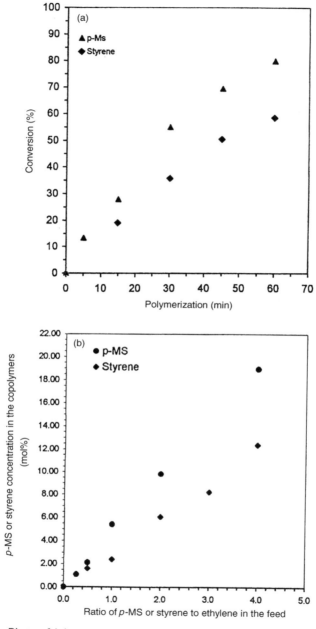

Figure 4 Plots of (a) comonomer conversion vs reaction time and (b) comonomer concentration in copolymer vs comonomer ratio, during the copolymerization reactions of ethylene with *p*-methylstyrene and with styrene using $[C_5Me_4(SiMe_2N^tBu)]TiCl_2$ catalyst

2.3 COMPARISON AMONG STYRENE DERIVATIVES

In the evaluation of p-MS comonomer, it is very interesting to compare p-MS with styrene and methylstyrene isomers. The consumption of comonomer during the reaction is useful way to understand the dynamics of copolymerization reactions. Figure 4(a) shows comparative plots of p-MS and styrene incorporation vs reaction time in the batch copolymerization reaction of ethylene (29 psi) and comonomer (0.356 mol/l), using $[C_5Me_4(SiMe_2N^tBu)]TiCl_2$ catalyst at 60°C.

Significantly better p-MS incorporation started at the beginning of copolymerization reaction. After 30 min, the difference became relatively constant owing to the depletion of p-MS. After about 1 h, >80 % of p-MS and <60 % of styrene were incorporated in the ethylene copolymers. Figure 4(b) compares the incorporated comonomer concentration (mol%) in copolymer with the comonomer ratio. Each copolymerization reaction was carried out at 40°C for 15 min by using $[C_5Me_2(SiMe_2N^tBu)]TiCl_2$ catalyst. At every monomer feed ratio, p-MS consistently shows more than a 30% higher incorporation than the corresponding styrene. The significantly higher p-MS incorporation must due to the electron donation by the p-methyl group, which is favorable in the 'cationic' polymerization mechanism [26]. Sterically, the methyl group at a para-position does not affect the monomer insertion. The p-MS incorporation was also compared with that of its derivatives, o-MS and m-MS, with consistently better incorporation. Both isomers (o-MS and m-MS) do not receive the full benefits of the electronic and steric effects that exist in p-MS.

2.4 p-MS-CONTAINING POLYOLEFIN ELASTOMERS

As discussed for poly(ethylene–co-p-methylstyrene) copolymers, despite the completely amorphous structure the lowest T_g observed with this type copolymer was about −5 °C, which is too high to be useful in most elastomer applications. For many commercial applications elastomers with low $T_g < -45$ °C and having 'reactive' sites (such as p-MS units), which can effectively form cross-linking networks and produce stable residues, are very desirable. It is certainly very interesting to expand the poly(ethylene–co-p-methylstyrene) system to polyolefin elastomers. In ethylene–propylene cases, This means preparing a random terpolymer containing a close to equal molar ratio of ethylene–propylene and some p-MS 'reactive' units. With the unprecedented capability of metallocene technology in copolymerization reactions, it is also very interesting to expand the polyolefin elastomer to new classes containing high α-olefins, such as 1-octene (instead of propylene), which can effectively prevent the crystallization of small consecutive ethylene units and provide low T_g properties. In chemistry, the terpolymerization reaction involving ethylene, 1-octene (high α-olefin) and p-methylstyrene (aromatic olefin) simultaneously, which is very difficult to achieve in Zieglet–Natta polymerization, will be an ultimate test of the metallocene technology.

2.5 POLY(ETHYLENE–TER-PROPYLENE–TER-p-METHYLSTYRENE)

The polymerization reactions were carried out in a Parr reactor. The desirable proportions of ethylene and propylene were mixed in a steel reservoir before piping into the reactor containing a mixed solution of p-MS, MAO and toluene. The polymerization reaction was initiated by charging $[C_5Me_4(SiMe_2N^tBu)]TiCl_2$ catalyst into the monomer mixture. A constant mixed ethylene–propylene pressure was maintained throughout the polymerization process. To ensure a constant comonomer ratio, the polymerization was usually terminated after 15 min by adding dilute $HCl–CH_3OH$ solution.

Figure 5 compares two GPC curves between poly(ethylene–ter-polypropylene–ter-p-methylstyrene) terpolymer (with ethylene: propylene mole ratio $\approx 54:44$ and 2 mol% p-MS) and the corresponding poly(ethylene–co-polypropylene) copolymer. Both polymers were prepared with $[C_5Me_4(SiMe_2N^tBu)]TiCl_2/MAO$ catalyst under similar reaction conditions.

Similar molecular weights and molecular weight distributions were observed, which indicates that the addition of p-MS did not significantly alter the polymerization process (similar propagation rate, no additional termination reaction). The high molecular weight ($M_w \approx 237\,700$ and $M_n \approx 107\,500$ g/mol) and narrow molecular weight distribution ($M_w/M_n \approx 2.2$) are indicative of an ideal single-site polymerization mechanism.

All experimental results with various monomer feed ratios are summarized in Table 2. Overall, the $[C_5Me_4(SiMe_2N^tBu)]TiCl_2/MAO$ metallocene catalyst shows excellent activity $[(3.4–5.4) \times 10^6$ g polymer mol Zr h at 50°C] in all reactions, with comparative reactivities between ethylene and propylene and good incorporation of

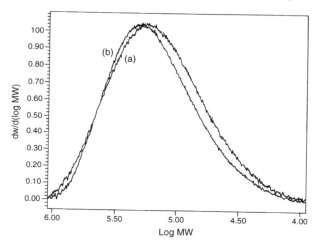

Figure 5 Comparison of two GPC curves of (a) poly(ethylene–ter-polypropy-lene–ter-p-methylstyrene) terpolymer (run p-120) and (b) poly(ethylene–co-poly-propylene) copolymer (run p-116)

Table 2 Summary of terpolymerization^a of ethylene, propylene and p-MS using [(C$_5$Me$_4$)SiMe$_2$N(^{t}Bu)TiCl$_2$]/MAO

Run No.	Ethylene:propylene mixing ratio (psi/psi)	Monomer concentration in the feed (mol/l)			Catalyst activity (kg mol/h)	Copolymer composition (mol%)			T_g (°C)	M_w (g/mol)	M_n (g/mol)	PD
		Ethylene	Propylene	p-MS		[E]	[P]	[p-MS]				
p-116	80:40	0.13	0.28	0	4.9 × 10³	53.9	46.1	0	−49.4	300 444	134 285	2.2
p-117	70:50	0.12	0.35	0	4.7 × 10³	41.8	58.2	0	−43.7	284 280	120 813	2.4
p-107	50:50	0.10	0.43	0	4.8 × 10³	36.2	63.8	0	−35.5	–	–	–
p-115	30:70	0.06	0.60	0	4.9 × 10³	13.0	87.0	0	−16.4	–	–	–
p-110	60:60	0.10	0.43	0.1	5.4 × 10³	37.0	58.5	4.5	−20.3	184 149	90 634	2.0
p-111	60:60	0.10	0.43	0.3	4.9 × 10³	39.2	52.9	7.9	−19.1	194 245	97 634	2.0
p-112	60:60	0.10	0.43	0.5	5.3 × 10³	40.3	48.6	11.1	−9.1	188 861	80 855	2.3
p-118	70:50	0.12	0.35	0.3	4.0 × 10³	46.4	43.6	10.0	−20.7	198 199	74 890	2.7
p-113	50:70	0.08	0.50	0.3	3.5 × 10³	32.4	59.2	8.4	−12.4	183 180	91 786	2.0
p-120	80:40	0.13	0.28	0.05	4.1 × 10³	54.4	43.8	1.8	−45.8	237 702	107 519	2.2
p-119	70:50	0.12	0.35	0.05	4.0 × 10³	46.1	52.3	1.6	−41.0	195 760	75 028	2.6
p-128	85:35	0.14	0.25	0.03	4.4 × 10³	56.3	43.1	0.6	−48.6	269 400	104 335	2.6
p-127	80:40	0.13	0.28	0.03	3.8 × 10³	50.7	48.6	0.7	−45.9	244 331	85 506	2.9

^{a} Polymerization conditions: 100 ml of toluene; [Ti] = 2.5 × 10^{-6} mol; [MAO]/[Ti] = 3000; 50 °C; 15 min.

p-MS. Such an effective incorporation of an aromatic monomer in polyolefins is very difficult to achieve with traditional Ziegler–Natta catalysts. In the control set (runs p-116, p-117, p-107, and p-115), only involving ethylene and propylene, the copolymer composition is basically governed by the monomer feed ratio. However, with *p*-MS, the incorporation of propylene seems to slow. This trend is very clearly observed in the comparative set (runs p-110, p-111 and p-112), with constant ethylene and propylene feeds and various *p*-MS concentrations: the higher the *p*-MS concentration in the feed, the lower is the propylene: ethylene ratio in the copolymer. As discussed in a previous paper [12], no consecutive *p*-MS incorporation was observed in $[C_5Me_4(SiMe_2N^tBu)]TiCl_2/MAO$ copolymerization reactions owing to steric hindrance at the propagating *p*-MS site. The same steric hindrance may also have some effect on the subsequent propylene insertion, preferring ethylene over propylene. On the other hand, the incorporation of *p*-MS seems insensitive to the ethylene: propylene feed ratio. In both comparative sets of runs p-120 vs p-119 and p-128 vs p-127, with constant *p*-MS concentration in each comparative run (0.05 and 0.03 mol/1, respectively) and varying ethylene:propylene feed ratio, the incorporation of *p*-MS is very constant at about 1.6–1.8 and 0.6–0.65 mol%, respectively.

In general, the molecular weights of these terpolymers are very high. Comparing runs p-116 vs p-120 and p-117 vs p-119, with the same amount of ethylene and propylene feeds and with and without *p*-MS, only a small reduction in molecular weight arises from the incorporation of *p*-MS. It is very interesting that the replacement of *p*-MS with styrene in the same reaction conditions significantly lowers the molecular weight of poly(ethylene–ter-propylene–ter-styrene). The results may be attributed to the comparable reactivities of *p*-MS and of ethylene and propylene. The electron donation from the *p*-methyl group in *p*-MS is favorable in this cationic coordination polymerization mechanism [26].

The glass transition temperature (T_g) was examined by DSC. Figure 6 shows several DSC curves for EP–*p*-MS terpolymers (samples p-112, p-118 and p-120 in Table 2). Each curve has only a sharp T_g on a flat baseline, without any detectable melting-point. The combination implies a homogeneous terpolymer microstructure with completely amorphous morphology. The same clean DSC curves were observed for all samples shown in Table 2, even the sample containing >87 mol% propylene, which may have the most atactic propylene sequences incapable of crystallization. The T_g is clearly a function of the propylene and *p*-methylstyrene contents. Comparing the ethylene–propylene copolymers (without *p*-MS units) (runs p-116, p-117, p-107 and p-115), the T_g are linearly proportional to the propylene content and level off at ca $-50\,^\circ$C with a composition containing ca 50 % propylene (similar results were reported for the EPDM case [27]). The T_g increases significantly with the incorporation of *p*-MS in ethylene–propylene copolymers. Comparing runs p-117 and p-118, both having containing ca 42 mol% ethylene and *p*-MS:propylene mole ratios of 0:58 and 10:48, respectively, the T_g increases from -43 to $-20\,^\circ$C. A similar result was observed in the pair of runs p-107 and p-110, with 37 mol% ethylene and a smaller difference in

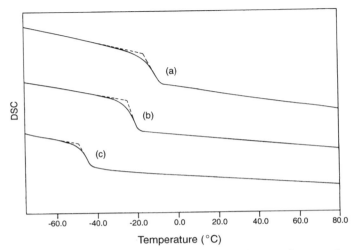

Figure 6 Comparison of DSC curves of EP–*p*-MS terpolymers from (a) run
p-112, (b) run p-118 and (c) run p-120

p-MS:propylene mole ratios of 0:63 and 4.5:58, respectively; the T_g change is
smaller, from −35 to −20 °C. It is very interesting to compare runs p-116 and p-120,
both with ideal ca 54 mol% ethylene and only a very small difference in *p*-
MS:propylene mole ratios (0:46 vs 1.8:44); the T_gs are −50 and −45 °C, respectively. The same T_g trend holds, although much smaller. Overall, the composition of
EP–*p*-MS material with low T_g < −45 °C is very limited, only up to a composition
with <2 mol% *p*-methylstyrene content. Despite the random terpolymer structure
and the ideal (ca 55:45) ethylene:propylene ratio, a further increase in *p*-MS raises
the T_g of the terpolymer to > −40 °C. Obviously, the high T_gs of both the propylene
(T_g of PP ≈ 40 °C) and *p*-MS [T_g of poly (*p*-MS) ≈ 110 °C] components preclude
EP–*p*-MS from achieving some desirable elastomers containing both a high content
of 'reactive' *p*-MS and a low T_g(< −45 °C).

2.6 POLY (ETHYLENE–TER-1-OCTENE–TER-p-METHYLSTYRENE)

In EP elastomers, the primary function of propylene units is to prevent the crystallization of ethylene sequences. In traditional Ziegler–Natta polymerization, propylene is a natural choice because it has the closest reactivity to ethylene. However, in
terms of effectiveness of preventing the crystallization of ethylene sequences and
obtaining low-T_g material, propylene is not the best comonomer, owing to (i) the
small CH_3 side group and (ii) the relatively high T_g(ca 0 °C) of the propylene
component. Concerning our objective to prepare polyolefin elastomers with low
T_g < −45 °C and containing a wide concentration range of 'reactive' *p*-MS units,
the EP system clearly shows the serious limitations as discussed above. It is very
interesting to replace propylene units with high α-olefins, such as 1-octene, which

can effectively prevent the crystallization of ethylene sequences (as known in LLDPE [19]) and is a low-T_g material [28] with no possibility of self-crystallization. Additionally, it is also very interesting to study the metallocene technology in a very complicated termonomer system, involving ethylene, 1-octene (high α-olefin) and p-MS (aromatic olefin).

The terpolymerization reaction of ethylene, 1-octene and p-MS was usually started by adding the catalyst mixture [$C_5Me_4(SiMe_2N'Bu)$]$TiCl_2$/MAO to the monomer solution, containing 1-octene and p-MS monomers and partially soluble ethylene (at constant pressure) in toluene solvent. A homogeneous solution was observed throughout the whole copolymerization reaction. Figure 7 compares the GPC curves of the terpolymers prepared with the same ethylene (0.4 mol/l) and 1-octene (0.8 mol/l) and different p-MS (0.1, 0.2 and 0.4 mol/l) concentrations.

Overall, the polymer molecular weight is fairly high ($M_w \approx 200\,000$ g/mol) and is not significantly dependent on the content of p-MS. The molecular weight distributions, $M_w/M_n < 2.5$, similar to those of most of metallocene-based homo-and co-polymers, indicate a single-site reaction with good comonomer reactivities. The detailed experimental results are summarized in Table 3.

Comparing runs p-471 and p-470, with similar 1-octene and p-MS concentrations and different ethylene contents, the molecular weight of the terpolymer is basically proportional to ethylene content. On the other hand, comparing runs p-471, p-472, p-473, p-474, p-475 and p-476, with the same ethylene concentration and different 1-octene and p-MS ratios, the molecular weights of all terpolymers are very similar with no clear correlation pattern. The molecular weight is clearly governed by ethylene concentration, which must be due to the significantly higher reactivity of

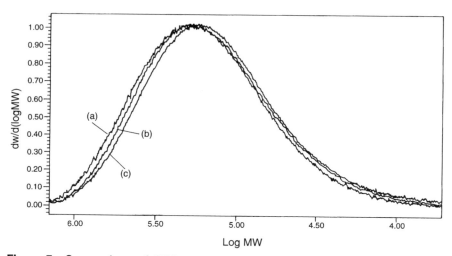

Figure 7 Comparison of GPC curves of three EO–p-MS terpolymers prepared from (a) run p-471, (b) run p-476 and (c) run p-472

Table 3 Summary of terpolymerization[a] of ethylene, 1-octene and p-MS using [(C$_5$Me$_4$) SiMe$_2$N(^{t}Bu)]TiCl$_2$/MAO catalyst

Run No.	Monomer concentration in feed (mol/l)			Yield (g)	Copolymer composition (mol%)			T_g (°C)	M_w (g/mol)	M_n (g/mol)	PD
	E[b]	1-Oct	p-MS		[E]	[O]	[p-MS]				
p-465[c]	0.25	0.89	0	7.0	41.4	58.6	0	−61.8	134 515	60 614	2.2
p-466[c]	0.25	0.89	0.13	5.2	40.0	54.5	5.6	−51.3	75 496	36 045	2.1
p-470	0.20	0.80	0.10	7.3	54.2	43.0	2.7	−56.2	173 989	74 703	2.3
p-471	0.40	0.80	0.10	10.1	61.1	36.0	2.9	−58.1	219 752	96 802	2.3
p-472	0.40	0.80	0.20	9.8	60.3	36.3	4.4	−55.7	182 185	77 497	2.4
p-473	0.40	0.40	0.20	8.2	59.6	34.0	6.4	−50.1	208 920	86 812	2.4
p-477	0.40	0.20	0.20	6.4	80.2	14.1	5.7	−37.3	227 461	91 490	2.5
p-476	0.40	0.80	0.40	9.0	63.4	29.3	7.3	−50.3	202 085	96 035	2.1
p-475	0.40	0.60	0.40	9.1	67.2	24.7	8.1	−48.2	205 124	93 763	2.2
p-478	0.40	0.40	0.40	7.9	73.3	18.5	8.1	−44.7	246 300	122 306	2.0
p-474	0.40	0.60	0.15	9.0	64.7	31.3	4.0	−55.7	224 476	102 617	2.2
p-396[d]	0.52	0.38	0.91	5.6	64.0	18.1	17.7	−25.8	106 954	55 825	1.9
p-383[e]	0.13	0	1.82	15.9	60.0	0	40.0	38.3	—	—	—

[a] Polymerization conditions (unless specified otherwise): 100 ml of toluene; [Ti] = 2.5 × 10^{-6} mol [MAO]/[Ti] = 3000; 50 °C; 30 min.
[b] Solubility of ethylene: 0.25 mol/l for 29 psi in hexane at 60 °C, 0.20 mol/l for 2 bar, 0.40 mol/l for 4 bar in toluene at 50 °C, 0.52 mol/l for 45 psi in hexane at 30 °C, 0.13 mol/l for 10 psi in hexane at 30 °C.
[c] Solvent: 100 ml of hexane; 60 °C.
[d] Solvent: 100 ml of hexane; 30 °C; [Ti] = 2.5 × 10^{-5} mol; [MAO]/[Ti] = 2000; ethylene pressure 45 psi.
[e] Solvent: 100 ml of hexane; [Ti] = 10 × 10^{-6} mol; [MAO]/[Ti] = 1500; 30 °C; 60 min.

ethylene among the three monomers. It is interesting that the incorporation of 1-octene is also accompanied by some reduction at high p-MS concentrations (comparative runs p-471 vs p-476 and p-475 vs p-474). Following the enchainment of p-MS, subsequent insertion of ethylene is faster than that of 1-octene, possibly owing to steric hindrance at the active site. In run p-478, using the same comonomer concentration, the resulting terpolymer, having an ethylene:1-octene:p-MS mole ratio of 9:2:1, clearly indicates the comonomer reactivity sequence ethylene > 1-octene > p-MS. In fact, the ratio is consistent with all results (runs p-470, p-471, p-472, p-473, p-474, p-475 and p-476), despite the very different comonomer feed ratios. This reactivity ratio is very useful for predicting the composition of EO–p-MS terpolymer. The thermal transition temperature of PO–p-MS terpolymer was examined by DSC studies. Figure 8 shows the DSC curves for two PO–p-MS terpolymers (runs p-472 and p-478) and one poly (ethylene–co-p-methylstyrene) copolymer (run p-383).

Comparing the curves for runs 472 and 383, both having the same ethylene (ca 60 mol%) content but different 1-octene:p-MS ratios, the T_g changes from >30 °C in run 383 (with no 1-octene) to < −55 °C in run 472 (with 35 mol% 1-octene). The T_gs of co- and terpolymers are summarized in Table 4. It is very interesting that the EO–p-MS sample with even up to 8 mol% of p-MS still shows T_g < −45 °C, which

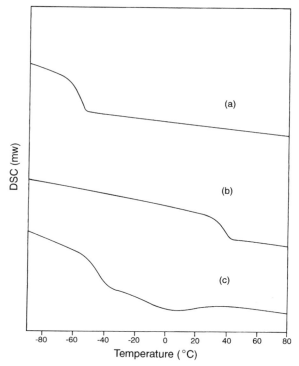

Figure 8 DSC curves of two poly(ethylene–ter-1-octene–ter-p-methylstyrene) terpolymers prepared from (a) run p-472 and (c) run p-478 and (b) poly(ethylene–co-p-methylstyrene) (run p-383)

Table 4 Summary of anionic graft-from polymerization

Lithiated polymer (g)	Monomer[a] (g)	Solvent	Temperature (°C)	Time (h)	Isolated polymer (g)	Comonomer in graft copolymer (mol%)
1.5	ST, 1.9	hexane	25	1	3.3	24.4
1.2	ST, 5.9	hexane	25	1	6.8	54.7
1.0	MMA, 3.7	THF	0	1.5	1.86	20.0
1.0	MMA, 3.4	THF	0	15	2.66	31.8
0.8	MMA, 4.0	hexane	25	5	3.08	44.4
0.8	MMA, 4.0	hexane	0	5	2.21	33.0
1.0	AN, 3.0	hexane	25	16	2.99	51.2
1.0	p-MS, 4.0	hexane	25	0.5	5.0	48.7

[a] ST = styrene; MMA = methyl methacrylate; AN = acrylonitrile; p-MS = p-methylstyrene.

is very different from the corresponding EP–*p*-MS copolymer, as discussed above. These results clearly show the advantages of 1-octene comonomer (over propylene), which ensures the formation of amorphous polyolefin elastomer with low T_g and high *p*-MS content. In Figure 8(c), the DSC curve of terpolymer (p-478) contains a very weak crystalline peak at ca 5°C. Apparently, a total concentration of 1-octene and *p*-MS of > 30 mol% may be necessary to eliminate completely the crystallization of ethylene sequences.

2.7 ANIONIC GRAFT-FROM REACTIONS

Our major research interest in incorporating *p*-MS into polyolefins is due to its versatility to access a broad range of functional groups. The benzylic protons are ready for many chemical reactions, such as halogenation, oxidation and metallation, as shown in Scheme 3. The metallated polymer can further be used in living anionic graft-from reactions, which offers a reatively simple process in the preparation of polyolefin graft copolymers. Usually, the lithiated polymer was suspended in an inert organic diluent before addition of monomers, such as styrene, MMA, vinyl acetate, acrylonitrile and *p*-methylstyrene. It is interesting that the heterogeneous condition allows the easy removal of excess reagent, which was impossible under homo-

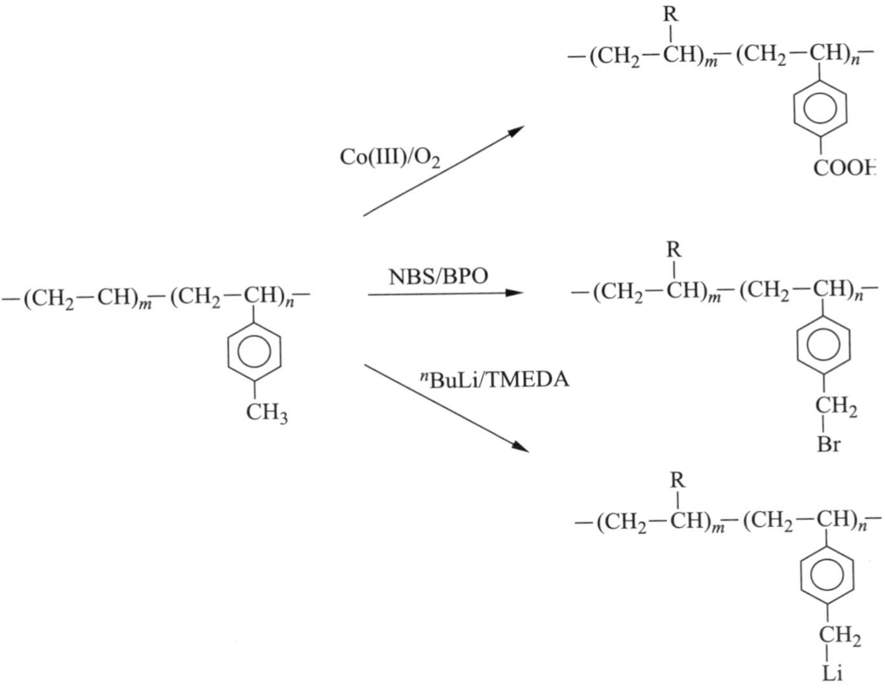

Scheme 3

geneous conditions, such as in the case of poly(isobutylene–co-*p*-methylstyrene) [29]. The unreacted alkyl lithium complex is much more reactive than benzylic lithium and can produce a lot of undesirable ungrafted homopolymers during the graft-from reaction.

To study the efficiency of the lithiation reaction, some of the lithiated polymer was converted to an organosilane-containing polymer by reaction, with chlorotrimethyl-silane. Figure 9 compares the ^{1}H NMR spectra of the starting P[E–co-(*p*-MS)], containing 0.9 mol% of *p*-MS, and the resulting trimethylsilane-containing PE copolymers, which had been metallated using either ^{s}BuLi/TMEDA^{n}BuLi/TMEDA, respectively, under the same reaction conditions.

In Figure 9(a), in addition to the major chemical shift at 1.35 ppm, corresponding to CH_2, there are three minor chemical shifts around 2.35, 2.5 and 7.0–7.3 ppm, corresponding to CH_3, CH and aromatic protons in *p*-MS units, respectively. After the functionalization reaction, Figure 9(b) and (c) show the reduction in peak intensity at 2.35 ppm and no detectable intensity change at both 2.5 and 7.0–7.3 ppm chemical shifts. In addition, two new peaks at 0.05 and 2.1 ppm, corresponding to $Si(CH_3)_3$ and ϕ-CH_2Si, are observed. Overall, the results indicate a 'clean' and selective metallation reaction at the *p*-methyl group. The integrated intensity ratio between the chemical shift at 0.05 ppm and the chemical shifts between 7.0 and 7.3 ppm and the number of protons that the two chemical shifts represent determines the efficiency of metallation reaction. The ^{n}BuLi/TMEDA converted only 24 mol% of *p*-methylstyrene to benzyllithium. On the other hand, ^{s}BuLi/TMEDA was much more effective, achieving 67 mol% conversion. Apparently, the metallation reaction was not inhibited by the insolubility of polyethylene; most of the *p*-MS units must be located in the amorphous phases which are swellable by the appropriate solvent during the reaction. Both DSC and GPC studies, comparing copolymer samples before and after the functionalization reaction, indicated no significant change in the melting-point and the molecular weight, respectively.

Most of the lithiated PE powder was suspended in cyclohexane before the addition of monomers. The living anionic polymerization took place at room temperature, similar to the well known solution anionic polymerization [30]. To ensure sufficient time for monomer diffusion in the heterogeneous conditions, the reaction was continued for 1 h before terminating it by the addition of methanol. The conversion of monomers (estimated from the yield of graft copolymer) was almost quantitative (>90 %) in 1 h. The reaction mixture was usually subjected to a vigorous extraction process, by refluxing THF through the sample in a Soxhlet extractor for 24 h, to remove any polystyrene or poly(*p*-methylstyrene) homopoly-mers. In all cases, only a small amount (<10 %) of THF-soluble fraction was obtained. The THF-insoluble fraction is mainly PE graft copolymer and is comple-tely soluble in xylene at elevated temperature. Figure 10 shows the ^{1}H NMR spectra of three PE–g-PS copolymers.

Compared with the ^{1}H NMR spectrum of the starting P[E–co-(*p*-MS)], three additional chemical shifts arise at 1.55, 2.0 and 6.4–7.3 ppm, corresponding to CH_2,

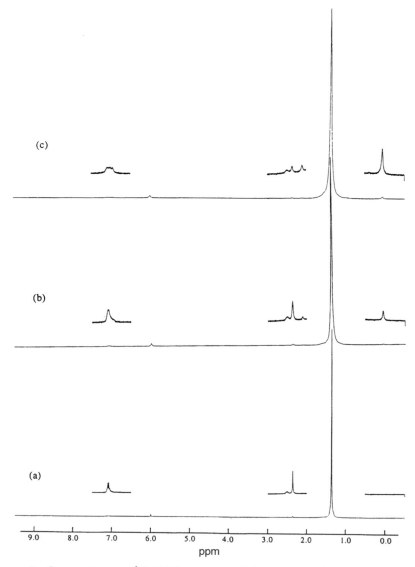

(c)

(b)

(a)

ppm

Figure 9 Comparison of ^1H NMR spectra of (a) poly(ethylene–co-*p*-methylstyrene) with 0.9 mol% of *p*-methylstyrene and two corresponding trimethylsilyl derivatives prepared via lithiation reactions using (b) ^{n}BuLi/TMEDA and (c) ^{s}BuLi/TMEDA reagents

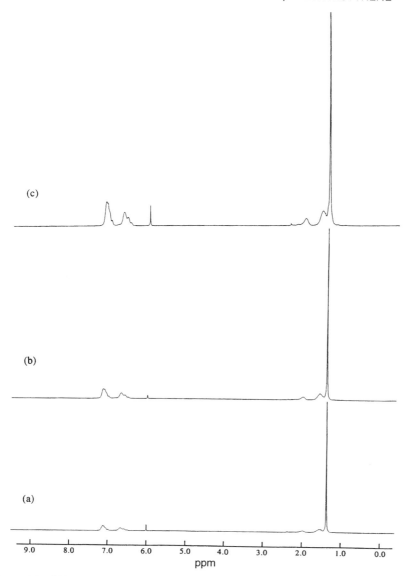

Figure 10 ^{1}H NMR spectra of PE–g-PS copolymers containing (a) 25.6 (b) 38.1 and (c) 43.8 mol% of polystyrene

CH and aromatic protons in polystyrene, respectively. The quantitative copolymer composition was calculated from the ratio of two integrated intensities between aromatic protons ($\delta = 6.4-7.3$ ppm) in the PS side-chains and methylene protons ($\delta = 1.35-1.55$ ppm) and the number of protons that the two chemical shifts represent. Figure 10(a), (b) and (c) indicate 25.6, 38.1 and 43.8 mol% of PS, respectively, in PE–g-PS copolymers. Table 4 summarizes the reaction conditions and the experimental results.

Overall, the experimental results clearly show a new class of PE graft copolymers which can be conveniently prepared by the tranformation of metallocene catalysis to anionic graft-from polymerization via p-MS groups.

2.8 PE–PS POLYMER BLENDS

It is interesting to study the compatibility of PE–g-PS copolymer in HDPE and PS blends. Polarized optical microscopy and the SEM were used to examined the surfaces and bulk morphologies, respectively. Two blends studied comprised an overall 50:50 weight ratio of PE and PS; one was a simple 50:50 mixture of HDPE and PS and the other was a 45:45:10 by weight mixture of HDPE, PS, and PE–g-PS (containing 50 mol% of PS). Figure 11 compares the polarized optical micrographs of two blends which were prepared by casting chlorobenzene solutions of the polymer mixtures on glass slides.

The optical patterns are very different. The gross phase separation in Figure 11(a) shows the spherulitic PE and the amorphous PS phases. The PS phases vary widely in both size and shape owing to the lack of interaction with the PE matrix. On the other hand, the continuous crystalline phase in Figure 11(b) shows the compatibilized blend. Basically, the large phase-separated PS domains are now dispersed into the inter-spherulite regions and cannot be resolved by the resolution of the optical microscope. The graft copolymer behaving as a polymeric emulsifier increases the interfacial interaction between the PE crystalline and the PS amorphous regions to reduce the domain sizes.

Figure 12 shows scanning electron micrographs obtained with secondary electron imaging, which show the surface topography of cold fractured film edges. The films were cryo-fractured in liquid N_2 to obtain an undistorted view representative of the bulk material.

In the homopolymer blend, the polymers are grossly phase separated, as can be seen by the PS component which exhibits non-uniform, poorly dispersed domains and voids at the fracture surface, as shown in Figure 12(a). This 'ball and socket' topography is indicative of poor interfacial adhesion between the PE and PS domains and represents PS domains that are pulled out of the PE matrix. Such a pull-out indicates that limited stress transfer takes place between phases during fracture. The similar blend containing graft copolymer shows a totally different morphology [Figure 12(b)]. The material exhibits flat mesa-like regions similar to pure PE. No distinct PS phases are observable, indicating that fracture occurred through both

(a)

(b)

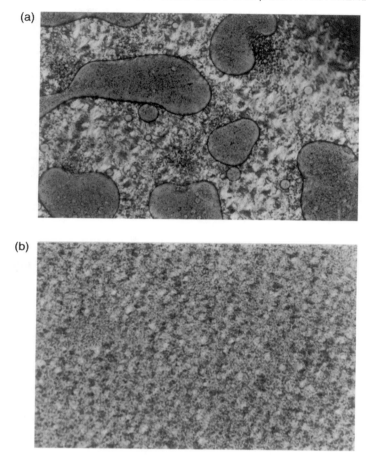

Figure 11 Polarized optical micrographs of polymer blends: (a) two homopolymer blend with PE:PS = 50 : 50(100×); (b) two homopolymers and PE–g-PS copolymer blend with PE : PE–g-PS : PS = 45 : 10 : 45 (100×)

phases or that the PS phase domains are too small to be observed. The PE–g-PS is clearly proved to be an effective compatibilizer in PE–PS blends.

3 CONCLUSION

A new class of reactive polyolefin co-and ter-polymers, containing *p*-methylstyrene groups, have been prepared using metallocene catalysts, with constrained ligand geometry. The combination of spatially opened catalytic sites and a cationic

(a)

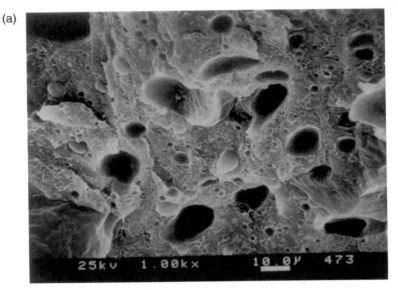

(b)

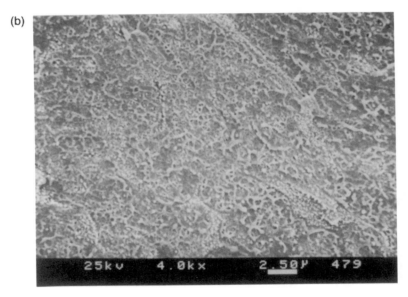

Figure 12 Scanning electron micrographs of the cross-sections of two polymer blends: (a) two homopolymer blend with PE:PS = 50:50(1000×); (b) two homopolymer and PE–g-PS copolymer blend with PE:PE–g-PS:PS = 45:10:45 (4000×)

coordination mechanism with a metallocene catalyst provides very favorable reaction conditions for *p*-methylstyrene incorporation to obtain high polyolefin co- and terpolymers with narrow molecular weight and composition distributions. The experimental results clearly show that *p*-methylstyrene performs distinctively better than styrene, *o*-methylstyrene and *m*-methylstyrene in the ethylene copolymerization reaction. In turn, the copolymers are very useful intermediates in the preparation of functional polyolefins and graft copolymers which can serve as effective compatibilizers in polyolefin blends.

4   ACKNOWLEDGMENTS

The authors thank the Polymer Program of the National Science Foundation for financial support.

5   REFERENCES

1. (a) Baijal, M. D., *Plastics Polymer Science and Technology*, Wiley, New York, 1982; (b) Carraher E. C., Jr, and Moore, J. A., *Modification of Polymers*, Plenum Press, Oxford, 1982; (c) Purgett, M. D, PhD, Thesis, University of Massachusetts (1984); (d) Clark, K. J. and City, W. G., *US Pat.*, 3 492 277 (1970).
2. Boor, J., Jr, *Ziegler–Natta Catalysts and Polymerizations*, Academic Press, New York, 1979.
3. Ruggeri, G., Aglietto, M., Petragnani, A. and Ciardelli, F., *Eur. Polym. J.*, **19**, 863 (1983).
4. (a) Eshuis, J. W., Tan, Y. Y., Meetsma, A., Teuben, J. H., Renkema, J. and Evens, G. G *Organometallics*, **11**, 362 (1992); (b) Johnson, L. K., Mecking, S. and Brookhart, M. *J. Am. Chem. Soc.*, **118**, 267 (1996).
5. Kesti, M. R., Coates, G. W. and Waymouth, R. M. *J. Am. Chem. Soc.*, **114**, 9679 (1992).
6. (a) Chung, T. C., Jiang, G. J. and Rhubright, D., *US Pat.*, 5 286 800 (1994); (b) Chung, T. C., Jiang, G. J. and Rhubright, D., *US Pat.*, 5 401 805 (1995).
7. (a) Chung, T. C. and Rhubright, D., *Macromolecules*, **26**, 3019 (1993); (b) Chung, T. C., Lu, H. L. and Li, C. L., *Polym. Int.*, **37**, 197 (1995).
8. (a) Chung, T. C. and Rhubright, D., *J. Polym. Sci., Polym. Chem. Ed.*, **31**, 2759 (1993); (b) Chung, T. C., Lu, H. L. and Li, C. L., *Macromolecules*, **27**, 7533 (1994).
9. Brown, H. C. *Organic Synthesis via Boranes*, Wiley–Interscience, New York, 1975.
10. (a) Chung, T. C., Lu, H. L. and Janvikul, W. *J. Am. Chem. Soc.*, **118**, 705 (1996); (b) Chung, T. C. and Jiang, G. J., *Macromolecules*, **25**, 4816 (1992); (c) Chung, T. C., Rhubright, D. and Jiang, G. J., *Macromolecules*, **26**, 3467 (1993); (d) Chung, T. C. and Rhubright, D., *Macromolecules*, **27**, 1313 (1994); (e) Chung, T. C., Janvikul, W., Bernard, R. and Jiang, G. J. *Macromolecules*, **27**, 26 (1994); (f) Chung, T. C., Janvikul, W., Bernard, R., Hu, R., Li, R. C., Liu, S. L. and Jiang, G. J., *Polymer*, **36**, 3565, (1995).
11. Chung, T. C., Lu, H. L. and Janvikul, W., *Polymer*, **38**, 1495 (1997).
12. (a) Chung, T. C. and Lu, H. L., *US Pat.*, 5 543 484 (1996); (b) Chung, T. C. and Lu, H. L., *J. Polym. Sci., Part A: Polym. Chem. Ed.*, **35**, 575 (1997).

13. (a) Frechet, J. M. J., *Crown Ethers and Phase Transfer Catalystsin Polymer Science*, Plenum Press, New York, 1984; (b) Powers, K. W., Wang, H. C., Chung, T. C. Jias, A. J. and Olkusz, J. A., *US Pat.*, 5 162 445 (1992).

14. (a) Mohanraj, S. and Ford, W., *Macromolecules*, **19**, 2470 (1986); (b) Jones, R. G. and Matsubayashi, Y., *Polymer*, **33**, 1069 (1992); (c) Pini, D., Settambolo, R., Raffaelli, A. and Salvadori, P., *Macromolecules*, **20**, 58, 1987; (d) Piirma, I. and Lenzotti, J. R., *Br. Polym. J.*, **21**, 45 (1989).

15. (a) Nagasaki, Y. and Tsuruta, T., *Makromol. Chem., Rapid Commun.*, **7**, 437 (1986); (b) Bonaccorsi, F., Lezzi, A., Prevedello, A., Lanzini, L. and Roggero, A., *Polymer Int.*, **30**, 93 (1993).

16. (a) Onopchenko, A., Schulz, J. G. D. and Seekircher, R. J., *J. Org. Chem.*, **37**, 1414 (1972); (b) Ferrari, L. P. and Stover, H. D., *Macromolecules*, **24**, 6340 (1991).

17. (a) Chung, T. C., Lu, H. L. and Ding, R. D., *Macromolecules*, **30**, 1272 (1997); (b) Bonaccorsi, F., Lezzi, A., Prevedello, A., Lanzini, L. and Roggero, A., *Polym. Int.*, **30**, 93 (1993).

18. (a) Kaminsky, W., Kulper, K. and Brintzinger, H., *Angew. Chem., Int. Ed. Engl.*, **24**, 507 (1985); (b) Ewen, J. A., *J. Am. Chem. Soc.*, **106**, 6355 (1984); (c) Slaugh, L. H. and Schoenthal, G. W., *US Pat.*, 4 665 047 (1987); (d) Turner, H. W., *US Pat.*, 4 752 597 (1988).

19. Canich, J. M., *US Pat.*, 5 026 798 (1991).

20. Lai, S. Y., Wilson, J. R., Knight, G. W., Stevens, J. C. and Chum, P. W., *US Pat.*, 5 272 236 (1993).

21. (a) Stevens, J. C., Timmers, F. J., Wilson, J. R., Schmidt, G. F., Nickias, P. N., Rosen, R. K., Knight, G. W. and Lai, S. Y., *Eur. Pat. Appl.*, 416 815 (1991); (b) Sernetz, F. G., Mulhaupt, R. and Waymouth, R. M., *Macromol. Chem. Phys.*, **197**, 1071 (1996).

22. (a) Longo, P., Grassi, A. and Oliva, L., *Makromol. Chem.*, **191**, 2387 (1990); (b) Oliva, L., Caporaso, L., Pellecchia, C. and Zambelli, A., *Macromolecules*, **28**, 4665 (1995); (c) Pellecchia, C., Pappalardo, D., D'Arco, M. and Zambelli, A., *Macromolecules*, **29**, 1158 (1996); (d) Oliva, L.; Longo, P., Izzo L. and DiSerio, M, *Macromolecules*, **30**, 5616 (1997).

23. Ewen, J. A., Jones, R. L., Razavi, A. and Ferrara, J. L., *J. Am. Chem. Soc.*, **110**, 6255 (1988).

24. Stevens, J. C. *Stud. Surf. Sci. Catal.*, **89**, 277 (1994).

25. Kelen, T., Tüdos, F., *React. Kinet. Catal. Lett.*, **1**, 487 (1974).

26. (a) Jordan, R. F., *J. Chem. Educ.*, **65**, 285 (1988); (b) Eshuis, J. J., Tan, Y. Y., Meetsma, A. and Teuben, J. H, *Organometallics*, **11**, 362 (1992); (c) Yang, X., Stern, C. L. and Marks, T. J., *J. Am. Chem. Soc.*, **116**, 10015 (1994).

27. (a) Baldwin, F. P. and Ver Strate, G., *Rubb. Chem. Technol.*, **45**, 709 (1972); (b) Ver Strate, G., *Encycl. Polym. Sci. Eng.*, **6**, 522 (1986).

28. Plate, N. A. and Shibaev, V. P., *J. Polym. Sci., Macromol. Rev.*, **8**, 117 (1974).

29. Powers, K. W., Wang, H. C., Chung, T. C., Dias, A. J. and Olkusz, J. A., *US Pat.*, 5 548 029 (1996).

30. Szwarc, M., *Adv. Polym. Sci.*, **47**, 1 (1982).

PART V

Commercial Metallocene Polymerization Methods

14

Application of Metallocene Catalysts to Large-scale Slurry Loop Reactors[†]

DARRYL R. FAHEY, M. BRUCE WELCH, ELIZABETH A. BENHAM, CARL E. STOUFFER, JOSE M. DIONISIO, DALE L. EMBRY, SYRIAC J. PALACKAL, R. W. HANKINSON, ROBERT W. BOHMER,[‡] MIKE C. CARTER,[‡] JOHN STEWART,[‡] ART ORSCHELN,[‡] L. W. BRENEK,[‡] ASHISH M. SUKHADIA, WILLIAM M. WHITTE AND LOUIS MOORE
Phillips Petroleum Company, Bartlesville, OK and[‡] Pasadena, TX, USA

1 INTRODUCTION

The slurry loop ethylene polymerization process is used by Phillips and its licensees in 14 countries to produce 5 million metric tons per year of polyethylene. During the past several years, Phillips Petroleum Company has been developing metallocene catalysts that will effectively operate in this slurry loop process. As the project progressed from laboratory through large-scale polymerization tests, the slurry loop reactor technology and metallocene catalysts displayed features that are well suited for each other. The following subjects will be discussed that support this contention.

- catalyst performance in bench, pilot plant and large-scale reactors for metallocene linear low-density polyethylene (m-LLDPE) production;
- relationship between catalyst performance and resin pricing;
- properties of m-LLDPE resins produced in the loop;
- next-generation, engineered resins produced with multiple metallocene catalysts in a single reactor.

[†] Dedicated to Professor Roy G. Miller in celebration of his retirement.

Metallocene-based Polyolefins Edited by J. Scheirs and W. Kaminsky
© 2000 John Wiley & Sons Ltd

2 COMPARISON OF PERFORMANCE CHARACTERISTICS IN VARIOUS SIZE REACTORS

Any catalyst used in a slurry loop process must meet several critical reactor performance criteria. The first is to be able to operate in the reactor continuously without 'fouling'. If the metallocene/methylaluminoxane (MAO) ethylene polymerization catalyst is partially soluble in the hydrocarbon polymerization medium, the dissolved catalyst species may produce polyethylene molecules which can precipitate not only onto solid particles in the slurry but also onto solid surfaces of the polymerization reactor. If a thin film of polymer plates out on the reactor wall (which is a heat exchange surface), removal of the heat of polymerization from the reactor is inhibited. In extreme cases, the heat transfer may become so poor that the heat of polymerization causes the internal reactor temperature to rise rapidly. If this temperature rise is not bated, the polymer in the reactor melts and 'fouls' the reactor, ultimately discontinuing operations.

A second criterion is that the metallocene catalyst must be capable of being fed uniformly and continuously to the reactor. For high-productivity catalysts, small amounts of catalyst are fed slowly. In practice, it is often difficult to deliver small quantities of air-sensitive solid catalysts consistently to the hydrocarbon-filled reactor at elevated temperatures and pressures. Metallocene catalysts can have high productivities, so their uniform and reliable feeding is a challenge.

A third criterion is that the catalyst should preferably produce a polymer fluff with a particle size distribution having very few particles smaller than 150 μm in diameter, termed 'fluff fines'. Fluff fines can cause post-reactor operating difficulties. Downstream from the reactor, the fluff fines may plug filters, cause fluff transport problems or be difficult to feed to the pelletizing extruder.

A fourth criterion is that the metallocene catalyst must be capable of providing reactor output rates equivalent to the highest rates attainable with conventional catalysts. Some of the factors that control reactor output are catalyst residence times, polyethylene fluff bulk densities, slurry viscosities and reactor solids levels.

Metallocene catalysts have been developed at Phillips that meet these critical reactor performance criteria. Table 1 compares the performance of one such catalyst in a bench batch reactor, a pilot plant slurry loop reactor (production scale-up factor 25) and two large-scale slurry loop (production scale-up factors 1000 and 40 000) reactors. The metallocene catalyst performance seen in the bench reactor translates very well to all three of the larger reactors.

Control of polyethylene fluff particle size (see Figure 1) and morphology is particularly satisfactory. Figure 2 shows that the fluff particles are fairly spherical and fairly uniform in particle size. The fluff particle size can be controlled by adjusting the catalyst synthesis method.

As scientists continually improve supported metallocene catalyst productivities to lower catalyst costs, the chosen polymerization process can impose production limitations owing to its macro and micro heat transfer capabilities. Ethylene

Table 1   Reactor performance experience

Performance criterion	Bench (batch)	Pilot plant (slurry loop)	Large-scale (slurry loop)
Fouling	No	No	No
Uniform catalyst feed	NA[a]	Yes	Yes
Fluff fines	Low	Low	Low
Relative fluff bulk densities[b]	0.7	0.9–1.1	1.0
Reactor throughput	NA[a]	Full rates	Comparable rates

[a] Not applicable.
[b] Compared with commercial standard.

polymerization is a highly exothermic reaction, and the heat generated within the growing catalyst/fluff particle must be effectively transferred to the surrounding reaction medium. Small high-productivity catalyst particles are the most troublesome. If the heat is not effectively transferred, the temperature of the particle may increase to the polymer melting-point. When this happens, sticky molten polymer will plate out on the internal heat exchange surfaces in a slurry loop reactor. In a gas-phase reactor, the sticky polymer can plate out on the walls or create large agglomerations of particles. Either scenario results in significant downtime and expensive clean-up procedures.

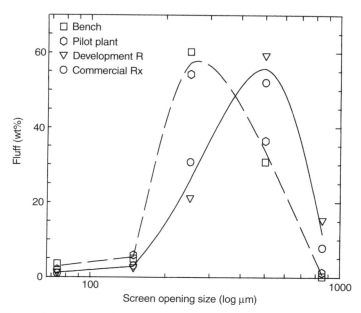

Figure 1   Polymer fluff particle size distribution by sieve analysis

Figure 2 Scanning electron micrograph of polymer fluff particles

A benefit of the slurry loop is its good heat transfer characteristics. Direct solid–liquid contact in the process promotes rapid dissipatation of heat from high-productivity catalyst particles at high reactor throughputs. This heat-dissipation phenomenon is illustrated in Figure 3, which shows the relationships between resin particle size and catalyst productivity that increase the temperatures of the resin particles to 10 °C above the reaction medium. These curves were calculated using a computer model, POLYRED [1], and assuming conventional slurry loop and super condensed gas-phase process conditions [2] for a 0.920 g/cm^3 density resin at a reaction temperature of 80 °C. At a productivity of 10 000 g/g-h, the fluff particle must have an average particle diameter larger than 170 μm to avoid the particle becoming more than 10 °C hotter than the reaction medium in the slurry loop process. At a given particle size and temperature above that of the reaction medium, the heat transfer characteristics of the slurry loop allow a much higher polymer production per unit volume of reactor per unit time than the gas-phase process. This special quality of the slurry loop process makes it uniquely suited for high-activity catalysts—a feature we call metallocene friendly.

3 RESIN PRICING

The penetration of metallocene polyolefins into the global market depends on the balance of polymer properties, processability and price. Of these factors, resin

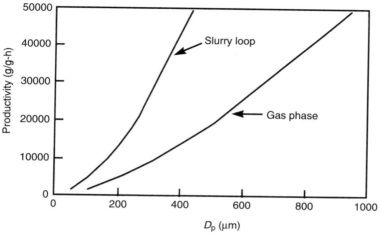

Figure 3 Minimum catalyst particle size–productivity relationship leading to a catalyst temperature rise of 10 °C

pricing historically has played the dominant role. There is a general rule of thumb that price and market size are inversely related (see Figure 4). Low-priced products have much larger markets than high-priced resins. In order for customers to realize the full potential of metallocene resins, it is imperative that resin prices be as low as possible.

Among the many factors that contribute to the final price of a resin are ethylene, comonomer, process, catalyst, additives, marketing and distribution costs. In this discussion, we will address only two of these costs—process and catalyst. Process costs associated with the use of metallocene catalysts are predicted to be competitive with those for conventional catalysts. Our experience with metallocene catalysts indicates that they can be used in existing large-scale reactors with minimal process changes and without additional capital investment. Such catalysts are considered to be 'process friendly'.

The contribution of the catalyst to polyethylene production costs is determined by two factors—the purchase price of the catalyst and the productivity of the catalyst. The purchase price of the catalyst is determined by raw material and production costs. Metallocene costs range from $200–1000/lb for simple structures to $1000–6000/lb for more complex structures [3]. Methylaluminoxane prices range from $100 to $300/lb and usually must be added in a 100–10000-fold molar excess to activate the metallocene. Alternatively, more expensive perfluorophenylborates can be used in much lower molar ratios. Production processes to convert an inherently hydrocarbon-soluble metallocene into an air-sensitive heterogeneous catalyst can require multiple steps and be expensive. When only small volumes of catalyst are being consumed, the lack of economy of scale is a major contributor to catalyst

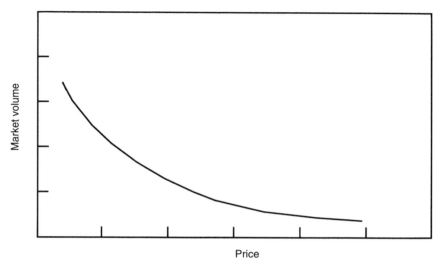

Figure 4   Relationship between market price and volume

costs. These four factors translate to high actual catalyst costs, especially during the initial stages of commercial use. To bring the catalyst cost down, major advances must be made in catalyst manufacturing, especially in developing inexpensive routes to metallocenes and their cocatalysts. Economies of scale also must be realized. At low volumes of metallocene catalyst usage, the cost of these catalysts per pound of resin will be relatively expensive (see Figure 5), but these costs will be quickly reduced as the volume of metallocene catalysts used increases. While such improvements are expected and necessary, this approach alone will not enable the actual catalyst costs to approach those of conventional catalysts.

Catalyst productivity is the other major lever in the economic equation. The higher the catalyst productivity, the lower is the catalyst cost on a unit weight of resin basis. Fortunately, some metallocenes can be made into high-productivity catalysts. Figure 6 shows the relative productivities, lb of polymer per lb of transition metal, of chromium/silica, titanium chloride/magnesium dichloride, and current metallocene catalysts. We predict that at relatively large production volumes and at high catalyst productivities, catalyst costs for conventional and metallocene catalysts will be approximately the same.

4   m-LLDPE RESIN PROPERTIES

Metallocene-produced LLDPE resins, at constant density and MI values, from bench-scale, pilot plant, developmental scale and commercial-scale reactors have nearly superimposable molecular weight distributions and complex rheology curves (see Figures 7 and 8). The bench-scale m-LLDPE has a slightly lower molecular

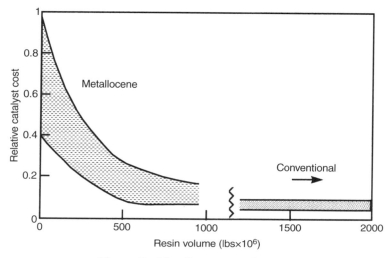

Figure 5   Metallocene catalyst costs

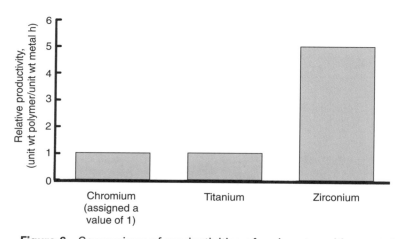

Figure 6   Comparison of productivities of various transition metals

weight, which accounts for the slight offset in its molecular weight distribution and rheology curves. Films 1 mil thick ($1/\text{mil} = 10^{-3}/\text{in}$) blown at high speeds using a 0.06 in die gap from the latter three resins also have identical (within the reproducibilities of the tests) optical and mechanical properties (see Figure 9). The consistency of the m-LLDPE resins produced in the different scale reactors is remarkable. The performance features of the metallocene catalyst site seem to be well preserved as the manufacturing scale increased.

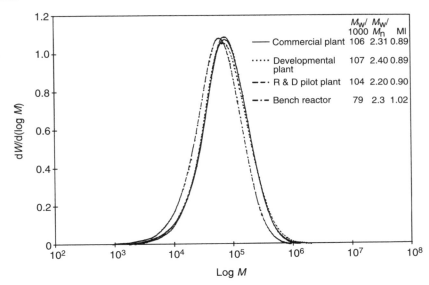

Figure 7 Molecular weight distributions of m-LLDPE resins (0.918 g/cm³, ca 1 MI) produced in different reactors

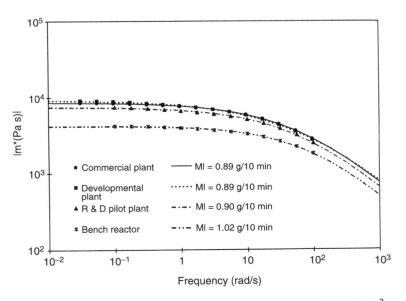

Figure 8 Complex viscosities of m-LLDPE resins (0.918 g/cm³, ca 1 MI) produced in different reactors

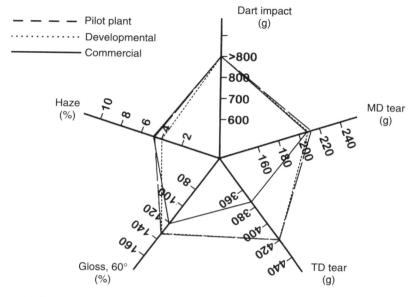

Figure 9 Film properties of m-LLDPE resins (0.918 g/cm^3, ca 1 MI) produced in different reactors

A comparison of selected film properties for the Phillips m-LLDPE resin with those of a some other LLDPE resins is shown in Table 2. The dart impact values for the m-LLDPE films are very high while the tear properties for all the films compare favorably. In addition, the clarity, gloss and level of extractables of the Phillips m-LLDPE are also very good. Overall, a commercial quality resin is obtained by using a metallocene catalyst in the Phillips slurry loop process.

5 ENGINEERED RESIN OPPORTUNITIES

Metallocene resins have often been referred to as tailored resins. In reality, they are much closer to 'molecularly pure' resins as each polymer molecule is nearly identical with every other molecule. Combining molecularly pure resins produced by two or more metallocene catalysts to produce a blend designed to provide specific combinations of optimized properties and processability is what we prefer to call an engineered resin. Engineered resins can be manufactured in a variety of ways, including physical blending, running two separate catalysts in one reactor, running one catalyst in two reactors in series and running two or more catalysts on the same support in one reactor. The most popular commercial process for producing these

Table 2 Comparisons of LLDPE and m-LLDPE blown film properties (processed at 250 lb/h)[a]

Sample	Density (g/cm^3)	Ml (°C/ min)	Dart drop (g)	MD tear (g)	TD tear (g)
Phillips m-LLDPE	0.918	1	770	230	400
Comparative resin A—octene m-LLDPE	0.902	1	650	160	320
Comparative resin B—octene LLDPE	0.920	1	120	280	475
Comparative resin C—hexene LLDPE	0.923	0.6	120	380	560

[a] 1 mil film thickness, 8 in die, 250 lb/h, 0.06 in gap, 2.5 BUR (Blow Up Ratio), 350 °F profile, 24:1 barrier screw, pocket bubble. Data sheets for the comparative resins may display different property values than those shown in this table.

engineered resins today uses one catalyst in two reactors, each under different conditions, in series.

The preferred molecular architecture of a resin depends on its intended application, but a particularly desired engineered resin has a rather broad molecular weight distribution with controlled comonomer distribution. At Phillips, a catalyst to produce such engineered resins has been developed and tested in the slurry loop pilot plant over a range of reactor conditions.

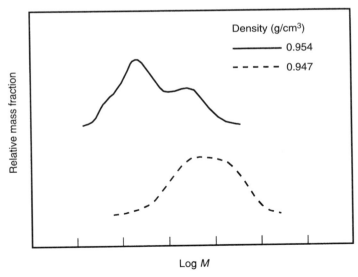

Figure 10 Molecular weight distributions of engineered resins

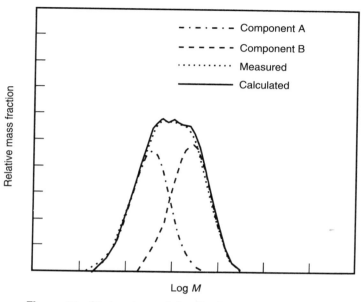

Figure 11 Molecular weight distributions of components

No significant operational problems were encountered during the test. The densities, melt indices and molecular weight distributions of the resins were controlled by changing the reaction conditions and catalyst formulation. In this way, we prepared resins ranging in density from 0.925 to 0.954 g/cm^3, in MI from 0.09 to 67 dg/min and in heterogeneity indices from 2.5 to 9.7. The molecular weight distributions of two of the resins are shown in Figure 10. In Figure 11, the molecular weight distributions of the two individual resins produced from two catalysts run separately, under reactor conditions comparable to those that produced the bottom molecular weight distribution in Figure 10, are shown. The sum of the two curves produces a curve that matches the molecular weight distribution of the engineered resin.

6 CONCLUSION

Based upon the performance of metallocene catalysts in the slurry loop process and the quality of the LLDPE resin that is produced, a technical foundation for an m-LLDPE business has been established. The ability to produce engineered polyethylene resins using metallocene catalysts in a slurry loop reactor has also been demonstrated. One might even think that when the slurry loop process was invented 40 years ago, it was for metallocene catalysts. Continued research is required to ensure the catalyst costs will be low and to develop the engineered resins that converters will demand in the future.

ACKNOWLEDGMENTS

Important contributions to this project by Drs Tim W. Johnson and David C. Rohlfing and Messrs Paul P. Barbee, Gary L. Glass and John E. Anderson are gratefully acknowledged.

8 REFERENCES

1. *POLYRED*, available from Process Research Corporation, Department of Chemical Engineering, University of Wisconsin, 1415 Johnson Drive, Madison, WI 53706, USA.
2. Exxon Chemical Patents, *U S Pat.*, 5 352 749.
3. Birmingham, J. M. and Hanna, G. J. presented at the International Congress on Metallocene Polymers, Brussels, April 26, 1995; Conference Proceedings available from Schotland Business Research, 16 Duncan Lane, Skillman, NJ 08558, USA.

15

Multi-catalyst Reactor Granule Technology for the Production of Metallocene Polymers

J. C. HAYLOCK, R. A. PHILLIPS AND M. D. WOLKOWICZ
Montell Polyolefins, Elkton, MD 21921, USA

1 INTRODUCTION

Metallocene-catalyzed polymerization of olefinic monomers is an important new technology that has started to make significant inroads into the polyolefins marketplace in recent years. Metallocene catalysts allow for an expansion of the property map of conventional, well established, Ziegler–Natta catalyst technology. Structural variables responsible for this property map expansion include resin molecular weight distribution, micro-tacticity, comonomer incorporation and new monomer combinations. The ability to tailor these structural variables in new ways with metallocene catalysts offers the potential for unique olefinic polymer and alloy structures not previously thought practical or even possible. The development of mixed catalyst systems based on both Ziegler–Natta and metallocene catalysts allows for added material design options and flexibility for the commercial production of many of these unique polymer structures.

Modern gas-phase polymerization processes require a method of supporting the metallocene catalyst to achieve good morphology control of the polymerized particle while retaining catalyst activity and the desired structural characteristics of the metallocene polymer. A variety of methods to support the metallocene catalyst have been reported [1–14]. In this chapter, a unique proprietary process for supporting the metallocene catalyst is discussed which utilizes a polymeric support derived from Ziegler–Natta catalysts [11,13,14]. This approach not only provides a means of supporting the metallocene catalyst, but also provides the opportunity for unique

Metallocene-based Polyolefins Edited by J. Scheirs and W. Kaminsky

material design options. These options are derived from the possibility of combining within the reactor the polymer structure versatility and potential product advantages of both Ziegler–Natta and metallocene catalysts. This multi-catalyst reactor granule (MRG) technology [15] is adaptable to all of Montell's process technologies (Spheripol, Catalloy, Spherilene) [16]. The qualitative features of the MRG technology are described while contrasting differences between Ziegler–Natta- and metallocene-catalyzed polymerization. Selected structure–property correlations highlight the potential and opportunities associated with the MRG approach to polymer structure variation.

2 MULTI-CATALYST REACTOR GRANULE TECHNOLOGY: POLYMERIZATION

2.1 OLEFIN POLYMERIZATION WITH ZIEGLER–NATTA AND METALLOCENE CATALYSTS

The structure and composition of Ziegler–Natta [17,18] and metallocene [18–21] catalysts has been extensively reviewed. Modern fourth-generation Ziegler–Natta catalysts are composed of an active δ-form [17,18,22] $MgCl_2$ support of $TiCl_4$ with an alkyl phthalate internal donor (D_i) used in conjunction with an aluminum cocatalyst (commonly triethylaluminum) and an alkoxysilane external donor (D_e) [18]. The external donor, D_e, can be eliminated when the internal donor, D_i, is composed of a diether compound [18,23]. In these modern catalysts, the $MgCl_2$ support is not inert, and plays a key role in active site determination [18] and the defining role in controlling the morphology of the polymerized particle. Owing to the complicated nature of these catalysts, details of the active site(s) and the mechanism for stereoregulation remain an active area of research [18]. It is generally accepted that there are multiple active sites which, while highly tunable, contribute to a broadened molecular weight distribution in addition to diffusional effects [24, 25], and for propylene polymerization a broadened inter-chain tacticity distribution [26–28] and the presence of a finite atactic content.

Metallocene catalysts are made up of cyclopentadienyl ligands coordinated to a Group IV metal center (Zr, Hf, Ti), with the ligands often joined by either a silyl- or carbon-based bridge [19–21]. The catalyst molecule generally either requires an alumoxane [19–21] cocatalyst or exists as a cationic complex with a weakly coordinating counteranion (generally containing boron) [19–21]. Both have the effect of creating a weakly associated alkylated cationic metal center for coordination with the incoming monomer [19–21]. Because the ligand–metal complex is a discrete molecule, the geometry and electronic environment of the active site are well defined. This contrasts with the case of Ziegler–Natta catalysts, and has allowed the detailed study of the active site for various metal centers and ligand structures [18–21]. A detailed understanding of the catalyst structure–polymer structure relationship has resulted [18–21]. This is perhaps best illustrated by the stereoregular polymerization of propylene. For this case, metallocene catalysts allow for the production of isotactic, hemi-isotactic, syndiotactic and atactic polypropylene by variation of the metallocene site symmetry [18–21].

Relative to Ziegler–Natta catalysts, the heterogeneity of active sites is eliminated in metallocene catalysts. This factor allows polymers produced by metallocene catalysts to approach the limiting molecular weight distribution ($M_w/M_n = 2$) predicted for identical catalyst centers with fixed chain propagation and termination rates [21]. In addition to a narrow molecular weight distribution, the homogeneity of active sites in metallocene catalysts can contribute to a narrowed interchain tacticity distribution and interchain composition distribution in copolymers [8,10,21,26]. Unlike high-activity Ziegler–Natta catalysts, metallocene catalysts are not intrinsically supported and a method of supporting the catalysts is required for morphology control of the polymerized product. Various methods of supporting the metallocene catalyst have been reported [1–14,19–21]. Most supported metallocene catalysts show a reduction in activity after fixation on a solid-support [2–7,12], and can also exhibit different requirements with respect to the type and amount of cocatalyst [1–3, 5–7,9,12]. With respect to polymer properties, the supported catalysts can be inert [4] or can substantially alter polymer properties such as tacticity/melting-point [2, 3, 5–9,12], molecular weight [5,7–10,12] and molecular weight distribution [2,6,8].

2.2 ZIEGLER–NATTA CATALYSIS: CATALYST SUPPORT AND REPLICATION

The key to Montell's proprietary Ziegler–Natta polyolefin processes, for both polyethylene and polypropylene, is the ability to control the morphology of the growing polymer particle through the control of catalyst particle morphology. The Ziegler–Natta 'catalyst architecture' refers to the complex of catalyst, donors and support [29]. The 'catalyst architecture' plays three major roles. The first role is to determine the intrinsic structural characteristics of the growing chain, such as molecular weight, molecular weight distribution, stereospecificity and comonomer reactivity ratios. The second role is to determine the intrinsic catalyst performance through such variables as catalyst activity and hydrogen response. The third role, inter-related with the second, is to determine the morphology of the growing polymer particle. It is this third role which is a critical factor determining the versatility of Ziegler–Natta catalysts in the MRG technology.

Figure 1 illustrates schematically the strategy of the MRG approach. The metallocene-catalyzed polymerization within the MRG framework is discussed briefly. For the present purposes, attention is focused on Ziegler–Natta homopolymerization and/or random copolymerization. As discussed earlier, The Ti active centers are on an activated $MgCl_2$ support within the catalyst–support–donor complex. The active $MgCl_2$ support is made up of very small primary crystallites. Wide-angle X-ray scattering (WAXS) measurements give typical crystal size

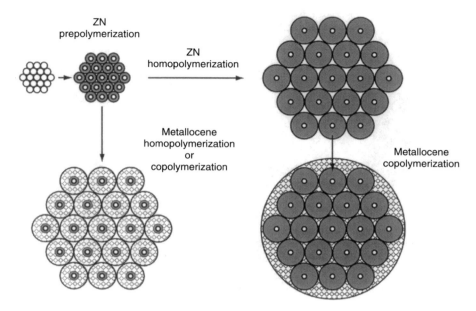

Figure 1 Schematic diagram of multi-catalyst reactor granule technology. See text for discussion

dimensions of 20–200 Å for highly disordered $MgCl_2$ [22]. The primary crystallites are organized on larger size scales in a loosely bound framework which gives the macroscopic catalyst high specific surface area and porosity [30]. Figure 2 shows scanning electron micrograph of macroscopic Ziegler–Natta catalyst particles that are representative of the catalysts used in this work. Because of the surface area and porosity characteristics, the initial stages of polymerization are characterized by polymer growth both on the exterior surface and within the interior of the macroscopic catalyst particle. This results in a uniform spatial distribution of catalyst fragments within the growing particle [30,31]. The subsequent growth of the polymer particle during polymerization has been extensively modeled [25,32–39]. Although quantitative aspects are beyond the scope of this chapter, the schematic diagram in Figure 1 adopts the qualitative features of the multi-grain model of polymer growth [33–37,39].

In the multi-grain model [33–37,39], the macroscopic catalyst undergoes fragmention into microparticles evenly distributed within the macroscopic particle. This fragmentation is due to the forces exerted by the growing polymer layers. In $MgCl_2$-supported Ziegler–Natta catalysts, fragmentation is initiated at very low polymer yields [30,39,40]. The microparticles behave as microreactors within the growing polymer particle. Monomer diffuses through the macroparticle and the growing polymer layer surrounding the microparticle. Subsequent polymerization results in

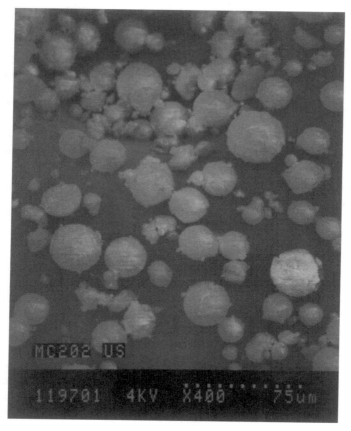

Figure 2 Scanning electron micrograph of porous $MgCl_2$-supported Ziegler–Natta catalyst

growth of the polymer layer at the microparticle level. The growth of the macroparticle results from the cumulative uniform expansion of microparticle layers. Heat and mass transfer during polymerization can be viewed at both the macro- and microparticle levels, and can be controlled to some extent by the 'catalyst architecture'. This growth of the macroparticle by uniform expansion of the microparticle layers is depicted in Figure 1 by the transition from catalyst particle to pre-polymerized particle to Ziegler–Natta homopolymer particle. The actual morphology of Ziegler–Natta polymerized particles shows a more complicated hierarchy than the 'two-level' approach of the multi-grain model [30,41]. However, a key feature of the multi-grain model relevant to the current discussion is the replication of the porous spherical catalyst morphology into a porous spherical polymer macroparticle during polymerization [29]. If polymerization is carried out to completion, the catalyst residues associated with microparticle growth are reduced to the ppm level in the final macroscopic porous polymer particle. This is due to the very high replication factors of modern catalysts.

Critical to the process of catalyst replication is the catalyst fragmentation associated with the early stage growth during prepolymerization. As mentioned above, for highly active Ziegler–Natta catalysts, fragmentation occurs at very low polymer yields [30,39,40]. This gives rise to a large active surface from the beginning of polymerization. For good replication to be achieved, the mechanical strength of the granule must be in balance with the catalyst polymerization activity [18]. If the reactivity is too high, an uncontrollable 'explosion' of the granule into a fine powder occurs due to the mechanical forces generated by the growing polymer chains. If the mechanical strength of the catalyst particle is too great, a low level of reactivity occurs because the internal active sites cannot be generated [39]. According to the summary of Albizzati *et al.* [18] and Galli and Noristi [42], good replication requires that the catalyst has:

  (i) high surface area;
 (ii) high porosity with a large number of cracks evenly distributed through the mass of the granule;
(iii) high enough mechanical resistance to withstand handling, but low enough to allow breakage into microscopic particles during polymerization;
 (iv) homogeneous distribution of the active centers;
  (v) free access of the monomer up to the innermost zones.

It is the ability to develop controllably and reproducibly the growing polymer particle that forms the basis of the MRG approach. As indicated in Figure 1, the porous Ziegler–Natta-catalyzed polymerization particle can provide a solid support medium for subsequent polymerization with metallocene catalyst. Two simplified scenarios are depicted. The first scenario depicts the use of a Ziegler–Natta prepolymerized support for metallocene-catalyzed homopolymerization and/or copolymerization. By this route, the metallocene polymer is overwhelmingly the dominant component of the polymerization. The second scenario depicts a Ziegler–Natta porous reactor particle for support of metallocene-catalyzed copolymerization. By this route, both Ziegler–Natta and metallocene polymers make up the final particle.

2.3 METHOD OF SUPPORTING METALLOCENE CATALYST

The method of supporting the metallocene by the MRG approach (Figure 1) utilizes either a prepolymerized particle or a polymerized reactor granule formed from Ziegler–Natta catalysts. The second approach in Figure 1, showing metallocene copolymerization within the reactor granule, has been described previously by Galli *et al.* [13] and Collina and co-workers [11,14]. Figure 3 shows a scanning electron micrograph of a polymerized polypropylene reactor granule at high and low

(a)

119728 2KV X20.0 1500um

Figure 3 Scanning electron micrograph of porous polymerization particle of Ziegler–Natta isotactic polypropylene. (a) The low-magnification micrograph shows the coarse spherical morphology

magnifications. The Ziegler–Natta reactor granule (Figure 3) shows spherical morphology at low magnification and 'globule-like' morphology at high magnification. This hierarchy of morphology qualitatively resembles earlier discussions of micro-grain growth within the macroscopic particle. The high magnification results resemble the 'secondary' structure of Kakugo *et al.* [41]. Following the Ziegler–Natta polymerization stage, the Ziegler–Natta catalyst is deactivated and the polymerized particle is impregnated with the activated metallocene catalyst [11, 13,14]. In contrast with other methods of supporting metallocene catalysts [1–10], the morphology of the metallocene polymerization stage is determined by the template established during Ziegler–Natta (pre)polymerization. Subsequent

(b)

HIGH POROSITY PP GRANULE
CORE
100727 6KV X5.00K 6.0um

Figure 3 (*continued*) (b) the high-magnification micrograph shows the fine-scale structure

polymerization with the metallocene catalyst produces polymer which grows on and in the polymeric support, replicating its shape.

2.4 ZIEGLER–NATTA PREPOLYMER SUPPORT OF METALLOCENE HOMOPOLYMERIZATION

Figure 4 shows (a) low- and (b) high-magnification scanning electron micrographs of an isotactic polypropylene granule from rac-Me$_2$Si(2-Me-1-indenyl)$_2$ZrCl$_2$ metallocene catalyst [8,10,43,44] after being impregnated on a Ziegler–Natta prepolymerized support. This corresponds to the first scenario depicted in Figure 1.

(a)

Figure 4 Scanning electron micrograph of isotactic polypropylene homopolymer polymerized with rac-Me$_2$Si(2-Me-1-indenyl)$_2$ZrCl$_2$ supported on a Ziegler–Natta prepolymer support. (a) The low-magnification micrograph shows the coarse spherical morphology.

The macroscopic morphology is spherical with bulk density of 0.42 g/cm^3. The macroscopic morphology of the metallocene iPP reactor granule shown in Figure 4(a) is qualitatively similar to the Ziegler–Natta case shown earlier (Figure 3). However, the high-magnification micrograph shown in Figure 4(b) does not show the secondary structure morphology seen in the Ziegler–Natta case (Figure 3). This indicates that the replication of the prepolymerized fragmented support to the porous metallocene iPP granule differs from that of conventional Ziegler–Natta polymerizations. Analysis of the polymer yield indicates that the metallocene homopolymer iPP can be produced with only residual Ziegler–Natta polymer from the prepolymer support retained in the final resin.

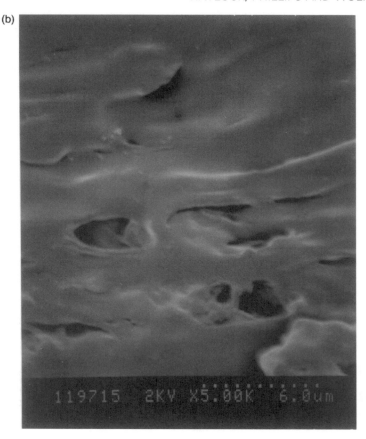

Figure 4 (*continued*) (b) the high-magnification micrograph shows the fine-scale structure

2.5 ZIEGLER–NATTA POLYMERIC SUPPORTS FOR METALLOCENE COPOLYMERIZATION

Previous studies have shown that the polypropylene reactor particle (Figure 3) can also be used as a polymeric support of metallocene copolymerization for the production of rubber-modified heterophasic copolymers [11,13,14]. In these studies, ethylene–propylene copolymer rubber from either *rac*-ethylene-bis(4,5,6,7-tetrahydroindenyl)$ZrCl_2$ (*r*-EBTHI) [45,46] or *meso*-ethylene-bis(4,7-dimethyl-indenyl)$ZrCl_2$ (*m*-EBDMI) metallocene catalyst [47] were grown within the polymer reactor particle. This strategy corresponds to the second scenario depicted in Figure 1, where the final product is made up of polymers from both Ziegler–Natta and metallocene catalysts. Depending on the polymerization conditions, the macroscopic heterophasic reactor particles were spherical and free-flowing, with the rubber dispersed within the particle morphology [13].

Alternatively, it has also been shown that polymerization of ethylene–propylene rubber can be carried out at yields far exceeding that of the polymeric support [48, 49], a situation more closely resembling the first scenario in Figure 1. Figure 5(a) shows a low-magnification scanning electron micrograph of a macroscopic granule containing 60 wt% of a metallocene rubber (80 mol% ethylene). The metallocene

(a)

Figure 5 (a) Low-magnification scanning electron micrograph of the coarse spherical morphology of a mixed catalyst reactor granule. Ethylene–propylene rubber is polymerized within a Ziegler–Natta porous polymeric support. The as-polymerized blend contains 60 wt% of a metallocene rubber (80 mol% ethylene) formed from *m*-EBDMI catalyst (see text).

(b)

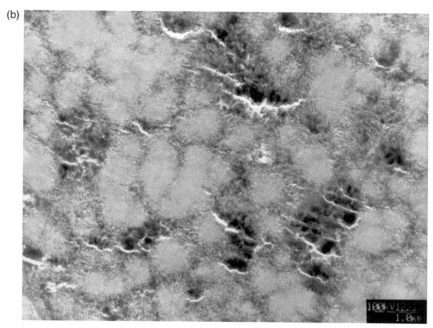

Figure 5 (*continued*) (b) High-magnification transmission electron micrograph. White globular regions are the ZN-iPP matrix, gray/dark regions are the dispersed metallocene ethylene–propylene rubber phase

rubber was grown after impregnation of the metallocene catalyst (*m*-EBDMI) within the polymerized Ziegler–Natta granule. The morphology is spherical and free-flowing. Figure 5(b) shows a higher magnification transmission electron micrograph of the internal morphology. The dark/gray areas are the metallocene rubber finely dispersed within the white granular Ziegler–Natta matrix morphology.

3   MULTI-CATALYST REACTOR GRANULE TECHNOLOGY: POLYMER STRUCTURE/PROPERTIES

The MRG technology represents an added approach for the production of metallocene polymers. In addition to developing viable commercial polymerization processes, it is also necessary to define the polymer structure–performance–application relationships. The characterization of new polymer structures, and correlation with properties, is a critical step in determining optimum cost–performance balances. It is this balance that metallocene polymer resins must obtain if metallocene polypropylene technology is to be justified in the highly competitive, cost-conscious polyolefin marketplace. Below, selected examples are presented which contrast the structure of metallocene and Ziegler–Natta polymers. Although metallocene polyethylene is a critically important area, emphasis is placed on polypropylene and modified polypropylene. Areas of opportunity for expanding the property map of Ziegler–Natta polymers with metallocene polymers are highlighted by selected structure–property correlations.

3.1 ISOTACTIC POLYPROPYLENE HOMOPOLYMER FROM METALLOCENE CATALYST

The unique feature of propylene polymerization, versus ethylene polymerization, is the symmetry of the monomer insertion into the growing polymer chain. It is the presence of the methyl group in the propylene monomer that is responsible for this difference. This gives the monomer insertion an orientation (the monomer has a 'head' and a 'tail') and a stereochemical configuration with respect to the other units in the chain backbone. The regularity with respect to monomer orientation is termed the 'regiospecificity' of the polymerization. Regularity of the methyl group placement relative to the other methyl groups along the chain backbone is termed the 'stereospecificity' of the polymerization. As mentioned earlier, metallocene catalysts, by variation of metallocene site symmetry, are capable of producing the major stereospecific classifications of homopolymer polypropylene: isotactic (iPP), hemi-isotactic, syndiotactic (sPP) and atactic (aPP) [18–21]. Ziegler–Natta polymerizations are generally restricted to the production of iPP and aPP. Discussion of these tacticity classifications will be returned to shortly.

Kinetic arguments suggest that iPP from both Ziegler–Natta catalysts (ZN-iPP) and metallocene catalysts (m-iPP) show a strong preference for primary {1,2}-insertion of monomer into the growing chain [50–52]. The {2,1}-insertion of monomer renders the active site kinetically dormant [18,50–55], and its incorporation into the growing polymer chain occurs infrequently, generally as an isolated regio misinsertion error [56]. The classification and formation mechanisms of regio misinsertion errors in iPP have been reviewed [21,26,52, 56–61]. In ZN-iPP, regio misinsertion errors are generally not present in the isotactic fraction [18]. In contrast, iPP from metallocene catalysts (m-iPP) can show appreciable levels of regio misinsertion errors in chains of high stereospecificity [21,26,53,56–59,61–66]. The dependence of regiospecificity on ligand structure has been reviewed [21]. This difference between Ziegler–Natta and metallocene catalysts is summarized in Figure 6 and Table 1.

Figure 6 compares the DSC heat of fusion and melting point as a function of single unit tacticity errors (SUE) per 1000 monomer units. In Figure 6, XSRT is the atactic weight fraction and $\Delta H/(1\text{-XSRT})$ represents the heat of fusion normalized by the mass of the isotactic fractions. This latter parameter is proportional to the crystallinity of the isotactic fractions. Calculation of SUE is based on two-site enantiomorphic modeling [67] of the solution ^{13}C NMR spectra. The SUE value for

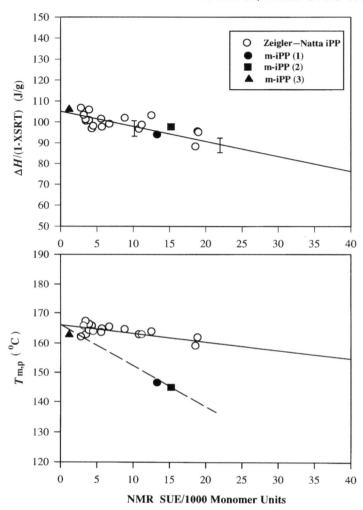

Figure 6 Relationship between DSC heat of fusion (ΔH) and melting-point ($T_{m,p}$) on NMR tacticity for metallocene (m-iPP) and Ziegler–Natta (ZN-iPP) isotactic polypropylene. XSRT is the room temperature xylene-soluble content, and is representative of the weight fraction of atactic polymer. $\Delta H/(1\text{-}XSRT)$ is an empirical parameter representing the heat of fusion normalized to the mass of isotactic fractions. The ^{13}C NMR single unit error content (SUE) is based on two-site entantiomorphic modeling [67]. Values of SUE per 1000 monomer units are shown

Table 1 Microstructure of m-iPP resins

Polymer	Catalyst type	XSRT (%)	M_W (GPC)	M_W/M_n (GPC)	% mmmm (^{13}C NMR)	$T_{m.p}$ (°C)	{2,1}-*Erythro* insertion (mol %)
ZN-1	Ziegler–Natta	1.44	210 000	5.4	95.1	163.6	0.00
m-iPP (1)	Metallocene	0.30	190 000	2.6	93.1	146.6	0.39
m-iPP (3)	Metallocene	0.50	154 000	2.1	98.1	162.9	0.00

ZN-iPP is applicable to the isotactic component in the context of the two-site model. Data are shown in Figure 6 for selected ZN-iPP and m-iPP resins.

When compared at similar molecular weight, Figure 6(a) shows that ZN-iPP and m-iPP resins both show decreasing crystallinity with increasing concentration of stereo-defects. This is due to the role of stereo-defects on disrupting the average length of isotactic sequences in a manner which is conceptually analogous to the incorporation of a comonomer in random copolymers. Figure 6(b) shows an analogous comparison for the melting-point of ZN-iPP and m-iPP resins. This comparison shows that the m-iPP resins deviate, relative to the relationship of ZN-iPP resins, towards lower melting structures. The data in Table 1 indicate that a distinguishing characteristic of the low-melting m-iPP resin is a measureable concentration of 2,1 *erythro* [58,61] regio misinsertions. Measurement of these misinsertions is quantified by methods detailed elsewhere [61]. Figure 6 shows that the stereospecificity of the m-iPP sample is within the range commonly observed for ZN-iPP.

Although metallocene catalysts can give other types of regio misinsertions [21, 26,52,56–61], these data suggest that the presence of these misinsertions contributes to the low melting-point observed for m-iPP (Figure 6). This conclusion is in agreement with literature results [10,21,26,59]. When the concentration of regio misinsertions is lowered by variation of the ligand structure [sample m-iPP(3) in Table 1 and Figure 6], the melting-point of the m-iPP is similar to that typical of ZN-iPP. Although the role of regio misinsertions on the melting-point depression has been debated [63,64], these results suggest that the control of defect structures into the iPP chain with supported metallocene catalysts can substantially alter, relative to ZN-iPP, the property balances associated with crystallinity and melting-point in the final product. Expanding the property map of ZN-iPP to lower melting structures can either be a drawback (for applications requiring high T_m) or a benefit. Areas of opportunity include applications which require a balance of room temperature properties with elevated temperature processing. Examples include heat sealing films and solid-state forming operations.

Despite the fact that m-iPP resins expand the ZN-iPP crystallinity–melting-point property map towards lower melting structures, Table 1 shows that the m-iPP resins also contain low room temperature xylene-soluble (%XSRT) levels. This is due, in part, to a narrowing of the inter-chain tacticity distribution of m-iPP relative to

ZN-iPP. With conventional ZN-iPP, copolymerization can give a melting-point which matches that of the low-melting m-iPP(1). However, ZN copolymerization is generally accompanied by an increase in room temperature extractables. The low extractable levels shown for the m-iPP resins, relative to ZN-iPP, may have potential benefits in food contacting and medical applications.

In addition to crystallinity and melting characteristics, metallocene catalysts are known to produce iPP with narrow molecular weight distribution (MWD). Values of the polydispersity index, M_w/M_n, can approach the limiting value of 2 [21]. Table 1 compares M_w/M_n of m-iPP with an as-polymerized ZN-iPP. Generally, Ziegler–Natta catalysts have difficulty producing iPP with M_w/M_n below 3.5–4. A relatively narrow M_w/M_n is observed for the as-polymerized m-iPP. This gives the melt rheology of m-iPP an increased Newtonian character relative to ZN-iPP, leading to less shear thinning at high melt shear rates.

One application area where a narrow MWD is thought to be valuable is in the area of melt spun fibers. Figure 7 shows single-filament fiber properties of an m-iPP resin compared with a series of ZN-iPP resins with different M_w/M_n. Fibers of 2–10 denier were prepared by melt spinning at relatively low line speeds (250–1000 m/min) in a single stage, without subsequent drawing. The melt flow rate (MFR) is held constant (MFR = 30 g/10 min). The characteristics of the m-iPP [m-iPP(1)] and broad M_w/M_n as-polymerized ZN-iPP (ZN-1) are given in Table 1. ZN-iPP samples of intermediate M_w/M_n were obtained by chemical visbreaking to 30 MFR. Relevant characteristics of these resins are given in the caption of Figure 7. Figure 7 shows that at higher line speeds, slight improvements in fiber tenacity are observed with decreasing M_w/M_n. This improvement occurs at the expense of the elongation at break. Very little difference in tenacity is observed at slow spin speeds for fibers with varying MWD. In this sense, the property differences observed for the m-iPP when compared with visbroken and as-polymerized ZN-iPP appear to be a natural consequence of the narrow M_w/M_n. Higher maximum spin speeds are attainable in the visbroken ZN-iPP ($M_w/M_n = 3.26$) versus the m-iPP resin (4560 versus 3990 m/min).

The microstructure of m-iPP also influences the morphology of melt spun fibers. Figure 8 compares the WAXS patterns of fibers with different M_w/M_n (at MFR = 30 g/10 min). iPP homopolymer is known to crystallize into the meso-morphic crystallographic form at the high cooling rates typical of melt spinning operations [68]. This crystallographic form retains the 3_1-helical conformation of the most common α-form, but with disrupted inter-chain order [68]. The mesomorphic form is identified in Figure 8 by patterns containing only two diffuse maxima. The more highly ordered and most common α-form [68] is identified by multiple scattering maxima. The m-iPP(1) resin and narrow MWD visbroken ZN-iPP resin ($M_w/M_n = 3.26$) both crystallize into the mesomorphic form at 250 and 1000 m/min. The visbroken ZN-iPP with intermediate MWD ($M_w/M_n = 3.78$) crystallizes into the mesomorphic form at low spin speeds and the dominant α-form at high spin speeds. The broad MWD ($M_w/M_n = 5.38$) as-polymerized

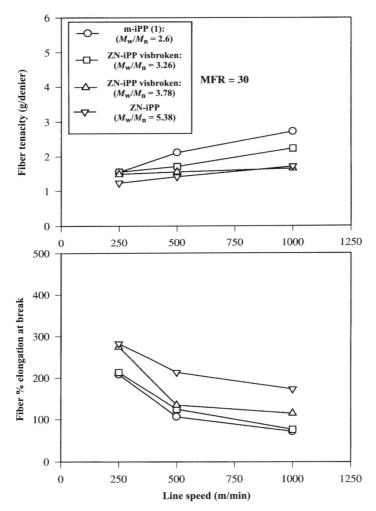

Figure 7 Single-filament tenacity and % elongation at break as a function of spinning speed for melt-spun fibers of metallocene (m-iPP) and Ziegler–Natta (ZN-iPP) isotactic polypropylenes with varying polydispersity index (M_W/M_n) determined by GPC. The m-iPP is sample m-iPP (1) in Table 1. All resins have a nominal melt flow-rate of 30 g/10 m

ZN-iPP crystallizes exclusively in the α-form at all spin speeds. These results indicate that crystallization into the α-form is favored by increased line speeds and broadened MWD. Strain-induced nucleation of the more stable α-form during melt spinning is promoted both by the high molecular weight tail in broad MWD ZN-iPP

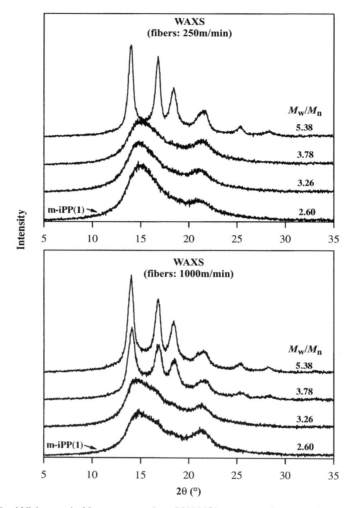

Figure 8 Wide-angle X-ray scattering (WAXS) patterns from melt-spun fibers of metallocene (m-iPP) and Ziegler–Natta (ZN-iPP) isotactic polypropylenes with varying polydispersity (M_W/M_n). The m-iPP is sample m-iPP(1) in Table 1. Fibers were randomized prior to data collection to minimize effects due to preferred orientation

and by increased line speed. These results imply that higher line speeds are required to nucleate m-iPP into the α-form relative to broadened MWD ZN-iPP. This is particularly relevant for processes which apply a subsequent solid-state drawing step, since different draw characteristics are expected for the mesomorphic and α-form morphologies. Generally, the morphology of m-iPP and narrow MWD visbroken ZN-iPP are very similar for all line speeds.

3.2 POLYPROPYLENE HOMOPOLYMER TACTICITY MIXTURES

The above discussion of Figure 6 and Table 1 indicates that the homopolymerization of iPP with metallocene catalysts (m-iPP) allows for an expansion of the property map relative to Ziegler–Natta iPP (ZN-iPP). Specifically, narrow molecular weight distributions are accessible without chemical visbreaking. Also, the balance of crystallinity and melting point is expanded to lower melting structures. This property map expansion is attributed, in part, to unique regio-specific microstructures from metallocene catalysts. The stereochemistry of polypropylene also strongly influences material properties. In addition to iPP homopolymer, metallocene catalysts have the potential to polymerize new tacticity microstructures, including high molecular weight (narrow distribution) atactic polypropylene (aPP) [69] and syndiotactic polypropylene (sPP) [68,70,71]. Homopolymer 'tacticity mixtures' refer to mixtures of polypropylene homopolymer with different intra-chain tacticity types. Examples include iPP–aPP, iPP–sPP, sPP–aPP and m-iPP–ZN-iPP. The properties of selected laboratory-prepared blends [72] are briefly discussed below to highlight the property map expansion associated with this class of mixtures.

High molecular weight aPP with a narrow molecular weight distribution is readily polymerized with metallocene catalysts [69]. aPP is non-crystalline, with a glass transition temperature (T_g) near $0\,°C$. While not necessarily attractive as a homopolymer, when used in conjuction with either semicrystalline sPP or iPP, a new balance of properties is possible relative to ZN-iPP. This is illustrated in Figure 9, which shows the stress–strain curves of ZN-iPP–aPP equimolecular weight $(M_w = 200\,000)$ mixtures as a function of atactic concentration. The aPP was prepared using dimethylsilylbis(9-fluorenyl)zirconium dichloride with methyl aluminoxane (MAO) cocatalyst [69,72]. Homopolymer iPP shows a well developed yield point, followed by subsequent drawing of the semicrystalline structure. As the aPP concentration varies from 0 to 80 wt %, a systematic reduction in Young's modulus (slope of stress–strain curve at zero strain), yield stress and draw stress is observed, with a corresponding broadening of the yield region and shift of the yield strain to higher values. These mechanical characteristics are consistent with a reduction in crystallinity with increasing aPP concentration due to the non-crystalline nature of aPP homopolymer. Relative to ZN-iPP homopolymer, iPP/aPP mixtures allow for a shift of the property map to 'softer', less crystalline, materials without a significant change in melting-point. This balance of properties differs from those of 'elastomeric' polypropylenes (ELPP) derived from metallocene catalysts [73–75]. In an idealized ELPP structure, blocks of aspecific and isospecific sequences alternate along the chain backbone. In this way, crystallinity is reduced owing to intra-chain incorporation of aspecific sequences. Although this idealized structure may not be realized in some ELPP catalysts [76], the low crystallinity structures resulting from intra-chain defect incorporation can exhibit unique mechanical recovery characteristics [73,75]. However, owing to the intra-chain defect incorporation, the melting-points are reduced [73–75] relative to the case of inter-chain incorporation of aPP in the iPP–aPP mixtures.

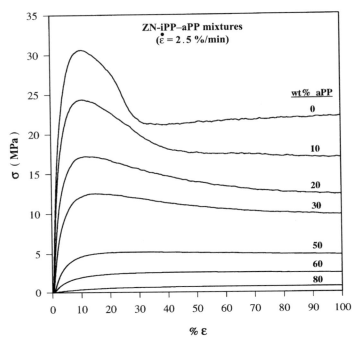

Figure 9   Stress–strain curves of iPP–aPP blends under tensile elongation. The strain rate is 2.5 %/min

Metallocene catalysts provide the only viable route for the production of sPP [18–21,70,71]. Figure 10 compares the Young's moduli of ZN-iPP–aPP mixtures with those of sPP–aPP mixtures. The sPP was prepared with isopropylidene(cyclopentadienyl)(9-fluorenyl)zirconium dichloride/MAO [70–72]. The volume fraction crystallinity (ϕ_c), as measured by the density gradient method, was varied with aPP addition. The highest crystallinity points in Figure 10 represent either the sPP homopolymer or ZN-iPP homopolymer. Each successive lower crystallinity point corresponds to increasing aPP concentration at 0, 10, 20, 30, 50, 60 and 80 wt % aPP. ZN-iPP–aPP and sPP–aPP blends show similar qualitative characteristics. Mechanical softening is observed with increasing aPP concentration. At any given aPP concentration, the sPP homopolymer and sPP–aPP blends show lower crystallinity (ϕ_c) and correspondingly lower young's modulus relative to ZN-iPP–aPP mixtures. Relative to ZN-iPP, sPP in general shows lower crystallinity, lower modulus and yield stress, lower crystallization temperature, lower melting-point and reduced flowability.

In the context of homopolymer tacticity mixtures, ZN-iPP–m-iPP blends highlight the potential of the MRG approach. Biaxial oriented polypropylene (BOPP) film is

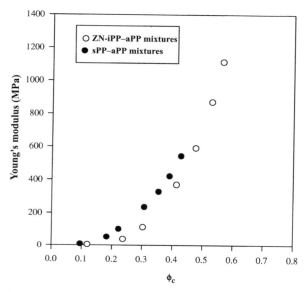

Figure 10 Young's modulus of iPP–aPP and sPP–aPP blends as a function of volume fraction crystallinity (ϕ_c). Volume fraction crystallinities were based on the measured mass density (ρ) determined by the density gradient method. Φ_c was reduced in the blends by increasing the concentration of aPP at 0, 10, 20, 30, 50, 60 and 80 wt% aPP

formed by either simultaneous or sequential biaxial stretching at elevated temperatures below the melting-point. Figure 11 shows the biaxial yield stress during BOPP film formation (simultaneous stretching) of three ZN-iPP resins (MFR = 4) chosen to represent a range of tacticities (high, medium, low) typical of Ziegler–Natta catalysts. Data at different draw temperatures are shown. Because partial melting has the effect of reducing the crystallinity at the draw temperature, the biaxial yield stress is lowered either by reducing the ZN-iPP stereoregularity or by increasing the draw temperature.

Superposed on Figure 11 is the biaxial yield stress of ZN-iPP–m-iPP mixtures. The m-iPP has the characteristics of m-iPP(1) in Table 1 and Figure 6. The ZN-iPP homopolymer matrix is chosen to have tacticity similar to the high- and medium-tacticity ZN-iPP homopolymers. However, the melt flow-rate (MFR) of the ZN-iPP matrix in the ZN-iPP–m-iPP blends was lowered to give similar MFR in the blends as the ZN-iPP homopolymers. Introduction of the low-melting m-iPP component also has the effect of reducing the biaxial yield stress relative to the ZN-iPP matrix at all draw temperatures. This can allow processing of BOPP film over a wider temperature range and at reduced draw temperatures. In addition, the molecular weight distribution can readily be tailored by the blend composition and individual

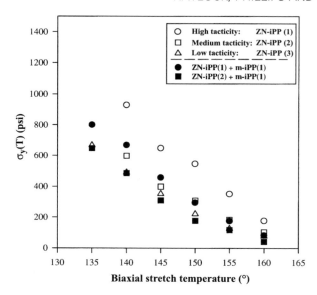

Figure 11 Biaxial yield stress, σ_y, as a function of stretch temperature for Ziegler–Natta isotactic homopolymer (ZN-iPP) and blends of ZN-iPP with metallocene isotactic homopolymer (m-iPP) during simultaneous biaxial drawing. The m-iPP is sample m-iPP(1) in Table 1. In the mixed ZN-iPP–m-iPP blends, the ZN-iPP MFR was modified to give the same MFR (4 g/10 m min) in the blend as the comparative ZN-iPP resins. The tacticity of the ZN-iPP component of the blends was similar to the corresponding comparative high-and medium-tacticity ZN-iPP resins

blend component characteristics. Figure 11 shows that a similar effect can be obtained by tuning the tacticity of conventional Ziegler–Natta homopolymers towards structures with reduced stereoregularity. However, this occurs with a corresponding loss of room temperature mechanical properties due to the reduction in crystallizability of resins with lower tacticity (Figure 6).

3.3 *MIXED CATALYST RUBBER MODIFIED ISOTACTIC POLYPROPYLENE*

Rubber-modified polypropylene is a commercially significant heterophasic polyolefin typically composed of Ziegler–Natta isotactic polypropylene (ZN-iPP) and a Ziegler–Natta ethylene–propylene elastomeric copolymer rubber (ZN-EPR) or ethylene–butene rubber (ZN-EBR). In subsequent discussions, EPR is used for illustrative purposes with the understanding that similar formulations can be based on EBR. The as-polymerized ZN-EPR weight fraction can range from 5 to 70 % while retaining a free-flowing particle morphology. Current process technologies allow for the use of bulk polymerization in combination with gas-phase or a series of gas-phase reactors using supported porous Ziegler–Natta catalysts [77]. Ziegler–Natta process technology allows for the independent control of molecular weight, comonomer composition, rubber phase volume fraction and particle morphology. As discussed earlier, the MRG technology allows for *in situ* polymerization within the porous ZN-iPP granule of EPR or EBR with metallocene catalysts (m-EPR, m-EBR). This mixed catalyst approach to rubber-modified heterophasic polypropylene allows for new polymer microstructures and property balances relative to Ziegler–Natta catalyst technology.

The EPR produced by classical Ziegler–Natta catalysts typically have a crystalline component at every composition. This is due to the broad inter-chain composition distribution. Whereas intra-chain incorporation of comonomer is often Bernoullian, the broad inter-chain composition distribution gives rise to an apparent 'blockiness' in unfractionated copolymer [18]. At an average copolymer composition of 30 wt % ethylene, a small amount of ethylene crystallinity is detected. As the concentration of ethylene increases beyond 50 wt %, the amorphous mass fraction of the EPR decreases owing to an increasing fraction of crystallizable chains rich in ethylene. Crystallinity in the rubber can be advantageous or detrimental [78–80]. Increased resistance to stress whitening can be achieved by incorporating high levels of ethylene crystallinity in the EPR phase [81–83], but the presence of crystallinity can impose restrictions on the effective sub-ambient glass transition temperature, T_g.

The statistics of comonomer incorporation, extent of comonomer uptake and molecular weight of copolymers produced by metallocene catalysts can vary over a wide range depending on catalyst structure [18–21, 84–88]. The statistics of intra-chain comonomer incorporation are often quantified by the product of reactivity ratios, $r_1 r_2$, where r_1 is the ratio of rate constants (k_{11}/k_{12}) for the addition of monomer 1 relative to monomer 2 to a site occupied by monomer 1 on the growing chain end. The value of r_2 is defined similarly (k_{22}/k_{21}) for a site occupied by monomer 2 on the growing chain end. Reported values of the product $r_1 r_2$ vary significantly depending on catalyst structure, but often are $\leqslant 1$ [18,19,21,84–86,88]. This indicates a tendency for alternation ($r_1 r_2 < 1$) or random ($r_1 r_2 = 1$) intra-chain incorporation statistics. Values of $r_1 r_2 > 1$, indicating increased 'blockiness', have also been reported [85,86,88]. The extent of comonomer uptake in ethylene–α-olefin copolymers is sensitive to catalyst structure and site symmetry [21,84], in accordance with the activities of ethylene–α-olefin homopolymerization [21]. Detailed analysis of monomer sequencing indicates that for some metallocene catalysts, a single catalyst center can exhibit a distribution of site selectivity due to the influence of the growing chain monomer sequencing on the site reactivity [87]. This factor can contribute to non-random intra-chain comonomer incorporation. While it is difficult to make generalizations regarding intra-chain comonomer incorporation, it is generally accepted that metallocene catalysts give copolymers with narrow inter-chain composition distribution relative to copolymerization with Ziegler–Natta catalysts [18,20,21,87].

With metallocene catalysts, it is possible to produce m-EPR with good elastomeric behavior and very low crystallinity at ethylene contents up to 75 wt%. The use of compositions with high ethylene content, which as discussed is required by existing technologies, can lead to desirable chemical cross-linking reactions (favorable reaction in polyethylene) when relevant physical–mechanical properties of cross-linked elastomers is desirable. The reaction may also be undesirable when thermally initiated during compounding or subsequent processing, and is a potential reason why targeted properties are altered. The m-EPR is conventionally used in melt mixing with the iPP component to yield a rubber-modified product. However, current metallocene technology lacks morphology control of the m-EPR. Therefore, for a given catalyst, the m-EPR must be polymerized with a sufficiently high ethylene content to provide some structure (low level of crystallinity) to allow for extrusion/pelletizing and minimization of cold flow. This limits the full utilization of the potential comonomer composition range possible from the metallocene catalyst. As discussed earlier, the MRG technology can, depending on polymerization conditions, produce macroscopic heterophasic reactor particles which are spherical and free-flowing, with the m-EPR dispersed within the iPP particle morphology [13]. It has also been shown that polymerization of m-EPR can be carried out with yields far exceeding that of the polymeric support [48,49]. This capability of the MRG technology provides distinct advantages over conventional technologies because it allows for the production of a wider range of m-EPR intra-chain compositions through the *in situ* polymerization and dispersion of m-EPR within the ZN-iPP particle morphology.

As discussed earlier, polymerization with metallocene catalysts can give polymers with narrowed inter-chain composition distribution and narrowed molecular weight distribution compared with polymerization with Ziegler–Natta catalysts [19–21,87]. These factors can be important contributing factors to polymer properties. Extractables can be reduced, blooming of low molecular weight species to the surface is reduced, thermal transition regions are sharpened (affecting impact, ductile–brittle transition temperature, thermo-mechanical properties) and the Newtonian rheological plateau is extended to higher melt shear rates. The last factor can lead to different melt processing requirements relative to blends with ZN-EPR (the m-EPR has reduced flowability owing to a higher effective viscosity at equivalent melt shear rates). Distinct differences in the dispersion of the rubber and iPP phases relative to Ziegler–Natta polymerization can also occur. The dispersion of the rubber phase is controlled by both rheological factors (through the imposed shear rate and viscosity ratio of dispersed and matrix fluids, η_r) and thermodynamic factors (through the interfacial tension, Γ).

The tendency for droplet break-up in the dilute Newtonian limit is characterized by the dimensionless capillary number (Ca) or Weber number (We) given by [89–92]

$$We = \frac{\dot{\gamma}\eta_m D}{2\Gamma} \qquad (1)$$

where $\dot{\gamma}$ is the shear rate, η_m is the viscosity of the matrix, D is the droplet diameter of the dispersed phase and Γ is the interfacial tension. We represents the ratio of the viscous force (or shear stress), $\eta_m\dot{\gamma}$, to the interfacial force (force which resists deformation), $2\Gamma/D$. The Taylor limit of the critical criteria for particle break-up, We_{cr}, is a function of the viscosity ratio, η_r, and the type of flow field [68]. In simple shear flow, no particle break-up occurs above a critical value of η_r, irrespective of shear rate. In extensional flow, droplet break-up is possible for all η_r, and We_{cr} is reduced. When η_r is changed to give a minimum value of We_{cr}, the droplets are most easily broken up. This occurs for η_r near unity. Hence differences in the extent of shear thinning with increasing shear rate, $\dot{\gamma}$, in m-EPR and ZN-EPR can be expected to give differences in rubber dispersion for the same matrix viscosity, η_m, even if the zero shear viscosities, η_0, are closely matched. Because of the viscoelastic nature of polymers, elasticity and yield stress also affect the deformation and rate of break-up of polymer droplets. The Taylor limit underpredicts the minimum particle size in this case [92]. The final particle size of the dispersed phase is also affected by droplet coalescence during processing. As concentration and inter-particle collisions increase relative to the idealized discussion above (as is the case for most commercial formulations), small particles can recombine to form larger particles, depending on the nature of the interface between them.

For incompatible iPP–EPR blends, the interfacial tension, Γ, is a factor which contributes to both (a) the dispersion of phases and (b) the interfacial adhesion between the respective components. Both the rubber dispersion and characteristics of the interfacial region govern mechanical properties such as strength and toughness. If the interfacial tension is large, the blend is highly immiscible, phase adhesion is minimal and little reinforcement between phases is observed. Changes in the composition of the EPR phase which favor interfacial tension must be balanced against the composition dependence of the glass transition temperature.

If a copolymer has a uniform inter-chain composition distribution and a narrow molecular weight distribution, then all molecules should contribute equally to surface tension. Polymerization of m-EPR by metallocene catalysts provides the opportunity to produce such uniform compositional parameters and thus to have more precise control of interfacial tension between phases. However, a narrow inter-chain composition distribution may not be beneficial for some aspects of the property balance. The broad inter-chain composition distribution of copolymers produced by Ziegler–Natta catalysts influences interfacial tension and interfacial characteristics. Even though the composition averaged interfacial tension may be equivalent (relative to a metallocene rubber with narrow inter-chain composition distribution), a distribution of surface tensions may favor interfacial adhesion on a molecular level due to an enrichment of favorable compositions at the interphase region (based on thermodynamic considerations) which differs from the bulk rubber composition.

The effect of different inter-chain composition distributions of metallocene and Ziegler–Natta-polymerized rubber on the resultant morphologies of rubber modified

(a)

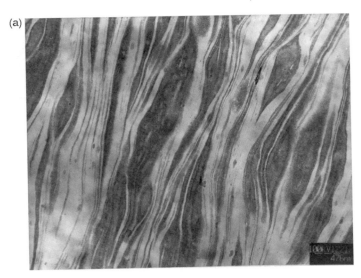

(b)

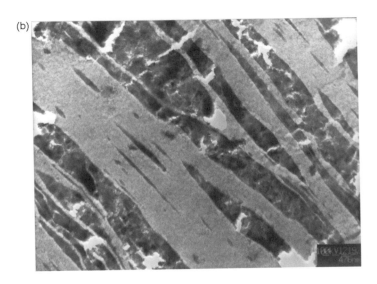

Figure 12 Transmission electron micrographs of iPP–EPR blown film morphologies. (a) Cross-section of a Ziegler–Natta polymerized heterophasic film (60 wt% ZN-EPR with 78:28 ethylene:propylene mole ratio, 40 wt% ZN-iPP). (b) Cross-section of a mixed catalyst (Ziegler–Natta iPP and metallocene EPR) polymerized heterophasic film (60 wt% m-EPR with 80:20 ethylene:propylene mole ratio, 40 wt% ZN-iPP). Dark regions are the rubber phase, light regions are the ZN-iPP phase

iPP is illustrated in Figure 12. Figure 12(a) shows a cross-section of a Ziegler–Natta-polymerized heterophasic blown film material (60 wt % EPR and 72:28 E:P mole ratio) with a broad inter-chain composition distribution in the rubber phase. Figure 12(b) shows the blown film morphology of a metallocene-polymerized (60 wt % EPR and 80:20 E:P mole ratio) rubber–ZN-iPP composition. The metallocene rubber has narrow inter-chain composition distribution. Both the Ziegler–Natta [Figure 12(a)] and mixed catalyst [Figure 12(b)] formulations have the same rubber phase volume fraction and similar viscosity ratios and film processing conditions. This comparison demonstrates that the Ziegler–Natta formulation shows a more complicated rubber phase morphology. A significant fraction of ethylene crystallinity, as indicated by lamellar textures, is observed within the EPR phase. The amorphous fraction (presumably propylene rich) of the EPR phase tends to reside at the interface. The metallocene-polymerized EPR shows a uniform phase morphology. No crystalline textures are observed in the rubber phase. These morphological differences can lead to differences in film properties. The blends based on Ziegler–Natta catalysts can show increased stiffness due to a significant fraction of ethylene crystallinity, and improved toughness resulting from rubber dispersion and effects of interface enrichment on interfacial adhesion. In contrast, blends based on metallocene catalysts allow for more precise control of interfacial tension by changes in rubber composition.

Hence some aspects of the property balance of ZN-iPP–rubber mixtures can be adversely affected by the use of metallocene rubber. These aspects can include the blend processability and phase morphology. However, the same molecular attributes of the metallocene rubber which lead to these difficulties, mainly the narrow inter-chain composition and molecular weight distribution, can also expand the property map of Ziegler–Natta formulations in beneficial ways. Extractables and low molecular weight blooming can be reduced, thermal transition regions sharpened (affecting impact, ductile–brittle transition temperature and thermo-mechanical properties) and copolymer composition can be more precisely controlled. These beneficial property attributes can be fully exploited by the MRG technology by allowing for the full utilization of the potential comonomer composition range possible from the metallocene catalyst. This capability is derived from the *in situ* polymerization and dispersion of m-EPR within the ZN-iPP particle morphology.

4 CONCLUSIONS

The understanding and control of Ziegler–Natta catalyst architecture has enabled Montell to develop a unique, practical and economic route to supported metallocene polymerization. The MRG technology allows additional freedom for the design of new resins and materials polymerized by metallocene catalysis. In this proprietary technology, the porous polymer or prepolymer produced by Ziegler–Natta catalysts becomes a polymeric support for a metallocene–catalyzed polymerization. The metallocene polymer then grows on and in that support, reproducing its shape. The potential for material synergies arises naturally from a process which allows for the *in situ* combination of Ziegler–Natta and metallocene polymers with widely varying polymerization strategies. If a porous Ziegler–Natta prepolymer is used as the support, then a pure metallocene polymer is produced with only residual Ziegler–Natta polymer remaining in the resin. If a porous Ziegler–Natta polymerization particle is used as a support, polymeric blends which combine Ziegler–Natta and metallocene polymers can be produced by *in situ* polymerization of metallocene catalysts. Free-flowing heterophasic copolymers can be produced by the *in situ* polymerization of metallocene rubber to very high rubber contents without reactor fouling. Application opportunities are diverse and include isotactic polypropylene homopolymer, propylene homopolymer tacticity mixtures, metallocene rubber and rubber-modified isotactic polypropylene.

5 ACKNOWLEGMENTS

The authors acknowledge the assistance of Dr Daniele Bugada with fiber spinning experiments, Tingh Nguyen with BOPP film experiments and Jean News with mixed catalyst film formulations. Characterization assistance was provided by Ken Klinger, Dr William Long, Deborah Morgan and Helena Rychlicki.

6 REFERENCES

1. Chien, J. C. W. and He, D., *J. Polym. Sci., Part A: Polym. Chem.*, **29**, 1603 (1991).
2. Kaminaka, M., and Soga, K., *Makromol. Chem., Rapid Commun.*, **12**, 367 (1991).
3. Soga, K., and Kaminaka, M., *Makromol. Chem., Rapid Commun.*, **13**, 221 (1992).
4. Collins, S., Kelly, W. M. and Holden, D. A., *Macromolecules*, **25**, 1780 (1992).
5. Kaminsky, W., and Renner, F., *Makromol. Chem., Rapid Commun.*, **14**, 239 (1993).
6. Soga, K., and Kaminaka, M., *Makromol. Chem.*, **194**, 1745 (1993).
7. Soga, K., Kim, H. J. and Shiono, T., *Macromol. Chem. Phys.*, **195**, 3347 (1994).
8. Langhauser, F., Kerth, J., Kersting, M., Kolle, P., Lilge, D. and Muller, P., *Angew. Makromol. Chem.*, **223**, 155 (1994).
9. Soga, K., *Macromol. Symp.*, **89**, 249 (1995).
10. Hungenberg, K. D., Kerth, J., Langhauser, F., Marczinke, B. and Schlund, R., in Fink, G., Mulhaupt, R. and Brintzinger, H. H. (Eds), *Ziegler Catalysts*, Springer, Berlin, 1995, p. 363.
11. Collina, G., Dall'Occo, T., Galimberti, M., Albizzati, E. and Noristi, L., *PCT Int. Appl.*, WO 96/11218 (1996).
12. Jungling, S., Koltzenburg, S. and Mulhaupt, R., *J. Polym. Sci. Polym. Chem.*, **35**, 1 (1997).
13. Galli, P., Collina, G., Albizzati, E., Sgarzi, P., Baruzzi, G. and Marchetti, E., *J. Appl. Polym. Sci.*, **66**, 1831 (1997).
14. Collina, G., Braga, V. and Sartori, F., *Polym. Bull.*, **38**, 701 (1997).
15. Galli, P., in *A Bimonthly Global Review of Technologies and End Use Trends*, Chemical Marketing Resource, Houston, TX, 1996, Vol. 2, No. 1, p. 46.

16. Haylock, J. C., and Galli, P., in *Metallocenes Asia '97*, 1997 p. 39.
17. Barbe, P. C., Cecchin, G. and Noristi, L., *Adv. Polym. Sci.*, **81**, 1 (1987).
18. Albizzati, E., Giannini, U., Collina, G., Noristi, L. and Resconi, L., in Moore, E. P., Jr (ed.), *Polypropylene Handbook*, Hanser, Munich, 1996, p. 11.
19. Gupta, V. K., Satish, S. and Bhardwaj, I. S., *J. Macromol. Sci., Rev. Macromol. Chem. Phys.*, **34**, 439 (1994).
20. Soares, J. B. P. and Hamielec, A. E., *Polym. React. Eng.*, **3**, 131 (1995).
21. Brintzinger, H. H., Fischer, D., Mulhaupt, R., Rieger, B. and Waymouth, R. M., *Angew. Chem., Int. Ed. Engl.*, **34**, 1143 (1995).
22. Giunchi, G. and Allegra, G., *J. Appl. Crystallogr.*, **17**, 172 (1984).
23. Albizzati, E., Giannini, U., Morini, G., Galimberti, M., Barino, L. and Scordamaglia, R., *Macromol. Symp.*, **89**, 73 (1995).
24. Zucchini, U. and Cecchin, G., *Adv. Polym. Sci.*, **51**, 103 (1983).
25. Galvan, R. and Tirrell, M., *Chem. Eng. Sci.*, **41**, 2385 (1986).
26. Tsutsui, T., Ishimaru, N., Mizumo, A., Toyota, A. and Kashiwa, N., *Polymer*, **30**, 1350 (1989).
27. Paukkeri, R. and Lehtinen, A., *Polymer*, **34**, 4075 (1993).
28. Kioka, M., Makio, H., Mizuno, A. and Kashiwa, N., *Polymer*, **35**, 580 (1994).
29. Galli, P., Barbe, P. C. and Noristi, L., *D. Angew. Makromol. Chem.*, **120**, 73 (1984).
30. Noristi, L., Marchetti, E., Baruzzi, G. and Sgarzi, P., *J. Polym. Sci., Polym. Chem. Ed.*, **32**, 3047 (1994).
31. Ferrero, M. A., Sommer, R., Spanne, P., Jones, K. W. and Conner, W. C., *J. Polym. Sci., Polym. Chem. Ed.*, **31**, 2507 (1993).
32. Schmeal, W. R. and Street, J. R., *AIChE J.*, **17**, 1188 (1971).
33. Nagel, E. J., Kirillov, V. A. and Ray, W. H., *Ind. Eng. Chem., Prod. Res. Dev.*, **19**, 372 (1980).
34. Floyd, S., Choi, K. Y., Taylor, T. W. and Ray, W. H., *J. Appl. Polym. Sci.*, **32**, 2935 (1986).
35. Ferrero, M. A. and Chiovetta, M. G., *Polym. Eng. Sci.*, **27**, 1436 (1987).
36. Ferrero, M. A. and Chiovetta, M. G., *Polym. Eng. Sci.*, **31**, 904 (1991).
37. Hutchinson, R. A., Chen, C. M. and Ray, W. H., *J. Appl. Polym. Sci.*, **44**, 1389 (1992).
38. Hoel, E. L., Cozewith, C. and Byrne, G. D., *AIChE J.*, **40**, 1669 (1994).
39. McKenna, T. F., Dupuy, J. and Spitz, R., *J. Appl. Polym. Sci.*, **57**, 3731 (1995).
40. Ferrero, M. A., Koffi, E., Sommer, R. and Conner, W. C., *J. Polym. Sci., Polym. Chem. Ed.*, **30**, 2131 (1992).
41. Kakugo, M., Sadatoshi, H., Yokoyama, M. and Kojima, K., *Macromolecules*, **22**, 547 (1989).
42. Galli, P., and Noristi, L., presented at the 5th European Plastics and Rubber Conference, Paris, June 12– 15, 1978.
43. Spaleck, W., Antberg, M., Rohrmann, J., Winter, A., Bachmann, B., Kiprof, P., Behm, J. and Herrmann, W. A., *Angew. Chem., Int. Ed. Engl.*, **31**, 1347 (1992).
44. Spaleck, W., Antberg, M., Aulbach, M., Bachmann, B., Dolle, V., Haftka, S., Kuber, F., Rohrmann, J. and Winter, A., in Fink, G., Mulhaupt, R. and Brintzinger, H. H. (Eds), *Ziegler Catalysts*, Springer, Berlin, 1995, p. 83.
45. Kaminsky, W., Kulper, K., Brintzinger, H. H. and Wild, F. R. W. P., *Angew. Chem., Int. Ed. Engl.*, **24**, 507 (1985).
46. Kaminsky, W., *Angew. Makromol. Chem.*, **145/146**, 149 (1986).
47. Resconi, L., Piemontesi, F., and Galimberti, M., *Eur. Pat.*, EP643078, 1995.
48. Galimberti, M., Ferraro, A., Baruzzi, G., Sgarzi, P., Camurati, I., Piemontesi, F., Mingozzi, I. and Vianello, M., presented at MetCon '97, Houston, TX, June 4–5, 1997.
49. Ferraro, A., Galimberti, M., Baruzzi, G. and Di Diego, M., presented at MetCon '97, Houston TX, June 4–5, 1997.

50. Busico, V., Cipullo, R. and Corradini, P., *Makromol. Chem.*, **194**, 1079 (1993).
51. Busico, V., Cipullo, R. and Corradini, P., *Makromol. Chem., Rapid Commun.*, **14**, 97 (1993).
52. Busico, V., Cipullo, R., Chadwick, J. C., Modder, J. F. and Sudmeijer, O. *Macromolecules*, **27**, 7538 (1994).
53. Tsutsui, T., Kashiwa, N. and Mizuno, A., *Makromol. Chem., Rapid Commun.*, **11**, 565 (1990).
54. Chadwick, J. C., Miedema, A. and Sudmeijer, O., *Macromol. Chem. Phys.*, **195**, 167 (1994).
55. Chadwick, J. C., van Kessel, G. M. M. and Sudmeijer, O., *Macromol. Chem. Phys.*, **196**, 1431 (1995).
56. Cheng, H. N. and Ewen, J. A., *Makromol. Chem.*, **190**, 1931 (1989).
57. Soga, K., Shiono, T., Takemura, S. and Kaminsky, W., *Makromol. Chem. Rapid Commun.*, **8**, 305 (1987).
58. Grassi, A., Zambelli, A., Resconi, L., Albizzati, E. and Mazzocchi, R., *Macromolecules*, **21**, 617 (1988).
59. Toyota, A., Tsutsui, T. and Kashiwa, N., *J. Mol. Catal.*, **56**, 237 (1989).
60. Mizuno, A., Tsutsui, T. and Kashiwa, N., *Polymer*, **33**, 254 (1992).
61. Resconi, L., Fait, A., Piemontesi, F., Colonnesi, M., Rychlicki, H. and Zeigler, R., *Macromolecules*, **28**, 6667 (1995).
62. Roll, W., Brintzinger, H.-H., Rieger, B. and Zolk, R., *Angew. Chem., Int. Ed. Engl.*, **29**, 279 (1990).
63. Rieger, B., Mu, X., Mallin, D. T., Rausch, M. D. and Chien, J. C. W., *Macromolecules*, **23**, 3559 (1990).
64. Chien, J. C. W. and Sugimoto, R., *J. Polym. Sci., Polym. Chem. Ed.*, **29**, 459 (1991).
65. Fischer, D. and Mulhaupt, R., *Macromol. Chem. Phys.*, **195**, 1433 (1994).
66. Stehling, U., Diebold, J., Kirsten, R., Roll, W., Brintzinger, H. H., Jungling, S., Mulhaupt, R. and Langhauser, F., *Organometallics*, **13**, 964 (1994).
67. Doi, Y., *Makromol. Chem., Rapid Commun.*, **3**, 635 (1982).
68. Phillips, R. A. and Wolkowicz, M. D., in Moore, E. P., Jr, *Polypropylene Handbook*, Hanser, Munich, 1996, p. 113.
69. Resconi, L., Jones, R. L., Rheingold, A. L. and Yap, G. P., *Organometallics*, **15**, 998 (1996).
70. Ewen, J. A., Jones, R. L., Razavi, A. and Ferrara, J. D., *J. Am. Chem. Soc.*, **110**, 6255 (1988).
71. Ewen, J. A., Elder, M. J., Jones, R. L., Haspeslagh, L., Atwood, J. L., Bott, S. G. and Robinson, K., *Makromol. Chem., Macromol. Symp.*, **48/49**, 253 (1991).
72. Phillips, R. A., Wolkowicz, M. D., and Jones, R. L., in *SPE ANTEC-97*, Toronto, 1997, Vol. XLIII, No. 2, p. 1671.
73. Llinas, G. H., Dong, S-H., Mallin, D. T., Rausch, M. D., Lin, Y. -G., Winter, H. H. and Chien, J. C. W., *Macromolecules*, **25**, 1242 (1992).
74. Hauptman, E., Waymouth, R. M. and Ziller, J. W., *J. Am. Chem. Soc.*, **117**, 11586 (1995).
75. Gauthier, W. J., Corrigan, J. F., Taylor, N. J. and Collins, S., *Macromolecules*, **28**, 3771 (1995).
76. Gauthier, W. J. and Collins, S., *Macromolecules*, **28**, 3779 (1995).
77. Lieberman, R. B. and LeNoir, R. T., in Moore, E. P., Jr, *Polypropylene Handbook*, Hanser, Munich, 1996, p. 287.
78. Collina, G., Bruga, V. and Sartori, F., *Polym. Bull.*, **38**, 701 (1997).
79. Gilbert, M., Briggs, J. E. and Omana, W., *Br. Polym. J.*, **11** (1979).
80. Corbelli, L., Martini, E. and Milani, F., presented at the 9th International Rubber Conference, Gottwaldow, 1987.

81. Fernando, P. L. and Williams, J. G., *Polym. Eng. Sci.*, **21**, 1003 (1981).
82. Karger-Kocsis, J. and Kuleznev, V. N., *Polymer*, **23**, 699 (1982).
83. Kolarik, J., Agrawal, G. L., Krulis, Z. and Kovar, J., *Polym. Composites*, **7**, 463 (1986).
84. Zambelli, A., Grassi, A., Galimberti, M., Mazzocchi, R. and Piemontesi, F., *Makromol. Chem. Rapid Commun.*, **12**, 523 (1991).
85. Chien, J. C. W. and He, D., *J. Polym. Sci., Polym. Chem. Ed.*, **29**, 1585 (1991).
86. Uozumi, T. and Soga, K., *Makromol. Chem.*, **193**, 823 (1992).
87. Galimberti, M., Martini, E., Piemontesi, F., Sartori, F., Camurati, I., Resconi, L. and Albizzati, E., *Macromol. Symp.*, **89**, 259 (1995).
88. Galimberti, M., Dall'Occo, T., Piemontesi, F., Camurati, I., Collina, G. and Barristi, M., presented at MetCon '96, Houston, TX, 1996.
89. Han, C. D., *Multiphase Flow in Polymer Processing*, Academic Press, New York, 1981.
90. Grace, H. P., *Chem. Eng. Commun.*, **14**, 225 (1982).
91. Han, C. D., in *'Polymer Blends and Composites in Multiphase Systems'*, ACS Symposium Series, Vol. 206, American Chemical Society, Washington, DC, 1984.
92. Sundararaj, U. and Macosko, C. W., *Macromolecules*, **28**, 2647 (1995).

16

Metallocene Catalyst Technology in a Bimodal Polymerization Process

HILKKA KNUUTTILA, ARJA LEHTINEN AND HANNU SALMINEN
Borealis Polymers Oy, Porvoo, Finland

1 INTRODUCTION

Polyethylenes (PE) produced using single-site catalysts have firmly entered the polyolefin market in the last 2 years. The growth projections for the global consumption of metallocene PE (mPE) have been widely debated. The growing mPE markets will compete not only with PE products made with conventional chromium and Ziegler–Natta catalysts, but also with LDPE and a range of other polymers such as PP, PVC and polar HP copolymers. The demand for mPE has been forecast by different consultants to be from 7.5 to 35 % of the polyethylene market in 2005. However, owing to high prices, unforeseen problems in production, resistance by processors and fear of patent infringement, until now the market penetration has been slower than predicted, especially in Europe [1–4].

There will be heavy competition between materials produced with metallocene and non-metallocene second-generation PE technologies, especially in film applications. The first-generation commercial metallocene polymers were polyethylene plastomers and elastomers having low density (below $910 \, kg/m^3$) and targeted at specialty markets [5]. Today, many companies are developing metallocene LLDPE and MDPE for replacement of LDPE and conventional LLDPE in order to gain larger sales volumes [6–11]. At the same time, bimodal LLDPE grades made with

Metallocene-based Polyolefins Edited by J. Scheirs and W. Kaminsky
© 2000 John Wiley & Sons Ltd

Ziegler–Natta catalysts and with excellent processability have great potential to replace large volumes of LDPE film. A bimodal process has shown its ability to improve the processability of LLDPE material considered very poor compared with LDPE [12–16]. The bimodal approach has been in existence already for some time, and has, e.g., helped producers address the problem of how to combine good processability with better mechanical properties in HDPE resins [17–19]. Especially in the case of mLLDPE, when the material offers superior mechanical properties such as excellent toughness, dart impact strength, clarity and sealability, but poor processability, bimodal technology brings many benefits including enhanced product processability and better overall balance of properties beneficial for many packaging and molding applications.

2 BIMODAL POLYETHYLENE AND TWO-STAGE PROCESSES

One of the important targets in material development is improved strength. The improvement of strength with increased molecular weight is apparent and, therefore, molecular weights of, e.g., HDPE resins have been creeping up over the years. At the same time, the molecular weight distribution (MWD) has broadened to minimize the negative effects of high molecular weight (as molecular weight increases, properties improve, but processability decreases).

Bimodal polyethylene consists of two fractions which usually differ from each other with respect to molecular weight and comonomer content. The high molecular weight component gives the material good mechanical strength while the low molecular weight component gives good processability. Each fraction should have a relatively narrow molecular weight distribution, with fairly different average molecular weights to generate a bimodal polymer with broad MWD. The key difference of the bimodal resin concept, when compared with conventional broad MWD polymers produced, for example, using chromium catalysts, is the possibility to control the shape of the MWD by concentrating the polymer chains needed for critical properties of the target resin: no very low molecular weight tail giving odor and smoke formation, more material that acts as a lubricant in processing, less 'average' polymer chains that are merely acting as a matrix, and again more high molecular weight material that is essential for strength of the polymer in its final application, e.g. as film and pipe (Figure 1).

Since the crystallization behavior and therefore the morphology and density of polyethylenes are to a large extent controlled by the amount, distribution and type of comonomer and short-chain branching, two-stage processes bring a completely new dimension in tailoring polymer structure: controlled comonomer distribution as a function of molecular weight combined with control of MWD. Normally when, e.g., Ziegler–Natta catalysts are used in unimodal processes, comonomers are incorpo-

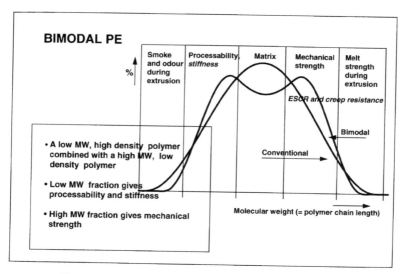

Figure 1 Control of molecular weight distribution

rated more readily in the short polymer chains. In a bimodal (two-stage) process, the comonomer incorporation can be flexibly controlled by producing short chains in one reactor and long chains in another. To optimize the material properties, the high molecular weight component usually has also a higher comonomer content (Figure 2).

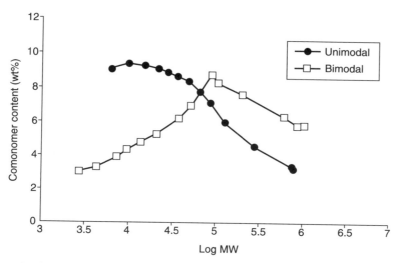

Figure 2 Comonomer distribution as a function of molecular weight for unimodal and bimodal Ziegler–Natta LLDPE

Bimodal polyethylene is produced in two cascaded reactors (Figure 3). The catalyst is fed into the first reactor where the first polymer fraction is produced. The polymer is taken out and transferred into the second reactor, where the second polymer fraction is produced. Between the reactors there may be a separation unit to remove the unreacted monomer, comonomer or hydrogen from the reaction mixture before the polymer is passed into the second polymerization stage.

If the process has a separation unit to remove at least the volatile components from the polymer (slurry) after the first polymerization reactor, either one of the two components can be produced in the first stage. If there is no separation unit between the stages, then the high molecular weight component has to be made in the first reactor, otherwise hydrogen would be carried over into the second reactor and a sufficiently high molecular weight would not be reached.

2.1 SLURRY–SLURRY PROCESS

The first commercial processes to produce bimodal polyethylene were combinations of two slurry reactors. In slurry processes the polymerization takes place at a temperature below the melting-point of the polymer in a liquid in which the polymer is essentially insoluble. The cascade reactor system can consist of two identical stirred tank reactors (this type of process is operated, e.g., by Mitsui Petrochemical and Hoechst [19]) or two loop reactors (operated or patented e.g., by Solvay [19] and Fina [20]). The monomer, comonomer and hydrogen are dissolved in a liquid diluent, where the catalyst and polymer are also suspended. Generally a high-boiling hydrocarbon, such as isobutane or hexane, is used as a diluent. Catalyst is fed into

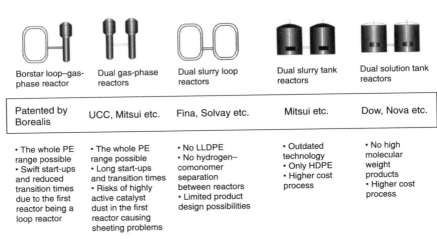

Possible bimodal process configurations

Borstar loop–gas-phase reactor	Dual gas-phase reactors	Dual slurry loop reactors	Dual slurry tank reactors	Dual solution tank reactors
Patented by Borealis	UCC, Mitsui etc.	Fina, Solvay etc.	Mitsui etc.	Dow, Nova etc.
• The whole PE range possible • Swift start-ups and reduced transition times due to the first reactor being a loop reactor	• The whole PE range possible • Long start-ups and transition times • Risks of highly active catalyst dust in the first reactor causing sheeting problems	• No LLDPE • No hydrogen–comonomer separation between reactors • Limited product design possibilities	• Outdated technology • Only HDPE • Higher cost process	• No high molecular weight products • Higher cost process

Figure 3 Two-stage processes to produce bimodal polyethylene

the first reactor with ethylene, the eventual comonomer, hydrogen and diluent. The polymer slurry is passed into the second reactor to produce the second component. The polymer slurry is then removed from the reactor and the polymer is separated from the liquid (e.g. by centrifuging). The liquid components are then recovered and recycled back to the process. The polymer is dried and extruded to pellets.

The solubility of polymer in the diluent increases with decreasing density. For this reason, LLDPE cannot be produced in slurry–slurry processes.

2.2 GAS-PHASE–GAS-PHASE PROCESS

Research on processes consisting of two cascaded gas-phase reactors has been conducted by Union Carbide [21], BP Amoco [22], Mobil [23], Montell [24] and Mitsui Petrochemical [25]. Catalyst, ethylene, comonomer and hydrogen are fed into the first reactor, where typically the high molecular weight component is produced. Polymer is then taken out of the reactor, the gases are removed and the polymer is passed to the second reactor, where typically the low molecular weight component is produced. The gases are removed by reducing the pressure and the polymer is dried and extruded.

The dual gas-phase reactor process offers a wide product range, from HD to LLD. Since there is no solvent or diluent, the solubility of the polymer is no problem. The softening of the polymer at low densities sets the lowest limit to the density that can be produced.

In theory, the process can be operated in either of the two modes: either the low molecular weight component is produced in the first reactor and the high molecular weight in the second reactor, or the high molecular weight component is produced first and then the low molecular weight component. In practice, the latter mode yields a more stable process operation.

2.3 SLURRY–GAS-PHASE PROCESS

A process consisting of a cascaded loop and gas-phase reactor is operated by Borealis (Figure 4) [26]. A catalyst, diluent, ethylene, hydrogen and the eventual comonomer are fed into the loop reactor. Usually the reactor is operated under supercritical conditions, i.e. the temperature and pressure in the reactor exceed the critical temperature and pressure of the reaction mixture. The polymer slurry is withdrawn from the reactor, after which the fluid is flashed off and recycled. The polymer is passed into a gas-phase reactor, where ethylene, hydrogen and comonomer are also added. The product is then taken out of the gas-phase reactor and the residual gases are recovered. The polymer is subsequently extruded.

The slurry–gas-phase process covers a wide product range from HD to LLD products. In theory, this process can also be operated in either of the two modes described earlier, but a more flexible operation is allowed if the low molecular

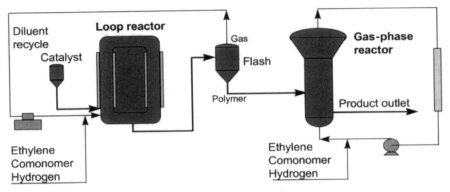

Figure 4 Borstar process flow chart

weight component is produced in the loop reactor and the high molecular weight component in the gas-phase reactor.

Compared with a gas-phase reactor, a slurry loop allows a higher production rate per unit volume owing to more effective cooling. Also, since the heat transfer between the surrounding fluid and polymer particles is more effective in a loop than in a gas-phase reactor, the presence of hot spots is more unlikely. On the other hand, a gas-phase reactor allows one to produce polymer with a higher comonomer content than a loop reactor, where the copolymer may dissolve in the fluid phase.

A loop reactor is ideal for production of the low molecular weight homopolymer for following reasons: (i) a lower hydrogen to ethylene ratio is needed to produce a specified molecular weight than in a gas-phase reactor; (ii) a shorter residence time of the polymer; (iii) a higher productivity of the catalyst; (iv) in general, the polymer does not significantly dissolve into the diluent; and (v) no hot spots in the reactor.

A gas-phase reactor is used to produce the high molecular weight copolymer, because there are no problems with solubility, it allows a wide product range and it is possible to produce a higher molecular weight at a fixed hydrogen to ethylene ratio than in slurry.

The combination of loop and gas-phase reactors offers, in addition to those mentioned above, the advantages such as no fresh catalyst feed into the gas-phase reactor is needed, with the result of more stable operation. There is the possibility of producing a wide range of products, from HD to LLD, and it is possible to influence the molecular weight and comonomer distributions of the final polymer.

3   SINGLE-SITE CATALYSTS IN MULTI-STAGE PROCESSES

In a bimodal process, the conditions in the two reactors differ radically from each other. This is very demanding for the catalyst performance, setting heavy requirements to be fulfilled: a high total activity balanced in all process conditions, a long polymerization lifetime, good hydrogen and comonomer response and good polymer morphology. Therefore, to have the right catalyst is a key issue in a bimodal process (Figure 5).

Already in the early stages of their development, metallocene catalysts were thought to be suitable for the production of bimodal polyethylene, either in multi-stage reactor systems or as ingredients of dual-site catalysts to create a bimodal resin in a single reactor. The main challenges that the metallocene catalysts would face in a bimodal role in any process were thought to be: (i) the catalysts were relatively low on the development curve; (ii) their adaptability to more economical low-pressure units had not yet been demonstrated; (iii) their narrow-MWD characteristics might not be suited to bimodals; (iv) de-ashing steps might be needed; and (v) legal conflicts over patents could retard commercial use [27].

High activity under all reactor conditions is necessary in order to have good economy in a bimodal process. Hydrogen which is used for molecular weight control strongly affects the catalyst activity. Conventional Ziegler–Natta catalysts usually lose a great part of their polymerization activity when the hydrogen concentration in the reactor is increased to produce a low molecular weight polymer (Figure 6). This restricts the overall productivity of the catalyst. However, Borealis has developed an advanced Ziegler–Natta catalyst which retains its activity and has a very balanced behavior throughout the whole operating range of hydrogen concentrations [28].

Some metallocene catalysts show an opposite behavior to normal Ziegler–Natta catalysts, with a higher activity when the hydrogen concentration is increased. Processwise, the high sensitivity of most single-site catalysts to hydrogen is also a feature which is useful in a bimodal process. This means that the low molecular

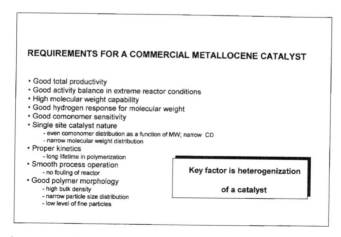

Figure 5 A commercial metallocene catalyst has to fulfil many severe requirements set by the process, catalyst manufacturing and the polymer product itself

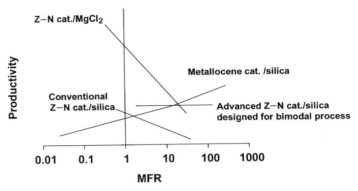

Figure 6 Metallocene catalysts possess very different behavior in varying process conditions compared with Ziegler–Natta catalysts

weight component in bimodal polyethylene can be produced at low hydrogen concentration. Owing to its high conversion, hydrogen is not carried over into the subsequent polymerization stages. This makes it possible to employ processes without separation of polymer and hydrocarbons between the polymerization stages. In Table 1, the hydrogen conversion in the production of a similar product using a single-site catalyst and a Ziegler–Natta catalyst is presented. Table 1 also shows that a lower comonomer to ethylene ratio is needed when a specific density is produced with a single-site catalyst compared with a Ziegler–Natta catalyst.

The high comonomer sensitivity of some single-site catalysts allows the production of a desired copolymer composition at a lower comonomer concentration. Also, the composition distribution produced by these catalysts is usually narrow. This makes it possible to produce a polymer with higher comonomer content while the flowability of the polymer powder is still good enough not to cause problems in the process. A material with a higher comonomer content and, therefore, with lower density can also be produced in the loop reactor while the solubility of the polymer remains at an acceptable level.

Table 1 Hydrogen and comonomer conversion with single-site and Ziegler–Natta catalysts

	Single site catalyst	Ziegler–Natta catalyst
MFR_2 (g/10 min)	90	89
Density (kg/m^3)	937	942
H_2/C_2 (mol/kmol)	0.3	171
C_4/C_2 (mol/kmol)	212	642
Conversion of H_2 (%)	96	19

With single-site catalysts, the effects of hydrogen and comonomer are not separate, but are strongly coupled (Figure 7). Thus an increase in the hydrogen concentration affects not only the MFR, but also the comonomer response of the catalyst and, therefore, the density of the material produced. On the other hand, comonomer concentration has an effect on the MFR, but this is less significant.

Hence the following advantages can be obtained by using SS catalysts:

- lower hydrogen concentration needed to produce a desired molecular weight;
- lower comonomer concentration needed to produce a desired density;
- possibility of producing a lower final density;
- possibility of producing narrower molecular weight and comonomer distributions for improved mechanical and optical properties.

One problem with the single-site catalysts is that many of them fail to produce a high enough molecular weight as required in some applications. As a consequence, the processability and the mechanical properties may then not be at the required level.

4 UNIQUE TAILORING OF MATERIAL

One of the major benefits of metallocene resins has been the potential for improving property performance over conventional polyethylene resins in many applications. Depending on the application and properties needed, low-pressure polyethylenes are usually produced using heterogeneous chromium or Ziegler–Natta catalysts. One of the distinctive characteristics of these resins is that they have both a broad molecular

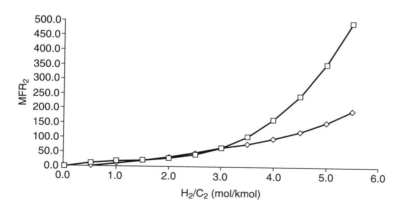

Figure 7 Effect of hydrogen on the performance of a metallocene catalyst in ($\Diamond$) homopolymerization and ($\square$) copolymerization of ethylene: MFR versus hydrogen to ethylene feed ratio

weight distribution and a broad, often multimodal, chemical composition distribution. This diffuses the material properties somewhat. Metallocene catalysts are best known for their single sitedness. Owing to more precise control of the length and structure of polymer chains, they produce structurally reasonably uniform polymers with narrow molecular weight distributions (Figure 8). Detailed structural studies have recently shown compositional inhomogeneity also in metallocene polyethylenes [29,30], but at least theoretically they should still be optimal catalysts for tailoring polymer structure and properties in a bimodal process. Structural inhomogeneity is partially dependent on the polymerization process used. For example, with the same catalyst, the slurry loop process seems to produce more homogeneous resins than the gas-phase process [31].

Since metallocene catalysts can incorporate large amounts of comonomer in the polymer chains, elastomeric materials with density as low as 855–860 kg/m^3 have been produced by solution and high-pressure processes. The upper usable density limit for mPE has not yet been defined but today a flood of linear low-and medium-density grades (910–940 kg/m^3) produced by slurry and gas-phase processes are supplementing the initial selection of elastomeric mPE grades [5–11,32]. The metallocene plastomers and elastomers have found markets, e.g. in food packaging, shoe soles, flooring, coated fabrics, medical products and cable insulation materials. Higher density materials are often targeted at fiber, rotational molding and injection molding applications.

However, a large-volume application of mLLDPE, mMDPE and mHDPE will be in film products. Consultant Chem Systems estimates that 18 % of conventional LLDPE will be replaced by mLLDPE by 2005. The excellent physical properties and there by the down-gauging possibilities make mLLDPE film well suited for a broad

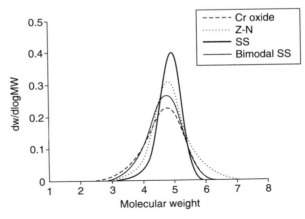

Figure 8 Typical molecular weight distributions produced by major polyethylene catalysts

range of applications. Fabrication of mLLDPE film is generally effected by the blown and cast film processes. The resulting film is usually characterized by lower haze, lower blocking tendency, higher gloss, higher dart impact strength, excellent puncture resistance and improved sealability and organoleptics compared with conventional Ziegler–Natta LLDPE resins of equivalent MFR and density. New property combinations can also be achieved with the new emerging grades, e.g. greater stiffness and impact strength or better sealability and improved heat resistance.

At the same time, however, the first-generation metallocene resins have poor processing characteristics compared with conventional polyethylene. Owing to their narrower molecular weight distribution, metallocene LLDPEs show less pronounced shear thinning behavior than the conventional LLDPEs. This means that they are more viscous at normal extrusion shear rates, generate more shear heating, melt faster, pump at slightly higher specific rate and generate higher motor torque [33]. Consultants use a processability rule of thumb: LDPE ranks 10, bimodal HDPE 8, conventional LLDPE 4 and typical mPE 1 [34]. Usually some processing modifications on the existing equipment have to be introduced when mLLDPEs are to be utilized.

One way to minimize the processability problems of mPE is to blend or coextrude it with conventional PE or to use processing aids (e.g. fluoroelastomers). However, the main trend is to develop a 'look-alike LDPE', i.e. a high-clarity, tough, easy-to-process LLDPE that would run well on LDPE equipment. One way to achieve this is to incorporate long-chain branching (LCB) in the polymer chains and at the same time maybe slightly broaden the MWD. With metallocene catalysts the amount of LCB can be controlled so that improved extruder processability and melt elasticity are gained without sacrificing toughness [5,7–10,35].

Another possibility for improving processability is to broaden the MWD with a dual-site catalyst or a multi-reactor process; i.e. make bimodal mPE with or without LCB. Piloting of bimodal single-site resins has confirmed the potential for the production of polymers with accurately tailor-made molecular structure. Since the metallocene catalysts produce a reasonably narrow compositional distribution, the comonomer (short-chain branching) can be inserted with greater precision. In two-step polymerizations it is also possible to adjust the comonomer distribution by feeding different concentrations of comonomer into the first and second reactors and in this way control the polymer morphology and especially formation of tie molecules to obtain optimum balance of properties. By changing the comonomer distribution as a function of molecular weight, at the same density more comonomer can be incorporated in the bimodal PE resins than in conventional unimodal resins. This tailoring of comonomer distribution also means that resins with a reversed comonomer distribution can be produced (more comonomer in the high molecular weight fraction as shown in Figure 2). This possibility has been widely utilized in the development of HDPE pipe grades [19].

Melt rheology is a useful tool in studying of molecular structure and comparing materials with each other. Some examples of the rheological behavior of unimodal and bimodal metallocene PE resins are presented in Figure 9. As mentioned earlier, bimodalism is a way to make products which are easier to process and have a good balance of material properties. With a broader MWD, the molecular weight of the polymer can also be increased and this usually enhances toughness and other mechanical properties.

Many companies have applied for patents on bimodal single-site catalyst PE materials [36–41], but Mitsui Petrochemical was the first to present semi-commercial bimodal metallocene LLDPE grades with the trade name Evolue [25]. These resins are produced in cascaded gas-phase reactors with commercial production started in 1998. According to Mitsui, their bimodal mPEs run at reduced die pressure and fast rates and down-gauge by 30 % over conventional LLDPE. Since the bimodal metallocene PE grades are still mainly in the pilot phase, all the potential improvements in material properties have not yet been realised. Nova Chemicals has developed both bimodal Ziegler–Natta and recently also new bimodal single site resins which can give some indications of the properties that it is possible to obtain with bimodality. These polymers are produced in an Advanced Sclairtech solution process and the components produced in each reactor have narrow or reasonably narrow MWD and compositional distribution. These new bimodal polymers are said to have an outstanding balance of physical, optical and sealing characteristics compared with unimodal metallocene resins having narrow MWD [13,42].

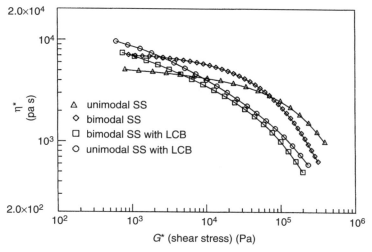

Figure 9 Rheological behavior of some unimodal and bimodal metallocene resins

5 CONCLUSION

Bimodal metallocene catalyst technology is still mainly in the pilot phase. The first commercial products came on the market in 1998.

The use of metallocene catalyst technology gives many benefits compared with advanced Ziegler–Natta catalysts. A lower hydrogen concentration is needed to produce a desired molecular weight. Lower comonomer concentrations are needed to reach a desired density, which means both reduced comonomer costs and shorter transitions. When a catalyst has a high activity also at high MFR production we can achieve increased overall productivity and better economy. It is also possible to produce a lower final density than with Ziegler–Natta catalysts without fouling the reactor. Also, of course, it is possible to have a narrower comonomer distribution, reflecting excellent material properties. A bimodal process combined with single-site/metallocene catalyst technology is a very powerful tool for advanced poly-ethylenes. It provides an opportunity for the unique design of materials where it is possible to tailor carefully the molecular weight distribution and also the chemical composition distribution in a polymer chain and overall in the material.

6 REFERENCES

1. Montagna, A. A., Burkhart, R. M. and Dekmezian, A. H., *Chem Tech* **27**, (12), 26 (1997).
2. *Eur. Chem. News*, 20–26 October, 16 (1997).
3. *Eur. Chem. News*, 21–27 April, 16 (1997).
4. Schut, J. H. *Plast. World*, **53**, (1), 33 (1995).
5. Schut, J. H. *Plast. World*, **54**, (4), 41 (1996).
6. Bailey P. N. and Varrall, D. C. presented at Metallocenes '96, 6–7 March 1996, Düsseldorf.
7. Murphy, M. W. in *Metallocenes Europe '97*, 8–9 April 1997, Düsseldorf, pp. 459–478.
8. Lastovica, J. E. in *Proceedings of the 7th International Business Forum Spec. Polyolefins SPO '97*, 1997, pp. 245–253.
9. Simpson, D. M. Whaley P. D. and Liu, H. T. in *Proceedings of the 7th International Business Forum Spec. Polyolefins SPO '97*, 1997, pp. 255–269.
10. Fraser, W. A. Adams J. L. and Simpson, D. M. in *4th International Conference on Metallocene Polymers, Metallocenes Asia '97*, pp. 69–95.
11. Everaert J. and Dewart, J.-C. in *Metallocenes Europe '97*, 8–9 April 1997, Düsseldorf, pp. 423–438.
12. Takakarhu, J. *Petrol. Technol. Q.*, Winter, 124 (1997/98).
13. Kelusky, E. C. in *Metallocenes Europe '97*, 8–9 April 1997, Düsseldorf, pp. 111–127.
14. Panagopoulos, G. Jr, and Kamla, R. D. in *Polymers, Laminations and Coatings Conference 1996*, pp. 241–259.
15. Yi K. C. H. and Michie, W. J. Jr, in *SPO '93*, pp. 7–25.
16. Ealer, G. E. Buehler-Vidal, J. O. Kupperblatt, S. A. Moy F. H. and Rickman-Davis, D. J. in *Polymers, Laminations and Coatings Conference 1997*, pp. 577–608.
17. Böhm, L. L. Enderle H. F. and Fleissner, M. *Adv. Mater.*, **4**,, 234 (1992).
18. Böhm, L. L. Enderle H.-F. and Fleissner, M. *Stud. Surf. Sci. Catal.*, **89**, 351 (1994).
19. Scheirs, J. Böhm, L. L. Boot J. C. and Leevers, P. S. *Trends Polym. Sci.*, **4**, 408 (1996).
20. Debras, G. *Eur. Pat.*, EP 0 580 930 (1994).

21. Daniell, P. T. Tilston, M. W. Spriggs, T. E. Wagner B. E. and Rammurthy, A. V. Rammurthy, *Eur. Pat.*, EP 0 691 353 (1996).
22. Jenny, C. and Lalanne-Magne, C. *Eur. Pat.*, EP 0 570 199 (1993).
23. Hussein, A. Hagerty R. and Ong, S. *Eur. Pat.*, EP 0 503 791 (1997).
24. Covezzi, M. Galli, P. Govoni G. and Rinaldi, R. *Eur. Pat.*, EP 0 517 183 (1992).
25. Hamada, N. in *Metallocenes Europe '97*, 8–9 April 1997, Düsseldorf, pp. 289–314.
26. Ahvenainen, A. Sarantila, K. Andtsjö, H. Takakarhu J. and Palmroos, A. *US Pat.*, 5 326 835 (1994).
27. Leaversuch, R. D. *Mod. Plast. Int.*, **21**, (10), 34 (1991).
28. Garoff, T. Johansson, S. Palmqvist, U. Lindgren, D. Sutela, M. Waldvogel P. and Kostiainen, A. *Eur. Pat.*, EP 0 688 794 (1995).
29. Mingozzi I. and Nascetti, S. *Int. J. Polym. Anal. Charact.*, **3**, 59 (1996).
30. Hsieh, E. T. Tso, C. C. Byers, J. D. Johnson, T. W. Fu Q. and Cheng, S. Z. D. *J. Macromol. Sci. Phys.* **B36**, 615 (1997).
31. H. Knuuttila, A. Lehtinen, H. Hokkanen and K. Kallio, presented at Metallocenes '96, 6–7 March 1996, Düsseldorf.
32. I. Sörum Melaaen, in *Metallocenes '96*, 6–7 March 1996, Düsseldorf, pp. 87–95.
33. W. Hellmuth, in *Metallocenes '96*, 6–7 March 1996, Düsseldorf, pp. 239–248.
34. R. D. Leversuch, *Mod. Plast. Int.* **24**, 38 (1994).
35. G. N. Foster and S. H. Wasserman, presented at *MetCon '97: Polymers in Transition*, 4–5 June 1997, Houston, TX.
36. R. L. Bamberger, *PCT Int. Appl.*, WO 96/18679 (1995).
37. P. M. Stricklen, *PCT Int. Appl.*, WO 92/15619 (1992).
38. G. N. Foster, D. E. James and F. J. Karol, *Eur. Pat.*, EP 0 770 629 (1995).
39. T. Tsutsui and T. Ueda, *Eur. Pat.*, EP 0 447 035 (1991).
40. M. Takahashi, A. Toda, S. Matsunaga and T. Tsutsui, *Eur. Pat.*, EP 0 587 365 (1993).
41. N. Toshimi, T. Housaki, J. Matsumoto, T. Okamoto, M. Watanabe and N. Ishihara, *Eur. Pat.*, EP 0 572 034 (1993).
42. Kelusky, E. C., in *Proceedings of the 8th International Business Forum Spec. Polyolefins SPO '98*, 1998, pp. 113–129.

PART VI

Rheology and Processing of Metallocene-based Polymers

17

Rheological Properties of Metallocene-catalyzed Polyolefins in the Melt and Solid States and Comparison with Conventional Polyolefins

JOSÉ MARÍA CARELLA
INTEMA (UNMdP–CONICET), Mar del Plata, Republica Argentina

LIDIA MARÍA QUINZANI
PLAPIQUI (UNS–CONICET), Bahía Blanca, Republica Argentina

1 INTRODUCTION

Since the discovery by Ziegler and Natta, during the 1950s, of the powerful catalyst used in the low-pressure polymerizations of ethylene and propylene, remarkable progress has been made in polyolefin technology. Continuing efforts have been made to design and produce new generations of highly active and stereoselective catalysts to achieve better tailoring of the macromolecular structure distributions, therefore allowing better control of the melt and solid-state rheological properties of polyolefins. Novel types of polyolefins are nowadays being produced following the introduction of metallocene catalysts, such as very low density polyethylenes, polyethylenes with controlled type and amount of short-chain branching sequences, syndiotactic polypropylenes and or polypropylenes with controlled tacticity sequences.

The rheological behavior of polyolefins has been investigated and reported in a very large number of papers during the last 30 years, attesting to the interest and

Metallocene-based Polyolefins Edited by J. Scheirs and W. Kaminsky
© 2000 John Wiley & Sons Ltd

progress in research in this field. Phenomenological and molecular theories have been developed to a high degree of refinement, tested with the aid of model polymers, and in some cases applied to model polyolefins, to verify predicted structure–property relationships [1–8]. However, there have been very few reports on the rheological characterization of the recently generated metallocene-catalyzed polyolefins. Chien *et al.* [9], Llinas *et al.* [10] and Eckstein *et al.* [11] studied the viscoelastic properties of several polypropylenes with different microstructures synthesized using metallocene catalysts. Vega *et al.* [12], Carella [13], Muñoz Escalona *et al.* [14], Sukhadia [15] and Rohlfing and Janzen [16] analyzed the melt rheological properties of metallocene-catalyzed polyethylenes. Woo *et al.* [17] and Lehtinen *et al.* [18] obtained dynamic mechanical data for metallocene-catalyzed copolymers of ethylene with α olefins. A few other studies [19–23] have contributed rheological measurements of ethylene copolymers. No reasons can be found in the publications cited above to suggest that the rheological behavior of these polymers will not fall within the framework of the rheological characteristics of previously existing types of polyolefins. This conclusion can be drawn when the rheological measurements are analyzed in the light of a complete molecular characterization of the materials.

The topic that is presented with greatest emphasis in this chapter is linear viscoelasticity. This is a very important subject that has been extensively studied in the past [5,24]. A brief analysis of the current phenomenological and molecular theories will permit an interpretation of the response of polymeric materials to a restricted class of flow, and relate it to the molecular structure of the polymers. We begin by presenting the necessary background followed by a discussion of experimental and theoretical results for the linear viscoelasticity of polyolefins.

2 GENERAL THEORETICAL ASPECTS

2.1 MELT RHEOLOGICAL PROPERTIES

The study of linear viscoelastic properties is the first and basic step in the rheological characterization of polymer melts. Linear viscoelastic parameters may be related to the molecular structure of the materials since they are very sensitive to the molecular weight, molecular weight distribution and long-chain branching of the molecules. On the other hand, linear viscoelastic data are also used in the quality control of industrial products and in pure rheological characterization for the determination of the coefficients of different constitutive equations.

Linear viscoelastic behavior is observed for simple fluids when the flow produces sufficiently small displacements and displacement gradients. In that case the response of the materials is associated with the distortion of the polymer chains from their equilibrium conformations and results in a reversible but time-dependent behavior. The viscoelastic response of amorphous polymers is successfully described by the general linear viscoelastic model, which is well described in several standard references [5,24].

The relaxation modulus $G(t)$ is the time function that captures the nature of the fluid. It measures the transient stress per unit strain in a step-strain experiment performed with a very small deformation. This modulus is a positive function which decreases monotonically to zero as time goes to infinity. The expression of the relaxation modulus most extensively used is that given by the generalized Maxwell model [5]:

$$G(t) = \sum_{k=1}^{\infty} G_k e^{-t/\lambda_k} = \sum_{k=1}^{\infty} \frac{\eta_k}{\lambda_k} e^{-t/\lambda_k} \tag{1}$$

where the infinite set of constants λ_k and η_k form the spectrum of relaxation times and viscosity coefficients. In practice, the spectrum is reduced to a discrete number of elements by organizing the relaxation times in a decreasing order and setting λ_k and η_k to zero for k greater than some finite number N. The relaxation modulus is then dominated by the largest relaxation time, λ_1.

The linear viscoelastic parameter most used in the frequency (ω) domain is the complex shear modulus $G^*(\omega)$, which characterizes the behavior of polymeric materials under small-amplitude oscillatory shear flow. The in-phase and out-of-phase components of the dynamic modulus are known as the storage or elastic modulus, $G'(\omega)$, and the loss or viscous modulus, $G''(\omega)$, respectively. Two other frequently used linear viscoelastic properties are the dynamic viscosity $[\eta'(\omega) = G''/(\omega)]$ and the dynamic rigidity $[\eta''(\omega) = G'/\omega]$.

The dynamic material parameters, calculated according to the generalized Maxwell model, are

$$\eta'(\omega) = \frac{G''(\omega)}{\omega} = \sum_{k=1}^{N} \frac{\eta_k}{1 + \lambda_k^2 \omega^2} \qquad \frac{\eta''(\omega)}{\omega} = \frac{G'(\omega)}{\omega^2} = \sum_{k=1}^{N} \frac{\eta_k \lambda_k}{1 + \lambda_k^2 \omega^2} \tag{2}$$

In the limit of small frequencies, the dynamic viscosity approaches the zero-shear-rate viscosity, $\eta_0 = \sum_k \eta_k$, as expected according to experimental results, and the function η''/ω approaches the number $\sum_k \eta_k \lambda_k$. These limits provide a way to measure the characteristic zero-shear-rate relaxation time ($\lambda_0 = \sum_k \eta_k \lambda_k / \eta_0$) and the zero-shear-rate viscosity experimentally.

The typical frequency dependence of G' and G'' for polymers with linear flexible chains and narrow molecular weight distribution is sketched in Figure 1. These are universal functions [24,25]. The dynamic viscosity is found to approach the constant zero-shear-rate viscosity (η_0) at low frequencies (terminal region), which corresponds to a loss modulus proportional to ω. The storage modulus is found to be proportional to ω^2 in this region. At intermediate frequencies both η' and η''/ω show power-law behavior. Narrow molecular weight distribution linear polymers exhibit a well defined maximum of G'' in this region, while the storage modulus shows a constant value which is known as the plateau modulus, G_N^0. The plateau modulus is

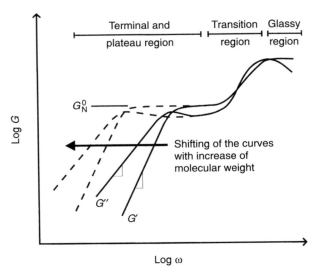

Figure 1   Typical frequency dependence of G' and G'' for polymers with linear flexible chains and a narrow molecular weight distribution

independent of the molecular weight of the polymer for high molecular weight materials, and it is related to the molecular weight of polymer segments between entanglements, M_e, which may be calculated from rubber elasticity theory as $M_e = \rho RT/G_N^0$ [24]. Graessley and Edwards [26] have shown that $G_N{}^0$ is proportional to temperature, chain contour length concentration and the characteristic ratio for the species, C_∞. The zero-shear-rate viscosity of melts and concentrated polymer solutions, and also the breadth of the plateau region, increase with increasing molecular weight as M^a, where a is typically in the range 3.4–3.6 for linear polymers. Broadening of the molecular weight distribution decreases the 1 and 2 slopes of G'' and G' in the terminal region, while an increase in the long-chain branching content increases the breadth of the transition from the terminal region to the plateau zone [25]. The terminal and transition regions are affected by the intermediate and long-time scale molecular dynamics, i.e. entanglement and flow behavior. The glass behavior at short times appears in the range of large frequencies.

Molecular theories developed from the basic ideas of Edwards [27] and de Gennes [28] analyze the origin of the $G(t)$ function starting from the ideas of the reptation model. If an entangled polymer molecule is suddenly deformed, the resisting stress decays with time. The rate of decay is a measure of the time required for the chains to disentangle from their neighbors. Doi and Edwards [3] considered three series of relaxation processes. Short-range interactions relax in the shortest times. Subsequently, Rouse-type relaxation takes place until the chain relaxes out to the distance between entanglements, where the $G(t)$ function takes the values predicted by the rubber elasticity theory [24]. Stresses continue to relax owing to the contraction of the contour length from its deformed length back to its equilibrium state. The final relaxation process corresponds to the chain diffusing out of its oriented tube and returning to an isotropic state. The above description is adequate for linear polymers, which diffuse by reptation-like and associated mechanisms.

It has been demonstrated that long branches in the molecules strongly suppress translational motions of the whole molecules which, however, can still change their configurations by retracting the long branches along their primitive path, pushing out unentangled loops into the surrounded matrix [29]. For similar molecular weights, this relaxation mechanism, which is not necessary for the relaxation of linear molecules, changes the flow activation energy toward higher values. A theoretical basis has been provided [29] for the observed thermorheological complexity of polymers with long-chain branching. The anomalous behavior is associated with long branches and the temperature coefficient of the chain dimensions ($K = \partial \ln\langle R^2 \rangle / \partial T$). It has been proposed and confirmed experimentally [29,30] that long branches in polymers with a negative K (such as polymethylene chains) produce an increase in the values of the flow activation energy, when calculated from zero-shear-rate viscosity measurements. This increase depends on the number of long branches per molecule, on the average branch length and on the value of K. Flow activation energies calculated from measurements in the plateau region are independent of the presence of long-chain branches. The value of the characteristic zero-shear-rate relaxation time, λ_0, of long-branched polymers is considerably larger than the corresponding value for linear molecules of similar (high) molecular weight, owing to the above-mentioned additional relaxation mechanism necessary for translational motions of the whole molecules.

Rheological data for linear polymers obtained at various temperatures may be combined into single master curves through the time–temperature superposition procedure [24]. This method considers that, when the same deformation is applied to a material at two different temperatures, then (a) the stresses generated in the polymer scale with a shift factor $b_T(T)$ which includes the effects of temperature, density and characteristic ratio for the species [26,31], and (b) the rate at which the deformation is applied scales with a shift factor $a_T(T)$. In consequence,

$$G(T_0) = G(T)b_T \qquad \eta(T_0) = \frac{\eta(T)}{a_T}b_T \qquad \omega(T_0) = a_T\omega(T) \qquad (3)$$

although the temperature and density dependence of stresses are usually neglected and only the effect of a_T is considered.

The shift factor a_T is generally determined from measurements of the zero-shear-rate viscosity at different temperatures. For materials far from the glass-transition temperature, such as polyolefins in the melt state, the temperature dependence of a_T is usually described by an *Arrhenius* type of exponential function of the form

$$a_T = \exp\left[\frac{\Delta H}{R}\left(\frac{1}{T} - \frac{1}{T_0}\right)\right] \qquad (4)$$

where ΔH is the flow activation energy and R is the universal gas constant. Typical values of $\Delta H/R$ are 3000 K for high-density polyethylenes, 4500 K for commercial low-density (high-pressure) polyethylenes and 5200 K for polypropylenes. These values are very dependent on the molecular parameters of the polymers. As mentioned above, the flow activation energy of long-chain branched materials is dependent on the relaxation mechanism. At low frequencies, where the long relaxation times dominate the relaxation process, the value of the flow activation energy is larger than at higher frequencies where the short-range interactions dominate the relaxation process. In consequence, it is not possible to apply the time–temperature superposition procedure in the rheological characterization of long-chain branched polymers unless small ranges of temperature and/or time are considered. This behavior is known as thermorheologically complex. Thermorheo-logical complexity has also been observed in blends of linear and long-chain branched polyethylenes of similar molecular weights [32]. The flow activation energy measured from zero-shear-rate measurements by Graessley et al. [32] was found to increase with the product of the molecular weight of the long branches times the volume fraction of the branched material.

2.2 SOLID-STATE RHEOLOGICAL PROPERTIES

Semicrystalline materials, such as linear and branched polyethylenes and ethylene copolymers, display different temperature-activated transitions. Below the melting temperature and in decreasing order, there are three temperature transitions which are conventionally designated as α, β and γ transitions or relaxations. It has been shown that the α relaxation corresponds to processes occurring in the crystallites. For polyethylenes this relaxation temperature is found between 0 and approximately 130 °C, and depends on the crystallite thickness [33]. The β transition corresponds to the relaxation of the crystalline–amorphous interphase [34,35]. For polyethylenes this relaxation temperature is observed between -15 and -35 °C and is a function of the molecular branching frequency [34]. For ethylene copolymers, the β relaxation temperature has been measured between -15 and approximately -70 °C as a function of the chemical nature of the comonomer, the comonomer concentration and the copolymerization statistics [34,36,37]. The β relaxation temperature of block copolymers is not affected by the comonomer concentration, whereas in the case of random copolymers this temperature decreases when the comonomer concentration increases [34, 35]. The γ transition corresponds to processes in the amorphous phase. The corresponding relaxation temperature has been detected at about -100 °C and it is expected to depend slightly on branching frequency for polyethylenes, and on the chemical nature of the comonomer and on comonomer concentration in the case of ethylene copolymers.

3 EXPERIMENTAL ASPECTS

To illustrate the above-mentioned rheological behavior of polyolefins, the following resins have been selected:

- three conventional polyethylenes: high-density polyethylene (HDPE) from a low-pressure Hoechst process [38], linear low-density polyethylene (LLDPE) from the gas-phase UNIPOL process [39] and a high-pressure low-density polyethylene (LDPE) [40];
- two metallocene-catalyzed polyethylenes, PE02 and PE03 [12], produced with technology developed by Kaminsky *et al.* [41] using Cp_2ZrCl_2 as catalyst;
- one conventional polypropylene (PP) made with a low-pressure BASF process;
- four metallocene-catalyzed polypropylenes, PP(50), PP(25), PP(0) and PP(−20) [10], produced with *rac-anti*-[ethylidene(1-η^5-tetramethylcyclopentadienyl)(1-η^5-indenyl)dichlorotitanium(IV)/methylaluminoxane (MAO).

3.1 POLYETHYLENE

Table 1 summarizes the weight-average molecular weight ($\overline{M}_w$) and molecular weight dispersity ($MWD = \overline{M}_w/\overline{M}_n$) of the five different polyethylenes, together with the corresponding zero-shear-rate viscosity (η_0), the characteristic zero-shear-rate relaxation time (λ_0) and the flow activation energy ($\Delta H/R$). The selected polymers differ mainly in the structure of the chains, which vary from very linear, as in the case of the HDPE, to long-branched, as in LDPE. All the polyethylenes have very similar molecular weight distributions. The metallocene-catalyzed polyethylenes PE02 and PE03 were chosen among several others because they have lower and higher average molecular weights, respectively, than the three conventional materials. This selection allows for the comparison of the rheological properties. The

Table 1 Molecular and rheological parameters of the analyzed polyethylenes (all the temperature-dependent properties are given at the same reference temperature of 190 °C)

Material	Reference	$\overline{M}_w \times 10^{-3}$	MWD	η_0 (Pa s)	λ_0 (s)	$\Delta H/R$ (K)
HDPE	Quinzani and Vallés [38]	86.3	4.3	2.48×10^3	0.61	3450
LLDPE	Quinzani and Vallés [39]	94.5	3.8	5.78×10^3	4.01	3320
LDPE (melt I)	Laun [40]	120.0	4.5	1.36×10^4	15.6	6500
PE02	Vega *et al.* [12]	69.0	3.5	3.77×10^3	$\sim 10^a$	4070/4320
PE03	Vega *et al.* [12]	194.0	2.9	3.50×10^6	976	4480

^{a} A much lower value, 0.347 s, is given by Vega *et al.* [12] which does not agree with the reported experimental results.

linear viscoelastic parameters measured for PE02 and PE03 are representative of the rheological behavior observed in the metallocene-catalyzed polyethylenes studied in the literature cited here [12,14–16].

The constants η_k and λ_k that conform the discrete relaxation time spectra of the three conventional polyethylenes, as determined by Quinzani and Vallés [38,39] and Laun [40] from small amplitude oscillatory shear flow experiments, are given in Table 2. The reported spectra were determined by fixing the number of modes and the relaxation times and calculating the viscosity coefficients by non-linear minimization procedures fitting the measured dynamic moduli of each material to equations (2). The most important relaxation times, i.e. those weighted by larger viscosity coefficients, are very different for HDPE (∼0.01s) and LLDPE (∼0.1s) than for LDPE (∼40s). Moreover, in the case of LDPE the importance of each mode (given by η_k) changes very rapidly. These results reflect the different rheological behaviors of materials that, although having similar microstructures, differ in their average large-scale molecular structures. LDPE, with its long-branched structure, has a dominant relaxation time much larger than those of HDPE and LLDPE, although the molecular weights and molecular weight distributions of the three resins are very similar, as predicted from theory [29].

Figures 2 and 3 show the parameters $\eta'(\omega)$ and $G'(\omega)$ at the reference temperature of 190 °C predicted by the multimode Maxwell model [equations (2)] using the relaxation time spectra of Table 2 and temperature shift factor coefficients a_T calculated using equation (4) with the corresponding flow activation energies listed in Table 1. Both the LLDPE and HDPE show qualitatively similar behavior, although the analyzed LLDPE has larger viscous and elastic parameters in the range of frequencies covered in the study. This is to be expected for a material with a higher average molecular weight and short-chain branching. The same effect has been observed by Carrot et al. [42] for similar materials. The LDPE, on the other hand, shows a qualitatively different linear viscoelastic behavior. The long-branched

Table 2 Range of relaxation times of the three conventional polyethylenes at the reference temperature reported in the original work

HDPE ($T_0 = 180\,°C$)		LLDPE ($T_0 = 200\,°C$)		LDPE ($T_0 = 150\,°C$)	
λ_k (s)	η_k (Pa s)	λ_k (s)	η_k (Pa s)	λ_k (s)	η_k (Pa s)
40	5.6	1000	4	1000	1000
10	45	100	50	100	18000
4	216	10	700	10	18900
1	350	1	930	1	9800
0.4	384	0.1	1650	0.1	2670
0.1	400	0.01	1200	0.01	586
0.04	432	0.001	330	0.001	94.8
0.01	650	0.0001	10.3	0.0001	12.9
0.001	450				

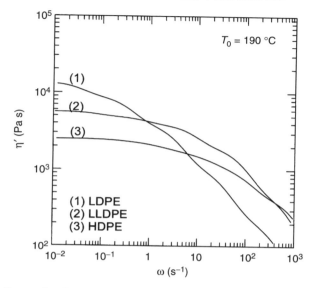

Figure 2 Dynamic viscosity versus frequency at the reference temperature of 190 °C for the three conventional polyethylenes listed in Table 1

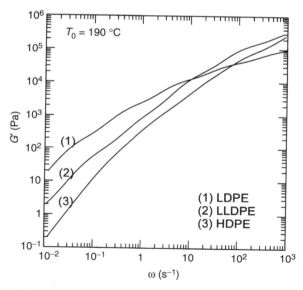

Figure 3 Elastic modulus versus frequency at the reference temperature of 190 °C for the three conventional polyethylenes listed in Table 1

molecules of this material induce high viscosity and elasticity at low frequencies ($\eta_0 = 13\,600\,\text{Pa s}$ and $\lambda_0 = 15.6\,\text{s}$) owing to the longer relaxation time originated by the additional relaxation mechanism necessary for translational motions of the whole molecules [29]. At high frequencies both η' and G' are smaller than the corresponding values measured for LLDPE and HDPE. The much smaller average radius of gyration of the molecules of LDPE is the reason for the large effect of 'shear thinning' seen in the dynamic parameters of this material.

Figure 4 shows the complex viscosity of the metallocene-catalyzed polyethylenes PE02 and PE03 at $T_0 = 190\,^\circ\text{C}$ [12]. The curves of the predicted complex viscosity of the three conventional polyethylenes at $190\,^\circ\text{C}$ have also been included to simplify the comparison. From this figure, and taking into account the molecular weights of the polymers, it is concluded that both metallocene-catalyzed resins behave as long-branched polyethylenes. A linear polyethylene with a molecular weight similar to that of PE02 should have a smaller zero-shear-rate viscosity and a Newtonian behavior up to frequencies larger than in the case of HDPE and LLDPE. However, on the contrary, PE02 shows a qualitative behavior that resembles that of LDPE. The calculated range of relaxation times of PE02 and PE03, which are given in Table 3, agree with this conclusion. Figure 5 shows the whole range of relaxation times in Tables 2 and 3 shifted to the same reference temperature of $190\,^\circ\text{C}$ and plotted as $\eta_k/\Sigma_k\eta_k$ to emphasize the importance of each mode of relaxation. This plot shows clearly the similar rheological behaviors of LDPE, PE02 and PE03. The

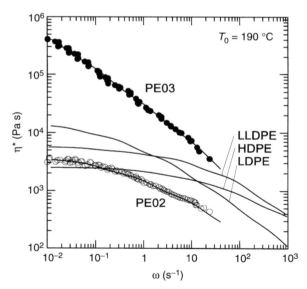

Figure 4 Complex viscosity versus frequency at the reference temperature of $190\,^\circ\text{C}$ for the five polyethylenes listed in Table 1

Table 3 Range of relaxation times of the two metallocene-catalyzed polyethylenes at the reference temperature of 190 °C

PE02		PE03	
λ_k (s)	η_k (Pa s)	λ_k (s)	η_k (Pa s)
1000	50	1000	450000
100	550	100	450000
10	1400	10	92000
1	1100	1	22000
0.1	550	0.1	5100
0.01	180	0.01	500
0.001	23	0.001	75

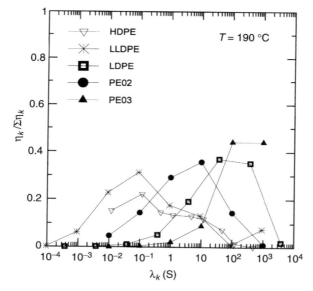

Figure 5 Range of relaxation times (λ_k, η_k) at the reference temperature of 190 °C for the five polyethylenes listed in Table 1. The values are listed in Tables 2 and 3 at the original temperatures

three materials have very few important modes of relaxation which correspond to large relaxation times. These relaxation times ($\sim$10 s for PE02, $\sim$100 s for LDPE and $\sim$400 s for PE03) increase as the molecular weight increases (69 000 g/mol for PE02, 120 000 g/mol for LDPE and 194 000 g/mol for PE03).

This is not the only result in the literature that shows evidence that metallocene-catalyzed polyethylenes are materials with a long-chain branched structure. For example, if the curves of η^* of PE02 and PE03 in Figure 4 are examined in detail, it can be seen that the superposition of the data on master curves is not good over the whole range of frequencies. This is a typical situation in the case of long-branched polymers that have a stress-dependent flow activation energy [32]. Rohlfing and Janzen [16] reached the same conclusion by analyzing the shape of the curves of dynamic viscosity of several metallocene-catalyzed polyethylenes modeled using the Carreau–Yasuda model. They are in full agreement with Carella [13] that neither 'new rules' nor anomalous molecular weight–viscosity relationships are needed to explain the rheological behavior of the new polyethylenes. Only the breadth of the molecular weight distribution and the effect of long-chain branches need to be considered. Even when usually the literature does not mention any systematic long-branching step in the polymerization mechanisms proposed for metallocene catalysts, there are some reports showing C—C terminal double bonds which may participate in the transfer reactions in solution polymerizations, giving rise to some sporadic long-chain branching [43]. In this case the result would be some long-branched molecules in a matrix of predominantly linear molecules which may be the cause of the observed behavior of the viscosity and the increase in the flow activation energy of these materials [29].

Woo et al. [17] used dynamic mechanical spectroscopy to study the β relaxation temperature for a series of metallocene-catalyzed copolymers of ethylene and α-olefins with solid densities ranging from 0.87 to 0.90. An ethylene–octene copolymer made in a solution process with a Ziegler–Natta catalyst with density of 0.912 is used for comparison. The β relaxation temperature measurements obtained for these copolymers show a gradual decrease with the increase in short-branching level, from -25 to $-44\,^\circ\text{C}$, indicating that the branch sequence distribution along the chains is random [35,36]. This is expected for catalysts with uniform site activity ('single-site' catalysts). Similar results have been obtained by Lehtinen et al. [18] and Starck [21].

3.2 POLYPROPYLENE

With the development of metallocene catalysts in the 1980s, it became possible to obtain polypropylenes with higher steric purity than the conventional materials. Polypropylenes obtained with ansa-zirconocene catalysts [41,44] activated with MAO, at polymerization temperatures between -10 and $20\,^\circ\text{C}$, have been shown to be 99+% iPP. Llinas et al. [10] and Chien et al. [9] investigated the structure, morphology and rheological properties of polypropylenes synthesized using two 'dual-site' catalysts, i.e. rac-anti-[ethylidene(1-η^5-tetramethylcyclopentadienyl)(1-η^5-indenyl)dichlorotitanium(IV)/MAO and rac-anti-[ethylidene(1-η^5-tetramethylcyclopentadienyl)(1-η^5-indenyl)dimethyltitanium(VI)/MAO, respectively. These two studies and that of Eckstein et al. [11] are the only ones in which the rheological behavior of new metallocene-catalyzed polypropylenes were analyzed.

Llinas et al. [10], performing polymerizations at different temperatures ($T_p = -20$, 0, 25 and $50\,^\circ\text{C}$), obtained polypropylenes with different microstructures. Table 4 summarizes the main characteristics of the synthesized polymers.

Table 4 Molecular and rheological parameters of the analyzed polypropylenes

Material	$\overline{M}_w \times 10^{-3}$	MWD	η_0 (Pa s)	Notes
PP(−20)	>500	—	>1.0×10^6	Amorphous
PP(0)	215	2.1	2.7×10^4	Evidence of a crystalline fraction
PP(25)	170	1.75	1.0×10^4	Thermoplastic elastomer [(cry-PP)$_{20}$(am-PP)$_{50}$]$_{30}$ $T_m = 67{-}71\,°C$
PP(50)	120	1.9	3.0×10^3	Thermoplastic elastomer [(cry-PP)$_{50}$(am-PP)$_{100}$]$_{10}$ $T_m = 67{-}71\,°C$
PP	293	4.1	2.0×10^4	Conventional $T_m \approx 165\,°C$

PP(50) and PP(25) have melting temperatures around 65 °C whereas PP(0) and PP(−20) do not exhibit a melting transition. Figures 6 and 7 show the viscous and elastic moduli of the four polymers at 50 and ~100 °C, respectively. At 50 °C, the materials PP(50) and PP(25) behave as a network displaying an elastic modulus larger than the viscous modulus. The melting transitions cause the lowering of the G' and G'' values by two and one orders of magnitude, respectively. At temperatures above the melting temperature both polymers show the typical liquid-like behavior of a melt. Both moduli decrease with decrease in frequency reaching, at low frequencies, slopes very near 2 and 1, respectively, in logarithmic plots, as they correspond to low molecular weight distribution polymers. The authors reported an equilibrium shear modulus (G_∞) of 1.16 and 0.56 MPa for PP(50) and PP(25), respectively.

PP(0) does not behave as a three-dimensional network at low temperature (it does not exhibit a low-frequency plateau of G') but, on the other hand, it is not possible to apply time–temperature superposition of the dynamic moduli at 50 and 120 °C. This indicates that this material still has a small amount of a crystalline fraction in this temperature range.

PP(−20) behaves as a typical liquid polymer over the whole range of temperatures (30–170 °C). The time–temperature superposition principle is applicable and master curves of the moduli can be built extending over nine decades of frequency.

At high temperature (approximately 100 °C), the elastic and viscous moduli of the four polymers studied present typical terminal and transition regions. The largest values of the elastic and viscous moduli are those of PP(−20) and they decrease for polypropylenes obtained at increasing polymerization temperature. This result is in accord with the measured molecular weights of the polymers, which decrease with increase in the T_p used. It is interesting that the zero-shear-rate viscosity of these polymers is proportional to $\overline{M}_w^{3.5}$, as expected for linear polymers [24].

The solid lines included in Figure 7(a) and (b) correspond to the elastic and viscous moduli of the conventional polypropylene (PP) at the reference temperature

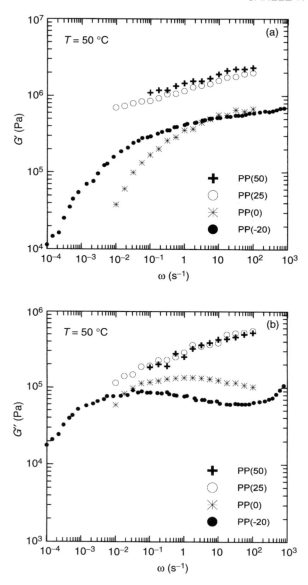

Figure 6 (a) Elastic modulus and (b) Viscous modulus versus frequency at the reference temperature of 50 °C for the four metallocene-catalyzed polypropylenes listed in Table 4

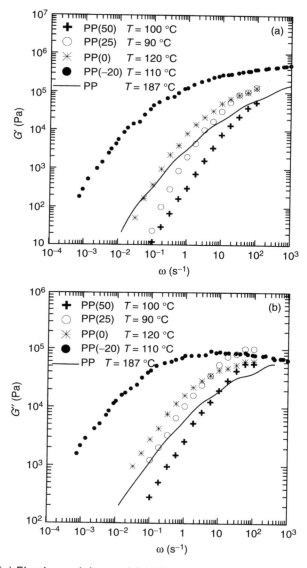

Figure 7 (a) Elastic modulus and (b) Viscous modulus versus frequency at the reference temperature of ~100 °C for the four metallocene-catalyzed polypropylenes listed in Table 4. The solid line corresponds to the conventional polypropylene at 187 °C

of 187 °C. The melting temperature of this resin is 165 °C. It can be observed that the rheological behavior is similar to that of the metallocene-catalyzed polypropylenes. The limiting slopes for G' and G'' are lower than 2 and 1, respectively, reflecting the broader molecular weight distribution.

Chien *et al.* [9] synthesized equivalent polypropylenes [PP′(25), PP′(0) and PP′(−20)] using the above titanium(VI)/MAO metallocene catalyst. These polymers have lower molecular weight than the corresponding materials produced with titanium(IV)/MAO catalyst but the rheological behavior is similar although shifted in T_p. Both PP′(0) and PP′(−20) are amorphous materials with liquid-like rheological behavior. The time–temperature superposition procedure is applicable to these polymers. They do not exhibit a phase transition in the 30–200 °C temperature range. On the other hand, PP′ (25) contains a significant fraction of crystallizable sequences and it has a melting transition at 63 °C. At low temperature (30 °C), this polymer has a rheological behavior not far from that found in PP(0). No low-frequency plateau is found in G'. This indicates that the crystal domains are weaker cross-links in this polymer than in PP(50) and PP(25) and that PP′(25) has a lower content of crystallizable stereoregular sequence than the PP(25) produced with the titanium(IV)/MAO catalyst.

Based on the observed rheological behavior for the metallocene-catalyzed polypropylenes in the melt, no evidences has been found for the presence of long-chain branching. These materials have narrow molecular weight distributions and follow the general behavior expected for linear polymers.

4 REFERENCES

1. Graessley, W. W., *Adv. Polym. Sci.*, **16**, 1 (1974).
2. Graessley, W. W., *Adv. Polym. Sci.*, **47**, 68 (1982).
3. Doi, M. and Edwards, S. F., *J. Chem. Soc., Faraday Trans. 2*, **74**, 1789 (1978); **74**, 1802 (1978); **74**, 1818 (1978); **75**, 38 (1979).
4. Pearson, D. S., *Rubber Chem. Technol.*, **60**, 439 (1987).
5. Bird, R. B., Armstrong, R. C. and Hassager, O., *Dynamics of Polymeric Liquids. Vol 1: Fluid Mechanics*, Wiley, New York 1987, 2nd edn.
6. Bird, R. B., Curtiss, C. F., Armstrong, R. C. and Hassager, O., *Dynamics of Polymeric Liquids. Vol. 2: Kinetic Theory*, Wiley, New York, 1987, 2nd edn.
7. Larson, R. G., *Constitutive Equations for Polymer Melts and Solutions*, Butterworths, New York, 1988.
8. Piau, J.-M. and Agassant, J. F., *Rheology for Polymer Melt Processing*, Elsevier, Amsterdam, 1996.
9. Chien, J. C. W., Llinas, G. H., Rausch, M. D., Lin, Y. G., Winter, H. H., Atwood, J. L. and Bott, S. G., *J. Polym. Sci., Polym. Chem. Ed.*, **30**, 2601 (1992).
10. Llinas, G. H., Dong, S. H., Mallin, D. T., Rausch, M. D., Lin, Y. G., Winter, H. H. and Chien, J. C. W., *Macromolecules*, **25**, 1242 (1992).
11. Eckstein, A., Friedrich, C., Lobbrecht, A., Spitz, R. and Mulhaupt, R., *Acta Polym.*, **48**, 41 (1997).

12. Vega, J. F., Muñoz Escalona, A., Santamaria, A., Muñoz, M. E. and Lafuente, P., *Macromolecules*, **29**, 960 (1996).
13. Carella, J. M., *Macromolecules*, **29**, 8280 (1996).
14. Muñoz Escalona, A., Lafuente, P., Vega, J. F., Muñoz, M. E. and Santamaria, A., *Polymer*, **38**, 589 (1997).
15. Sukhadia, A. M., in *ANTEC Proceedings* SPE, p. 832 (1997).
16. Rohlfing, D. C. and Janzen, J., presented at Metallocene Technology '97, Chicago, IL, 16–17 June, 1997.
17. Woo, L., Ling, M. T. K. and Westphal, S. P., *Thermochim. Acta*, **272**, 171 (1996).
18. Lehtinen, C., Starck, P. and Lofgren, B., *J. Polym. Sci., Polym. Chem. Ed.*, **35**, 307 (1997).
19. Kim, Y. SD., Chung, C. I., Lay, S. Y. and Hyun, K. S., in *ANTEC Proceedings*, SPE, p. 1122 (1995).
20. Westphal, S. P., Ling, T. K. and Woo, L., *Thermochim. Acta*, **272**, 181 (1996).
21. Starck, P., *Euro. Polym. J.*, **33**, 339 (1997).
22. Rohlfing, D. C., Hicks, M. J. and Janzen, J., presented at the 68th Annual Meeting of the Society of Rheology, Galveston, TX, 16–20, February, 1997.
23. Aaltonen, P., Seppala, J., Matilainen, L. and Leskela, M., *Macromolecules*, **27**, 3136 (1994).
24. Ferry, J. D., *Viscoelastic Properties of Polymers*, Wiley, New York, 1980, 3rd edn.
25. Raju, V. R., Menezes, E. V., Marin, G., Graessley, W. W. and Fetters, L. J., *Macromolecules*, **14**, 1668 (1981).
26. Graessley, W. W. and Edwards, S. F., *Polymer*, **22**, 1329 (1981).
27. Edwards, S. F., *Proc. Phys. Soc.*, **92**, 9 (1967).
28. de Gennes P.-G., *J. Chem. Phys.*, **55**, 572 (1971).
29. Graessley, W. W., *Macromolecules*, **15**, 1164 (1982).
30. Carella, J. M., Gotro, J. T. and Graessley, W. W., *Macromolecules*, **19**, 659 (1986).
31. Carella, J. M., Graessley, W. W. and Fetters, L. J., *Macromolecules*, **17**, 2775 (1984).
32. Graessley, W. W. and Raju, V. R., *J. Polym. Sci., Polym. Symp.*, **71**, 77 (1984).
33. Popli, R., Glotin, M. and Mandelkern, L., *J. Polym. Sci., Polym. Phys. Ed.*, **22**, 407 (1984).
34. Popli, R. and Mandelkern, L., *Polym. Bull.*, **9**, 260 (1983).
35. Krigas, T. M., Carella, J. M., Struglingski, M. J., Schilling, F. C., Crist, B. and Graessley, W. W., *J. Polym. Sci. Polym. Phys. Ed.*, **23**, 509 (1985).
36. Baldwin, F. P. and Ver Strate, G., *Rubber Chem. Technol.*, **45**, 709 (1972).
37. Mandelkern, L., *Polym. J.*, **17**, 337 (1985).
38. Quinzani, L. M. and Vallés, E. M., *Lat. Am. J. Chem. Eng. Appl. Chem.*, **17**, 121 (1987).
39. Quinzani, L. M. and Vallés, E. M., *J. Rheol.*, **29**, 725 (1985).
40. Laun, H. M., *Rheol. Acta*, **17**, 1 (1978).
41. Kaminsky, W., Hahnsen, H., Külper, K. and Wöldt, R., *US Pat.*, 4 542 199 (1985).
42. Carrot, C., Guillet, J., Revenu, P. and Arsac, A., in Piau, J.-M. and Agassant, J. F. (Eds), *Rheology for Polymer Melt Processing*, Elsevier Amsterdam, 1996, p. 159.
43. Galland, G. B., Quijada, R. and Mauler, R. S., in *Proceedings of the 5th Latin American and 3rd Ibero American Polymer Symposium*, 1996, p. 93.
44. Kaminsky, W., Külper, K., Brintzinger, H. H. and Wild, F. R. W. P., *Angew. Chem., Int. Ed. Engl.*, **24**, 507 (1985); Kaminsky, W., *Angew. Chem., Makromol. Chem.*, **145/146**, 149 (1986); Kaminsky, W., in Keii, T. and Soga, K. (Eds), *Catalytic Polymerization of Olefins*, Kodansha Elsevier, Tokyo, 1986, p. 293.

18

Rheology and Processing of Metallocene-based Polyolefins

CHR. FRIEDRICH, A. ECKSTEIN, F. STRICKER
AND R. MÜLHAUPT
Institut für Makromolekulare Chemie der Universität Freiburg, Freiburg i.
Br., Germany

1 INTRODUCTION

The discovery of single-site metallocene catalysts has revolutionized polyolefin technology. These catalysts give excellent control of stereochemistry and molecular weight without sacrificing the narrow molecular weight distribution (MWD) [1]. Today, well defined atactic polypropylene (PP) and highly isotactic and highly syndiotactic PP are available covering the entire feasible stereoregularity and molecular weight range. Moreover, comonomers are incorporated randomly to afford controlled short- and long-chain branching. In contrast to many multi-site Ziegler–Natta catalysts, metallocene catalysts produce very uniform copolymers without wax-like byproduct formation. The polymer properties can be varied as a function of comonomer incorporation. For example, 1-olefins such as 1-butene and 1-octene copolymerized with ethylene give linear low-density polyethylene (LLDPE) or very low-density polyethylene (VLDPE). Novel families of stiff cycloaliphatic engineering resins are produced by copolymerizing ethene with cycloolefins such as norbonnene. Key to commercial application of metallocene-based polymers is their processability, which is reflected by the viscoelastic properties of polyolefin melts.

Metallocene-based Polyolefins Edited by J. Scheirs and W. Kaminsky
© 2000 John Wiley & Sons Ltd

In this chapter we analyze the linear viscoelastic properties of metallocene-based polyolefins in order to establish or revise, if necessary, structure–property relationships. Attention is also given to processing relevant properties.

2 RHEOLOGY OF METALLOCENE-BASED POLYPROPYLENES

Rheology is a powerful tool for the characterization of polymers owing to the pronounced correlation between viscoelastic behavior and molecular architecture of polymers. The well known 3.4 scaling relationship between zero shear viscosity η_0 and molecular weight M_w, which holds for almost all polymers, is such an example. Usually, the viscoelastic properties can be determined by mechanical spectroscopy, measuring the storage modulus $G'(\omega)$ and the loss modulus $G''(\omega)$ at small deformation amplitudes over a wide range of frequencies ω and temperatures T. In principle, it is possible to deduce from these moduli material parameters such as the zero shear viscosity, η_0, the terminal relaxation time, λ_0, and the plateau modulus G_N^0. The thermorheological properties characterized by temperature shift factors $a_T(T)$ can also be evaluated. These approaches give an insight into such properties as the fractional free volume and volume expansion coefficient as a function of temperature.

In the following sub-sections we derive such parameters from the material functions and correlate them with molecular weight, stereoregularity and copolymer composition of metallocene-based homo- and copolymers.

2.1 INFLUENCE OF STEREOREGULARITY AND MOLECULAR WEIGHT ON THE VISCOUS PROPERTIES OF PP MELTS

For a long time it was assumed that stereoregularity, which defines the conformational properties of a polymer chain on a small scale, has no influence on its terminal, large-scale properties. It was found for poly(methyl methacrylate) (PMMA) [2] that stereoregularity influences all properties, viscous and elastic, determined from the terminal relaxation zone. Using metallocene-based polypropylenes of controlled stereoregularity, we found that the same is true for polyolefins [3]. Figure 1 shows the dynamic viscosities η' of a series of highly isotactic [Figure 1(a)] and a series of highly syndiotactic [Figure 1(b)] polypropylenes as a function of the reduced frequency ωa_T.

All curves correspond to a common reference temperature $T_0 = 170\,^{\circ}\mathrm{C}$. The horizontal plateaus determine the zero shear viscosities of the polypropylenes of corresponding molecular weight. The presentation of these viscosities versus the molecular weight M_w in Figure 2 clearly establishes the strong influence of stereoregularity on viscosity, which is given by the following equation:

$$\eta_0 = K(T_0, St)M_w^{3.4} \tag{1}$$

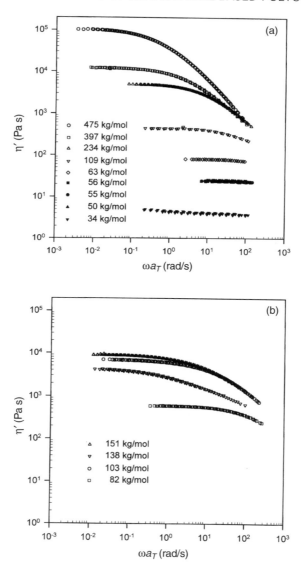

Figure 1 Dynamic viscosity η' as a function of reduced frequency, ωa_T, for (a) highly isotactic and (b) highly syndiotactic polypropylenes. The numbers indicate the molecular weight, M_w

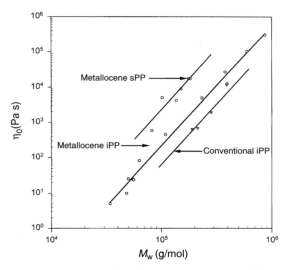

Figure 2 Zero shear viscosity η_0 of polypropylenes with different stereoregularities versus molecular weight M_w on a logarithmic scale

for $M_w > M_c$, where M_c is the critical molecular weight which indicates the onset of the 3.4 scaling behavior and St designates the stereoregularity characterized by percentage of corresponding triads.

The three series of different polypropylenes (including conventional PP) follow the well known scaling behavior with an exponent α of 3.4. For conventional isotactic polypropylenes similar exponents have been found [4,5]. Comparing metallocene polypropylenes with different stereoregularities at the same temperature (here 170 °C), the difference in absolute viscosities at the same molecular weight is the most striking feature. Metallocene sPP samples have zero shear viscosities approximately 10 times higher than that of metallocene iPP with the same molecular weight M_w. Conventional iPP exhibits a lower viscosity than metallocene iPP. This can be explained by the presence of low molecular weight fractions which are present in conventional PP as a consequence of the broad MWD.

Stereoregularity-dependent differences in behavior can be found for these polymers when analyzing the temperature shift factors a_T which were used to construct the master curves in Figure 1. We found Arrhenius behavior in all cases, with differences in activation energies of flow E_A for syndiotactic and isotactic materials. Figure 3 displays this behavior.

Metallocene-based sPP possesses about a 15 kJ/mol larger activation energy of flow than metallocene-based iPP, as is shown in Figure 3. The activation energy of flow of conventional iPP is about 5 kJ/mol smaller than that of the metallocene-based iPP. This can be explained by the presence of low molecular weight fractions in the conventional samples. These components are responsible for a plasticizer effect which gives the polymer more mobility at the same temperature. For

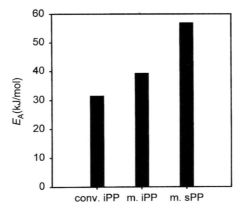

Figure 3   Activation energies of conventional and metallocene-based PP

molecular weights above M_c the activation energy of flow is independent of molecular weight.

For atactic PP, WLF behavior was found. This is not a contradiction of our previous findings. It reflects the fact that due to the onset of crystallization and the general low shift factors for stereoregular polyolefins, the few data points in the case of iPP and sPP can be identified only by the Arrhenius law, whereas for aPP a WLF relationship is valid. As can be seen from Figure 4, the few data points for iPP and

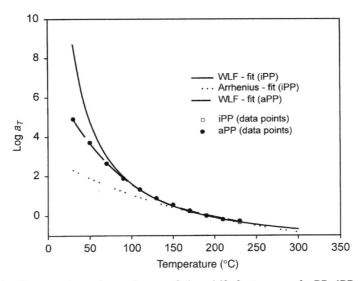

Figure 4   Temperature dependence of the shift factor a_T of aPP, iPP and sPP and comparison with calculations on the basis of Arrhenius and WLF relationships

sPP do not allow one to differentiate between the two relationships. In such a case the model with the smaller number of parameters is chosen: the Arrhenius model.

2.2 INFLUENCE OF STEREOREGULARITY ON ENTANGLEMENT BEHAVIOR OF PP MELTS

Another important rheological parameter is the plateau modulus, G_N^0, which defines the entanglement molecular weight M_e, the molecular weight between adjacent temporary entanglement points [2]. The following relationship between the two parameters holds:

$$G_N^0 = \frac{\rho RT}{M_e} \qquad (2)$$

where R is the universal gas constant (8.314 J/mol K) and ρ the density of the polymer at the temperature T at which the plateau modulus was measured.

There are two methods for the determination of the plateau modulus of which one, the tan δ minimum method, cannot be used for polyolefins. Owing to the crystallization of these polymers at temperatures which correspond to the frequency range where the minimum in tan $\delta (= G''/G')$ is found, this criterion is not applicable. The second method, the G'' integration method, is based on the following equation:

$$G_N^0 = \frac{2}{\pi} \int_{-\infty}^{+\infty} G_{FT}''(\omega) \mathrm{d}\ln \omega \qquad (3)$$

From the technical point of view, this method demands polymers of high enough molecular weight which yield G'' curves as presented in Figure 5. The existence of a distinct maximum in these curves which allows safe extrapolation is a necessary prerequisite to integrate the G'' data over the frequency range which is called flow transition (FT) (see, e.g., Ref. 6). Metallocene technology gives the opportunity for the synthesis of such polymers and the results are presented in Figure 5. The physical data for the polypropylenes used are given in Table 1. The application of the presented methods to all three polymers results in plateau moduli and corresponding entanglement molecular weights which are given in Table 2. From the M_w/M_e ratio it can be seen that these polymers are, indeed, highly entangled.

The plateau properties and all other rheological parameters characterizing the flow transition of all PP samples were markedly influenced by their stereoregularity. A reason may be the existence of different stereoregularity-dependent conformations of PP in the melt state. Whereas the solid-state conformations are well known for isotactic PP (3_1 helix) [7], syndiotactic PP (2_1 helix corresponding to a planar zig-zag conformation) [8] and atactic amorphous PP (random coil), the liquid-state conformations are not as clear, at least for sPP. The similarity of plateau moduli, entanglement molecular weights and activation energies of flow for iPP and aPP indicates that iPP is likely to form a random coil conformation in the molten state also.

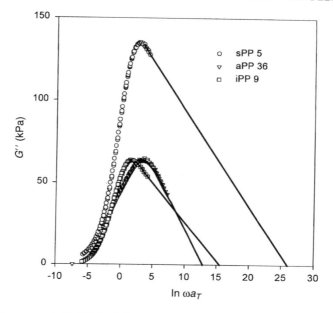

Figure 5 Loss moduli $G''(\omega)$ of the three PPs versus reduced angular frequency ωa_T in a linear–natural logarithm plot at the reference temperature $T_0 = 463$ K

Table 1 Physical data for high molecular weight PPs

Polymer	M_n (kg/mol)	M_w (kg/mol)	M_w/M_n	Density, ρ (g/cm^3)^a
iPP 9	365	871	2.4	0.766
sPP 5	231	483	2.1	0.762
aPP 36	1890	3670	1.9	0.765

^{a} Measured at 190 °C.

Table 2 Plateau moduli, G_N^0, entanglement molecular weights, M_e, and number of entanglements per chain, M_w/M_e, of metallocene-based polypropylenes

Polymer	$G_N{}^0$ (kPa)	M_e (g/mol)	M_w/M_e
iPP 9	427	6900	126
sPP 5	1350	2170	222
aPP 36	418	7050	520

The significant differences in the rheological properties of sPP with respect to those of isotactic and atactic PP may be due to the presence of predominantly all-*trans* conformations in the melt. This was reported by Loos *et al.* [9] and confirmed by molecular modeling investigations [10]. The observation of similar material characteristics of isotactic and atactic PP in the molten state in contrast to sPP is in accordance with similar observations in pressure–volume–temperature (*pvT*) experiments [11,12].

3 RHEOLOGY OF METALLOCENE-BASED ETHENE–1-OLEFIN COPOLYMERS

We shall now analyze the influence of copolymer composition on the rheological properties of new metallocene-catalyzed polymers. First we present results for polymers composed of ethene and 1-butene, then we analyze the rheological properties of ethene copolymers containing the more polar styrene units.

3.1 INFLUENCE OF COMPOSITION ON THE RHEOLOGICAL PROPERTIES OF ETHENE–1-BUTENE (EB) COPOLYMERS

The molecular data for polymers to be analyzed here are given in Table 3. In addition to our EB polymers we include two commercial polymers from BASF, Luflexen 94 and 95 (LU94, LU95). EB00 [neat poly(1-butene)] is an isotactic, crystalline polymer which displays all the attributes of a linear, relatively narrow distributed polymer. For EB00 and all copolymers we constructed a master curve at a reference temperature of $T_0 = 130\,°C$. For polymers with an ethylene content x_E of up to 80 mol% we found WLF behavior and beyond that limit both relationships, the WLF and Arrhenius equations, can be used. This behavior is similar to that which was observed for PPs (see Figure 4). For EB100 it was impossible to describe the shift

Table 3 Chemical composition, molecular weight and polydispersity of the polymers under investigation

Polymer	x_E (mol%)	M_w (kg/mol)	M_w/M_n
EB00	0	131	2.8
EB22	22.2	197	2.4
EB34	34.5	152	2.4
EB64	64.7	199	2.4
EB80	80.9	248	2.3
EB88	88.8	305	2.5
LU94	94.6	118	2.3
LU95	96.5	—	—
EB100	100	190	2.5

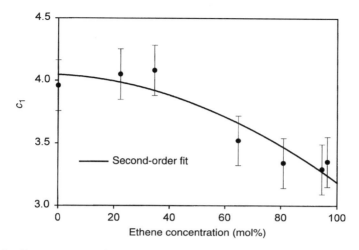

Figure 6 Parameter c_1 of the WLF equation versus the composition of different copolymers. The line represents a second-order fit to experimental data

factors by the WLF equation and, consequently, the c_1 value for this polymer is not available. To analyze the thermorheological properties of these polymers as a function of composition the c_1 parameters are presented in Figure 6.

We observe a decrease in c_1 values from around 4 for neat poly(1-butene) to 3.4 for polyethylene-like copolymers. This decrease corresponds to an increase in fractional free volume of about 25 %. Polyethylene is denser than poly(1-butene). Calculation of the WLF activation energy of flow at the reference temperature yields values of about 35 kJ/mol for polyethylene and about 60 kJ/mol for neat poly(1-butene). This tendency is in agreement with other measurements indicating an increase in activation energy with increasing length of short side-branches.

The master curves for some of the polymers in Table 3 are presented in Figure 7. Regardless of the fact that the polymers in Figure 7 do not have the same molecular weight, two observations can be made. First, a Newtonian plateau is only observed for EB00 and with increasing ethene content the possible onset of such a plateau, which of course, exists (see also Figures 12 and 13), shifts to lower frequencies. Except for EB00, none of the polymers reaches the terminal relaxation region which is characterized by $G'(\omega) \propto \omega^2$ and $G''(\omega) \propto \omega$ (which is equivalent to the plateau for η'). Because the polydispersity of all polymers is almost the same, this shift and the form of transition to the region where all curves merge into one is an indication of long-chain branching rather than molecular non-uniformity. Second, owing to the observed behavior we are not able to check the viscosity–molecular weight scaling relationship. The construction of a viscosity composition relationship is impossible because we are not able to separate the compositional and configurational aspects. However, from consideration of Figure 7 we can speculate that such a relationship

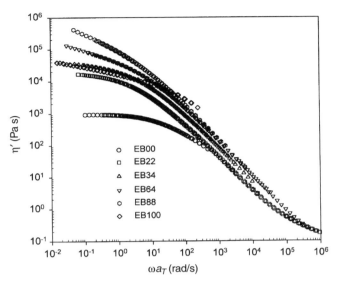

Figure 7 Dynamic viscosity η' of ethene–1-butene (EB) copolymers versus reduced frequency ωa_T

has a maximum on the ethene-rich side of compositions. Whether this is an effect due to composition or to molecular configuration (long-chain branching) is not clear at present. In section 3.3 we will analyze this problem in more detail. However, the analysis of the slopes of moduli at the smallest measured frequencies supports the idea of branching. The ethene-rich copolymers EB64 to EB88 show the smallest slopes, indicating that these polymers are far from the terminal region, although the molecular weights are comparable. Such a situation may occur if the relaxation is hindered owing to the presence of few but long branches.

The only one parameter which does not depend on molecular weight is the plateau modulus. We were able to determine this parameter for some of the copolymers using the tan δ method. The results are presented in Figure 8. These experimentally determined values (filled circles) match the plateau moduli of the neat components. The value for PE was taken from the literature [6] and the value for neat poly(1-butene) was determined from its zero shear viscosity, η_0 (980 Pa s) and the cross-over relaxation time, λ_x, which is the inverse of the frequency where G' and G'' are equal. The relationship

$$G_N^0 \approx G_{px} = \eta_0/\lambda_x \tag{4}$$

where G_{px} is the value of G' at the cross-over frequency, was found to be valid for the copolymers to be discussed in the next section [13].

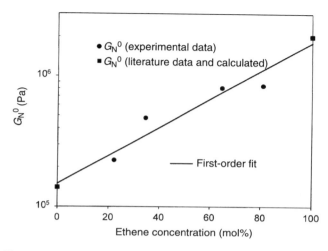

Figure 8   Plateau moduli of polyethylene (from the literature), poly(1-butene) and some copolymers versus the concentration (mol%) of incorporated ethene

3.2   INFLUENCE OF COMPOSITION ON THE RHEOLOGICAL PROPERTIES OF ETHENE–STYRENE (ES) COPOLYMERS

Little was known concerning the behavior of random ethene–styrene copolymers. This situation is a consequence of the difficulty residing in the synthesis of poly(ethene–co-styrene) by conventional Ziegler–Natta catalysis. Indeed, for a vinyl aromatic comonomer such as styrene, only a maximum of about 1 mol% styrene units in the copolymer was reached in the past. The polymers reported often comprised a mixture of homo- and copolymers [14,15].

Recently, the synthesis of mono-Cp-amido complexes [16,17] provided the basis for the new generation of metallocene catalysts which produce poly(ethylene–co-styrene) containing more than 30 mol% styrene. To date only a few papers [18–20] concerning the characterization of poly(ethylene–co-styrene) are available. We have studied the rheological properties of these new copolymers.

For the temperature dependence of the horizontal shift factors a_T we expect an interesting transition from Arrhenius behavior for ethene-rich copolymers to WLF behavior for styrene-rich polymers. Indeed, as shown in Figure 9, the shift factors of the copolymers containing up to 16.5 mol% styrene are very similar to these of polyethylene and fall within the shaded area. Moreover, the low absolute values for a_T obtained for the copolymers indicate that, in contrast to many homopolymer melts such as polystyrene and poly(methyl methacrylate), their rheology is hardly temperature dependent as for PP. Figure 9 also shows that owing to the linear relationship of a_T vs $1/T$ the rheological behavior of such polymers can be described by the empirical Arrhenius equation. A systematic deviation from this behavior starts

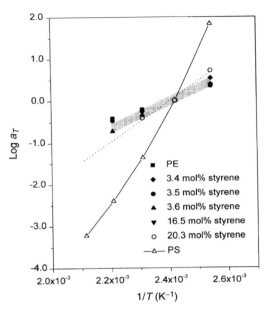

Figure 9 Shift factors a_T versus the reciprocal temperature $1/T$ for copolymers and neat polyethylene and polystyrene

with the copolymer containing 20.3 mol% styrene. The corresponding data points are connected by a dotted line to guide the eye. This onset of changing the shift properties from Arrhenius-like behavior for the polyethene and polyethene-like copolymers to the WLF behavior of pure polystyrene corresponds to the disappearance of crystallinity. Copolymers with a higher styrene content are not available at present as an increasing styrene concentration rapidly reduces the catalyst activity.

Figure 10 depicts the dynamic viscosity of copolymers with a molecular weight of about 190 kg/mol. What we observe is the rheological response of polymers with a polydispersity far from that of polymers prepared by living polymerization: a broad transitional region. The flow zone is not finally reached because η' shows, as for most other polymers discussed so far, a deviation from proportionality to ω^0 (which defines the zero shear viscosity plateau). Nevertheless, it is already possible to determine the zero shear viscosity η_0 either by extrapolation or by taking η' values corresponding to the lowest measured frequency.

We checked the viscosity–molecular weight relationship for three copolymers of almost identical comonomer composition ($x_s = 3.5$ mol%) and the results are presented in Figure 11. In this case the exponent is $\alpha = 3.34$, showing that the copolymers follow the empirical scaling relationship usually exhibited by homopolymers, and confirming the fact that these copolymers have a linear structure without any long-chain branching known for 'substantially linear olefin polymers'. Because both the plateau and the terminal zones could not be reached, the

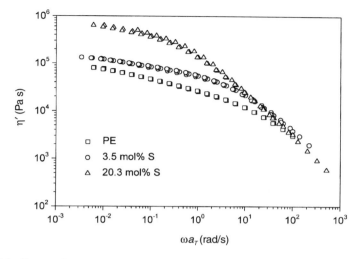

Figure 10 Dynamic viscosity η' of polyethene and two ethene–styrene (ES) copolymers versus reduced frequency ωa_T

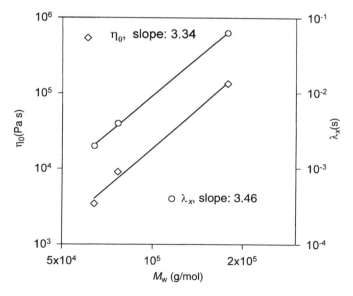

Figure 11   Zero shear viscosity η_0 and cross-over relaxation time λ_x for ethene–styrene (ES) copolymers with a styrene concentration $x_s = 3.5\,\text{mol}\%$ as a function of the molecular weight

determination of the terminal relaxation time λ_0 and the plateau modulus G_N^0 was not possible. However, the cross-point defined as the point were G' and G'' are equal can be evaluated. It was shown earlier that the coordinates of this cross-point are related to molecular parameters [13]. The time $\lambda_x = 1/\omega_x$, where ω_x is the abscissa of the cross-point, was estimated for each sample (see Figure 11) and can be considered as an approximation for the terminal relaxation time, as Dumoulin and Utracki [21] had already done for polyethylene blends. Furthermore, we will use this characteristic time to determine an approximation of the plateau modulus, G_{px}, using equation (4). Although λ_x is not the terminal relaxation time, interesting results have been deduced from this approximation and the values found for the polyethylene and polystyrene references ($G_{px}^{PE} = 32.8 \times 10^5$ Pa and $G_{px}^{PS} = 1.9 \times 10^5$ Pa) were also consistent with the plateau moduli in the literature [6,22]. The G_{px} values found are given in the literature [13].

The dependence of the materials parameters on the copolymer composition should also be mentioned. We found that for ES copolymers the zero shear viscosity and the plateau modulus follow logarithmic linear mixing rules. This relationship is expressed by

$$\log p_{CP} = x_S \log p_S + x_E \log p_E \qquad (5)$$

where p_{CP} is either the viscosity or the plateau modulus of the copolymers and the other subscripts indicate neat polystyrene (S) and neat polyethylene (E). Using such equations together with relationship $\eta_0 \propto M_w$, the properties of copolymers of arbitrary molecular weight and copolymer content can be predicted.

3.3 INFLUENCE OF LONG-CHAIN BRANCHING ON RHEOLOGICAL PROPERTIES

It is known that long-chain branching has a strong influence on the rheological properties of polymers. For model systems such as anionically synthesized poly-styrene or polyisoprene star- [22] or H-shaped polymers [23], it was found that the zero shear viscosity scales with molecular weight in an exponential manner:

$$\eta_0 \propto K(M_a, M_b) \exp(M_a/M_e) \qquad (6)$$

where M_a is the molecular weight of an arm (e.g. it is the fth part of the molecular weight of a star where f is the number of arms per molecule), M_b is the molecular weight of the strut of H and M_e is the entanglement molecular weight. The function K is of power law character. This equation explains why for molecular weights two to four times higher than the critical molecular weight the viscosity increases tremendously in comparison with a linear molecule of the same molecular weight. A schematic representation of this behavior is given in Figure 12, where the viscosity molecular weight characteristics of linear polymers according to equation (1) are also given. Because the plateau modulus is nearly the same for linear and branched species, this relationship also indicates that the relaxation times of branched systems

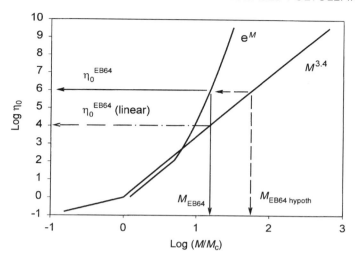

Figure 12 Viscosity–molecular weight relationships for linear and branched polymers

also increase exponentially with increasing molecular weight. Unfortunately, our polymers do not allow us to check this equation and to explain the appearance of long-chain branching by showing a change from power law behavior to exponential behavior. While it is easy to do this for molecules of defined architecture, the detection of similar relationships for polyolefins is more complicated. There are two reasons. First, it is difficult to synthesize model polyolefins of defined branched structure. Second, it is even more difficult to characterize the degree of branching by any physical method. Rheology itself is the only technique that reflects changes in degree of branching very well. However, rheology itself cannot detect the rheological properties and the structure of the polymers at the same time. The lack of an independent method for the characterization of the degree of long-chain branching, especially for polyolefins, makes it difficult to establish the desired relationships. The introduction of empirical relationships such as the Dow rheology index [25] (*DRI*) is the consequence. The definition of this index is based on the relationship combining the zero shear viscosity η_0, the terminal relaxation time λ_0 and the plateau modulus G_N^0 similarly to equation (4) and accounts for the deviation from that relationship due to long-chain branching in the following way:

$$DRI \propto G_N^0 \frac{\lambda_0}{\eta_0} - 1 \qquad (7)$$

Unfortunately, this equation does not allow us to discriminate between effects originating from polydispersity and long-chain branching. However, for similar molecular uniformity *DRI* increases owing to long-chain branching and helps to account for that effect, at least in principle. Knowing that all configurational peculiarities of a polymer are reflected at best in the low frequency or long time limits of material functions, *DRI* should reflect these peculiarities. However, the necessary parameters, the zero shear viscosity η_0 and the terminal relaxation time λ_0 are determined by fitting capillary data to a proposed fit equation. While this fit equation contains these parameters, the data are determined in the high shear rate range where the flow properties depend only marginally on structural aspects. This discrepancy my explain why some authors [26] find an unexpected dependence of the activation energy of flow on the degree of branching whereas others [27] do not.

Nevertheless, we shall try to discuss some qualitative features of the properties of our metallocene-based polymers to demonstrate alternative approaches to the characterization of long-chain branching. In our discussion we make use of the fact that all our polymers exhibit nearly the same polydispersity of $M_w/M_n \approx 2$. This allows us to separate some rheological properties which may arise from the interaction of polymers of very different lengths from those which are caused by long-chain branching. As already mentioned, we believe that some peculiarities (extremely high viscosity level at a comparable molecular weight for some polymers) in rheological response of EB copolymers can be understood qualitatively only when assuming long-chain branching for EB64 to EB88. To explain this, we use another feature of randomly branched polymers [27]: the appearance of an intermediate relaxation regime between the plateau zone and the terminal region. We found this region for the mentioned copolymers by combining two experimental techniques: oscillatory measurements and creep measurements. Figure 13 shows for

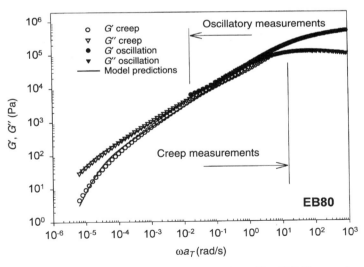

Figure 13 Storage modulus G' and loss modulus G'' of EB80 as a function of the reduced frequency ωa_T

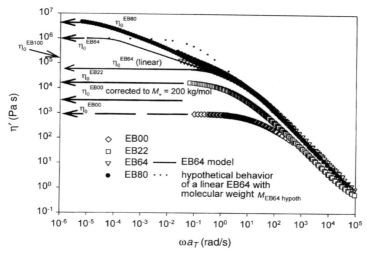

Figure 14 Dynamic viscosity η' of some ethene–1-butene (EB) copolymers versus reduced frequency ωa_T

the example of EB80 the intermediate power law region. Similar observations describing the appearance of such an intermediate region were made for branched EPDM [28]. In the literature models are available describing this behavior empirically. It was found [29] that rheological constitutive equations with fractional derivatives are suitable to describe branched polymers quantitatively correctly. Such an example is given in Figure 14 where the model calculations are given by the solid lines. The agreement between the data and the model confirms our hypothesis that branched polymers relax in a hierarchical manner as described by Friedrich et al. [29].

The function G'' can be transformed to dynamic viscosity ($\eta' = G''/\omega$) and we shall compare this material function for the polymers EB00, EB22, EB64 and EB80. The results are depicted in Figure 14, which presents the dynamic viscosity (symbols) of these polymers. To exclude the influence of molecular weight, we correct the zero shear viscosity of some polymers to a molecular weight of about 200 kg/mol. First, we start with the zero shear viscosity of EB00, η_0^{EB00}, by applying equation (1). The second arrow from the bottom shows the result: η_{0corr}^{EB00}. The viscosity of EB100 (this is a material from Ref. 13) was also corrected to the reference molecular weight and its value is depicted by the small arrow pointing from the left to the η'-axis. The zero shear viscosity of EB22 can be read directly from the data (the third arrow from the bottom). Next we discuss EB64, which is a branched polymer similar to EB80. Because we did not have enough material at our disposal, we could not extend the viscosity data to lower frequencies as shown for EB80. Calculations with a fractional derivative model [29] on the basis of our oscillatory data allow us at least to predict the course of the curve, and therefore the zero shear viscosity. The fourth arrow from the bottom represents the zero shear viscosity of linear EB64 of the same molecular weight (see also Figure 12). The dotted curve illustrates the behavior of a linear polymer with a molecular weight which would lead to the same zero shear viscosity of the branched EB64. The corresponding relationships between viscosity and molecular weight are also given in Figure 12. The difference in viscosity between the dotted and the full lines explains the improvement in processability of a branched product in comparison with linear polymers of same molecular weight. Moreover, the reduced results allow us to confirm the validity of equation (5) for linear EB polymers. The investigation of the intermediate relaxation region seems to be a promising way to establish a quantitative method which allows one to account for different degrees of long-chain branching.

At present the set of experimental data is too narrow to decide which method is correct and which conclusions are wrong. In presenting our qualitative results, we would like to support the idea that quantitatively correct results can be found if material functions are measured for well defined systems in the low-frequency range. Metallocene catalysis offer model systems to clarify some of these aspects.

4 OUTLOOK

Processing of metallocene-based polymers is influenced by the molecular architectures including molecular weight, molecular weight distribution, stereochemistry and short- and long-chain branching. In the future, the development of metallocene-based polymers will be aimed at polyolefins with bi- or multimodal molecular weight distributions and long-chain branching to enhance processing. Moreover, hybrid catalysts, containing two or more types of catalytically active centers, will be tailored to produce reactor blends of linear and branched homo- and copolymers, including block copolymers. The unprecedented control of polymer architectures achieved with 'single-site' metallocene catalysts will facilitate the investigation of basic correlations between molecular architecture, rheology and processing.

5 REFERENCES

1. Brintzinger, H. H., Fischer. D., Mülhaupt. R., Rieger, B. and Waymouth, R. M., *Angew. Chem., Int. Ed. Engl.*, **34**, 1143 (1995).
2. Fuchs, K., Friedrich, Chr. and Weese, J., *Macromolecules*, **29**, 5893 (1996).
3. Eckstein, A., Friedrich, Chr., Lobbrecht, A., Spitz, R. and Mülhaupt. R., *Acta Polym.*, **48**, 41 (1997).
4. Minoshima, W., White, I. J., and Spruiell, K., *Polym. Eng. Sci.*, **20**, 1166 (1980).
5. Hingmann, R. and Marczinke, B. I., *J. Rheol.*, **38**, 573 (1994).

6. Donth, E. J., *Relaxation and Thermodynamics in Polymers: Glass Transition*, Akademie Verlag, Berlin, 1992.
7. Wunderlich, B., *Macromolecular Physics*, Academic Press, New York, 1980.
8. Tonelli, A. E., *Macromolecules*, **24**, 3069 (1991).
9. Loos, J., Buhk, M., Petermann, J., Zoumis, K. and Kaminsky, W., *Polymer.*, **37**, 387 (1996).
10. Ito, M. and Kobayashi, N., *Trans. Mater. Res. Soc. Jpn.*, **16A**, 559 (1994).
11. Maier, R. D., Thomann, R., Kressler, J., Mülhaupt. R. and Rudolf, B., *J. Polym. Sci., Part B: Polym. Phys.*, **35**, 1135 (1997).
12. Walsh, D. J., Graessley. W. W., Datta. S., Lohse, D. J. and Fetters. L. J., *Macromolecules*, **25**, 5236 (1992).
13. Lobbrecht, A., Friedrich. Chr., Sernetz. F. G. and Mülhaupt. R., *J. Appl. Polym. Sci.*, **65**, 209 (1997).
14. Soga, K., Lee, D. and Yanagihara. H., *Polym. Bull.*, **20**, 237 (1988).
15. Lu, Z., Líao, K. and Liu. S., *J. Appl. Polym. Sci.*, **53**, 1453 (1994).
16. Shapiro, P. J., Cotter, W. D., Schaefer, W. P., Labinger, J. A. and Bercaw, J. F., *J. Am. Chem. Soc.*, **116**, 4623 (1994).
17. Okuda, J., *Chem. Ber.*, **123**, 1649 (1990).
18. Ren, J. and Hatfield, G. R., *Macromolecules*, **28**, 2588 (1995).
19. Sernetz, F. G., Mülhaupt, R. and Waymouth, R. M., *Macromol. Chem. Phys.*, **197**, 1071 (1996).
20. Cheung, Y. W. and Guest, M. J., *ANTEC SPE Tech. Pap.*, **96**, 1634 (1996).
21. Dumoulin, M. M. and Utracki, L. A., in Utracki. L. A. (Ed.), *Two-Phase Polymer Systems*, Hanser, Munich, 1991.
22. Ferry, J. D., *Viscoelastic Properties of Polymers*, Wiley, New York, 1980.
23. Roovers, J., *Polymer*, **26**, 1091 (1985).
24. McLeish, T. C. B., *Macromolecules*, **21**, 1062 (1988).
25. Lai, S., Plumley, T. A., Butler, T. I., Knight, G.W. and Kao, C. I., *ANTEC SPE Tech. Pap.*, **94**, 1814 (1994).
26. Kim, Y. S., Chung, C. I., Lai, S. Y. and Hyan. K. S., *Korean. J. Chem. Eng.*, **13**, 294 1996.
27. Kasehagen, L. J. and Macosko, C. W., *J. Rheol.*, **40**, 689 (1996).
28. VanGourp, M. and Palmen, J., *Rheol. Bull.*, **67**, 5 (1998).
29. Friedrich, Chr., Schiessel, H. and Blumen, A., in Siginer, D. A., Chhabra, R. P. and DeKee, D. (Eds), *Advances in the Flow and Rheology of Non-Newtonian Fluids*, Elsevier, Amsterdam, 1999, pp. 429–466.

19

Melt-rheological Characteristics of Metallocene-catalyzed Polyethylenes

DAVID C. ROHLFING AND JAY JANZEN

Phillips Petroleum Company Research Center, Bartlesville, OK, USA

1 INTRODUCTION

With the introduction of metallocene ('single-site')-catalyzed polyethylene (PE) resins into the marketplace, the industry has been confronted with the need to integrate a new and different class of resins into its understanding of melt behavior. Since many producers and processors of PE deal with limited subsets of the available varieties (LLDPE, MDPE and HDPE based on titanium halide- and chromium oxide-type catalysts, high-pressure LDPE, and now metallocene-based varieties), the temptation is to treat each of these types separately, as a special case, and to accumulate local correlations and folklore about each one in order to help develop, produce and process these resins successfully.

We prefer to presume that the various resin types are specific instances of a more general model. We shall show that the rheological characteristics of the metallocene-catalyzed polyethylenes can be attributed to, and expected from, the relatively narrow molecular weight distributions that these resins usually exhibit, sometimes with long-chain branching (LCB) effects included [1]. In our view, much of the treatment of the different resin types as independent special cases [2–3] arises from the widespread practice of regarding resins with the same melt index (MI) and

Metallocene-based Polyolefins Edited by J. Scheirs and W. Kaminsky
© 2000 John Wiley & Sons Ltd

density as equivalents. While this may have served the industry in the past and may be useful when dealing with a single subfamily of polyethylene, it only clouds understanding when a broader range of types is considered.

2   A COHERENT MODEL

Since 1992, the authors' laboratories have made extensive use of the Carreau–Yasuda (CY) model [4,5], which the monograph by Bird *et al.* [6] classifies as a 'useful empiricism', and which Hieber and Chiang [7,8] have demonstrated to be capable of closely fitting viscosity data for several kinds of polymer melts. After examining several alternatives (power-law, Eyring, Ellis and Sabia [9] models), which proved for various reasons to be far less satisfactory, we adopted the CY equation for the routine representation of experimental data on polymer melts, having confirmed that it is capable of closely fitting data on a considerable variety of materials of interest to us, including all the various types of polyethylenes mentioned above.

The simplified form of the Carreau–Yasuda non-Newtonian viscosity model that we use for fitting isothermal complex viscosity data is

$$|\eta^*(\omega)| = \eta_0[1 + (\tau_\eta\omega)^a]^{(n-1)/a} \tag{1}$$

where $|\eta^*(\omega)|$ is the scalar magnitude of the complex viscosity [10] $\eta^*(\omega)$, η_0 is its low-frequency limiting value, called the zero-shear viscosity, ω is the angular frequency of an oscillating shearing deformation, τ_η is a characteristic viscous relaxation time, a is a parameter inversely related to the breadth of the transition from Newtonian to power-law behavior, and n fixes the final slope (the limiting power-law behavior of the viscosity at large ω). If the Cox–Merz [11] rule holds, the analogous form

$$\eta(\dot{\gamma}) = \eta_0[1 + (\tau_\eta\dot{\gamma})^a]^{(n-1)/a} \tag{1a}$$

where $|\eta^*(\omega)|$ is replaced by $\eta(\dot{\gamma})$ and ω is replaced by the shear rate $\dot{\gamma}$, can also be applied to constant shear and shear flow situations, short of shear rates where wall slip occurs.

Figures 1 and 2 illustrate the general form of the CY function and the roles of the various parameters in defining its shape in detail. η_0 is the vertical [Figure 1(a)] scaling parameter; it defines the vertical (viscosity) placement of the curve. Similarly, τ_η is the horizontal [Figure 1(b)] scaling parameter, defining the horizontal (frequency or shear rate) location of the 'knee' or transition from the Newtonian plateau to power-law behavior at large ω. The parameter a defines the breadth of the transition from Newtonian to power law behavior, as illustrated in Figure 2; a is

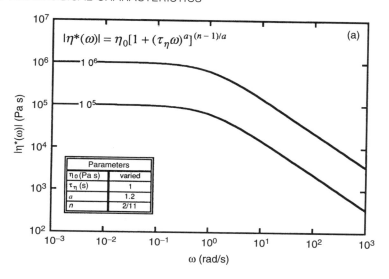

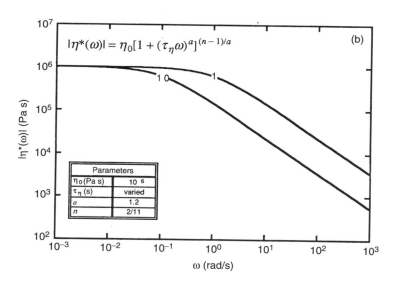

Figure 1 Carreau–Yasuda viscosity function. Parameter varied: (a) η_0; (b) τ_η

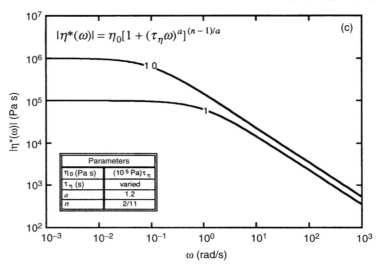

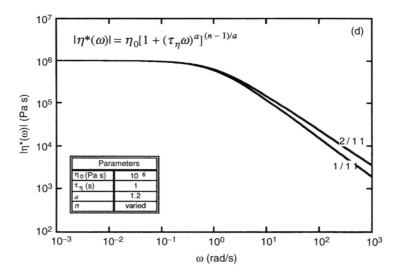

Figure 1 (*continued*) Carreau–Yasuda viscosity function. Parameter varied: (c) τ_η (fixed η_0/τ_η); (d) n

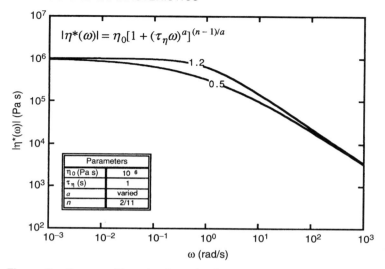

$$|\eta^*(\omega)| = \eta_0[1 + (\tau_\eta\omega)^a]^{(n-1)/a}$$

Figure 2 Carreau–Yasuda viscosity function. Parameter varied: a

an inverse measure of the breadth of this transition, decreasing in value as the transition becomes broader. A rheological breadth index h_a that is a direct measure of the relaxation spectrum breadth is given by the following (inverse) function of a:

$$h_a = e^{(2-a)^3} \qquad (2)$$

In the case of the narrowest conceivable relaxation spectrum, a single Maxwell element, the limiting value of a obtained by fitting the viscosity is 2, for which $h_a = 1$.

We have found that it is possible to fit adequately the experimentally accessible range of melt viscosity data for polyolefin resins with n fixed at 2/11, a value suggested by Graessley [12,13] on theoretical grounds. In fact, such a constraint is necessary to obtain coherent numerical results such that the other three CY fit parameters take on consistently comparable, useful and physically meaningful values. The authors of some previous attempts [14,15] at similar analyses, evidently unaware of theoretical limitations [16] on the actual information content of their experimental data, have unfortunately overlooked this point and have consequently been led to erroneous conclusions. We also read the survey of viscosity models by Elbirli and Shaw [17] as having been strongly influenced by the same considerations that led us to recognize the necessity to fix n at some judiciously selected constant value. [As a practical matter, the best value to use for n is very difficult to determine, because quality-of-fit statistics are insensitive to what value of n is assumed, in the range $0 < n \leqslant 2/11$. However, the other three CY parameter values are sensitive to the assumed value of n, so it is important to choose a constant value and stick with it. One of the authors (J.J.) has explored the rheological data of Schausberger $et\ al.$ [18] and of Rubinstein and Colby [19] on narrow-distribution model polystyrenes and polybutadienes for clues about the optimum value of n. For the highest molecular weights available, the results were consistent with the range found by fitting Curtiss–Bird model [20,21] viscosities with the CY model, namely $0.039 \leqslant n \leqslant 0.074$].

A supplementary model used in our unified approach to polyethylene is the commonly observed proportionality between a power of the mass-average relative molecular mass (IUPAC nomenclature and notation [22]) ('weight-average molecular weight') and the zero-shear viscosity:

$$\eta_0 = k(\overline{M}_{r,w})^p \tag{3}$$

where for linear chains the value of p is known empirically to be usually near 3.4. Carreau–Yasuda fitting provides plausible and useful estimated values for η_0 even when Newtonian plateaus are not directly observable experimentally. Average molecular weights are determined using size-exclusion chromatography.

3   RESULTS AND DISCUSSION

Figure 3(a) shows the results of viscosity and molecular weight measurements on sets of polyethylene resins produced using three different metallocene catalysts. Viscosity measurements were made at 190 °C using a Rheometrics RMS-800 instrument and molecular weight data were obtained with Waters 150CV and

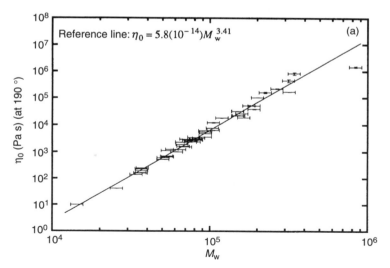

Figure 3   Dependence of η_0 on M_w for (a) metallocene-catalyzed polyethylenes

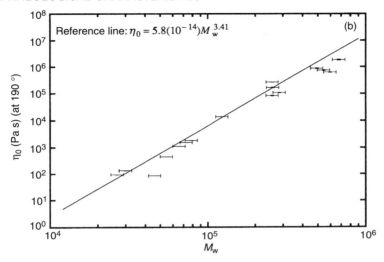

Figure 3 (*continued*) Dependence of η_0 on M_w for (b) titanium halide-catalyzed polyethylenes

150CV Plus gel-permeation chromatographs. The viscosity data were fitted using the CY equation described above. The error bars on the zero-shear viscosities are (1σ) internal estimates obtained from goodness-of-fit statistics for each respective sample. Error estimates accounting for replicability would be somewhat larger. A standard relative error of 9 % [23] was assumed for the molecular weight values. The line on the plot represents the results of Arnett and Thomas [24] for unbranched hydrogenated polybutadienes (narrow-distribution model polyethylenes), and it also matches the results of Mendelson et al. [25] on HDPE fractions. The metallocene-catalyzed resins are obviously not statistically distinguishable from this classical reference line. Figure 3(b) shows data for several resins made with titanium halide-type catalysts. On average, these data are also well described by the conventional relationship [equation (3)] with the same values of k and p.

In Figure 4 are plotted zero-shear viscosities vs relaxation times. These two quantities are strongly covariant. As the molecular weight increases, the zero-shear viscosity also increases and the knee in the transition between Newtonian and power-law behavior shifts to lower frequency (τ_η increases). This covariance is expected since it is through an increase in relaxation times of the system of entangled polymer chains that an increase in molecular weight raises the viscosity.

The relationship of the CY parameter a to molecular structure is much more difficult to quantify. It provides (via h_a) a measure of the breadth of the polymer's relaxation time spectrum. This breadth reflects the breadth of the molecular weight distribution, but is also influenced by the degree of entanglement, especially that associated with long-chain branching in the polymer molecules.

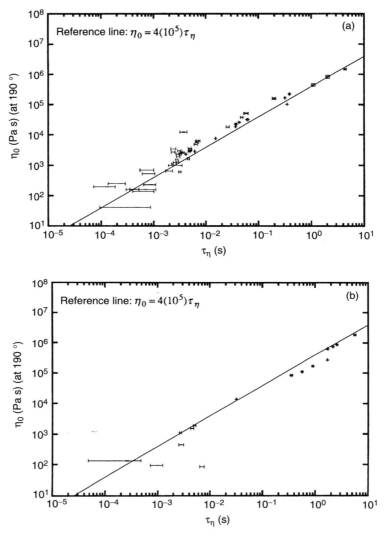

Figure 4 Co-variation of η_0 and τ_η for (a) metallocene-catalyzed polyethylenes, (b) titanium halide-catalyzed polyethylenes

Breadth-index data for the samples represented in Figures 3 and 4 are shown in Figure 5, where h_a is plotted against $\overline{M}_{r,w}/\overline{M}_{r,n}$ (for which we shall hereafter use the abbreviated notation M_w/M_n). In addition, data from several titanium halide-catalyzed resins with generally broader molecular weight distributions are shown. It is obviously not possible in general to predict accurately the rheological breadth

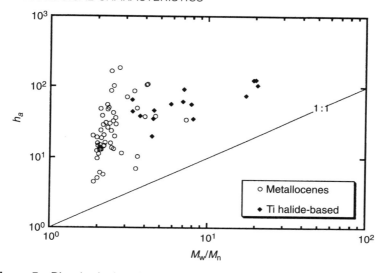

Figure 5 Rheological and chromatographic breadth indexes compared

from chromatographic (molecular weight distribution) breadth, nor is the converse possible, either. Effects such as the presence of very large molecules at chromatographically undetectable levels or some slight long-chain branching must also be playing a confounding role. Our observations are summarized in Table 1.

3.1 TRANSLATION INTO MELT-INDEX TERMS

Using the data in Figures 3 and 4 represented by the correlation lines shown on the plots and the general observations about the relationships between the parameter a and the breadth of the molecular weight distribution from Table 1, it is possible to translate the observations into the language of melt indexes. Figure 6 shows the CY model of six polymers of the same M_w but with different rheological breadths as given by the different a values. The broader the distribution, the lower is the

Table 1 Usual ranges of parameters for different types of catalyst

Catalyst type	Usual range of M_w/M_n	Usual range of a	Usual range of h_a
Metallocene	2–3	0.7–0.5	9–29
Titanium halide	3–5	0.5–0.35	29–89
Chromium oxide	> 8	0.25–0.12	213–769

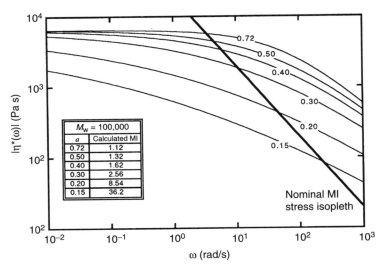

Figure 6 MI dependence on breadth, with M_w constant. Parameter varied: a

viscosity across the range of frequencies typically measured. The diagonal straight line across the plot is a constant-stress isopleth depicting the nominal stress of the standard (2.16 kg load) melt flow measurement. The shear rate where that line crosses a viscosity curve marks where the shear rate in the melt index measurement would be, assuming no pressure losses in the barrel, in the converging region above the capillary, or at the capillary exit. A more realistic model that accounts for these effects has been described elsewhere [26]; this was used to calculate the melt indexes tabulated on the plot for the indicated values of a. The point is, because the polymers modeled here are identical in M_w, that when dealing with resins significantly different in breadth, the melt index is a poor measure of molecular weight.

Figure 7 shows the calculated relationships between melt index (MI) and M_w for model polymers with three different values of a. Thus each family appears to have its own special MI–M_w relationship, with these three curves coming close to describing the behavior of PEs with narrow, intermediate and broad terminal relaxations. From a more general perspective, however, these relationships are necessary consequences of the breadths of the (relaxation time) distributions, and are not necessarily peculiar to the types of catalyst used to make the resins, since similar consequences could be reproduced via blending.

Our model can also be used to illustrate how resins can be very different even though having the same melt index. Figure 8 shows model viscosity curves for three resins with nominally the same melt index but different values of a, corresponding to narrow, medium and broad rheological characteristics, as above. The isopleth of

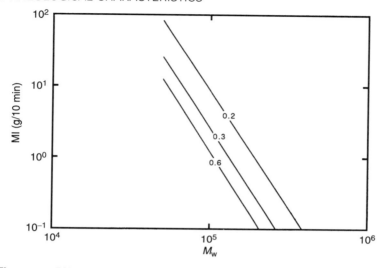

Figure 7 Effect of breadth on MI–M_w correlation. Parameter varied: *a*

constant MI stress is shown to indicate that the frequency where these curves coincide is to the left of this line and hence that there are indeed appreciable losses in the melt-flow plastometer. At the shear rate or frequency where these curves coincide, they should process (flow) similarly, but that is the only place where this is true. The weight-average molecular weights are very different for the three

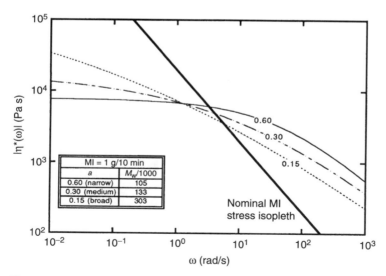

Figure 8 Effect of breadth, with MI constant. Parameter varied: *a*

resins and any properties or process behaviors that depend on M_w will reflect this. The low-shear-rate viscosities, for example, are very different and the melt strengths of these resins would hence be expected to differ appreciably.

The curves in Figure 8 are replotted in Figure 9 to show the corresponding shear stresses as functions of frequency or shear rate. At the higher rates that are more representative of practical processing conditions, it can be seen that for a fixed shear rate (*i.e.* for a process where the limiting factor is rate), the stresses, and therefore the processing pressures, will be greater for the narrow distribution resin. If the process is limited by stress (pressure), that limiting pressure will be reached at lower rates for the narrower distribution resin. Melt fracture phenomena generally occur at some critical shear stress and provide examples of stress limitations on processes. This is an illustration of why processing aids are needed for narrower distribution resins to prevent melt fracture but less so for broad distribution resins.

3.2 EFFECT OF LONG-CHAIN BRANCHING

Up to now the discussion has centered on linear polymers, meaning little or no long-chain branching. We will now enlarge the scope of the discussion to include gross effects of LCB. We have already indicated that a in the CY model depends not only on the breadth of the molecular weight distribution, but also on the effects of entanglement, especially when enhanced by even a small amount of LCB. Figure 10

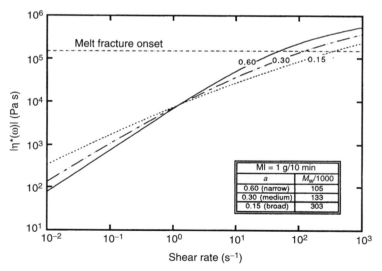

Figure 9 Effect of breadth on stress vs rate, with MI constant. Parameter varied: *a*

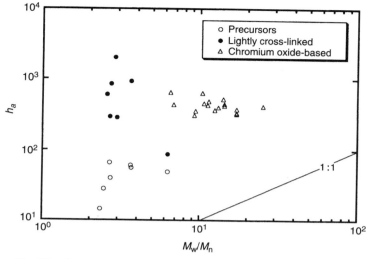

Figure 10   Rheological and chromatographic breadth indexes compared

shows breadth-index data for a few metallocene resins both before (open circles) and after (filled circles) small amounts of LCB were introduced intentionally by the addition of a peroxide during pelletization or unintentionally by extruding an under-stabilized resin in the presence of oxygen. Representative data for several chromium oxide-catalyzed materials are also included in Figure 10. Figures 11 and 12 compare the viscosities and the molecular weight distributions of one of the metallocene pairs—a parent (no LCB) and a treated derivative (with LCB). Although the molecular weight distributions are essentially the same, the viscosity curves are very different. For the resin with LCB, a is smaller and the zero-shear viscosity and viscous relaxation times are larger.

Figure 13 compares data for the samples in Figure 10 with the η_0 vs M_w reference line shown in Figure 3. Although the molecular weight analysis has not been corrected for the branching and the M_ws are therefore slightly underestimated, it is obvious that the zero-shear viscosities for resins with LCB are greatly enhanced over those of the linear polymers of comparable M_w.

The processing consequences of the presence of LCB can be seen by comparing Figure 11 with Figure 8. The resin with LCB has the viscosity curve of a resin with a broad relaxation time distribution, and therefore a higher zero-shear viscosity and a lower viscosity at high rates than would an unbranched resin of the same MI. It for the most part retains its narrow molecular weight distribution and its lower weight-average molecular weight compared with a resin with a broad molecular mass distribution but without LCB. It will reach a critical stress at lower shear rates than the no-LCB version.

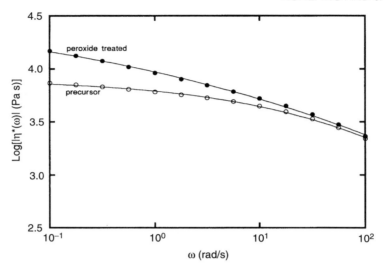

Figure 11 Effect of LCB introduced via peroxide treatment

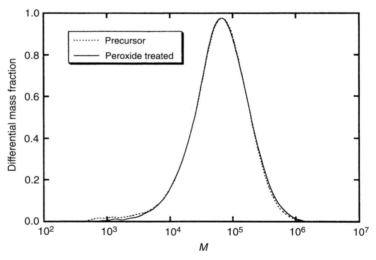

Figure 12 Effect of LCB introduced via peroxide treatment

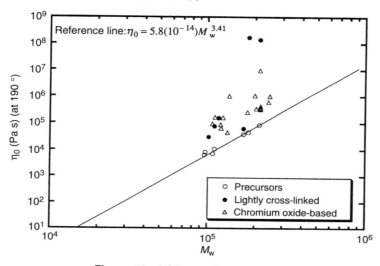

Figure 13   LCB contributions to η_0

4   CONCLUSION

Our model accounts for the melt-rheological characteristics of metallocene-catalyzed polyethylenes as natural consequences of their (usually) narrow molecular weight distributions. We have explored the translation of the viscosity model into terms of melt flow measurements to show how the different breadths of distribution translate into different families of models of melt flow behavior for narrow, medium and broad distribution resins. The effect of long-chain branching is to broaden the relaxation time distribution and as a result to increase the zero-shear viscosity and the viscous relaxation time while decreasing a to produce a resin that behaves rheologically like a broad linear resin while retaining its narrow molecular weight distribution.

While the melt-rheological characteristics of metallocene-catalyzed polyethylenes are dominated by, and expected from, the narrow molecular weight distributions and the absence or presence of long-chain branching, this is not to say that more subtle features of their molecular architecture, especially as relating to solid-state behavior, are not important or interesting. They just do not play a large role in the linear viscoelastic properties of the molten polymers.

5   ACKNOWLEDGMENTS

The authors gratefully acknowledge the contributions of colleagues far too numerous to mention, all of whom participated in synthesizing and characterizing the many materials forming the basis for this study, but special thanks are due to Michael J. Hicks, without whose skills in the making of rheological measurements this work would not have been possible, and to Jerold D. Wood and Delores J. Henson, who made similarly indispensable contributions to the gel permeation chromatographic data we used.

6 REFERENCES

1. Carella, J. M., *Macromolecules*, **29**, 8280 (1996).
2. Lai, S. and Knight, G. W., in *ANTEC '93*, New Orleans, 1993, pp. 1188–1192.
3. Vega, J. F., Muñoz-Escalona, A., Santamará, A., Muñoz, M. E. and Lafuente, P., *Macromolecules*, **29**, 960 (1996).
4. Yasuda, K., Ph.D Dissertation, Massachusetts Institute of Technology (1979).
5. Yasuda, K., Armstrong, R. C. and Cohen, R. E., *Rheol. Acta*, **20**, 163 (1981).
6. Bird, R. B., Armstrong, R. C. and Hassager, O., *Dynamics of Polymeric Liquids*, Wiley, New York, 1987, 2nd edn.
7. Hieber, C. A. and Chiang, H. H., *Rheol. Acta*, **28**, 321 (1989).
8. Hieber, C. A. and Chiang, H. H., *Polym. Eng. Sci.*, **32**, 931 (1992).
9. Sabia, R., *J. Appl. Polym. Sci.*, **7**, 347 (1963).
10. Dealy, J. M., *J. Rheol.*, **39**, 253 (1995).
11. Cox, W. P. and Merz, E. H., *J. Polym. Sci.*, **28**, 619 (1958).
12. Graessley, W. W., *J. Chem. Phys.*, **47**, 1942 (1967).
13. Graessley, W. W., *Adv. Polym. Sci.*, **16**, 1 (1974).
14. Plumley, T. A., Lai, S. Y, Betso, S. R. and Knight, G. W., in *ANTEC '94*, San Francisco, 1994, Vol. I, pp. 1221–1225.
15. Wood-Adams, P. M. and Dealy, J. M., *J. Rheol.*, **40**, 761 (1996).
16. Latimer, P., *J. Colloid Interface Sci.*, **39**, 497 (1972).
17. Elbirli, B. and Shaw, M. T., *J. Rheol.*, **22**, 561 (1978).
18. Schausberger, A., Schindlauer, G. and Janeschitz-Kriegl, H., *Rheol. Acta*, **24**, 220 (1985).
19. Rubinstein, M. and Colby, R. H., *J. Chem. Phys.*, **89**, 5291 (1988).
20. Bird, R. B., Saab, H. H. and Curtiss, C. F., *J. Chem. Phys.*, **77**, 4747 (1982).
21. Öttinger, H. C., in Roe, R. J. (Ed.), *Computer Simulation of Polymers*, Prentice Hall, Englewood Cliffs, NJ, 1991.
22. Kratochvil, P. and Suter, U. W., *Pure Appl. Chem.*, **61**, 211 (1989).
23. Lederer, K. and Aust, N., *J. Macromol. Sci., Pure Appl. Chem.*, **A33**, 927 (1996).
24. Arnett, R. L. and Thomas, C. P., *J. Phys. Chem.*, **84**, 649 (1980).
25. Mendelson, R. A., Bowles, W. A., and Finger, F. L., *J. Polym. Sci., Polym. Phys. Ed.*, **8**, 105 (1970).
26. Rohlfing, D. C. and Janzen, J., presented at ANTEC '97, Toronto, 1997.

20

Extrusion Characteristics of Metallocene-based Polyolefins

C. Y. CHENG
Exxon Chemical Company, Baytown, TX, USA

1 INTRODUCTION

The most significant revolution in polyolefins in the past 20 years was the discovery and commercialization of metallocene-based polyolefins. Almost all of the major producers of polyethylene and polypropylene (collectively called 'polyolefins' in this chapter) resins are engaged in research, development and manufacture of this new family of polymers.

Polyolefins are used in a wide variety of applications, covering all major consumer products. The extrusion process is used almost exclusively in the fabrication of these products, ranging from films to fibers to injection-molded products.

Extrusion is the process of converting polymers from the solid to the molten state and delivering it to the forming device, or die, at a constant rate and in the desired quality. With the new family of metallocene-based polyolefins, extrusion becomes more challenging [1]. The first generation of metallocene polymers possess unique rheological and morphological characteristics which will be the main focus of this chapter. Some of these first-generation polyethylene resins may require modification of even the present 'state-of-the-art' equipment and processing conditions. Ongoing processability improvements are occurring in the industry through machinery changes and resin development programs [2].

Metallocene-based Polyolefins Edited by J. Scheirs and W. Kaminsky
© 2000 John Wiley & Sons Ltd

2 EXTRUSION SYSTEM DESIGN FOR METALLOCENE POLYOLEFINS

2.1 MECHANICAL DESIGN OF A SINGLE-SCREW EXTRUDER

Figure 1 illustrates the major components of a single-screw extruder. The key components and their functions are outlined below.

2.1.1 Motor

The electric motor provides the main source of energy to melt the polymer. The energy is transmitted to the polymer through viscous shear heating generated by the rotation of the screw. Metallocene resins do not necessarily require more energy than conventional resins of the same molecular weight to reach the same melt temperature. However, a higher horsepower (HP) drive may be required to provide a higher torque capability for the extruder when processing metallocene polyolefins.

2.1.2 Gear Box

The purpose of the gear box, or reducer, is to step down the high rotational speed of the motor to a lower level to drive the screw. When the drive is directly coupled to the gear box, the speed ratio in the gear box determines the relationship between the motor and the screw speed. For example, assuming that the maximum speed of a direct current (d.c.) motor is 1750 rpm, a 17.5:1 gear reduction ratio would provide a maximum screw speed of 100 rpm while a 10:1 reduction would have a higher screw speed of 175 rpm.

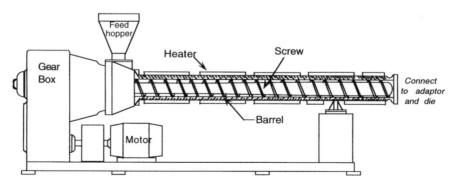

Figure 1 Schematic diagram of an extruder and major components

Table 1 Typical extruder screw speed and torque (24:1 L/D)

Extruder size (in)[a]	Maximum screw speed		Torque requirement (HP/rpm)				
	Blown film	Cast film	LDPE	LLDPE	mLLDPE	Plastomer	mPP
2.5 (65)	105–125	125–150	0.25	0.35	0.40	0.45	0.20
3.5 (90)	90–115	110–140	0.65	1.10	1.20	1.25	0.50
4.5 (115)	75–105	105–125	1.40	2.20	2.40	2.50	1.30
6.0 (150)	60–85	90–110	3.30	5.10	5.60	6.10	3.00

[a] Size in mm in parentheses.

A higher reduction ratio (lower maximum screw speed) provides a higher torque to the screw since full horsepower can be delivered at a lower screw speed. Basically, for a d.c. drive system, the available screw torque is calculated by

$$\text{Torque (HP/rpm)} = \text{rated horsepower/maximum screw rpm}$$

Higher extruder torque can be achieved by using a larger motor and/or a higher reduction ratio gear box (lower screw speed). Although the lower maximum screw speed provides a higher torque capability of the screw, too low a maximum speed may limit the output capacity of the extruder.

Typical extruder torque requirements for metallocene polyolefins are shown in Table 1. The range of rpm represents highly viscous plastomers at the low end and more shear-sensitive conventional high-pressure low-density polyethylene (LDPE) or lower viscosity polypropylene (PP) at the high end. In general, the maximum extruder screw speed for PP is higher than that for PE because PP has a lower specific output rate (70–80 % of PE) and a lower melt viscosity. Simply multiply the torque value by the maximum screw speed to determine the recommended horsepower.

2.1.3 Extruder L/D

The extruder screw length to diameter ratio (L/D) is a very important extruder design parameter. A shorter L/D extruder such as 24:1 is generally used to obtain lower melt temperature and lower extruder torque. For longer L/D extruders, the advantages are better mixing, higher output per rpm and better melt temperature uniformity. For processes where a higher melt temperature is desirable, such as a cast film process, a long L/D of 30:1 is recommended.

2.2   EXTRUSION SYSTEM FOR VARIOUS PROCESSES

2.2.1   Blown film

Extruder L/D

A large percentage of metallocene polyethylene (mPE) is used in blown film processes. Since the resins used in this application are generally of low melt index (MI) or high molecular weight, the extruder design is most challenging. A shorter L/D (e.g. 24:1) is preferred over a longer L/D extruder. However, a longer L/D (e.g. 30:1) can be used successfully when extruding at a higher rate, provided that the screw design is very low shear in nature and the drive train (motor and gear box) is capable of handling the high torque.

Barrel temperature control

To control the melt temperature better, barrel heaters equipped with a water cooling jacket (as opposed to forced air cooling) are strongly recommended for blown film extruders. The water cooling on the barrel is more efficient and precise in controlling the melt temperature. However, a new air-cooled barrel with efficient cooling capability has been used satisfactorily in small extruders (up to 115 mm). For mPE extrusion, it is very common to apply cooling to the last two barrel temperature control zones, especially at high screw speed.

Gear box

Extrusion of mPE film is characterized by a high extrusion torque. Hence the mechanical strength of the gear box needs to have a high torque rating. A higher torque rating gear box allows the installation of a larger drive or a greater speed reduction ratio. If an extruder is not capable of processing mPE, it is most likely due to the gear box torque capability limitation.

2.2.2   Cast Film Process

Extruder L/D

For the cast film process, it is desirable to have a high melt temperature to improve the optical properties and reduce draw resonance [3]. The MI or melt flow-rate (MFR) of polyolefins used for this process is generally higher than that for the blown film process. Therefore, it is advantageous to have a longer L/D to achieve high melt temperatures through viscous shear heating.

Extruder torque

The torque requirement for the cast film process is lower than that for the blown film process because of the higher MI resins used. However, an extruder designed for today's metallocene PE cast films should still have a higher torque capability than a similar line designed for Ziegler–Natta-catalyzed resins.

2.2.3 Fiber Spinning Process

The MFR or MI of the polyolefin used for the fiber spinning process is generally fairly high (low molecular weight). Therefore, the extruder torque is well within the capability of a typical extruder. A long L/D extruder such as 30:1 is strongly recommended to achieve good temperature uniformity and extrusion stability.

3 PLASTICATING EXTRUDER PROCESS ANALYSIS

Inside the extruder, the polymer advances down the screw channel as the screw rotates. While advancing, the polymer temperature increases and part of the polymer starts to melt owing to heat transfer from the barrel and from shear heating. Eventually, all of the polymer should become molten as it reaches the end of the screw. The output rate and melt temperature uniformity of the extrudate are dictated by the screw design.

3.1 FEEDING SECTION

To accomplish effective melting, a single screw has at least three zones, each with a specific function. The first zone is the feed section, where the solid polymer is introduced to the screw, conveyed, and compressed against the heated barrel. Most extruders have a smooth-bore barrel in the feed section and the pellet conveyance is dependent on friction between the pellet bed and the barrel wall. Metallocene PE and PP can be fed very efficiently with a smooth-bore feed section.

3.2 TRANSITION SECTION

The second zone is the transition or compression section, where the compacted polymer begins to melt. In the compression section, the frictional forces against the barrel on the compacted polymer bed steadily increase as the screw depth decreases. The associated mechanical energy is dissipated as heat, which raises the temperature of the polymer and ultimately causes it to melt. For lower density mPE, the polymer can be 'compressed' easily and the compression ratio can be as low as 3–3.5. The low compression ratio and gradual compression can reduce shear heating of the highly viscous metallocene resins.

3.3 METERING SECTION

The third functional zone of a screw is the metering section, where homogenization and pumping of the melt take place. The metering depth and metering section length play an important role in controlling the melt temperature. The metering section determines the amount of shear heating of the molten polymer and hence the melt temperature level and homogeneity. The metallocene polyolefin screw tends to have a deeper metering section than the screw for conventional resin of the same MI or MFR.

4 SCREW DESIGN FOR METALLOCENE POLYOLEFINS

4.1 TYPES OF SCREW DESIGN

It is important to keep in mind that the objective of the screw is to deliver a homogeneous melt at a constant target melt temperature and steady flow-rate. There are no differences in basic screw design principles for metallocene polyolefins and conventional resins. The amount of mechanical energy applied to the resin, and hence the melt temperature, are a result of resin properties and screw geometry [4]. The most commonly used screw designs for metallocene polyolefins are shown in Figure 2. Extrusion of polyolefins does not require a two-stage vented extruder since venting is not required except for the color compounding process.

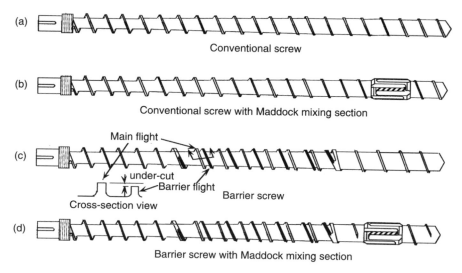

Figure 2 Types of screw design for polyolefins

4.1.1 Conventional Screw [Figure 2(a)]

This traditional design is suitable for low output rate extrusion. At higher output rates, melt temperature uniformity deteriorates. Unlike the mixing screws described below, there is no physical constraint to prevent the unmelted polymer from exiting the screw. For an extruder operating at medium to high screw speed, this type of screw does not adequately deliver the desired melt quality.

4.1.2 Screw with Fluted Mixing Section [Figure 2(b)]

To improve melt quality, a fluted mixing section is often placed near the end of the metering section of a conventional screw. The most common type of fluted mixing section is the Union Carbide or Maddock mixing section shown in Figure 2(b). This is the most commonly used design for polyethylene since its invention. The Egan mixing section is similar to the Maddock mixing design except that the flute is at an angle with the screw axis which incorporates some pumping action to the mixing section.

The advantage of this design is that the unmelted solid is trapped in the inlet channel until it is melted to a small enough size to pass through the barrier clearance. Hence a greater metering depth can be used, resulting in a lower melt temperature with good uniformity. The high shear rate encountered in the barrier clearance also incorporates dispersive mixing, which improves the homogeneity of minor components such as pigments or additives.

4.1.3 Screw with Barrier Flight Design [Figure 2(c)]

The barrier design is based on the principle of confining the unmelted solid polymer in the solid channel while conveying the molten polymer in a separate channel. The dual-channel screw ensures that the polymer particles remain compacted and exposed to frictional forces until they melt. This effect enhances the melting efficiency of the screw. By segregating the solid bed from the molten polymer, the melt channel can be made deeper to reduce shear while maintaining good melt temperature homogeneity, which is advantageous for metallocene polymers.

Figure 3 compares the temperature uniformity of the conventional, fluted, and barrier screws [5]. The melt temperature is measured at the discharge of the screw before entering the die. As the screw speed increases, the melt temperature fluctuation increases substantially for the conventional screw, while the Maddock and barrier screws continue to have minimal variations in melt temperature.

There are several variations of the barrier design. The major differences are in the flight pitch of the main and barrier flights, the depth profiles of the solid and melt channels and how the barrier flights start and end. For metallocene polyolefin extrusion with limited mixing requirements, a well designed barrier screw can offer the advantages of low melt temperature, high output rate and good melt quality.

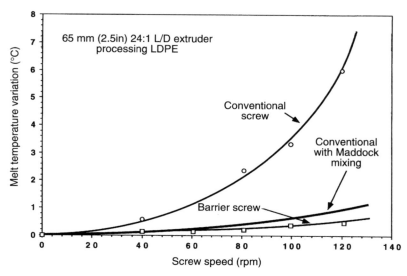

Figure 3   Melt temperature variation vs screw speed for various types of screw design

4.1.4   Barrier Screw with Fluted Mixing Section [Figure 2(d)]

By combining the barrier section with a fluted mixing section, the screw has the capability of a high output rate with good melt homogeneity. With the addition of a fluted mixing section, good melt homogeneity is assured, but at the expense of a slightly higher melt temperature.

4.2   MIXING IN A SINGLE SCREW

Mixing can be as simple as color or temperature homogenization or as complex as compounding multiple resin systems and additives. Mixing in a single-screw extruder is accomplished by laminar flow due to the high viscosity of polymers. A combination of shear and reorientation is required to accomplish adequate distributive mixing. Mixing effectiveness is often counter to other desirable extrusion characteristics such as output and melt temperature.

The type of screw design shown in Figure 2(d) is adequate for mixing high-viscosity metallocene resins. The clearance over the barrier flight and the flute section provide dispersive mixing to shear the agglomerate and the fluted mixing section provides an excellent reorientation of the melt stream for distributive mixing.

4.3   *SCREW DESIGN CONSIDERATIONS FOR METALLOCENE POLYOLEFINS*

The melt viscosity of the metallocene polyolefins is generally much less shear sensitive than those of conventional LDPEs and LLDPEs. They are more viscous at extrusion shear rates, generate more shear heating, melt faster, pump at a slightly higher specific rate and require a higher screw torque. Hence a barrier screw offers more design flexibility and can be used for a wider range of resin molecular weights. A fluted mixing section can be added to ensure good mixing capability.

To control melt temperature and minimize screw torque, the screw should be such that it imparts lower shear to the material. This is done by

- deepening the metering or melt channel depth;
- increasing barrier flight clearances;
- increasing fluted mixing section clearances.

The barrier clearance depends on the resin type and whether other mixing sections exist. For example, mPP will require a smaller clearance of 0.75 mm (0.030 in)– 1 mm (0.039 in). For mPE, the clearance can range from 0.9 mm (0.035 in) to 1.25 mm (0.050 in), regardless of the screw size. A larger clearance is used when there is more than one mixing section on the screw.

5   POLYMER PROPERTIES AFFECTING EXTRUSION

5.1   *SOLID-STATE PROPERTIES*

The density, crystallinity and melting characteristics of polyolefins are controlled by their composition and molecular structure. Resins with a higher density have a higher crystallinity and rigidity, which leads to a higher melting peak. The key solid-state properties affecting extrusion, such as bulk density, thermal conductivity and specific heat, may be influenced by the type of catalyst used. Hence the feeding, solid conveying and melting initiation properties of the metallocene resin may be different to those of conventional resins of comparable molecular weight and density.

5.2   *THERMAL PROPERTIES*

The thermal properties of metallocene polyolefins are very similar to those of conventional resins of the same MI and density. However, the melting-point and crystallization temperature of metallocene polyolefins are generally lower than those of Ziegler–Natta-catalyzed resins. Before replacing an existing resin, it is advisable to compare the density and thermal properties (e.g. melting-point, crystallization temperature and heat of fusion, using differential scanning calorimeter) of the new resin with those of the existing resin to assess the possible extrusion and physical property differences. Specific thermal property differences will be discussed later.

5.3 RHEOLOGICAL PROPERTIES

The rheological properties of the resins have the most profound effect on extrusion. Shear viscosity determines the torque required to turn the screw and the extent of viscous shear heating. For polyolefins, the polymer melt exhibits a non-Newtonian behavior and the viscosity is a function of shear rate. Figure 4 illustrates the effect of shear rate and molecular weight distribution (MWD) for two polymers of the same MI or MFR.

For typical metallocene polyolefins, the shear viscosity is higher than that of conventional resins at the extrusion shear rate. As a result, a higher extrusion torque is required, more shear heating is generated and a higher pressure drop is observed compared with conventional resins. Since the two polymers have the same MI or MFR, the viscosity is the same at the shear rate where the MI or MFR measurement is made (where the two viscosity curves cross each other). At low shear rates, the narrow MWD metallocene resin shows a lower viscosity as a result of its lower molecular weight. Since melt strength is related to viscosity at very low shear rates, the metallocene resin exhibits a lower melt strength.

The shear rate in the screw channel can be approximated by the peripheral velocity of the screw root divided by the channel depth. For the shear rate in the flight clearance, it is the peripheral velocity of the flight tip divided by the flight clearance. Hence it can be easily calculated that the shear rate in the extruder ranges from $10\,s^{-1}$ at very low screw speeds on a small screw to well over $1000\,s^{-1}$ for high screw speeds on a larger extruder. Within this range, the shear dependent melt

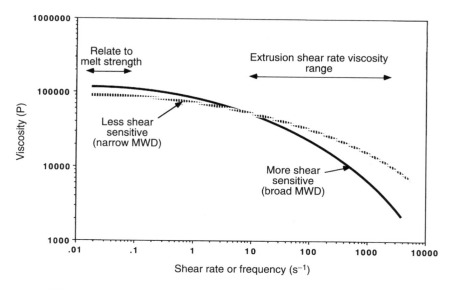

Figure 4 Comparison of shear sensitivities of different resins

viscosity (η) as a function of the shear rate and temperature can be reasonably approximated with the following constitutive equation:

$$\eta = m_0 \exp[-b(T - T_\text{b})]\dot{\gamma}^{n-1}$$

where m_0, b and n are polymer constants and $\dot{\gamma}$ is the shear rate. The constant n is also called the power law index and is the slope of the viscosity curve when viscosity is plotted against the shear rate on a log–log scale. The viscosity is independent of shear rate when $n = 1.0$ (Newtonian fluid). The smaller the value of n, the more shear dependent the polymer melt is. Hence the n value for narrow MWD mPE is higher than those for conventional LLDPE or LDPE. It should be noted that the equation cannot be extrapolated to low shear regions where the melt behaves as a Newtonian fluid.

The shear rate at the flow path downstream of the extruder is also relatively high. Hence the metallocene resin will have a higher extruder head pressure.

6   KEY ATTRIBUTES OF METALLOCENE POLYETHYLENE

6.1   SOLID-STATE PROPERTIES

The feeding or solid conveying efficiency in a plasticating screw is determined by the coefficient of friction (COF) of the pellets with the barrel. For metallocene polyethylene, the COF varies with the density of the polymer. The higher the COF, the greater is the feeding capability, and the output is unlikely to be feed limited. For a low-COF material, such as high-density PE, solid conveying by pellet to barrel friction may not be sufficient and a grooved feed barrel may be required.

Studies have shown that polyethylene with a density above about 0.908 g/cm³ behaves as a rigid polymer. The friction with the barrel wall exhibits the sliding mechanism. The COF increases with decreasing density. For example, the COF increased from 0.186 for 0.963 g/cm³ density HDPE to 0.76 for 0.908 g/cm³ density low-density PE in one study of constant pressure and sliding speed [6,7]. For metallocene-based polyethylene with a density below about 0.908 g/cm³ (mPE plastomers), the resin behaves as a soft elastomer. The resin is easy to feed and a screw with less feeding depth and a lower compression ratio may be used. For very low-density plastomer mPE (density below 0.870 g/cm³), the resin may become tacky and 'dusting' with an inert powder such as talc may be required.

6.2   THERMAL PROPERTIES

The single-site metallocene catalyst produces narrow MWD polymers with very narrow composition distributions of comonomers. Hence mPE has no high molecular weight molecules with few comononers. The resin exhibits a controllable and lower peak melting-point. The peak melting-point of mPE and conventional PE over

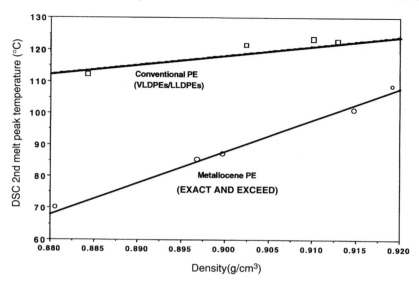

Figure 5 DSC melting point comparison

a broad density range is shown in Figure 5. Note that the amount of comonomer incorporation, as indicated by the density, has a strong effect on the melting-point of the resin. The resin exhibits lower heat-seal initiation temperatures, very good heat sealability and excellent hot tack strength compared with conventional PE. Figure 6 compares the DSC second melt of 0.900 vs 0.917 g/cm^3 density EXACT metallocene resins.

6.3 MOLECULAR WEIGHT DISTRIBUTION

Linear low-density polyethylenes are copolymers of ethylene and α-olefins (e.g. 1-butene, 1-hexene). All branches are short-chain branches (<7 carbons) because of the presence of comonomers. Branching controls density and crystallinity in polyethylene. Since metallocene catalysts have only one type of reaction site, the polymer has a narrower MWD and a narrower composition distribution than conventional linear ethylene polymers [8]. Typical MWDs of EXCEED and EXACT metallocene PE resins are shown in Figure 7.

6.4 RHEOLOGICAL PROPERTIES

Metallocene PE such as EXCEED LLDPE have a narrower MWD, thus exhibits lower shear sensitivity. All other things being equal, they are more viscous than conventional resins at typical extrusion shear rates [9]. The melt viscosity

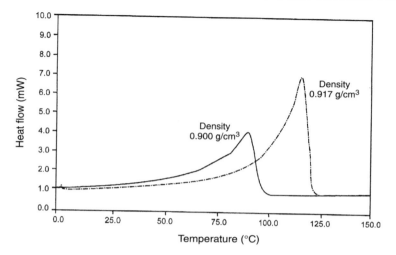

Figure 6 DSC melting peaks of mPE

differences between EXCEED mLLDPE and conventional LLDPE of the same density and MI are shown in Figures 8 and 9 for 190 and 220 °C, respectively. Since melt fracture occurs when measuring highly viscous mPE in a capillary rheometer at high shear rate, the complex viscosity using a cone-and-plate viscometer was used. The frequency in the graphs is equivalent to shear rate in s^{-1} (Cox–Merz rule).

Figures 8 and 9 also illustrate the lower 'zero shear' viscosity of the metallocene PE, which manifests itself in lower melt strength during the blown film process. The low melt strength is favorable for the cast film process, but it reduces the bubble stability of the blown film process and may limit the amount of cooling air that can

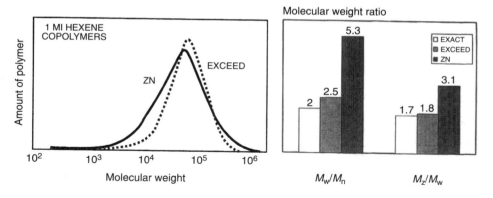

Figure 7 Molecular weight distribution comparison

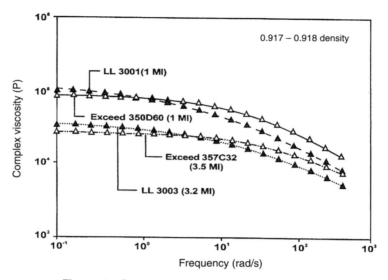

Figure 8 Complex viscosity comparison at 190 °C

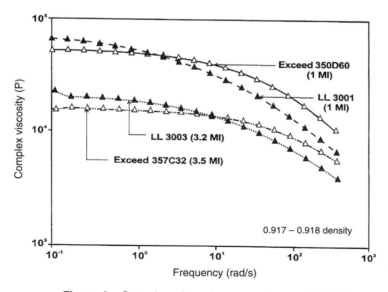

Figure 9 Complex viscosity comparison at 220 °C

be applied to the bubble. One simple way of improving low melt strength is a 3–10 % addition of LDPE. This will improve the production output owing to better bubble stability with little change in film properties.

To overcome the unfavorable processability of mPE in blown film applications, resin producers are aggressively developing new resins with better processability. The objective is to modify the viscosity curve so that the viscosity–shear rate curve is closer to that of conventional LLDPE or LDPE. This is done by using different catalyst systems to incorporate some branching or broaden the MWD. However, the challenge is to improve the processability without sacrificing the physical properties of the fabricated products.

7 KEY ATTRIBUTES OF METALLOCENE POLYPROPYLENE

7.1 THERMAL PROPERTIES

The current generation of mPP homopolymers have melting-points of 147–158 °C compared with 160–165 °C for conventional polypropylene. This is a result of occasional reversals of the propylene molecules as they attach to the growing polymer chain. A reversal of the head-to-tail polymerization (regio-defects) prevents the growth of the crystalline structure, thereby suppressing the melting-point. The melting-point of a particular metallocene polypropylene depends on the catalyst system used.

The low melting-point provides advantages in various aspects of the extrusion process. For example, the melt temperature of the extrudate can be lowered to reduce the energy of extrusion and the subsequent cooling process. This would be beneficial for processes such as blow molding and thermoforming where low melt temperatures provide an improved melt strength.

Figures 10 and 11 compare the melting and crystallization behaviors of an ACHIEVE metallocene homopolymer PP sample and a conventional narrow MWD PP of comparable molecular weight. The metallocene PP shows a lower heat of fusion and crystallization. The combination of lower melting-point and lower heat of fusion are beneficial to the extrusion process. The lower melting temperature leads to earlier melting and therefore the initiation of viscous shear heating occurs sooner than with conventional PP.

7.1 MOLECULAR WEIGHT DISTRIBUTION

Polypropylene polymers produced by the current metallocene catalyst have an MWD, as measured by M_w/M_n, of 2.0. This is in contrast to a broad range of MWD of 3–6 from conventional catalysts. The broad MWD of the polymer produced by the conventional catalyst is a result of multiple active sites of the catalyst solid [10]. For applications where broad MWD is required, the catalyst

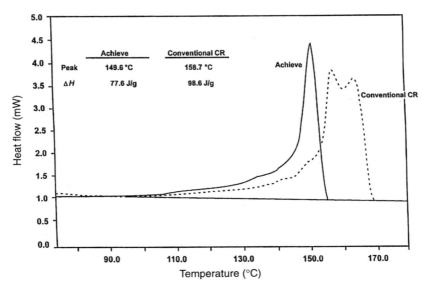

Figure 10   DSC melting behavior comparison of polypropylene homopolymer

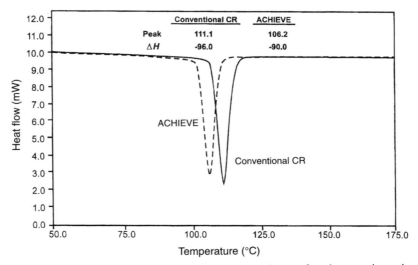

Figure 11   DSC crystallization behavior comparison of polypropylene homo-polymer

system can be modified to produce specific MWD and composition distributions to meet the process and product needs.

7.2 MELT ELASTICITY AND ELONGATION VISCOSITY

Owing to the narrow MWD, the metallocene-based propylene polymer has a low melt elasticity and elongation viscosity. The low melt elasticity reduces die swell at the die exit and the melt can be drawn down more easily to form thinner film, finer fibers, etc. The low elongation viscosity reduces spin line stress, and the polymer can be more easily extended [11].

Figure 12 compares the complex viscosity of a conventional 35 MFR polypropylene and the mPP (ACHIEVE) of the same MFR. Although the MFR is similar, the conventional resin shows a higher shear thinning effect owing to the broader MWD.

7.3 CONTROLLED RHEOLOGY

Currently, PP made from a conventional catalyst has a broad MWD and must be thermally or chemically broken down in the post-reactor extrusion to obtain narrow MWD polymer grades for applications such as films and fibers. When peroxide-initiated breakdown of the PP is employed, it is generally referred to as controlled rheology (CR process).

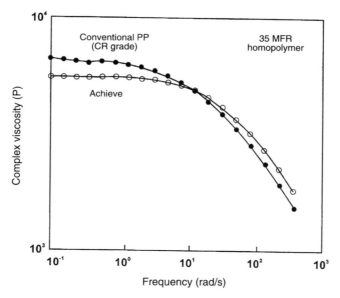

Figure 12 Viscosity comparison

Since mPP is more uniform in MW and has a narrower MWD than a typical CR product, it does not require post-reactor modification. In fact, the granules from the reactor (after the addition of stabilizer) can be fed directly into the extrusion system without pelletizing. This eliminates the extrusion step and preserves the molecular attributes.

7.4 VOLATILE AND EXTRACTABLE COMPONENTS

The CR process produces small amounts of low and very low molecular weight species. Being a non-CR product, the metallocene PP has minimal low-MW volatile components or fumes at the die exit. These attributes have significant environmental benefits, and at the same time reduce equipment downtime for equipment cleaning and housekeeping. The low extractable content is especially attractive for food packaging applications.

8 EXTRUSION PERFORMANCE—METALLOCENE POLYETHYLENE

8.1 OUTPUT AND TORQUE

Using the same 90 mm (3.5 in) screw of the type shown in Figure 2(d) and the same barrel temperature setting, EXCEED mLLDPE and conventional LLDPE have the same output rates. However, owing to the viscosity differences, the torque or power consumption is approximately 10 % higher, as shown in Figure 13. As a result, the

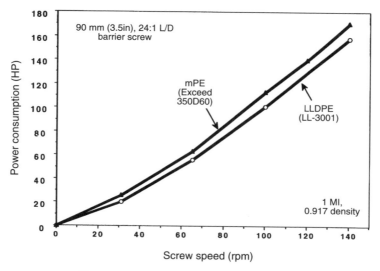

Figure 13 Power consumption comparison

melt temperature of the mLLDPE is 10–15 °C higher, depending on the screw speed. In some cases, blending of conventional LLDPE or LDPE is required to obtain desirable film properties and at the same time improve processability.

As the screw speed increases, the total output rate increases but the specific output rate may decrease.

8.2 LOW SHEAR SCREW

To reduce the melt temperature, a screw with low shear can be used. The low shear is obtained by deepening the metering section and increasing the clearance of the mixing sections. The deeper screw increases the specific output rate and reduces the shear rate at the metering and mixing section.

Figure 14 shows the output rate differences of a 90 mm (3.5 in) medium shear screw and a low shear screw running 3.4 MI mLLDPE. The metering depth of the low shear screw is 13 % deeper and the mixing section clearance 0.4 mm (0.015 in) larger than that of the medium shear screw. The power consumption for both screws is the same with screw torques of 0.87 Hp/rpm at 60 rpm and 1.1 Hp/rpm at 120 rpm. Figure 15 shows the melt temperature differences in the two screws. The combination of higher specific output rate and lower viscous shear heating contributes to the lower melt temperature.

8.3 DIE PERFORMANCE

To optimize extrusion performance, changes in die design should be considered when running mLLPE. A conventional die will have a higher pressure when running

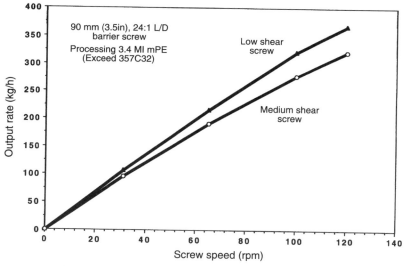

Figure 14 Output rate of low vs medium shear screw

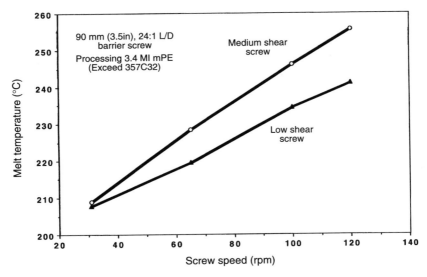

Figure 15   Comparison of melt temperatures

mLLDPE. The high back-pressure has two negative impacts on output performance: it reduces the specific output of the screw and the high pressure eventually dissipates into heat, resulting in a higher melt temperature. As with a low shear screw, the flow channel should be increased when running narrow MWD, low shear sensitivity mPE.

9   EXTRUSION PERFORMANCE—METALLOCENE POLYPROPYLENE

9.1   OUTPUT RATE

Earlier studies have indicated that, in general, the narrower the MWD, the higher is the achievable extrusion output rate [12]. The output rate is more stable and is less susceptible to surging.

Figure 16 shows the output differences of ACHIEVE mPP versus conventional PP for a 65 mm (2.5 in) extruder. With the particular screw design and the temperature profile used, the mPP shows a higher output rate than the conventional PP. Note that the output rate does not increase linearly with screw speed and the specific output rate decreases significantly with screw speed. This may be the combined effect of the type of screw design and the temperature profile used.

For a larger machine such as a 115 mm (4.5 in) extruder, polypropylene may require a reversed temperature profile to improve the specific output rate. With the

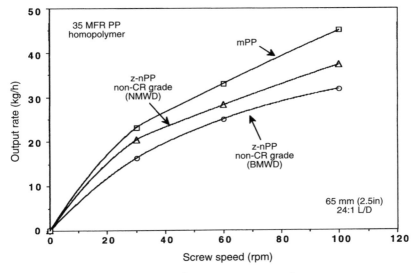

Figure 16 Output rate comparison

reversed temperature profile, the output rate of metallocene resins of 35 and 25 MFR are also found to have a higher output rate than conventional resins.

These results indicate that the metallocene-based polypropylene can be processed under the same conditions as conventional resin with the same or a higher output rate. Such characteristics make it easier for the processor to convert resin usage from conventional to metallocene-based resin. Since the extrusion rate is controlled by the screw design, the output differences between the two resins will depend on the type of the screw used. The extrusion performance may be further enhanced by optimizing the temperature profile for each resin to account for melting behavior differences.

9.2 MELT TEMPERATURE AND EXTRUSION TORQUE

While less shear sensitive owing to the narrower MWD, the melt temperature of the ACHIEVE resin is expected to be higher than that of conventional PP with a broader MWD under the same barrel temperature settings and the same specific output rate. Figure 17 compares the melt temperature of metallocene vs conventional resins when the specific output rates are the same. When the specific output rate (output per rpm) of the ACHIEVE resin is higher, the melt temperature will be lower owing to the lower specific energy input.

For most commercial-grade polypropylene, the MFR is relatively high and therefore, the melt viscosity is relatively low compared with blown film-grade

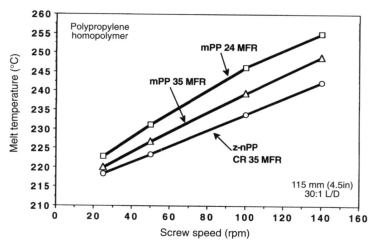

Figure 17   Melt temperature comparison

polyethylene. A narrow MWD does not pose extrusion difficulties in terms of extruder torque or excessive melt temperature. The melt temperature of a metallo-cene resin can be controlled through proper screw design and barrel temperature setting [13].

The extruder torque is reflected in the extruder drive motor amperage draw. The narrow MWD metallocene resin is expected to have a higher torque than a conventional resin. This is not an issue with PP owing to the generally high MFR and the extrusion torque of most processes are well within the capacity of a typical extruder.

10   EXTRUSION PROCESS OPTIMIZATION

Productivity is often the most critical parameter of an extrusion system. Although the output rate of some extruders may be limited by the speed of the haul-off system, such as winder speed, the output limitation is often related to the extruder performance. For metallocene polyolefins, the limitations are often due to high melt temperature, poor melt quality, excessive amperage draw of the motor (excessive torque) and output instability.

10.1   HIGH MELT TEMPERATURE

A high melt temperature causes poor bubble stability in the blown film process because of the lower melt viscosity (lower melt strength) and limits the productivity.

Unless the process calls for higher melt temperatures to improve physical properties such as clarity, an excessively high melt temperature increases the downstream cooling load and is not energy efficient. To reduce the melt temperature, the following approach may be considered:

(i) Increase the barrel cooling capability: this approach can reduce the melt temperature somewhat (5–10 °C) if the existing cooling is not efficient.

(ii) Lower the extruder back-pressure by reducing downstream resistance: this often involves shortening the adapter or enlarging the flow path of the die to match the rheology of the metallocene polyolefin. One can expect to lower the melt temperature by approximately 2–5 °C per 1000 psi of head pressure reduction. The lower temperature is due to slightly higher output per rpm and lower shear heating in the downstream system.

(iii) Use a lower shear rate screw: a new screw design of lower shear can significantly decrease the melt temperature by increasing the specific output of the extruder (output/rpm) and reducing the viscous shear heating in the screw. It should be noted that increasing the specific output may also increase the torque of the screw. Therefore, it may be necessary to increase the torque capability of the extruder at the same time.

10.2 POOR MELT QUALITY

Output may be limited by the fact that at certain screw speeds, the melt temperature non-uniformity becomes excessive and product uniformity, such as gauge control, is beyond the acceptable level. Poor melt quality may be a result of insufficient mixing of additives or minor components. To improve melt quality, the following approaches may be considered.

(i) Increase the extruder back pressure: this approach should be used only in special cases to improve mixing. A finer screen pack is the easiest way of increasing back-pressure on a temporary basis.

(ii) Incorporate a mixing section on the screw: the most efficient way is to put a removable fluted mixing section at the end of an existing screw. The clearance can be made small initially (e.g. 0.75 mm) and enlarged later if needed. For metallocene polyolefins, it may be necessary to increase the metering section depth of the screw at the same time to reduce shear heating.

10.3 EXCESSIVE EXTRUDER MOTOR AMPERAGE DRAW

For metallocene PE blown film processes, this is often the case. The solution requires mechanical modifications.

(i) Increase the speed reduction ratio: this can be achieved by changing the gears inside the gear box to the desired ratio, or the pulley ratio of the motor and the gear box, if the motor is connected to the gear box via a belt-driven arrangement. The extruder torque is increased because full power can now be applied at a lower screw speed.

(ii) Increase the drive motor size: this approach has the advantage of maintaining the same screw speed so that the output will not be limited by the screw speed.

The above two approaches require that the gear box has a sufficient torque rating to accept the higher torque. Otherwise, a new gear box will be required. Approaches such as increasing the barrel temperature setting can also be used at the expense of a higher melt temperature.

10.4 OUTPUT INSTABILITY

Output instability may occur at some high screw speeds which limit the production rate. Output rate fluctuation is evidenced by head pressure variations. A head pressure variation of less than 2 % (which translates to 2–3 % output rate change) is achievable even at high screw speeds with a good screw design. For mPE, extruder surging is rare and pressure variations are caused by the presence of poor melt quality (melt viscosity fluctuation). For mPP, it is possible to have severe surging (large-scale output rate fluctuations), which is accompanied by motor amperage variations. The solutions to the output instability are:

(i) Optimize the barrel temperature settings: this is especially true for polypropylene. A reversed temperature profile can reduce output instability by accelerating melting in the compression section of the screw.

(ii) Modify the screw design: for mPE, a mixing section can be added or mixing section clearance reduced to improve melt quality and hence reduce head pressure variations. Modification of the barrier flight design may be necessary to balance the melting rate. This is especially true for mPP.

11 CONCLUSIONS

Metallocene polyolefins have unique rheological and thermal properties which require modifications of processing conditions and equipment. The processability differences between metallocene-catalyzed resins and conventional resins depend on the shear viscosity curve differences.

In general, metallocene polyolefins have higher viscosity at the extrusion shear rate. Therefore, the polymers require a higher torque extruder to process the less shear-sensitive materials. A lower shear screw with mixing sections may be required to obtain the desired melt temperature and good melt quality. With the knowledge of resin attributes and their contribution to extrusion characteristics, a high-productivity extrusion system is achievable with metallocene polyolefins.

12 REFERENCES

1. Knights, M., *Plast. Technol.* **42**, (6) 49 (1996).
2. Sukhadia, A. M., in *SPE ANTEC Conference Proceedings*, 1997, pp. 2–7.
3. Hellmuth, W., in *Metallocene '96 Proceedings*, 1996, pp. 139–248 (available from Schotland Business Research, Skillman, NJ).
4. Wong, C. M., Shih, H. H. and Huang, C. J., in *SPE ANTEC Conference Proceedings*, 1997, pp. 1522–1526.
5. Cheng, C. Y., *Plast. Eng.*, November, 32 (1978).
6. Spalding, M. A., Hyun, SK. S. and Cohen, B. R., in *SPE ANTEC Conference Proceedings*, 1997, pp. 201–210.
7. Hyum., K. S. and Spalding, M. A., in *SPE ANTEC Conference Proceedings*, 1997, pp. 211–218.
8. Malpass, G. D., *Plast. World*, **54**, (10), 41 (1996).
9. Chung, C. I., in *SPE ANTEC Conference Proceedings*, 1997, pp. 121–126.
10. Blechschmidt, D., Fuchs, H., Vollmar, A. and Slemon, M., *Chem. Fibers Int.*, **46**,, 332 (1996).
11. Cheng, C. Y., *Plast. Eng.*, **44**, (4) 55 (1988).
12. Cheng, C. Y. and Christensen, R. E., in *SPE ANTEC Conference Proceedings*, 1991, pp. 74–78.
13. Cheng, C. Y. and Kuo, J. W. C., in *SPE ANTEC Conference Proceedings*, 1997, pp. 1942–1949.

PART VII

Applications of Metallocene-based Polyolefins

21

Film Applications for Metallocene-based Propylene Polymers

ASPY K. MEHTA, MICHAEL C. CHEN AND CHARLIE Y. LIN
Exxon Chemical Company, Baytown, TX, USA.

1 INTRODUCTION

Isotactic polypropylene (iPP) finds usage in a variety of end applications, many of them consumer oriented. Film is one of the top-tier end-use markets, accounting for about 18 % of worldwide polypropylene usage [1]. Polypropylene (PP) films utilize both homopolymers (HPPs) and random copolymers (RCPs). PP films can be biaxially oriented (BOPP) or non-oriented (NOPP). BOPP films, with their crisp feel, outstanding clarity, moisture barrier (low water vapor transmission rate, WVTR) and high stiffness, are widely used in tape and food packaging, especially snack foods and bakery products. Other uses include shrink wrap, for the form-fitting display of packaged articles as varied as pizza and hardware items, and capacitor film, where the cleanliness, good electrical insulation and low moisture absorption of PP are valued attributes. NOPP films are also used in food packaging. Two examples are the packaging of individually wrapped processed sliced cheese 'singles' and twist-lock wrappers for hard sweets and candies. The packaging of textile products (hosiery, shirts, blankets) and stationery products is another end-use market for non-oriented films. RCP films are used as heat seal layers and where enhanced toughness or softer films are desired. These films are generally laminated or co-extruded on to other substrates, but can also be used as stand-alone films.

How will metallocene propylene polymers (mPP) impact this large and diverse market? Although the commercial utilization of mPP is only in its infancy today,

Metallocene-based Polyolefins Edited by J. Scheirs and W. Kaminsky
© 2000 John Wiley & Sons Ltd

developments are under way to set in place the necessary product building blocks to launch commercial grades for film. These developments include:

- the making of high molecular weight, i.e. low melt flow-rate (MFR), film forming products;
- the capability to tailor molecular parameters, e.g. molecular weight distribution (MWD), to optimize film properties and film fabrication;
- the making of a broad range of RCPs to meet market needs on heat sealing and film toughness;
- the introduction of syndiotactic polypropylene (sPP) to the marketplace, in addition to the standard isotactic form, to broaden the available range of film product properties; and
- film fabrication trials of mPP on existing PP film lines to establish film processability.

These and other developments will be reviewed in this chapter.

2   METALLOCENE PROPYLENE POLYMERS FOR FILM APPLICATIONS

The study of new metallocene-based catalysts and of propylene polymers derived from them is the subject of intense investigation today in both industrial and academic research circles [2–4]. As a consequence, the development of new products is in a state of rapid flux. For application to films, high molecular weight, e.g. MFR in the range 1–40 dg/min, high levels of crystallinity to maximize film stiffness and moisture barrier, low levels of atactic material and the controlled incorporation of a range of comonomer(s) in RCPs represent what would appear to be a minimum set of product requirements needed of mPP to compete effectively against today's highly optimized slate of Ziegler–Natta (Z–N)-based PP film products.

2.1   ATTRIBUTES OF METALLOCENE PROPYLENE POLYMERS

2.1.1   Isotactic Polypropylene (iPP)

The current state of metallocene iPP product development, involving high molecular weight, film-forming propylene polymers of high isotacticity, can be gauged from the representative polymer property profiles on HPP and RCP shown in Table 1. These products were derived from supported catalysts, the basis for economically viable commercial production.

The property comparisons in Table 1 show some differences between the mPPs and their Z–N counterparts. Four key structural features that differentiate the metallocene-based from Z–N polymers are the narrower MWD, narrower composi-

Table 1 Comparison of polymer properties for metallocene and Ziegler–Natta iPPs[a]

	Homopolymer		Random copolymer	
Polymer property	Metallocene	Z–N	Metallocene	Z–N
Melt flow rate (dg/min)	5	5	5	5
mmmm pentad fraction	0.95	$\geqslant 0.95$	—	—
Comonomer (type/wt%)	—	—	$C_6/2.8$	$C_2/5.0$
$T_m(°C)$	151	160	128	132
n-Hexane extractables (wt%)	0.5	2.8	0.7	3.3
MWD (M_w/M_n)	2.0	5.0	2.0	3.0
Composition distribution (qualitative)	—	—	Narrow	Broad
Tacticity distribution (qualitative)	Narrow	Broad	Narrow	Broad

[a] Examples of Z–N polymers: PP 1012 (homopolymer) and PD9282 E2 (random copolymer) from Exxon Chemical.

tion distribution (CD), narrower tacticity distribution and enhanced capability to incorporate higher α-olefins (HAO) as comonomers. Each of these attributes influences film forming and film properties in a complex manner. An attempt to rationalize this is shown in Table 2, which indicates that, by and large, the film implications of these attribute differences are favorable. Of particular note are the easy drawability, high film clarity, low extractables and good thermal processing stability anticipated for mPP.

The metallocene RCP example shown in Table 1 is a 1-hexene-based polymer, to illustrate the capability of incorporating HAO comonomer, among many others. Conventional Z–N RCPs are based predominantly on ethylene as comonomer, with some products utilizing 1-butene. Figure 1 shows the depression of DSC peak melting-point for ethylene, 1-butene and 1-hexene RCPs [5]. The data show 1-hexene to be a very efficient comonomer, versus ethylene or 1-butene, in lowering the melting temperature. Recent work on a variety of metallocene RCPs indicates that the 1-hexene trend line in Figure 1 is followed by many other (longer) α-olefin comonomers [6].

2.1.2 Syndiotactic Polypropylene (sPP)

Metallocene-based catalysts, especially those containing the combination of a cyclopentadienyl ligand and a bulky fluorenyl ligand, have transformed sPP from a laboratory curiosity to a commercially polymerizable addition to the family of propylene polymers [7–9]. Products having syndiotactic pentad fractions about 0.80 have been introduced to the marketplace [10]. An assessment of selected HPP properties for these metallocene sPP resins versus metallocene iPP and Z–N iPP is shown in Table 3. The property profiles outlined for the metallocene sPP and iPP polymers reflect their current state of commercial product development. The data show metallocene sPP to have a different balance of polymer properties versus

Table 2 Structural features of metallocene propylene polymers: film implications

Structural feature		Film implication[a]
Narrow MWD	−	Potentially more difficult extrusion at same MFR (higher head pressure and motor load)
	+	Higher cast film line speeds without draw resonance
	+	Less low molecular weight extractables/ migratory polymer and less volatiles
	+	Greater thermal processing stability
Narrow composition distribution	+	More efficient use of comonomer in depressing crystallinity → lower seal initiation temperature (SIT)
	+	More uniform comonomer incorporation → less low molecular weight/high comonomer-containing species → less sticky polymer plating-out on film lines
	+	Narrower crystal size distribution → lower haze
	+/−	Narrower melting range
Narrow tacticity distribution	+	Low FDA extractables → wider range of packaging opportunities
	+	Potentially stiffer films at same T_m or SIT
Higher α-olefin (HAO) incorporation	+	HAO effects analogous to LLDPE → favorable toughness and stiffness balance
	+	Faster crystallization → higher hot tack

[a] + indicates benefit; − indicates debit; +/− indicates benefit or debit depending on the application.

metallocene iPP or Z–N iPP, the two iPP property sets being closer together. The key feature that emerges for metallocene sPP, separate from the expected single-site properties of narrow MWD, CD, tacticity distribution and improved comonomer incorporation, is that of a polymer of low stereoregularity and, as a consequence low crystallinity [8,13,14]. This feature affects many of the attributes shown in Table 3, such as poorer property stability on physical aging [15].

The influence of these characteristics on film forming and film properties is outlined in Table 4. The combination of film toughness, clarity, flexibility or softness and low heat seal temperature in a syndiotactic HPP appears to be unique, matching more the behavior of a conventional isotactic RCP, but with potentially lower extractables. The poorer property stability on physical aging, however, could be a concern, with the potential for unacceptable changes in film properties (e.g. permeability).

It must be cautioned that the crystallinity of metallocene sPP (like that of metallocene iPP) is dependent on the level of defects in the polymer. The low crystallinity of today's offerings (with a syndiotactic pentad fraction about 0.80) will

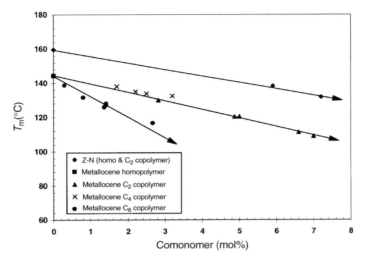

Figure 1 Depression of iPP melting temperature by various α-olefin comono-mers. (Metallocene polymer prepared from supported catalyst that provides homopolymer T_m of 145 °C. Recent catalyst advances have raised the homo-polymer T_m to about 151 °C.)

Table 3 Comparison of polymer properties for sPP versus iPP[a]

Polymer property	Metallocene sPP	Metallocene iPP	Z–N iPP
Stereoregularity (pentad fraction mmmm or rrrr)	0.8	0.95	$\geqslant 0.95$
T_m(°C)	130	151	160
Crystallinity (%)	– – –	–	○
Crystallization rate (min^{-1})	– – –	–	○
Breadth of MWD (M_w/M_n)[b]	–	–	○
Entanglement MW (M_e)[c]	– –	○	○
n-Hexane extractables (%)	–	– –	○
Radiation stability[d]	+	○	○
Property stability on aging[e]	– –	○	○

[a] Data shown are representative values for current commercial products (sPP from Fina Oil and Chem, iPP from Exxon Chem). Comparative assessment based on Z–N iPP as 'control': ○, 'control' property value; assigned neutral; +, ++, progressive property value increase; –, – –, – – –, progressive property value decrease.
[b] Assessment for tacticity distribution, will be similar to that for MWD.
[c] M_e is average molecular weight between entanglements, reflecting entanglement density in melt.
[d] Radiation stability data are on sPP from Donohue [11,12], on low pentad fraction polymers.
[e] Property stability on aging reflects transient effects, separate from effectiveness of stabilizer package.

Table 4 Film implications of sPP polymer properties

Polymer property		Film implication[a]
Lower crystallinity	+	Tougher film (tear and impact)
	+	Higher film clarity
	+/−	Softer, more flexible film
	+	Lower heat sealing temperature
Slower rate of crystallization	−	Potential for film processing difficulties on conventional polypropylene film lines (e.g. roll blocking, sticking)
		Modifications (e.g. blending with iPP or different processing conditions) may be needed to process on existing equipment
Lower property stability on aging (film aging under ambient conditions)	−	Potentially unacceptable changes in film properties (e.g. permeability, heat sealability, additives bloom)
Higher level of melt entanglements (versus iPP)	+/−	Different rheological behavior (e.g. lower intrinsic viscosity at same melt flow rate)
	+	Potentially higher hot tack and heat seal strength

[a] + indicates benefit; − indicates debit; +/− indicates benefit or debit depending on the application.

increase, as catalyst and reactor process developments raise the pentad fraction to the level of today's metallocene iPPs ($\geqslant 0.95$). Certainly, the implications for film will change from the snapshot presented in Table 4. Laboratory characterizations on such higher pentad fraction sPPs have been reported [16,17], which provide some insight into how the properties profile might change. For example, the crystalline melting temperature is increased to about 150 from 130 °C, indicating a higher sealing temperature and an increase in film stiffness.

2.1.3 Atactic Polypropylene (aPP)

Metallocene-based catalysts have also extended the property envelope of atactic PP. Z–N aPP is typically of low molecular weight, broad MWD and with some low level of residual crystallinity. Recent metallocene catalyst developments, however, have produced a high molecular weight product that is narrow in MWD and amorphous [9,18–20]. This polymer shows interesting elastomeric properties. Similar properties have also been obtained in propylene–ethylene copolymers made with aspecific metallocene catalysts [21], but with a lower glass transition temperature. One area of use of metallocene aPP for film applications is likely to be in blends, with iPP for example. In such a blend composition, the aPP should render enhancements in clarity, toughness and flexibility.

2.2 POTENTIAL FOR TAILORING

As is well known, one of the positive attributes of single-site, metallocene-based products is their homogeneity or narrow property distribution, be it of molecular weight or composition or other polymer parameters. This feature makes them ideal individual components to combine, for the purpose of tailoring selected properties. This ability to tailor metallocene propylene polymers in a controlled fashion allows, in principle, the design and production of polymers to meet any reasonable set of film performance criteria [22]. An example of such tailoring would be the controlled broadening of MWD, beyond the 'most probable' M_w/M_n value of 2.0. The narrow MWD is beneficial for properties, such as stiffness, but is a drawback in some film forming processes, such as BOPP film fabrication. As discussed later, controlled broadening of the MWD does indeed render a substantial improvement in film processability [23]. Another form of tailoring involves combining metallocene iPP with metallocene sPP. The individual property profiles of today's offerings for these two polymers being so different [24], some unique film attributes are attained on blending [25]. Combining metallocene aPP with metallocene iPP should also offer film property benefits [18,21] as noted above. Finally, tailoring can also involve combinations of metallocene-derived propylene polymers with those from conventional Z–N catalysts. For example, the combination metallocene RCP with Z–N RCP offers some novel film property benefits in NOPP films, as will be shown later [26,27].

Selected examples demonstrating the ability to tailor molecular parameters and the implications of these for film forming or film properties are outlined in Table 5. As mPPs strive to find their niche in today's diverse film marketplace, it is anticipated that polymer tailoring will play an increasingly important role in the design of new, differentiated products.

2.3 OTHER FILM-RELATED ATTRIBUTES

Single-site mPPs (HPPs and RCPs) show lower levels of volatile materials in addition to reduced levels of low molecular weight solvent extractables, *vis-à-vis* conventional Z–N PPs of similar composition and melt flow rate. This is favorable for film packaging of food items, since the original aroma and taste of the packaged food will be less affected by the ingress of volatile chemicals from the film. There is also a positive influence during film fabrication, where lower levels of fumes are released at the die exit and in orientation ovens. The reduced level of low molecular weight species also helps lower the tendency for films to block, providing better machinability. At the other end of the molecular weight spectrum, the lower level of high molecular weight species in single-site mPP, which inherently has a narrow MWD, contributes to an improvement in resin stability (less thermal oxidative breakdown) during multiple film processing steps. This has favorable implications for recyclability and the maintenance of target physical properties in the end-films.

Table 5 Implications of polymer tailoring on film properties

Description of tailored feature		Film implication[a]
Tailored MWD	+	Increased melt shear sensitivity for easier film processing
	+	Increased melt strength
	+	Reduced film neck-in
Tailored CD and/or tacticity distribution	+	Broader melting range for heat sealability at lower temperatures
	+	Broader film orientability window to lower temperatures
	+	Controlled film stiffnes/flexibility
	+	Controlled film shrinkage
Tailored compositions		
Example 1		
Metallocene sPP added to	+	Lower film haze
metallocene iPP (major component)	+	Increased film toughness (dart impact, tear)
	+	Upgrade of overall film properties profile
Example 2		
Metallocene C_6 RCP added to	+	Increased film toughness (dart impact, tear)
Z–N C_2 RCP (major component)	+	Lower film haze
	+	Lower coefficient of friction

[a] + indicates benefit; − indicates debit; +/− indicates benefit or debit depending on the application.

As for additive packages for mPPs, the optimum make-up will depend on the details of the product tailoring and design; however, the attributes inherent in 'single-sitedness' provide some general guidelines. A lower requirement of antiblock additive would be anticipated, based on the reduced amount of low molecular weight species. On the requirement of slip additive, crystallization data suggest that many mPPs crystallize more slowly than conventional Z–N polymers [28]. In such cases, an increase in the level of slip additive (versus the comparable Z–N product) might be required, particularly in cases where fast slip bloom is desired. Levels for stabilizer and neutralizer additives are generally lower, reflecting the high efficiencies of metallocene catalysts, high resin stability and the overall cleanliness of the resulting products. Between metallocene iPP and sPP of roughly similar molecular weight and pentad fraction, the data suggest that sPP has greater thermo-oxidative stability [29].

Finally, since the blending of mPPs with conventional Z–N PPs (and also other polyolefins) will undoubtedly be practised in the film marketplace, the miscibility of these polymers is of importance. Metallocene aPP and iPP [30], in addition to Z–N aPP and iPP [31], are melt miscible. However, metallocene aPP and sPP are melt immiscible [30]. Conventional iPP and metallocene iPP should be melt miscible. Metallocene sPP and iPP (conventional Z–N or metallocene) however, appear to be melt immiscible [30].

For metallocene RCPs, miscibility with conventional Z–N ethylene-containing RCPs will depend on the choice and the amount of comonomer. Ethylene RCPs, metallocene and conventional Z–N, of similar comonomer content appear to be melt miscible. Blends of metallocene 1-hexene RCPs with conventional ethylene RCPs, of similar comonomer content, also appear to be miscible. Novel cocrystallization behavior has been observed for this blend, with a film properties profile different from that displayed by the individual components, as alluded to earlier [26,27].

3 BIAXIALLY ORIENTED FILMS

3.1 BOPP FILM RESINS

Typical resins for BOPP film, either HPP or RCP, are in the MFR range 1–8 dg/min and MWD range, expressed as M_w/M_n, 3–10. A preferred BOPP film resin should provide satisfactory processability during the film orientation steps and produce quality film with good properties and gauge uniformity. With regard to processability, a high melt strength to allow film stretching to thin gauges without sagging, and the ability to be evenly stretched at low temperatures, are important considerations. Key resin requirements to provide good processability and film properties include a balanced crystalline and amorphous content and a broad MWD. Propylene polymers with controlled levels of long-chain branches would also offer improved processability during film stretching. Because of the inherently broader MWD of Z–N-catalyzed iPP, versus single-site metallocene, the processability is usually better. However, the more heterogeneous nature of the Z–N product, with relatively higher amounts of atactic and syndiotactic species, leads to a reduction in stiffness and an increase in surface blocking, among other undesirable characteristics.

As mentioned previously, the metallocene catalyst system promotes molecular tailoring of polymers for specific application needs. Taking advantage of this molecular design feature, development activities using tailored metallocene iPP to extend the film property envelope and improve film processability are becoming a focus of interest [23,27,32]. One example of molecular tailoring of iPP is the use of a mixture of different, metallocene catalysts, which offers the capability of controlling the extent of MWD broadening [22,33–35]. Typical resin properties of standard single-site metallocene iPP polymers, tailored mPPs and Z–N iPPs for BOPP film applications are listed in Table 6. The tailored metallocene iPP polymers are seen to exhibit broader MWD and higher recoverable compliance in the melt, reflecting the presence of high molecular weight species. As will be seen, an attractive feature of the tailored product is its easier stretchability, while maintaining the characteristic low extractables level of the standard single-sited metallocene iPP.

3.2 BOPP FILM PROCESSABILITY

There are two major production methods for biaxially oriented film: the double-bubble and tenter frame processes. The double-bubble process is less capital intensive. The tenter frame process is more widely used because of its higher

Table 6 Properties of metallocene iPP for BOPP applications

Type	MFR (dg/min)	Ethylene (wt%)	M_w/M_n	Melting temperature (°C)	Recoverable compliance ($Pa^{-1} \times 10^{-4}$)	Xylene solubles (wt%)
Metallocene	2.0	0	2.1	151	2.1	0.4
Metallocene	4.0	3.4	<2.5	127	~1	<1.0
Metallocene[a]	1.7	<1	2.9	147	4	0.4
Metallocene[a]	3.4	1.6	2.9	142	4.3	0.5
Ziegler–Natta[b]	2.1	<1	4.3	157	3.7	2.9
Ziegler–Natta[b]	3.6	4	3.3	138	2.4	8.7

[a] Metallocene propylene polymers by molecular tailoring.
[b] Examples of Ziegler–Natta polymers: PP4782 (2.1 MFR) and PD9302 E1 (3.6 MFR) from Exxon Chemical.

productivity. In this section, discussions of film processability will center on the sequential stretch tenter frame process.

In the tenter method (Figure 2), polymer is extruded through a flat die, quenched by a chill roll which may be half immersed in a water-bath, stretched in the machine direction orienter (MDO) over a set of S-wrap heated rolls rotating at differential speeds, and finally forwarded to the transverse direction orienter (TDO), where the sheet is grasped by clips which travel along the tenter frame. In the TDO, the sheet is preheated, stretched by diverging the film width along the transverse direction (TD), and finished with a heat set. The film is then edge trimmed, treated if necessary and wound on to a roll.

The TD stretching step is usually the most critical in the sequential stretching process. Film breaks occur more often at this point than any other, causing the line to be brought down. These breaks are particularly disruptive on large commercial lines, which operate at line speeds typically 250 to 300 m/min, with film webs 8–10 m wide. At TD stretching temperatures lower than the optimum, the high yield strength and cold-drawability of the film give rise to stretch bands and eventually film breakage. Conversely, at stretching temperatures higher than the optimum, partial melting occurs at the film surface, resulting in gauge deterioration, film sagging and

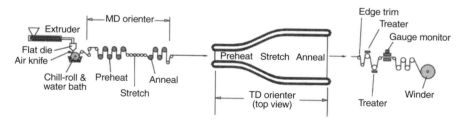

Figure 2 Schematic diagram of BOPP tenter frame sequential stretching process

finally film breakage. A processing window can be defined in terms of the range of TDO temperatures over which film quality and gauge uniformity are judged to be acceptable. Preferred BOPP film resins are those that provide the broadest possible processing window with competitive film properties. Evolutionary developments in BOPP film fabrication technology, involving faster line speeds, wider webs and thinner film gauges, will pose additional demands on BOPP film resins [36]. Thus resins having acceptable processing windows today may not be adequate, say, 10 years from now. Tailored metallocene resins show potential to address these stringent demands, as will be discussed below.

For standard single-site metallocene iPP, the processing window is relatively narrow, owing to the narrow MWD and melting range. Using a batch biaxial stretching apparatus, e.g. a T. M. Long machine, the resin processability was determined in terms of the range of oven temperatures over which uniform stretching could be achieved. It is seen in Figure 3 that good film formation for standard metallocene iPP is limited to only 5 °C (152–157 °C) versus about 10 °C (156–164 °C) for the Z–N iPP. The data suggest that a tightly controlled BOPP film process will be necessary, in order to stretch this metallocene iPP polymer into a uniform film.

To determine true BOPP film processability, a more realistic approach is to fabricate film on a pilot scale, sequential stretch tenter frame line. This method employs an in-line gauge monitor to measure stretched film quality and gauge uniformity over a range of TDO temperatures. As mentioned before, the processing window is determined by the range of TDO temperatures over which high-quality

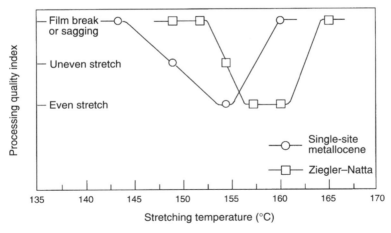

Figure 3 Comparison of BOPP processability for isotactic propylene polymers (MD × TD stretching ratio = 6 × 6)

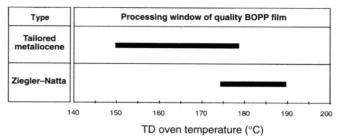

Figure 4 Comparison of processing window for producing quality BOPP film on sequential tenter frame

film is obtained, per a selected quality criterion. Typically, gauge uniformity across the film web is selected as the quality criterion. The processability improvement possible with a tailored metallocene iPP, beyond that of the standard single-site product is seen in Figure 4. It demonstrates a broad processing window of more than 26 °C. By comparison, the Z–N iPP exhibits good processing only over a 15 °C range. The optimal processing temperature is usually selected at the midpoint of the processing window. The optimal processing temperature for the tailored metallocene iPP is below 170 °C versus 183 °C for the Z–N iPP. Metallocene resins by molecular tailoring demonstrate not only greater processing latitude, but also the capability to operate at significantly reduced TDO temperatures. The tailored metallocene iPP offers the potential capability [23,27,32] to meet the demands of higher speed BOPP film lines, either in the sequential stretch tenter process or in the emerging technology of simultaneous stretch tenter processing for BOPP film [37].

3.3 BOPP FILM PROPERTIES

Processability in oriented film operation is a critical factor in judging a good BOPP film resin. An equally important factor is the ability to produce film with competitive properties. Since the start of metallocene PP development in the early 1990s, there have been many published data on BOPP film properties (e.g. Refs 23, 27, 32 and 38–48). Some typical properties from these sources are reported in Table 7, which compares selected BOPP film properties for mPPs versus Z–N controls. It should be noted that the different evaluations referenced in Table 7 utilized different mPP and Z–N polymers, making it difficult to draw a meaningful comparison of film properties. Still, on balance, the data do show the mPP film properties profiles to be competitive. Since BOPP film properties are greatly influenced by resin characteristics and the type of fabrication process and orientation conditions used,

Table 7 Comparison of BOPP film properties[a]

Film property	Testing method	Peiffer et al. [39] Metallocene	Z-N	Bidell et al. [41] Metallocene	Z-N	Saito and Ushioda [44] Metallocene	Z-N	Wheat and Hanyu [48] Metallocene	Z-N	Lin et al. [32] Tailored metallocene	Z-N[b]
MFR (dg/min)	DIN 53–735 or ASTM D-1238	4	–	8	2.5	1.6	1.7	2.5	2.8	1.7	2.1
Gauge average (µm)	–	20	20	12	18	23	23	18	18	21	20
Haze (%)	ASTM D-1003	1.0	2.8	0.1	2.5			<1	<1	0.6	0.2
Gloss	DIN 67–530	140	100	116	85–90	124	112	>90	>90	92	95
WVTR (g/m² day)	ASTM D-2457 DIN 53–122	1.1	1.3								
(g/m² day)	ASTM F-372							2.5	3.5		
(g/m² day per 25.4 µm at 37.8 °C, 100 % RH)	ASTM F-372									7.2	5.8
Oxygen transmission rate (cm³/m² day bar)	DIN 53–380	1600	1900					2200	2600		
(cm³/m² day)	ASTM D-1434										
E-Modulus (N/mm²) MD/TD	DIN 53–457	2600/4700	2200/4100	1700/3110	2000/2800	2150/5190	1900/4250			1906/2690	2144/3449
Secant modulus (N/mm²)	ASTM D-882							1500/2400	1200/1900		
1 % Secant modulus (N/mm²)	ASTM D-882							1200/1900			
Tensile strength at Break (N/mm²) MD/TD	DIN 53–455 ASTM D-882	160/320	140/300	134/362	120/260	137	118			152/251	164/264
Elongation at break (%) MD/TD	DIN 53–455 ASTM D-882	125/70	160/60	210/37	200–240/40–50					116/43	106/36
Shrinkage (%) MD/TD at 135 °C, 180 s	DIN 53–455 ASTM D-882							120/40	150/50	15/20	7/9
at 140 °C						1.0	1.8	10/20	7/10		
Scratch resistance (Δ haze)		8	28								
COF, kinetic film-to-film								0.4	0.6		

[a] Both metallocene and Ziegler–Natta polymers are isotactic polypropylene.
[b] Example of Ziegler–Natta polymer: PP4782 from Exxon Chemical.

a precise comparison of film properties can only be performed within the same experimental design.

Because of the features of narrow MWD and uniform stereoregularity, BOPP films from standard single-site metallocene iPP are distinguished, in general, by their improved tensile strength, stiffness and barrier along with good clarity and gloss. Both tensile strength and stiffness are important properties in the conversion of films to final packaging products. Improved stiffness offers the opportunity for down gauging, an industry demand to reduce the cost and volume of packaging material. In the application of metalized BOPP film, higher stiffness provides the ability to withstand flexing without developing cracks during the normal shipping and handling of packaged goods; this helps preserve barrier integrity and package appearance. Improved barrier properties mean better flavor preservation and therefore longer shelf-life of the final packaged item. One particular deficiency of metallocene iPP at this stage of development is its lower melting temperature versus conventional Z–N iPP. For this reason, the high melting temperature metallocene iPP reported by Saito and co-workers [44,46] is of interest ($T_m = 162\,°C$). A high melting temperature provides good dimensional stability, i.e. less shrinkage, during elevated temperature film conversion processes. As metallocene catalyst technology continues to evolve, it is anticipated that novel iPP products will become available in the area of BOPP film.

Other segments of the BOPP film market where mPP films show potential are shrink film and heat sealable film. In the shrink film market, improved shrinkage performance, low heat sealing temperature and good hot tack properties accompanied by balanced BOPP physical properties are desired. Typical polymers employed in this niche market segment are RCPs with modest levels of ethylene. In the conventional Z–N catalyst system, the distribution of comonomer usually varies with molecular weight. Also, there is a significant amount of atactic polymer, which causes die drool, fuming and plate-out during film extrusion. Opportunities for metallocene RCPs in shrink films relate to their uniform distribution of comonomer and the ability to incorporate HAOs [23,27,49–52]. It is seen from the data in Table 8 that the metallocene resins provide control over a greater range of shrinkage, shrink tension and seal strength. In addition, metallocene RCPs show a favorable balance among the shrinkage level, shrink tension and film physical properties.

In the seal film market, easy sealability and high hot tack strength are desired attributes. RCPs with lower melting temperatures are widely used as seal layer resins. Metallocene RCPs with their uniform comonomer incorporation, low extractables and good organoleptic behavior appear well matched to this opportunity. Product performance data reported to date [34,41,49,50], reinforce this expectation and are encouraging. This can be gauged from Figure 5, where a monolayer cast film comparison is made of the seal properties profile for a metallocene RCP and a 5 wt% ethylene Z–N RCP used widely as a BOPP seal layer resin [34].

Table 8 Comparison of BOPP shrink film properties[a]

Film property	Metallocene RCPs		Z-N RCP[b]
	1.1 wt% C_2	3.4 wt% C_2	4.0 wt% C_2
MFR (dg/min)	3.9	4.0	3.8
$T_m(°C)$	139	127	139
Hexane extractables (%)	0.4	1.4	3.0
Film thickness (μm)	15	27	17
Shrinkage at 135 °C, 180 s (%)	41	65	41
Shrink tension at 110 °C (g force)	125	215	86
WVTR at 37.8 °C, 100 % RH (g/m^2 day per 25.4 μm)	7.1	11.0	10.5
1 % Secant modulus (N/mm^2)	2096	1290	1240
Haze (%)	0.5	0.7	0.4
Gloss at 45°	93	85	92
Seal strength (g/25.4 μm) at 130 °C	50	440	64
135 °C	104	1694	123
140 °C	204	4073	123

[a] Films prepared on T. M. Long biaxial stretching apparatus; Z–N RCP and low C_2 metallocene RCP stretched at 138 °C; high C_2 metallocene RCP at 124 °C.
[b] Example of Ziegler–Natta RCP: PD9302 E1 from Exxon Chemical.

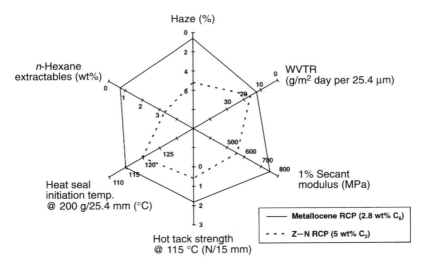

Figure 5 Comparison of heat sealing properties in monolayer cast film

4 NON-ORIENTED FILMS

4.1 NOPP FILM RESINS

For general-purpose NOPP films, isotactic HPPs and RCPs in the MFR range 4–15 dg/min are generally used. HPPs are selected for their stiffness, moisture barrier and higher use temperature, while RCPs are used for their flexibility, clarity, heat sealability and impact strength.

Resins for NOPP films are chosen depending on the specifics of the film fabricating equipment and conditions, and the targeted end-use application(s). Within this framework, the main processing and product requirements are high output, high line speed and draw-down, ability to co-extrude, efficient quench, ease of surface treatment or embossment, good film properties such as clarity, moisture barrier, toughness and machinability and good organoleptic properties.

Both iPP and sPP polymers from metallocene catalyst systems are candidates for NOPP films. Z–N-based iPP has traditionally been used in NOPP applications. There appear to be good opportunities for metallocene iPP to penetrate this market with its favorable product properties and tailoring potential. Syndiotactic PP is a recent entrant and is in a very early stage of development. At this point, the benefits of sPP appear to be primarily as a modifier in blends with iPP, to produce tough, clear films with good impact, tear strength, heat sealability and low coefficient of friction [25,47].

4.2 NOPP FILM PROCESSABILITY

NOPP film can be produced by the chill-roll cast (flat die) process or blown bubble (air- or water-quenched) process. The chill-roll cast process is widely used and produces high-quality film with good clarity and uniform gauge at high production rates. The air-quenched blown process has the advantage of lower capital cost, but it produces hazy and brittle PP film owing to the slow quenching rate and so is rarely used. The water-quenched blown process, employing chilled water to quench a downwardly blown bubble, does produce a clear and glossy film. However, it is less flexible in its capability to produce films of different width and is limited in line speed capability. For these reasons, chill-roll casting dominates the manufacture of NOPP film. Film casting is well suited to high-speed printing and other downstream conversion processes.

The typical casting process comprises polymer extrusion, melt feeding through a slot die, melt draw-down in the air gap, chill-roll casting, edge-trim slitting, surface treating if necessary and winding (Figure 6). Polymer structure development and morphology formation take place in the air gap region and during contact with the chill roll. As the melt leaves the die lip, the flow is mainly elongational and draw-down takes place in the air-gap region between die exit and chill roll. At the air gap, the dimensions of the polymer melt curtain become reduced laterally, a phenomenon

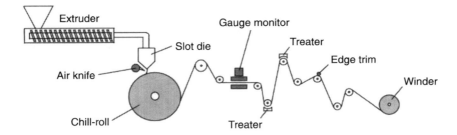

Figure 6 Schematic diagram of chill roll casting proces

termed edge neck-in. The edge neck-in causes an undesirable tear drop-shaped edge bead, the edges being thicker than the rest of the melt curtain. Edge trimming is therefore a necessary step to eliminate winding difficulties due to the uneven film thickness. It is a common practice to utilize quenching air in the air gap region (delivered by an air knife) to provide firm contact of the melt curtain on to the chill roll and at the same time minimize the edge neck-in.

With the evolutionary development of faster, more robust winding technology, cast film line speeds have consistently increased in recent years. This has led to improvements in productivity and manufacturing economics [53]. In this highly competitive segment of the film market, a versatile resin capable of being processed at high line speeds, drawn-down to a thin and uniform web, efficiently quenched to a clear film and with a good profile of film properties is very desirable. The key concern in processing standard single-site metallocene polymer at high output is its lower shear sensitivity and therefore higher melt viscosity during film extrusion, compared with broad-MWD Z–N iPP of similar MFR. This results in higher extruder torque and creates more energy transfer to the polymer melt by shear heating, which gives rise to excessive melt temperatures and consequent flow instability problems. These extrusion challenges may be alleviated by using higher MFR metallocene iPPs to compensate for the lower shear sensitivity. Another potentially useful approach is tailoring of the MWD, aiming to reduce melt viscosity at high shear rates while maintaining sufficient elasticity for high draw-down to a uniform web thickness.

Table 9(a) compares the processabilities of four different types of metallocene and Z–N iPPs using a monolayer casting process. The metallocene iPP (at 24 MFR) exhibits the highest specific output, the lowest head pressure and the lowest shear heating (extrusion temperature increase from a set-point) among all samples run in the extrusion study. This should be of benefit in those extrusion situations that can handle higher extruder throughput. The tailored metallocene iPP is seen to require a lower motor load for screw rotation than the metallocene iPP, despite the fact that the tailored metallocene iPP has a lower MFR (12 vs 24 dg/min). This enables an existing extruder with low motor load capability to run metallocene resins

Table 9   Comparison of cast film processability[a]

(a) Extrusion

Processing performance[b]	Metallocene 24 MFR	Tailored metallocene 12 MFR	Z–N HPP 7 MFR	Z–N RCP 7 MFR/2.8 wt% C_2
Head pressure (N/mm^2)	15	18	22	22
Melt temperature (set/actual) (°C)	232/242	232/249	232/257	232/256
Specific output (kg/rpm/h)	1.8	1.5	1.4	1.5
Motor load (A)	102	91	107	107
Edge neck-in (cm)	32	31	31	40

(b) Draw resonance tendency

Processing performance[c]	Metallocene 24 MFR		Z–N HPP 12 MFR	Z–N RCP 7 MFR/2.8 wt% C_2
Melt temperature (set) (°C)	216	232	216	232
Onset of draw resonance (m/min)	70	160+	58	82
	(36 μm)	(23 μm)	(43 μm)	(41 μm)
Specific output (kg/rpm/h)	1.6	1.4	1.3	1.3

[a] Examples of Ziegler–Natta polymers: PD 4443 (7 MFR HPP), PD 3284 (12 MFR HPP) and PP 9513 E2 (7 MFR RCP) from Exxon Chemical.
[b] Using 90 mm extruder, 107 cm die at 0.64 mm die gap, 193 kg/h output, 46 draw-down ratio, 244 m/min line speed and 27 °C chill roll.
[c] Using 90 mm extruder, 107 cm die at 0.64 mm die gap, 153 rpm screw speed and 27 °C chill roll.

successfully at high output. In general, the data in Table 9(a) show both metallocene polymer types to display favorable cast film processability. This is in line with the positive marketplace experience of metallocene ethylene polymers in cast film.

Analysis of edge neck-in behavior [Table 9(a)] was performed by measuring the total width reduction of the melt curtain from the die exit to the chill roll. During film extrusion, the air knife pressure is usually adjusted to ensure a stable melt curtain that provides firm chill roll contact and minimum edge neck-in. The maximum air knife pressure that can be applied on to the melt curtain is limited by the onset of melt instability due to air impingement on the flowing melt. Therefore, for each material, there is an optimal range of air knife pressures to achieve the best balance of gauge uniformity, film clarity and stable web dimensions. For Z–N iPPs, it is generally more difficult to process high-MFR, low melt strength materials for cast film, owing to the narrow range of usable air knife pressures and the greater sensitivity to air disturbance. It is seen from the data that both the metallocene iPPs, despite their higher MFRs, have the same degree of edge neck-in as the lower-MFR Z–N HPP.

Another important processing criterion for a good cast film resin is its capability to withstand high line speed and high draw-down without the problems of gauge variation or draw resonance. The draw resonance phenomenon, periodic gauge variation along the web length, is a function of draw-down ratio, aspect ratio of the melt curtain (the ratio of air-gap length to the die width) and polymer viscoelasticity [54]. For high draw-down processes, an increase of melt elasticity will normally help to stabilize the process and minimize the tendency for draw resonance. A comparison of line speed capability can be made by maintaining the same extrusion conditions for each material and progressively increasing line speeds to the maximum limit before the onset of draw resonance. The data in Table 9(b) indicate that the metallocene iPP has significant processing advantages over the Z–N polymers. Specifically, line speed and resin throughput could be substantially increased, before the onset of draw resonance. This is particularly evident at the melt temperature of 232 °C. In addition, the metallocene iPP was drawn to a thinner film gauge: 23 versus 41 µm for the Z–N polymer.

In addition to the above-mentioned processing advantages, an improvement in manufacturing efficiency, through reductions in die drool, undesired fuming and plate-out, can also be anticipated from the superior cleanliness of metallocene polymers. It is apparent that processing advantages can give metallocene iPP a competitive edge in the high line speed, cast film market.

4.3 NOPP FILM PROPERTIES

In this section, NOPP film property comparisons of iPP and sPP with conventional Z–N iPPs will be discussed. It is these film properties that bring value to end-use performance. As would be expected from metallocene iPPs, the stiffness and clarity should be greatly improved and these benefits have been observed [32,46]. Metallocene iPPs show good maintenance of clarity and gloss during high line speed processing, which is a desirable attribute. At high line speeds, a deterioration of clarity is often encountered owing to the shorter quench time for the extrudate and the less firm contact of the melt curtain on the chill roll.

As shown in Table 10, metallocene HPP exhibits excellent film clarity and gloss, even better than those of Z–N RCPs. Z–N RCPs are often used for applications requiring high film clarity. The narrower MWD and tacticity distribution of the single-site metallocene iPP results in films having a more uniform distribution of crystal sizes, which reduces the surface roughness and diminishes light scattering from the film surfaces. As a consequence, the film clarity and gloss are excellent.

In addition to favorable optical properties, metallocene iPPs feature advantages in film moisture barrier, stiffness, tensile, tear, puncture, dart impact and heat seal properties. The film combines good stiffness, tensile, puncture and moisture barrier of Z–N homopolymer with good clarity, tear, impact and heat seal of Z–N RCP. For applications with more stringent requirements on heat seal temperature, hot tack strength and impact resistance, metallocene iPP offers the efficient incorporation of ethylene or HAO comonomers to attain these advantages [34,44]. Furthermore, novel film properties, including impact, tear and heat sealability, as shown in Table 11, can be obtained from blends of metallocene HAO RCP (1-hexene) and Z–N ethylene RCP [26].

Table 10 Comparison of isotactic homopolymer and random copolymer cast film properties[a]

Film property	Metallocene 9 MFR	Z–N HPP 7 MFR	Z–N RCP 7 MFR/2.8 wt% C_2
Thickness (μm)	43	41	41
Haze (%)/gloss	1.3/88	4.1/78	1.6/86
Heat seal temp. at 10 N/15 mm (°C)	143	150	137
WVTR (g/m^{2}day per 25.4 μm)	12.7	13.8	16
1 % Secant modulus (N/mm^2) MD/TD	846/787	783/768	555/516
Tensile strength at yield (N/mm^2) MD/TD	24/24	22/22	16/15
Elongation at yield (%) MD/TD	3.9/4.0	3.9/3.9	4.1/3.9
Tensile strength at break (N/mm^2) MD/TD	68/60	68/50	59/47
Elongation at break (%) MD/TD	683/746	678/718	651/721
Puncture resistance (kg/25.4 μm)	2.0	2.1	2.1
Dart impact at 23 °C (g force/25.4 μm)	153	108	224
Elmendorf tear strength (g force/25.4 μm) MD/TD	46/66	31/78	33/94

[a] Examples of Ziegler–Natta polymers: PD 4443 (7 MFR HPP) and PP 9513 E2 (7 MFR RCP) from Exxon Chemical.

Table 11 Comparison of isotactic random copolymer cast film properties[a]

Film property	Metallocene RCP 4 MFR/ 2.8 wt% C_6	50:50 blend[b]	Z–N RCP 5 MFR/ 5 wt% C_2
Thickness (μm)	36	45	51
Haze (%)/gloss	0.6/76	0.3/73	5.2/67
Heat seal temp. at 10 N/15 mm (°C)	119	119	124
1 % Secant modulus (N/mm^2) MD/TD	738/765	467/447	556/562
Tensile strength at yield (N/mm^2) MD/TD	24/24	17/17	19/19
Tensile strength at break (N/mm^2) MD/TD	63/48	59/48	52/44
Elongation at break (%) MD/TD	535/635	595/630	635/740
Dart impact at 23 °C (g force/25.4 μm)	51	393	60
Elmendorf tear strength (g force/25.4 μm) MD/TD	26/54	35/149	15/43

[a] Example of Ziegler–Natta RCP: PD9282 E2 from Exxon Chemical.
[b] 50:50 blend of metallocene RCP (4 MFR/2.8 wt% C_6) and Z–N RCP (5 MFR/5 wt% C_2).

In addition to iPPs, metallocene catalysts also make it commercially practical to produce sPPs. Syndiotactic PPs obtained today by metallocene catalysis exhibit low melting temperatures, much lower stiffness and slower crystallization than their isotactic counterparts. The current use of sPP in NOPP film appears to be primarily as a modifier in blends with iPP. As seen from the data in Table 12(a) and (b), the addition of sPP to an isotactic Z–N RCP improves film clarity, tear and impact strength, but at the expense of stiffness and tensile properties. Owing to this limitation of film mechanical properties, today's sPP is expected to be used in NOPP for specialty applications where its attributes provide advantages.

Table 12 Comparison of isotactic/syndiotactic blend PP cast film properties [25][a]

(a) Improved impact and high clarity

Film property	Z–N isotactic HPP	Z–N RCP 2 % C_2	10 % ULDPE– 90 % Z–N RCP	10 % sPP– 90 % Z–N RCP
Thickness (mm)	0.1	0.1	0.1	0.1
Haze (%)/gloss	12.2/51	1.9/81	43/21	0.7/85
Tensile strength at yield (N/mm^2) MD/TD	18.4/18.6	15.3/15.1	13.2/12.5	13.6/12.9
Elongation at yield (%) MD/TD	11.9/10.6	15.0/13.6	14.6/13.7	15.6/15.3
Elmendorf tear strength (g force) MD	190	170	430	410
Dart impact at 1.52 m (g force)	140	158	236	189

(b) Improved COF and heat sealability

Film property	Z–N isotactic RCP	90 % RCP– 10 % sPP	80 % RCP– 20 % sPP	70 % RCP– 30 % sPP
Thickness (mm)	0.03	0.03	0.03	0.03
MFR (dg/min)	7.1	7.1	7.7	7.7
COF at 23 °C, 3 days	0.34	0.09	0.09	0.07
35 °C, 1 day	0.38	0.11	0.09	0.09
35 °C, 3 days	0.32	0.11	0.09	0.07
Heat seal temp. at 300 g/25 mm (°C)	132.3	132.3	128.6	125.0
500 g/25 mm (°C)	135.8	135.8	131.7	129.3
Tensile strength at yield (N/mm^2)	20.1	18.8	17.4	16.7
Tensile modulus (N/mm^2)	680	610	560	520

[a] Example of Ziegler–Natta iPP: 3576X (HPP) from Fina Oil and Chemical. ULDPE: ultra-low-density polyethylene at 0.895 g/cm^3.

Although still in the early stages of development, metallocene PPs have already demonstrated some promising results in NOPP film applications. It is anticipated that further optimization of metallocene catalyst technology will take NOPP processability and film property performance to a level beyond current limits set by Z–N PPs, opening up new opportunities for NOPP films.

5 CONCLUSION AND OUTLOOK

Metallocene-based propylene polymers for film applications are in their infancy today. Catalyst technology, the backbone to being able to provide differentiated products, is still being developed and will continue to be the focus of considerable effort. The products of tomorrow will likely be polymers designed for specific end-use applications. An outlook on overall potential application opportunities for metallocene-based propylene polymers can be found in an article by Sinclair [55], while current projections on the penetration and growth of these polymers in the marketplace can be found in studies by Chem Systems [1,56], among other database sources. Their assessment appears to be one of limited presence in the marketplace until after the year 2000. Most of the demand is expected to be in the regions of North America and Western Europe, having large markets for disposable and specialty items.

In assessing the outlook for mPP, the performance of these polymers in general-purpose BOPP films, NOPP cast films and seal layer films are reviewed. It is recognized that, as in the case of metallocene ethylene polymers, the different balance of properties inherent in these products will bring forward additional film opportunities heretofore untapped. At this point, the status today (based primarily on efforts, referenced in the literature, on developmental products) is compared against the main end-use property requirements for each of these applications in Table 13.

Table 13(a) outlines the performance of mPP in BOPP film. The data indicate a fair match with the requirements of the application. In particular, there appears to be good potential for easier processing candidates, as has been highlighted previously. The lower melting temperatures typically observed for today's metallocene HPPs from commercially viable supported catalysts (5–10 °C lower than for Z–N-based HPP), however, contribute to higher levels of film shrinkage at elevated temperatures, which are undesirable. This shortcoming is well recognized, and the development and introduction of higher crystallinity, higher melting products is anticipated. Such products, constituting the metallocene equivalent of so-called 'high-crystallinity PP', should offer less shrinkage and increased stiffness and moisture barrier in this application, to go along with the projected easier processing advantage.

In Table 13(b), NOPP films, both HPPs and RCPs, with relatively low levels of comonomer are addressed. The match-up of mPP performance with application requirements appears strong. HPP films show high stiffness, excellent clarity and sparkle and easy sealability. The narrow MWD allows high drawability to thin films, with a reduced tendency for draw resonance. A 'high-crystallinity' version, when developed, would afford even higher film stiffness and downgauging potential.

Table 13(c) assesses seal layer films. The balance of properties displayed by today's narrowly distributed metallocene-based products (lower extractables, more efficient use of comonomer to depress crystallinity), is encouraging, although seal initiation temperature and associated seal strength are defensive to top-of-the-line Z–N terpolymer sealants available today. As previously mentioned, the access to HAO-based RCPs affords a new dimension to use in mPP polymer design for seal layer films [23,27].

In terms of potential new film opportunities for metallocene polymers, two building blocks are highlighted for attention: the development of novel RCPs and the development of sPP. The capability of metallocene catalysts to incorporate a wide array of comonomers in addition to the α-olefins (e.g. aromatics, cyclic structures) is

Table 13 Assessment of fit of today's mPP in selected film markets[a]

(a) Biaxially oriented film for general-purpose packaging (typical construction: homopolymer core with heat seal layers on one or both sides, e.g. 1/20/1 μm)

Film requirement[b]	Productivity				Optical		
	Output	Line speed	Stiffness	WVTR	Haze	Gloss	Shrinkage
mPP fit	O	O	O	O	O	O	–

(b) Non-oriented film for general-purpose packaging (typical construction: monolayer homopolymer or random copolymer of 3 wt% C_2, e.g. 30 μm)

Film requirement[c]	Productivity		Optical					Sealability	
	Output	Line speed	Haze	Gloss	Stiffness	WVTR	Impact strength	Seal initiation temp.	Seal strength
mPP fit	+	+	+	+	O	O	O	+	+

(c) Seal layer film (typical construction: homopolymer core with random copolymer skins on one or both sides, e.g. 1/20/1 μm)

Film requirement[d]	Sealability				Optical		
	Seal initiation temp.	Seal strength	Hot Tack	Extractables	Haze	Gloss	Blocking
mPP fit	–	–	–	+	+	+	+

[a] Assessment of performance based on today's best Ziegler–Natta polypropylene in this application; 'O', Control' property value, assigned neutral; '+', property benefit; '–', property debit.
[b] Film properties refer to the 20 μm core layer produced on typical tenter frame line.
[c] Film properties refer to 30 μm cast film.
[d] Film properties refer to seal layer(s).

well known. An example, in metallocene ethylene polymers, is ethylene–styrene copolymer, a new entrant in the field [57]. It displays elastomeric film behavior, with good recovery after extension. Similar types of new polymers, covering the range from semi-crystalline to elastomeric, are also possible based on propylene. These polymers are expected to show unique balances of toughness, stiffness, clarity, low extractables and thermal properties (e.g. seal) that will considerably broaden the properties envelope afforded by today's RCPs. Application in medical and health care film could potentially be commercial opportunities. As regards sPP, the potential is harder to gauge. Today's commercial offerings are of relatively low crystallinity (syndiotactic pentad fraction only about 0.8) and have been targeted primarily in blends with iPP and other polyolefins, to impart clarity and toughness benefits. The properties of higher crystallinity polymers (syndiotactic pentad fraction $\geqslant 0.95$) are likely to be completely different from today's offerings. Syndiotactic PP improvements, as they become available, will be followed with great interest. The opportunities are properties and property balances not seen today.

The entry of metallocene products into PP film is at the stage where potential in the various market segments is only now being assessed. The necessary building blocks to make viable commercial products are still under development. Once these are in place, penetration should occur in selected film areas (e.g. cast NOPP film). Beyond this, the growth of metallocene products is likely to proceed at a measured pace, displacing conventional Z–N polymers or creating new opportunities when differentiable products that offer added value to the customer are developed.

6 REFERENCES

1. *The Global Polyolefins Industry: 1996, A Period of Structural and Technological Change*, Chem Systems, Tarrytown, NY, 1996.
2. Horton, A. D., *Trends Polym. Sci.*, **2**, 158 (1994).
3. Kaminsky, W., *Macromol. Chem. Phys.*, **197**, 3907 (1996).
4. Olabisi, O., Atiqullah, M. and Kaminsky, W., *J. Macromol. Sci., Rev. Macromol. Chem. Phys.*, **C37**, 519 (1997).
5. McAlpin, J. J. and Stahl, G. A., in *MetCon '94 Proceedings*, Houston, TX, 1994.
6. Arnold, M., Henschke, O. and Knorr, J., *Macromol. Chem. Phys.*, **197**, 563 (1996).
7. Ewen, J. A., *J. Am. Chem. Soc.*, **110**, 6255 (1988).
8. Fong, W. S., *Syndiotactic Polypropylene*, Report No. 128B, SRI International, Menlo Park, CA, 1995.
9. Moore, E. P., *Polypropylene Handbook*, Hanser, New York, 1996.
10. Rotman, D., *Chem. Week*, **152**, 7 (1993).
11. Donohue, J., in *SPO '95 Proceedings*, Houston, TX, 1995, p. 59.
12. Donohue, J., in *MetCon '96 Proceedings*, Houston, TX, 1996.
13. Shamshoum, E. S., Kim, S., Sun, L., Paiz, R., Goins, M. and Bartol, D., in *SPO '93 Proceedings*, Houston, TX 1993, p. 207.
14. Shamshoum, E. S., Sun, L., Reddy, B. R. and Turner, D., in *MetCon '94 Proceedings*, Houston, TX, 1994.
15. Wheat, W. R., in *SPE ANTEC '97*, 1997, p. 1968.

16. Galambos, A., Wolkowicz, M. and Ziegler, R., *ACS Symp. Ser.*, **496**, 104 (1991).
17. Balbontin, G., Dainelli, D., Galimberti, M. and Paganetto, G., *Macromol. Chem. Phys.*, **193**, 693 (1992).
18. Resconi, L., Jones, R. L., Albizzati, E., Camurati, I., Piemontesi, F., Guglielmi, F. and Balbontin, G., *ACS Polym. Prepr.*, **35**, 663 (1994).
19. Silvestri, R., Resconi, L. and Pelliconi, A., in *Metallocenes '95 Proceedings*, Brussels, 1995, p. 207.
20. Resconi, L., Piemontesi, F. and Jones, R. L., in *SPE Polyolefins X Conference Proceedings*, Houston, TX, 1997, p. 71.
21. Galimberti, M., Martini, E., Sartori, F. and Albizzati, E. in *MetCon '94 Proceedings*, Houston, TX, 1994.
22. Speca, A. N. and McAlpin, J. J., in *SPE Polyolefins X Conference Proceedings*, Houston, TX, 1997, p. 33.
23. McAlpin, J. J., Chen, M. C. and Mehta, A. K., in *SPO '96 Proceedings*, Houston, TX, 1996, p. 429.
24. Wheat, W. R., in *SPE ANTEC*, 1995, p. 2275.
25. Schardl, J., Sun, L., Kimura, S. and Sugimoto, R., *Plast. Film Sheeting*, **12**, 157 (1996).
26. Chen, M. C. and Mehta, A. K., *PCT Int. Appl.*, WO 97 10300 (1996).
27. Mehta, A. K., Chen, M. C. and McAlpin, J. J., in *SPE Polyolefins X Conference Proceedings*, Houston, TX, 1997, p. 417.
28. Bond, E. B. and Spruiell, J. E., TANDEC 6th Annual Conference, University of Tennessee, Knoxville, TN, 1996.
29. Osawa, Z., Kato, M. and Terano, M., *Macromol. Rapid Commun.*, **18**, 667 (1997).
30. Maier, R-D., Thomann, R., Kressler, J., Mühlhaupt, R. and Rudolf, B., *J. Polym. Sci., Phys.*, **35**, 1135 (1997).
31. Lohse, D. J. and Wissler, G. E., *J. Mater. Sci.*, **26**, 743 (1991).
32. Lin, C. Y., Chen, M. C., Kuo, J. W. C. and Mehta, A. K., in *SPO '97 Proceedings*, Houston, TX, 1997, p. 203.
33. Winter, A., Kelkheim, D. and Spaleck, W., *US, Pat.*, 5 350 817 (1994).
34. McAlpin, J. J., Mehta, A. K., Plank, D. A. and Stahl, G. A., in *SPO '95 Proceedings*, Houston, TX, 1995, p. 125.
35. Seelert, S., Langhauser, F., Kerth, J., Müller, P., Fischer, D. and Schweier, G., *US, Pat.*, 5 483 002 (1996).
36. Gabriele, M. C., *Mod. Plast.* **74** (7), 82 (1997).
37. Colvin, R., *Mod. Plasti.*, **73** (3), 26 (1996).
38. Peiffer, H., Busch, D., Dries, T., Schlögl, G. and Winter, A., *Eur. Pat.*, EP 747 211 (1996).
39. Peiffer, H., Busch, D., Dries, T., Schlögl, G. and Winter, A., *Eur. Pat.*, EP 747 212 (1996).
40. Bidell, W., Hingmann, R., Jones, P., Langhauser, F., Marczinke, B., Müller, P. and Fischer, D., in *Metallocene '96 Proceedings*, Düsseldorf, 1996.
41. Bidell, W., Fischer, D., Hingmann, R., Jones, P., Langhauser, F., Gregorius, H. and Marczinke, B., in *MetCon '96 Proceedings*, Houston, TX, 1996.
42. Jones, P. J. V., Hingmann, R., Langhauser, F., Marczinke, B. L., Horton, M., Fischer, D. and Schweier, G., *PCT Int. Appl.*, WO 97 19980 (1997).
43. McAlpin, J. J., Kuo, J. W. C. and Hylton, D. C., *US Pat.*, 5 468 440 (1997).
44. Saito J. and Ushioda, T., in *Metallocenes Asia '97 Proceedings*, Singapore, 1997.
45. Saito, J., Kawamoto, H., Kageyama, T., Hatada, K., Oki, Y. and Tanaka, S., *Jpn. Pat.*, JP 948 858 (1997).
46. Ushioda, T., Fujita, H. and Saito, J., in *SPO '97 Proceedings*, Houston, TX, 1997, p. 101.
47. Shamshoum E. S. and Schardl, J., in *SPE Polyolefins X Conference Proceedings*, Houston, TX, 1997, p. 31.
48. Wheat, W. R. and Hanyu, A., in *SPO '97 Proceedings*, Houston, TX, 1997, p. 193.

49. Dries, T., Spaleck, W. and Winter, A., *Eur. Pat.*, EP 668 157 (1995).
50. Winter, A., in *SPO '97 Proceedings*, Houston, TX, 1997.
51. Chen, M. C. and Mehta, A. K., *PCT Int. Appl.*, WO 97 11115 (1997).
52. Fischer, D., Bidell, W., Grasmeder, J., Hingmann, R., Jones, P., Kersting, M., Langhauser, F., Marczinke, B., Moll, U., Rauschenberger, V., Süling, C. and Popham, N., in *SPO '97 Proceedings*, Houston, TX, 1997.
53. Gabriele, M. C., *Mod. Plast.*, **73** (9), 62 (1996).
54. Silagy, D., Demay, Y. and Agassant, J.-F., *Polym. Eng. Sci.*, **36**, 2614 (1996).
55. Sinclair, K. B., *Metallocene Technology '97*, Chicago, IL, 1997.
56. *Impact of New Technology on the Global Polyolefins Business*, Chem Systems, Tarrytown, NY, 1994.
57. Hoenig, S., Turley, R. and Van Volkenburgh, W., in *SPO '97 Proceedings*, Houston, TX, 1996, p. 263.

22

Medical Applications of Metallocene-catalyzed Polyolefins

ROBERT C. PORTNOY AND JOSEPH D. DOMINE
Exxon Chemical Company, Baytown, TX, USA

1 INTRODUCTION

Since the commercial introduction of metallocene-catalyzed, ultra-low-density ethylene polymers (plastomers) by Exxon Chemical Company in 1991, the list of available metallocene-catalyzed polyolefins has grown to include also linear low-density polyethylene and isotactic and syndiotactic propylene homopolymers. From the beginning these novel materials have intrigued those who select construction materials for medical devices and for the packaging of medical devices and solutions and drugs. Principally because of their narrow molecular weight and composition distributions (including comonomer composition distributions in all of the polymers and also the distributions of stereo- and regio-defects in the propylene polymers), the new materials have promised unique and heretofore unknown combinations of properties which could be brought to bear for the solution of design and manufacturing problems in this very specialized industry. Many of the applications for which these materials have been studied are described below. This overview includes successful commercial applications, areas of interest which are under study and development and proposals for additional uses of the metallocene-catalyzed polymers, which are suggested by a thorough understanding of their properties. The information is largely organized by application type; the different polymers are discussed separately or together as appropriate for a particular use. There is, of

Metallocene-based Polyolefins Edited by J. Scheirs and W. Kaminsky
© 2000 John Wiley & Sons Ltd

course, considerable overlap between the applications in the material technology employed, but a conscientious attempt has been made to avoid redundancy in descriptions of the technical aspects of the materials. Repeated mention of similar applications, but approached by different material strategies, has been unavoidable.

2 PLASTICIZED PVC ALTERNATIVE

2.1 KEY POLYMER ATTRIBUTES

The first commercially available (1991) metallocene-catalyzed ethylene polymers were positioned between traditional linear low-density polyethylene (LLDPE) plastics and elastomers such as ethylene–propylene rubber/ethylene–propylene–diene monomer rubber (EPR/EPDM). Generically, they were referred to as plastomers because they possessed many of the characteristics of both plastics and elastomers [1]. Some of the key attributes of plastomers which make them of interest as alternatives to plasticized PVC (pPVC) in medical applications are:

- clarity, especially in thick sections (>250 μm);
- softness/low modulus/flexibility;
- superior low-temperature toughness;
- radiation sterilizability;
- competitive raw material costs on a density-adjusted basis;
- low level of extractable material and no extractable plasticizers;
- USP Class VI, tripartite, ISO 10993 compliance;
- ready disposability by incineration.

Plastomer haze levels, measured on nominal 1 mm thick plaques, range from as low as about 8 % at 0.90 density down to 5 % and less at 0.88 density and lower. Hence, plastomers can be considered for applications requiring high clarity.

Product developers have available to them plastomers with flexural modulus values down to about 800 psi (5.5 MPa) and with Shore A hardnesses down to about 60, in the range of all but the softest pPVC compounds.

Plastomers have significantly better low-temperature impact strength than pPVC. The low-temperature impact performance of pPVC is dependent on both the amount and type of plasticizers used in their formulation. Brittle/ductile transition temperatures [2], below which the impact energy is essentially zero, are typically about -15 to $-35\,°C$. For plastomers, on the other hand, brittle/ductile transition temperatures ranging from -60 to $-70\,°C$ have been reported by Woo and co-workers [3]. In addition, they found in the same study that the temperature above which falling dart impact failure was zero was about 0–$5\,°C$ for the pPVC films and -35 to $-45\,°C$ for the plastomers.

Plastomers have a high level of radiation tolerance (Table 1). They do not degrade and usually do not discolor after the typical 25 kGy of sterilizing gamma radiation. In fact, higher levels of radiation cause cross-linking, resulting in increased heat resistance and strength without any increase in stiffness or hardness. Thus, with plastomers, radiation can actually improve performance.

With densities in the range from 0.90 down to less than $0.87 \, g/cm^3$, plastomers offer about a 40–45 % increase in yield of parts per unit weight compared to pPVC with a density of $1.25 \, g/cm^3$. For many applications this more than offsets any difference in price per unit weight.

In contrast with pPVC, plastomers do not contain any plasticizers which could migrate into the liquids they hold or transport. Their aliphatic hydrocarbon structure and low level of total extractables are the reasons why plastomers with densities even as low as $0.865 \, g/cm^3$ have been found to be compliant with USP Class VI requirements [4].

As is the case with any polyolefin, plastomers can be readily disposed of by incineration. When pPVC is incinerated, it generates several undesirable by-products such as chlorinated dibenzofurans, chlorinated dibenzodioxins and hydrogen chloride [5]. Since plastomers are not made from chlorine-containing monomers, none of these by-products are produced when they are burned [6].

Plastomers offer product developers the first cost-competitive polyolefin with the required level of softness and clarity to constitute an alternative to pPVC in many health care applications. The applications and potential uses of plastomers discussed below will usually require a combination of two or more of these key attributes.

As in any material replacement, however, there are some tradeoffs. In the case of plastomers, these tradeoffs have included differences in bonding and welding techniques, heat resistance and strength.

Plasticized PVC is readily bonded to itself and some other polymers with simple, room temperature solvent welding and/or radiofrequency (r.f.) welding. Plastomers will not bond to themselves or other polymers with either of these techniques. They are, however, readily heat sealed using equipment designed for traditional PE which can also be used to bond plastomers to some other polymers, especially to other polyolefins. Bonding and welding of plastomers have been a challenge for most fabricators of components and devices who have an installed base of assembly lines

Table 1 Effect of high levels of gamma radiation

Property	Nominal dose (kGy)			
	0	50	100	150
Melt index (dg/min)	2.7	0.05	–	–
Vicat SP (°C)	45.3	48.1	48.4	49.7
Compression set (% after 24 h at 60 °C)	81.1	68.7	66.7	64.6

Plastomer: 3MI, 0.878 density, ethylene–butene copolymer.

for pPVC based parts which use solvent and/or r.f. welding. While no simple solvent welding technique, which works well with plastomers, is available, liquid adhesive bonding techniques have been developed, including UV-curable adhesives [7]. In addition, technology has been developed which allows fabricators to bond plastomer films and fittings using r.f. welding [8].

PVC contains about 10–15 % crystallinity and has a peak melting-point of about 420 °F (215 °C); in addition, PVC retains much of its crystallinity even when heavily plasticized [9]. In contrast, plastomers have less crystallinity and significantly lower peak melting points. This results in differences in heat resistance between plastomers, especially the lower density, lower modulus versions, and most pPVC compounds, leading to more deformation set unless adequate precautions are taken. When components and articles made from these lower density plastomers are to be exposed for one or more extended periods to temperatures in excess of 120 °F (49 °C) they should be packaged or held in an undeformed condition. Parts which show excellent elastic recovery from short-term deformations at room temperature will assume the deformed conformation if held in that condition for an extended period at an elevated temperature. In addition, the lower density, lower modulus plastomers are not steam sterilizable whereas many pPVC compounds are.

Plastomers, especially the lower density, lower modulus versions, frequently do not yield tubing with the same level of kink and rekink resistance that pPVC based tubing has. For most applications, this is not an issue. For some applications, including some IV tube applications, plastomers have been modified with other polymers to increase their kink and rekink resistance [10,11]. In addition to yielding tubing with improved kink performance, the blends described by these workers are expected to also have better melt processability, including higher melt strength for dimensional stability of the extrudate and the potential for being processed at higher speed because of their reduced sensitivity to melt fracture.

In the following examples some of the plastomer-based pPVC alternatives described are blends in which the plastomer is functioning as a modifier for another polymer or is itself modified by it. The emphasis of these discussions is, however, on the utility of these materials as pPVC alternatives. More detail on the role of metallocene-catalyzed polyolefins in these blends is given in Section 5. Some additional, incidental description of the use of the blends as pPVC alternatives is also provided there.

2.2 INJECTION MOLDED PARTS

Plastomers with melt indices in the range from 4 to over 30 dg/min are usually suitable for injection molding into a variety of medical device components ranging from small parts such as caps, closures, fittings and connectors to larger articles such as face masks [12]. For many of these applications, clarity is essential. Plastomers with densities below about 0.85 g/cm^3 are usually clear enough even in thick sections to meet the requirements for the product. In order to mask the yellow tinge which results from gamma radiation sterilization [13,14] or melt processing, pPVC is usually tinted a light green or violet. Plastomers, however, do not discolor after gamma radiation and can be molded either colorless ('water white') or tinted to look like their pPVC counterparts. Face masks molded from high-clarity plastomers permit observation of fluids expelled into the mask by the patient.

Components molded from these plastomers can be assembled into devices or articles ready for use, then packaged and either sterilized by gamma radiation, which penetrates the package, or by ethylene oxide (EtO), which diffuses through special packaging designed for this purpose. Gamma radiation sterilization appears to be growing at the expense of EtO sterilization because it is fast and effective and there is no residual gas which has to be expelled from the packages.

Plastomers are non-corrosive and therefore neither the molding machines nor the molds designed for plastomer components need to be protected from corrosion. Mold shrinkage of low density plastomers can range from as low as about 0.5 % to about 1.5 % [15]; therefore, molds need to be designed with generous draft angles to facilitate part release. Plastomer parts also release more easily from molds with honed surfaces, reducing the desirability of highly polished mold surfaces except in those regions of the part requiring the highest level of clarity. Because the modulus and strength of high-clarity plastomers are low, especially at the elevated temperatures expected at the time of part ejection, knockout pins and other ejection mechanisms should be larger than those which are used with LDPE in order to contact a larger area of the part; this reduces the deformation of the part by the pins and the potential for knockout pins to punch through the parts.

2.3   EXTRUDED FILM

Plastomers, like most other ethylene polymers and/or copolymers, can be extruded into film using conventional fabrication equipment. Both blown and cast films are available and are used in a wide range of applications. Plastomers, especially the narrow molecular weight distribution (MWD), linear versions, are more difficult to extrude than conventional LLDPE and LDPE, requiring higher torque and horsepower, generating higher head pressure and melt temperature, developing sharkskin melt fracture at lower output rates, and exhibiting more bubble instability. However, equipment designed to handle LLDPE efficiently will usually not have any problem even with the narrow MWD, linear plastomers [16].

For a variety of both technical and economic reasons, plastomers are usually not extruded into or used as neat monolayer films. As a result, many of the processing issues associated with their narrow MWD and linear structure are obviated because the plastomer is either co-extruded with other polymers which have better bubble stability, etc., and/or is extruded into a film as a blend with other polymers. In addition, polymer processing aids are available to allow conversion at commercially attractive output rates without sharkskin melt fracture [17].

The largest volume film applications are in packaging where plastomers offer superior toughness (especially at sub-ambient conditions), excellent heat sealability, high hot tack strength, high cling for improved stretch wrap performance and a favorable balance of moisture barrier and gas barrier compared with other low-melting polymers such as EVA. As stated before, the plastomers are almost invariably used in blends or in multilayer films.

Plastomer films, however, are used in both packaging and non-packaging medical applications. Packaging applications include both device packaging and drug or solution packaging such as IV bags. The focus of this section, however, is on medical applications where plastomers are being used as alternatives to pPVC and not on conventional packaging where plastomers are either competing with or used in conjunction with other polyolefins.

For IV bags plastomers are used in blends and/or co-extrusions with other polyolefins, usually polypropylene (PP). The PP is needed to provide the burst strength for the bag, especially for systems which are steam sterilized, and to improve the barrier properties of the film. Plastomers are sometimes blended into the PP to improve the toughness, especially at sub-ambient temperatures which the bag might experience from storage either in a refrigerator or freezer. For multilayer films PP random copolymer modified with plastomers can be used as the heat seal layer. Because the sealing of IV bags is critical, IV bags made from polyolefins are usually fabricated on a line designed specifically to handle and seal these materials and not on a line which typically is used to make pPVC IV bags.

For this application, high clarity of the film is essential. Fortunately, the plastomers which appear to have the optimum balance of improving the toughness of the PP without significantly reducing the strength and stiffness are those around $0.90\,g/cm^3$ density; blends of these plastomers with PP do not adversely affect the clarity as other types of impact modifiers do. Thus films made from blends of PP and nominally $0.90\,g/cm^3$ density plastomer are usually as clear as the film made from unmodified PP, even for propylene RCPs [18].

In comparison with pPVC plastomer films have lower water vapor transmission rates and higher oxygen transmission rates. In the 0.895–$0.905\,g/cm^3$ density range, plastomer WVTRs are about one fifth those of typical medical grade pPVC; oxygen permeability of the plastomer is about four times higher [19].

Non-packaging applications for extruded plastomer film include collection bags and medical devices. Films extruded for these applications are intentionally made hazy to obscure the contents of the bags; they have sufficient contact clarity to allow one to determine whether the bag is empty. For in-hospital use, plastomer film-based colostomy and urine collection bags are readily disposed of by incineration. For colostomy bags worn by ambulatory patients under their clothing, bags made from plastomer films offer the softness, quietness and feel of pPVC, with slightly less weight.

Extruded plastomer films can also be used as alternatives to pPVC in medical devices. A non-PVC based, sequential compression device which uses plastomer based film to form the air chambers has been recently commercialized. The device is worn by a postoperative patient on the leg. The sequential pressurization of the three air chambers results in a sequential compression of the ankle, calf, and thigh, aiding the venous blood flow. This helps prevent deep vein thromboses and pulmonary emboli. The finished device is not only lighter in weight than the pPVC version, but also readily disposable by incineration.

2.4 EXTRUDED PROFILES

In addition to extruded film, plastomers can be used to make extruded profiles which can can serve as alternatives to pPVC in many applications. The most common extruded profile in medical applications is tubing of various kinds. Most respiratory tubing is made from either LDPE or EVA, depending on the softness needed for the application. Metallocene plastomer could be used for respiratory tube applications, but it is usually not cost-competitive against LDPE or EVA.

Low-density plastomers have a good balance of properties for use in IV tubing, including low modulus, clarity, drug and body fluid compatibility and kink resistance, especially in the smaller diameter tubes. As addressed above, processing and fabrication/assembly differences have been impediments to the broader use of metallocene plastomers in this application. Avoiding sharkskin melt fracture requires processing at relatively low line speeds, increasing the cost of production. Processing aids are available to increase output, but some of them might affect the compatibility with drugs in particular, and possibly also body fluids. In addition, the inability to solvent weld fittings on to the tubing is a major fabrication difference compared with pPVC tubing.

Plastomers can be used as alternatives to pPVC in tubing which carries air or water for devices such as the sequential compression device mentioned above or for a mattress pad with thermostatically controlled water circulating through it. In addition to not having the same issues of drug and body fluid compatibility, applications such as these frequently need a higher modulus plastomer than what is used to make IV tubing.

3 NON-WOVENS

In a totally different arena, metallocene-catalyzed polymers, both ethylene-based plastomers and propylene polymers, can be used to make non-woven fabrics. Linear, metallocene-catalyzed polymers are well suited for processing by both melt blown and spun bond techniques because they exhibit very low extensional viscosities as a result of their narrow molecular weight distribution. There are several patents describing the non-woven processes and fabrics made with those processes [20–22].

A recent paper by a non-wovens manufacturer [23] stated that metallocene-catalyzed PPs have a superior combination of performance advantages versus traditional peroxide broken PP in spunbonded fabrics:

- process stability–less sensitivity to melt temperature, fewer broken filaments, less die drool, less pluggage in the spin packs;
- line speeds—4000–4500 m/min vs <2000 m/min;
- stronger fibers—higher tenacity even at low denier per fiber.

In addition to these advantages, metallocene PPs also have lower melting-points, which allow fabricators to use lower temperatures on the calendering rolls that bond the non-woven mat into a fabric. Their smaller amount of volatiles and extractables leads to improved performance in breathing-air filtration applications and liquid filtration applications. It has also been reported that lower MFR grades of metallocene-catalyzed PPs (22 versus 35 dg/min) can be spun at high line rates versus conventional, peroxide-broken PP, yielding finer diameter fibers with higher tenacity and fabrics with higher tensile strengths [24].

For melt blown fabrics, metallocene-catalyzed PPs yield finer fibers than conventional, peroxide-broken PP. This improves filtration efficiency by about 60–80 % and results in a higher hydrostatic head (15–30 %) indicating a higher resistance to penetration by water [25]. The lower melting-point of these polymers allows melt blowing to be carried out at about a 50 °F(30°C) lower melt temperature, this not only saves energy, but also improves the service factor of the melt blowing line [21].

As a result of the processing and property advantages of metallocene-catalyzed PPs, their use in non-wovens for medical applications will initially be in replacing traditional peroxide-broken PP in these same applications; surgical gowns, wraps, drapes, diaper cover sheet and other skin-contact liners, filtration media, etc., and other disposable items. Other potential applications include films for adhesive and bandage tapes. In all of these areas the absence of volatile, extractable and oily chain fragments in the metallocene-catalyzed resin constitutes a significant improvement in the processing and medical application of the material over peroxide-broken PP.

With the advent of metallocene-catalyzed propylene RCPs, fabricators will soon have available to them polymers which will produce softer fabrics with even lower melting-points. Gowns and drapes made from these metallocene-catalyzed RCPs will feel more like traditional cloth than fabrics available today.

4 SYNDIOTACTIC POLYPROPYLENE

In addition to the advantages due to narrow molecular weight, tacticity, defect and comonomer distributions which are common to all metallocene-catalyzed PPs, additional positive features are claimed for the syndiotactic PP structure which metallocene catalysis has made readily available for the first time. Among these are enhanced toughness, clarity, heat sealability, tear strength and tolerance of high-energy radiation [26], all properties which are extremely important to the application of the material to medical uses [27].

On the other hand, there are drawbacks to the application of sPP to health care. It has been observed that the rate of crystallization of the syndiotactic metallocene PP is much lower than either Ziegler–Natta (Z–N) or metallocene-catalyzed isotactic PP (iPP). This means that conversion processing of syndiotactic resin would be much slower and obviously more expensive than processing of isotactic material [28]. The low immediate crystallinity leads to various long-term aging effects which cause the properties of sPP to change drastically over time [29,30]. Also, the peak melting-points of syndiotactic resins are much lower than the values commonly observed for isotactic propylene homopolymers [29]. This makes the syndiotactic resins much less resistant to softening at elevated temperature and unsuitable for sterilization by heat.

It is very important, however, for the reader to understand that sPP, as currently defined, is much lower in tacticity than the iPP with which it is being compared with respect to these properties. This lower tacticity translates directly to lower crystallinity in the sPP than in the isotactic iPP. Crystallinity is further reduced in sPP by the slow crystallization described above, owing to the higher flexibility and entanglement density of this structural form. These factors alone could be the cause of most or all of the differences favoring sPP over iPP for some medical applications. Until sPP comparable in tacticity and crystallinity to standard iPP is available, a true comparison of the materials will not be possible. The fullest application of sPP to medical applications will most likely await the appearance of these more comparable polymers.

5 POLYOLEFIN MODIFIERS

While the use of neat, metallocene-catalyzed, ethylene-based plastomers in medical applications has been limited by tacky surfaces on molded parts, blockiness in film, slow processing and the relatively low temperature softening characteristics of the polymers, the application of these materials to the modification of medical grades of polyolefins is essentially unaffected by these properties. Blends of the plastomers with various PEs and PPs have been formulated with much improved properties such as tear and impact resistance and resistance to sterilizing doses of high-energy radiation. In some cases blends of plastomer and propylene homopolymer or moderate ethylene content RCPs can substitute effectively for high-ethylene copolymers which are either difficult to manufacture, prohibitively expensive or have insufficient demand to make them economically attractive to manufacture.

5.1   PLASTOMER SELECTION AND GENERAL ADVANTAGES

Extensive background investigations on the modification of a wide range of polyolefins with ethylene-based plastomers have been conducted by Yu and co-workers [31–37]. They have verified both theoretically and experimentally that ethylene-based plastomers are miscible with the entire range of PEs no matter which common comonomer is used to provide the low densities, which define these species. This is in contrast to the situation for EPR, another useful modifier for polyolefins, which yields only compatible blends with PEs. Ethylene–butene-based plastomers are the logical choice for PE modification [31].

Blends of PPs with either the common types of plastomeric ethylene polymers or with EPR have also been shown to be at best only compatible, but the balance of compatibility (and the resulting resistance to blooming), low cost and ease of handling is maximized with the use of plastomeric ethylene–butene copolymers in combination with the propylene polymer [31].

The following is a compilation of specific examples of the use of plastomers to modify polyolefins and the medical uses to which the compounds can be applied.

5.2   MODIFICATION OF HIGH-DENSITY POLYETHYLENE

According to Yu and Wagner [31]:

> . . . It is interesting to note that plastomers and [other, general] polyolefins can be used alternatively to modify each other. Plastomers can be used to improve the tear and impact of polyolefins, and similarly, polyolefins can modify the stiffness and melt strength of plastomers.

The addition of a minor amount of plastomer to high-density polyethylene (HDPE) in a pellet blend greatly improves the machine direction (MD) tear strength of a blown overwrap film valued for its stiffness and impermeability to the transmission of water vapor. Although these crystallinity-dependent properties of HDPE are certainly reduced approximately in proportion to the level of dilution with plastomer, blends with a very useful compromise of properties are easy to obtain. Figure 1 shows that a blend of 75 % HDPE and 25 % plastomer has twice the MD tear strength of the neat HDPE and still retains a level of stiffness far greater than that of linear low density PE [31].

On the other hand, looking at blends which have plastomer as the major component, the miscibility of HDPE and plastomers has allowed the development of clear plastomer–PE alloys as alternatives to the use of PVC for medical applications such as flexible, medical bags. Currently, some of these devices are produced from PVC by extrusion blow molding, and an alternative which could be converted without production hardware modifications would be preferred. Such a product results from the fortification of plastomer with only 20–30 % of an HDPE. The HDPE boosts the melt viscosity and softening point of the blend giving it the required processability and durability which the application requires. The resulting

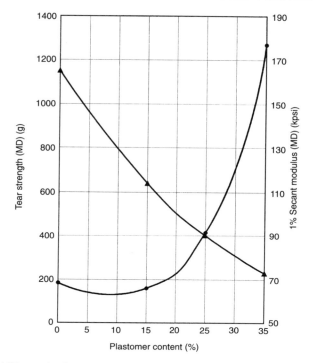

Figure 1 (●) Elmendorf tear strength (MD) and (▲) stiffness of 75 % HPDE–25 % plastomer blend

bag is clear, flexible, and like the current product, easily collapsible inside a protective canister [31].

5.3   RADIATION-RESISTANT BLENDS OF POLYPROPYLENE AND ETHYLENE-BASED PLASTOMERS

As one of these authors has written previously;

> Communication with the highest volume consumers of medical grades of PP indicates that the preferred PP resin for injection molded and radiation sterilized medical devices is a radiation tolerant, clear, autoclavable formulation.... Unfortunately, these three desired attributes have until recently been thought to be incompatible [38].
> ...Normally stabilized PPs are not suitable for sterilization by high-energy radiation at doses of 1.5–5.0 kGy because of the severe embrittlement and discoloration that occur in the plastic immediately after sterilization and worsen with aging. While the embrittlement of the plastic after irradiation is an inherent

property of the polymer, the discoloration is caused by reaction products of the phenolic antioxidants normally included in standard PPs. The modern, injection-molded resins that have been most successful in withstanding irradiation exhibit reduced crystallinity and narrow molecular weight distribution, are formulated with hindered amine light stabilizers (HALS), and contain none of the discoloring phenolic antioxidants. Ethylene-containing RCPs are useful substrates for building radiation tolerant formulations, as are homopolymers with low isotacticity and homopolymers to which hydrocarbon oils or greases have been added [39–41].

Of these methods for the protection of injection-molded PP from high-energy radiation, only the blends of clarified homopolymer with hydrocarbon oils or greases have been capable of providing all three of the desired properties, radiation resistance, clarity and resistance to softening at elevated temperature (autoclavability). As these formulations are protected by patents and not generally available for broad use in the medical device industry, alternative approaches to a similar result have been sought [38,42–45].

Recently, injection-molded blends of clarified PP homopolymers with metallocene-catalyzed, ethylene-based plastomers have been shown to have excellent tolerance of sterilizing doses of high-energy radiation (1.5–7.5 kGy) while retaining their homopolymer-like resistance to softening at elevated temperature and their clarity. These properties of blends of a clarified, hindered amine light stabilizer (HALS) stabilized, Z–N propylene homopolymer with 0–15 % of a 0.905 g/cm^3 density, 4.5 dg/min melt index (MI) metallocene-catalyzed, plastomeric, ethylene–butene copolymer are shown in Figures 2–6. At the minimum plastomer usage levels (7.5–10.0 %) which provide significant resistance to post-irradiation embrittlement, the blends exhibit clarity nearly equivalent to the neat, clarified homopolymer itself and stiffness and resistance to softening at elevated temperature (heat deflection temperature, HDT) similar to standard non-nucleated PP [38].

Figure 7 shows that the presence of the plastomer in the blends also has a very positive influence on their color, making them significantly whiter than the neat propylene homopolymer both before and after irradiation. The degree of whitening is directly related to the plastomer content in the blends [38].

The improved radiation resistance is imparted by any of a wide range of the plastomers. Likewise, retention of the homopolymer-like characteristics depends only on the homopolymer content of the blend. However, retention of clarity in the blend is highly dependent on the choice of plastomer. At moderate levels of plastomer usage, haze will be minimized by the choice of plastomer MI and density as indicated by the response surface shown in Figure 8. Optimum plastomer choices for PP–plastomer blends with minimum haze are in the ranges of 4–10 dg/min MI and 0.895–0.905 g/cm^3 density. These ranges of plastomer attributes appear to provide a low increase in haze to blends with a variety of PPs: homopolymers and RCPs with ethylene contents up to at least 3 % by weight and melt flow rates (MFRs) in the range 1–40 dg/min [36,38].

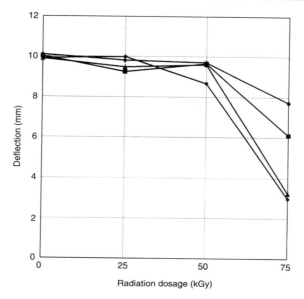

Figure 2 Deflection at peak flexural load of irradiated Ziegler–Natta polypropylene–plastomer blends. Plastomer content: (●) 0; (▲) 5; (■) 10; (◆) 15 %

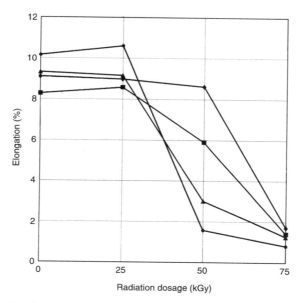

Figure 3 Tensile elongation at break of irradiated Ziegler–Natta polypropylene–plastomer blends. Plastomer content: (●) 0; (▲) 5; (■) 10; (◆) 15 %

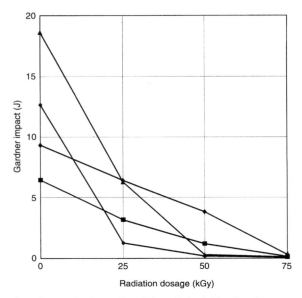

Figure 4 Gardner impact strength of irradiated Ziegler–Natta polypropylene–plastomer blends. Plastomer content: (●) 0; (▲) 5; (■) 10; (◆) 15 %

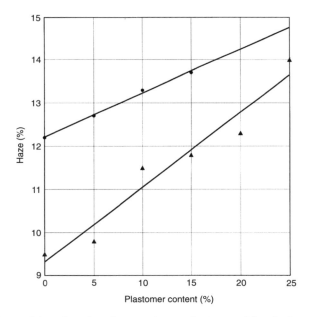

Figure 5 Haze of irradiated polypropylene–plastomer blends (extrapolated to equal ranges). (●) Ziegler–Natta PP; (▲) Metallocene PP

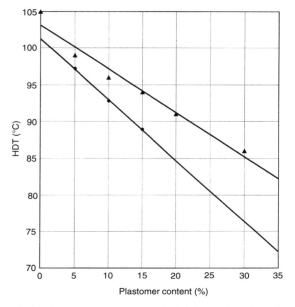

Figure 6   Heat deflection temperature at 455 kPa of irradiated polypropylene–plastomer blends (extrapolated to equal ranges). (●) Ziegler–Natta PP; (▲) Metallocene PP

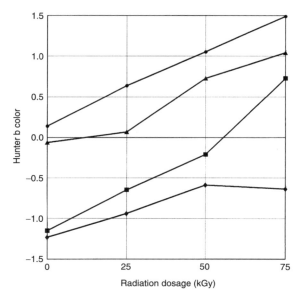

Figure 7   Color of irradiated Zielger–Natta polypropylene–plastomer blends. Plastomer content: (●) 0; (▲) 5; (■) 10; (◆) 15 %

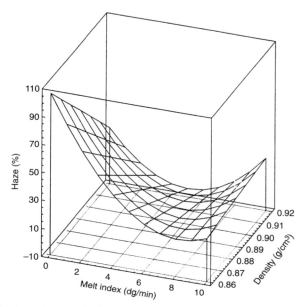

Figure 8 Effect on haze of plastomer density and melt index in blends with Ziegler–Natta propylene homopolymer (clarified)

Figures 5 and 6 and 9–11 show the same properties for blends of the same plastomer with a nucleated and clarified, 32 dg/min MFR, metallocene-catalyzed propylene homopolymer. In these blends also, incorporation of the plastomer protects the compound from post-irradiation embrittlement while having a negligible effect on the haze of the PP and causing only a small loss of the homopolymer-like stiffness and resistance to softening at elevated temperature. Because of the inherent higher level of crystallinity in the metallocene-catalyzed propylene polymer on which the blends are based, however, higher levels of plastomer are required in the blends to provide a given level of protection against post-irradiation embrittlement than with a normal Z–N PP. However, in these blends the balance of radiation and softening resistance is similar to that of the blends of plastomer with conventional PP. The higher levels of plastomer used to obtain a given level of radiation resistance balance the higher stiffness and softening point of the metallocene-catalyzed propylene polymer [46].

The most useful blends of plastomer with either Z–N or metallocene PP (mPP) are likewise similar to each other in stiffness, but there is a noticeable advantage in clarity for the blends made with mPP. The neat, clarified, metallocene-catalyzed propylene homopolymer is clearer than common Z–N analogs. Although a higher level of plastomer (12–15 %) is required to provide the needed level of radiation

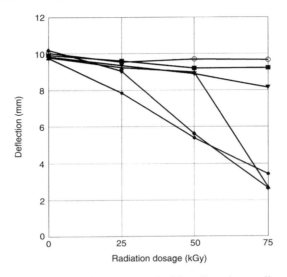

Figure 9 Deflection at peak flexural load of irradiated metallocene polypropy-lene–plastomer blends. Plastomer content: (●) 0; 5; (▲) 10; (▼) 15; (■) 20; (○) 30%

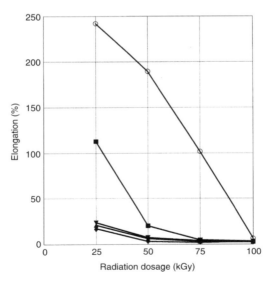

Figure 10 Tensile elongation at break of irradiated metallocene polypropylene–plastomer blends. Plastomer content: (●) 0; 5; (▲) 10; (▼) 15; (■) 20; (○) 30%

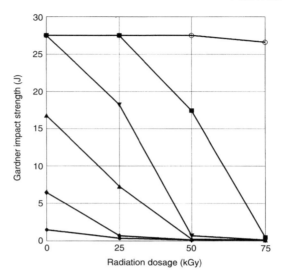

Figure 11 Gardner impact strength of irradiated metallocene polypropylene–plastomer blends. Plastomer content: (●) 0; 5; (▲) 10; (▼) 15; (■) 20; (○) 30 %

tolerance, the final haze of the useful blends is still lower than the blends based on Z–N PP (7.5–10 % plastomer). This is already shown in Figure 5 [38,46].

5.4 PLASTOMER-MODIFIED PP FILMS FOR MEDICAL PRODUCT PACKAGING

A recently published description of the difficulty in stabilizing thin PP articles to post-radiation embrittlement reads as follows [48,49]:

> Despite the rapid growth of PP as a construction material for radiation-sterilized medical devices, PP films have not achieved similar use in packaging of either devices or medical solutions. The high ratio of film surface area to mass, combined with the sensitivity of irradiated PP to oxygen-promoted degradation, causes them to be severely embrittled after normal sterilizing doses of radiation. Even those traditional resin formulations which yield highly radiation resistant injection molded devices are badly degraded after irradiation in thin film form.

These formulations would include, for example, both low-tacticity homopolymers and RCPs containing around 3 % ethylene, both of which are stabilized with hindered amines. Nucleation of a PP is especially damaging to its radiation tolerance when the material is converted to film [47–49].

Following up on the development of the blends of metallocene-catalyzed, ethylene-based plastomer with propylene homopolymers as radiation-tolerant materials for injection-molded medical devices, it has also been demonstrated that similar

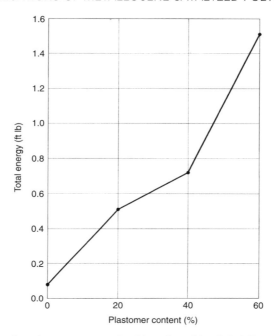

Figure 12   Dart drop impact strength of irradiated (50 kGy) films made from Z–N polypropylene–plastomer blends

compounds can be used to give films with the desired improved resistance to embrittlement after γ irradiation. The improvement increases in direct relationship to the amount of plastomer used in the blend. Other properties of the films such as ductility and flexibility also increase with increasing plastomer content. These developments are demonstrated in Figures 12–15, which depict the properties of irradiated films cast from blends of a metallocene-catalyzed, ethylene-based plastomer with a non-nucleated, HALS-stabilized, Z–N propylene homopolymer. This development makes possible the use of PP films for the packaging of radiation-stabilized medical devices [48,49].

5.5   PP–PLASTOMER BLENDS AS SUBSTITUTES FOR HIGH-ETHYLENE RANDOM COPOLYMERS

Another interesting application of the modification of PP with plastomers is the access to a wide range of polymer properties by blending the two radically different substrates in any proportions. Figures 16 and 17 demonstrate the results of this exercise in the case of a readily available, nucleated, 2.8 % ethylene, propylene–

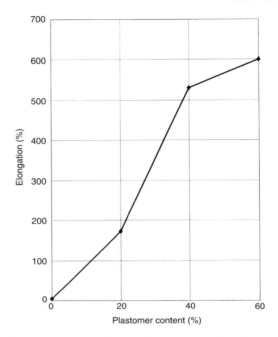

Figure 13 Tensile elongation at break (transverse direction) of irradiated (50 kGy) made from Z–N polypropylene–plastomer blends

ethylene RCP and the ethylene–butene, 4.5 dg/min MI, $0.905\,\text{g/cm}^3$ density plastomer described previously.

In physical properties such as tensile and impact strength and flexural modulus, the blends mimic a range of PPs from the neat 2.8% ethylene RCP to RCPs containing much higher levels of ethylene. The 80:20 RCP–plastomer blend, for example, is remarkably similar in tensile strength and modulus to a nucleated 4.5% ethylene RCP. Similarly, the 60:40 RCP–plastomer blend resembles a nucleated 6.3% RCP with respect to the same properties [50].

Such high-ethylene RCPs are not widely available, however, and are sold at significant premiums by those companies which do manufacture them. If an RCP with a different level of ethylene in this range might be ideal for a small-volume application, it might not be available at all, or at best might require months or years of pilot research and development to identify. Softer, more flexible propylene RCPs containing much higher ethylene levels cannot be produced owing to limitations in the polymer manufacturing process. However, in contrast, the corresponding blend of homopolymer or moderate ethylene RCP with plastomer that provides the required physical properties can be easily developed from the readily available components with a single day's work at the conversion machine. Also, the price of

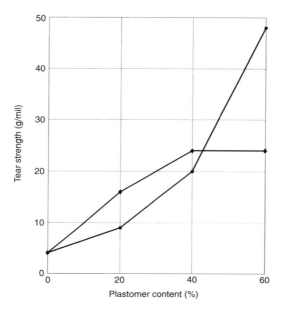

Figure 14 Elmendorf tear strength (machine direction) of irradiated (50 kGy) films made from Z–N polypropylene–plastomer blends. (●) MD; (◆) TD

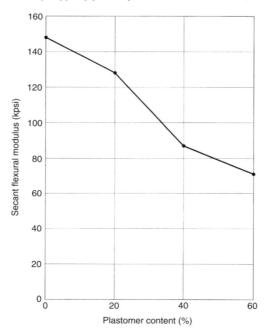

Figure 15 Stiffness of films made from Z–N polypropylene–plastomer blends

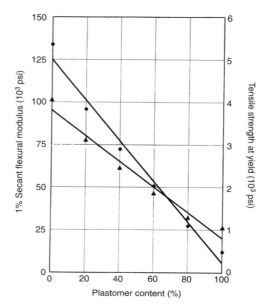

Figure 16 (●) Stiffness and (▲) strength of Z–N random copolymer–plastomer blends

the desired blend is fixed at the weighted average of the prices of the two components.

The most prominent example of a medical application which could utilize such blends (as in the case of the HDPE–plastomer blends above) is parenteral solution packaging (IV bags). Tough, heat-resistant, soft films prepared from a readily accessible blend of PP and plastomer would be suitable for conversion by heat sealing or r.f. welding into bags for the packaging and delivery of IV solutions and would provide a bona fide alternative to the use of pPVC for this application. In certain medical devices such as IV tubing kits, tubing clamp valves and drip chambers, a positive snap fit for the assembly of parts is desired, often balanced with other properties such as good impact resistance and flexural ductility. These opposed properties can be more easily combined in a custom PP–plastomer blend than in a RCP or impact copolymer drawn from a restricted selection of candidates.

5.6 CLEAR, IMPACT-RESISTANT PP

The work of Yu and co-workers on the general impact modification of PP with plastomers and the studies they have conducted on the optimization of clarity in radiation resistant PP–plastomer blends largely set the stage for the development of clear, impact-resistant PPs suitable for use in medical and related applications. As

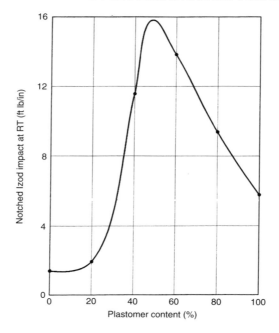

Figure 17   Notched Izod impact strength of Z–N random copolymer–plastomer blends

described above, the value of such materials is principally the improved balance they display between tensile strength and stiffness on the one hand and impact strength on the other.

Figure 18 presents some representative data on the impressive improvement in Gardner impact strength achieved with the progressive addition of the 4.5 dg/min MI, 0.905 g/cm^3 density ethylene–butene plastomer to clarified propylene homopolymers, both Z–N and metallocene catalyzed. Figure 19 likewise shows the improvement observed in instrumented, falling tup impact strength of clarified, 35 dg/min MFR, 3% ethylene RCP when it is blended with a 3.5 dg/min MI, 0.90 g/cm^3 MI ethylene–butene plastomer. The modifier improves the impact strength of the propylene copolymer both at room temperature and cold. The proportional decrease in stiffness and negligible increase in haze observed in blends with homopolymer is also seen in the case of the copolymer [36,38,46,51].

A promising role for blends of this type is to provide a margin of safety to medical devices against accidental breakage without the loss of the stiffness and clarity required by the application. Modification of the PP material of construction with plastomer can greatly improve the impact resistance of devices without greatly altering their other physical characteristics and clarity.

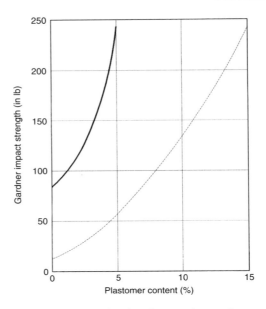

Figure 18 Gardner impact strength of polypropylene–plastomer blends. Solid line, Z–N PP; dashed line, metallocene PP

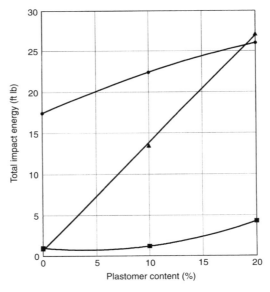

Figure 19 Instrumented impact strength of random copolymer–plastomer blends. (●) 23; (▲) 0; (■) −10 °C

An example of a conceptually similar use of the blends would be in infant feeding devices (baby bottles, disposable feeder holders, 'sip' cups for toddlers, etc.) and pacifiers commonly molded from clarified PP. These are used over a broad temperature range and are especially subject to breakage when dropped or struck while they are cold, as is often the case. The PP–plastomer blends would significantly improve the resistance of these devices to breakage.

6 PRECISION MOLDING

Despite the widespread use of PP in injection-molded medical applications, its relatively broad molecular weight distribution (MWD, as often indicated by poly-dispersity, the ratio of weight-average to number-average molecular weight) and semi-crystallinity have made its use in precisely molded parts very difficult. A fairly high polydispersity in the range of 4–5 signals the presence in the material of large, elastic polymer chains, which commonly result in molded-in orientation and stresses in parts made from it. The crystallinity and resultant shrinkage of PP are difficult to control precisely, as they depend on several temperature-related variables in the molding process and in the environment in which the finished parts are stored.

The use of narrower MWD, peroxide-broken, Z–N catalyzed resins with a polydispersity of about 2.8–3.5 can alleviate the orientation and molded-in stress problems to some extent, but at the expense of purity. The organic peroxide-treated materials contain measurable levels of peroxide decomposition products such as *tert*-butanol and acetone and products of oxidation of the polymer. These are undesirable in resins used for medical devices or device or drug packaging. However, the metallocene-catalyzed PPs exhibit a very low polydispersity of about 2.0 directly from the polymerization reactor without any peroxide treatment. Not only are they superior in providing stress-free, isotropic moldings, but also they are superior in purity to the peroxide-produced formulations.

Therefore, a more attractive and comprehensive solution is to use for this application nucleated, higher MFR, metallocene-catalyzed PP. This material has the requisite purity, chemical resistance and low molecular weight and ultra-narrow MWD to provide warp and stress-free parts. Furthermore, crystallization of this material is rapid and nearly complete in the molding press, giving a part with much more stable dimensions than obtained from non-nucleated materials.

The applications described above offer only a cursory glimpse at the possible ways in which metallocene-catalyzed polyolefins can be applied to medical devices and packaging. Emphasis in the early work with these materials has focused on providing ethylene- and propylene-based alternatives to pPVC, improving the impact and radiation resistance of PP, and exploiting the cleanliness and processing advantages of metallocene-catalyzed PP. Not only is the understanding of the properties of these material constantly growing, but also the list of available materials, themselves, is continually expanding. Properties which have been inherent in this class of materials as originally defined, such as ultra-narrow MWD, may be redefined through future research. This will, thereby, broaden the applicability of these materials to various uses and processes for which the current materials are not really well suited, such as extrusion blow molding in the narrow MWD case mentioned above. In another example, cross-linking of the low-melting, lowest density plastomers provides to them the heat resistance needed for sterilization by steam autoclaving. As this technology is developed and eventually incorporated in medically attractive plastomer formulations, the utility of the neat plastomers will increase markedly and compete more effectively with the blends and co-extrusions of plastomer and higher melting polymers which are more commonly used today. Also, of course, it is expected that new homo-and copolymers will be developed that will be of interest in the medical field. The growth of medical applications of metallocene polyolefins in all directions should continue at a rapid pace.

7 REFERENCES

1. Speed, C. S., Trudell, B. C., Mehta, A. K. and Stehling, F. C., in *Proceedings of Polyolefins VII*, Society of Plastics Engineers, Brookfield, CT, 1991, pp. 45–66.
2. Temperature for the transition from brittle to ductile failure in impact testing.
3. Woo, L., in *Proceedings of Metallocenes '95*, Schotland Business Research, Skillman, NJ, 1995, pp. 191–205.
4. Exxon Chemical Co., EXACT® Plastomer Grade Data Sheet on EXACT 5008 (10 dg/min MI, 0.865 g/cc density).
5. Erikson, D., *Plast. World*, September, 1989, pp. 39–43.
6. Switzer, W. G., *Labscale Incineration Analytical Study of Medical Polymers, Phase I*, Final Report No. 01-5099, Department of Fire Technology, Southwest Research Institute, Houston, TX, 1992.
7. Medical grades of UV-curable adhesives which can be used with plastomers are available from Loctite Corporation, 705 North Mountain Road, Newington, CT 06111, USA and Dymax Corporation, 51 Greenwoods Road, Torrington, CT 06790, USA; there may also be other suppliers as well.
8. Plastics Welding Technology, RF Systems, 6849 East 32nd Street, Indianapolis, IN 46226, USA.
9. Gilbert, M., *J. Macromol. Sci., Rev. of Macromol. Chem. Phys.*, **C34**, 77 (1994).
10. Fanselow, D. L., *et al.*, *US Pat.*, 5 562 127 (1996).
11. Ko, J. H. and Odegaard, L., in *ANTEC '96 Conference Proceedings*, Society of Plastics Engineers, Brookfield, CT, 1996, pp. 2804–2806.
12. *Res. Disclos.*, 732 (1993).
13. Hong, K. Z., in *ANTEC '95 Conference Proceedings*, Society of Plastics Engineers, Brookfield, CT, 1995, pp. 4192–4198.
14. Luther, D. W. and Linsky, L. A., in *ANTEC '95 Conference Proceedings*, Society of Plastics Engineers, Brookfield, CT, 1995, pp. 4203–4207.
15. Hoenig, S., Hoenig, W. and Parsley, K., in *ANTEC '96 Conference Proceedings*, Society of Plastics Engineers, Brookfield, CT, 1996, pp. 1970–1974.
16. *Technical Guidelines for Processing EXACT® Plastomers For Film*, Exxon Chemical Americas, Houston, TX, 1994, pp. 4–8.
17. Ref. 16, p.9.
18. Mehta, A. K. and Chen, M. C., *US Pat.*, 5 358 792 (1994).

19. Lipsitt, B., in *ANTEC '97 Conference Proceedings*, Society of Plastics Engineers, Brookfield, CT, 1997, pp. 2854–2858.
20. Davey, C. R., *et al.*, *US Pat.*, 5 322 728 (1994).
21. Reed, J. F. and Swan, M., *US Pat.*, 5 324 576 (1994).
22. Stahl, G. A. and McAlpin, J. J., *PCT Int. Pat. Appl.*, WO 94/28219 (1994).
23. Velasco, J. L. R., in *Proceedings of SPO '97*, Schotland Business Research, Skillman, NJ, 1997, pp. 317–333.
24. McAlpin, J. J., *et al.*, in *Proceedings of SPO '95*, Schotland Business Research, Skillman, NJ, 1995, pp. 125–144.
25. McAlpin J. J. and Stahl, G. A., in *METCON '94 Proceedings*, The Catalyst Group, Spring House, PA, 1994.
26. Donahue, J., in *METCON '96 Proceedings*, The Catalyst Group, Spring House, PA, 1996.
27. Schardl, J., *et al., Plast. Film Sheet.*, **12**, 157 (1996).
28. Shamshoum, E. S., *et al.*, in *Proceedings of SPO '93*, 1993, pp. 207–226.
29. Wheat, W. R., in *ANTEC '95 Conference Proceedings*, Society of Plastics Engineers, Brookfield, CT, 1995, pp. 2275–2278.
30. Wheat, W. R., in *ANTEC '97 Conference Proceedings*, Society of Plastics Engineers, Brookfield, CT, 1997, pp. 1968–1971.
31. Yu T. C. and Wagner, G. J., in *Proceedings of Polyolefins VIII*, Society of Plastics Engineers, Brookfield, CT, 1993, pp. 539–546.
32. Yu, T. C., in *ANTEC '94 Conference Proceedings*, Society of Plastics Engineers, Brookfield, CT, 1994, pp. 2439–2441.
33. Yu, T. C., in *ANTEC '95 Conference Proceedings*, Society of Plastics Engineers, Brookfield, CT, 1995, pp. 2358–2368.
34. Yu, T. C., in *ANTEC '95 Conference Proceedings*, Society of Plastics Engineers, Brookfield, CT, 1995, pp. 2374–2385.
35. Dharmarajan, N. R. and Yu, T. C., in *ANTEC '96 Conference Proceedings*, Society of Plastics Engineers, Brookfield, CT, 1996, p. 2006.
36. Yu, T. C. and Davis, D. S., in *Society of Plastics Engineers RETEC*, Rochester, NY, September 25–26, 1996, Society of Plastics Engineers, Brookfield, CT, 1996.
37. Yu, T. C., in *Proceedings of Polyolefins X*, Society of Plastics Engineers, Brookfield, CT, 1997, pp. 227–229.
38. Portnoy, R. C., *Med. Plast. Biomater.*, **4**, 40 (1997).
39. Portnoy, R. C., *Med. Plast. Biomater.*, **1**, 43 (1994).
40. Williams, J. L., in *Proceedings of the Medical Design and Manufacturing West 96 Conference and Exposition*, Anaheim, CA, Canon Communications, 1996, pp. 202-17–202-26, Santa Monica, CA (1996).
41. Westfal, J. C., Carman, C. J. and Layer, R. W., *Rubber Chem. Technol*, **45**, 402 (1972).
42. Williams, J., Dunn, T. and Stannett, V., *US Pat.*, 4 110 185 (1978).
43. Williams, J., Dunn, T. and Stannett, V., *US Pat.*, 4 274 932 (1981).
44. Williams, J., Dunn, T. and Stannett, V., *US Pat.*, 4 467 065 (1984).
45. Williams, J., Dunn, T. and Stannett, V., *US Pat.*, 4 845 137 (1989).
46. Portnoy, R. C., in *Proceedings of the Medical Design and Manufacturing Orlando 97 Conference and Exposition*, Orlando, FL, Canon Communications, 1997, pp. 303-17–303-35, Santa Monica, CA (1997).
47. Portnoy, R. C., Gulla C. T. and Kozimor, R. A., in *ANTEC '92 Conference Proceedings*, Society of Plastics Engineers, Brookfield, CT, 1992, pp. 230–232.
48. Portnoy, R. C., in *ANTEC '97 Conference Proceedings*, Society of Plastics Engineers, Brookfield CT, 1997, pp. 2844–2848.
49. Portnoy, R. C., *J. Plast. Film Sheet.*, **13**, 115 (1997).
50. Portnoy, R. C., unpublished results.
51. Portnoy, R. C., unpublished results.

23

Constrained Geometry-catalyzed Polyolefins in Durable and Wire and Cable Applications

S. BETSO
Dow Chemical Company, Horgen, Switzerland

L. T. KALE
Currently with Mobil Chemical Co., Edison, NJ, USA

J. J. HEMPHILL
DuPont Dow Elastomers L.L.C., Freeport, TX, USA

1 INTRODUCTION

Constrained geometry catalyst (CGC) technology ethylene–1-octene copolymers have demonstrated the ability to displace incumbent materials in a variety of rubber and plastic applications. Production capability ranges from ethylene-based homo-polymers to incorporation of over 18 mol% of 1-octene (density $<0.87 \mathrm{g/cm^3}$), with melt indices from fractional to over 100. Commercially, two product families have been developed; AFFINITY[†] polyolefin plastomers (POPs, <10 wt% 1-octene) and ENGAGE[‡] polyolefin elastomers (POEs, >10 wt% 1-octene). These products can be converted into end-use products in a number of ways. For example, the polyolefin pellets can be used 'as is' or compounded with other ingredients into a formulation. These products can also be used as a thermoplastic or cross-linked into a thermoset.

[†] Trademark of The Dow Chemical Company.
[‡] Trademark of DuPont Dow Elastomers L.L.C.

Metallocene-based Polyolefins Edited by J. Scheirs and W. Kaminsky
© 2000 John Wiley & Sons Ltd

In general, cross-linking methods include peroxide, irradiation and moisture (silane) processes. Similar to ethylene–propylene–diene terpolymers (EPDM), CGC technology ethylene–α-olefin–diene terpolymers can be cured with sulfur.

This chapter will focus on the performance of CGC technology copolymers in durable applications. More specifically, we will focus on the practical application and benefits of these products in wire and cable, flooring, geo-membranes, rotomolding and decorative applications.

2 WIRE AND CABLE APPLICATIONS

2.1 LOW-VOLTAGE FLEXIBLE CABLE INSULATION

CGC technology polyolefin elastomers (POEs) have demonstrated many advantages in the replacement of ethylene propylene rubbers (EPRs) for low-voltage flexible cable insulation. In general, flexible cable compounds are heavily loaded with fillers and plasticizers to meet flexibility and economic considerations and peroxide cross-linked to provide higher service temperatures. A typical compound is described in Table 1. Evaluations of this compound, containing a POE, has been carried out in both batch and continuous mixers with the compound normally being strained and extruded into a continuous strip for ease in feeding a rubber extruder. The rubber extruder feeds a crosshead die for extrusion onto wire followed by a continuous vulcanization line operating at temperatures between 180 and 200 °C.

Batch compounding with a Banbury mixer is widely practised in the rubber and plastics industry because of its versatility to melt blend a wide variety of raw materials. These ingredients may include polymers in pellet or baled form, and dry or liquid additives such as carbon black, mineral fillers, plasticizers, curatives, processing aids and antioxidants. For the wire and cable compound, the dry raw materials (including dry master batches containing the liquid vinylsilane and coagent) are generally weighed into batch loading carts for later introduction into the Banbury along with directly injected plasticizers. Experience has shown that

Table 1 Typical low-voltage flexible cable compound

Product	phr
Ethylene co(ter)polymer	100
Untreated mineral filler	250
Paraffinic oil	80
Dicumyl peroxide	4
Co-agent	3
Antioxidant	1.5
Vinylsilane	1.25

minor modifications in the mixing procedure will assist the compounder when converting from a baled EPR to the pelletized POE in this compound.

For example, mixing order of the ingredients may need to change when converting from a baled to a pelletized elastomer. Many manufacturers will begin their batch by introducing the baled product prior to other raw materials. This procedure ensures the breakdown of the bale for better dispersion in the compound. In contrast, POE pellets are readily dispersible and, as a general rule, the POE-based compound gives an improved mixer temperature and torque profile with an upside-down (polymer in last) mixing procedure.

Banbury loading factor changes can also improve the mixing performance of POEs. Compared with a baled elastomer, a 2–4 % increase from the normal 70–75 % load factor will ensure more comparable mixing profiles. Figure 1 illustrates the effect of load factor for an insulation compound comparing a POE and a baled EPDM product. The higher POE loading factor results in a mixing profile that more closely matches the baled EPDM based compound.

POE-based compounds will often have a lower melt viscosity than a compound using a high Mooney viscosity EPDM. Therefore, the compounder is encouraged to use a lower Banbury batch 'drop' temperature when possible, to optimize post-compound handling. Once again, POE pellets produce a good dispersion without having to generate higher temperatures in the Banbury. The low-voltage flexible cord compound and many other compounds have been mixed at temperatures below 100 °C with no adverse effects. This is a definite advantage when mixing compounds

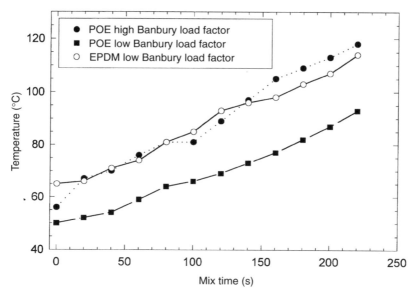

Figure 1 Effect of Banbury load for low-voltage flexible cable compounds

containing curatives which 'scorch' (premature cross-linking) when exposed to high mixing temperatures. Of course, the compounder should always consider the melt temperature of all compound ingredients before using lower drop temperatures, and might consider using alternative additives or master batches if necessary.

The low-voltage flexible cable compound was also mixed in an Farrel continuous mixer (FCM), which is essentially a counter rotating non-intermeshing twin screw [1]. POE pellets make it conducive to the continuous mixing equipment as opposed to a baled EPR, which would require a rubber grinder or a pelletization step to feed this type of equipment adequately. The FCM mixer was equipped with a loss-in-weight twin screw and volumetric feeders for all of the dry ingredients, including the polymer, filler, antioxidant and peroxide curative. The liquid ingredients, including paraffinic oil, vinylsilane and co-agent, were best introduced by pre-blending and volumetrically feeding just downstream of the dry feeder port. Some care was required for the raw material feeding operations to avoid bridging in the feeder port. Also, the compatibility of the liquid ingredients had to be determined to avoid potential segregation prior to injection. During process operations, early injection of the oil blend was required to maintain the compound temperature below 130 °C, which inhibited scorch.

A comparison of the processing and end-use performance for samples of the extruded strip from the Banbury and FCM mixes was conducted. Microscopic analysis was unable to detect any significant ingredient dispersive or distributive differences, and further testing shown in Table 2 demonstrated that both mixing methods gave virtually identical performance.

Other continuous compounding methods are available and have been evaluated using POEs as the base polymer for other complex rubber compounds. For instance, these compounds can be blended with co-rotating, intermeshing twin screw and reciprocating single screw mixers [2,3]. The key is to design the screw stack and feeding operations to obtain good dispersive and distributive mixing without shear heating the compound beyond the scorch and/or hydrolysis safety limits.

The processing and performance attributes of a POE compared with an EPR and EPDM are shown in Table 3. The POE has a higher cure rate and significantly

Table 2 Comparison of batch versus continuous mixing ($0.87 g/cm^3$, 0.5 dg/min POE; see Table 1 for formulation)

	Banbury	FCM
Processing Information:		
Mooney (ML $1 + 4/121$ °C)	25	24
200 °C ODR (3° arc) delta torque (N m)	1.47	1.41
200 °C ODR (3° arc) time to 90 % cure (min)	1.30	1.27
Mechanical properties (cured at 200 °C/2 min):		
Tensile strength (MPa)	7.9	7.7
Elongation (%)	250	275

Table 3 Comparison of POE and EPR in a typical low-voltage flexible cable compound (see Table 7 for formulation)

	ENGAGE CL800	EPR	EPDM
Comonomer	Octene	Propylene	Propylene
Diene level	None	None	High
Molecular weight	Medium	Medium	High
Density	0.87	0.87	0.87
Melt index	0.5	0.5	0.05
Mooney (ML 1 + 4/121 °C)	35	25	60
Base polymer MWD	Very narrow	Broad	Narrow
Processing information:			
Mooney (ML 1 + 4/121 °C)	25	23	29
200 °C ODR (3° arc) delta torque (MPa)	1.47	1.47	3.5
Time to 90 % cure (min)	1.3	1.5	1.3
Mechanical properties (Cured at 200 °C/2 min):			
Tensile strength (MPa)	7.9	4.1	6.6
Elongation (%)	250	450	250
Dielectric constant at 1 kHz	2.7	2.7	2.7
Dissipation factor at 1 kHz	0.002	0.002	0.002

exceeds the toughness of the ethylene–propylene-based elastomer when cross-linked. The EPDM-based compound has a higher cure state than the POE, as measured by oscillating disc rheometer (ODR) delta torque, but the cure rate is comparable. Note that the final POE compound exceeds the toughness of the EPDM compound as measured by the tensile strength. Thus, a wire and cable manufacturer can process a POE at line speeds comparable to an EPDM and still maintain good end-use performance characteristics. The improved thermal stability of the POE versus EPM and EPDM is shown in Figure 2. This improvement in heat-aging performance can benefit the end-user in either longer life cables or higher temperature applicability. The heat-aging performance of various CGC ethylene–α-olefins has been documented by Kale et al. [4]. More specifically, improved heat-aging performance was observed in ethylene–octene copolymers relative to ethylene–pentene and ethylene–butene copolymers. This was attributed to the fact that ethylene–octene copolymers have fewer tertiary carbons, i.e. the weak links in the thermo-oxidative process. The ethylene–octene, ethylene–pentene and ethylene–butene elastomers investigated contained 12.2, 14.3 and 15.8 mol% comonomer, respectively, based on ^{13}C NMR analysis.

2.2 HALOGEN-FREE FLAME-RETARDANT APPLICATIONS

Although space does not allow an in-depth discussion of flame-retardant (FR) technology, the plastics industry has used several approaches to address this issue [5]. Obviously, inherently FR polymers are frequently used in such applications;

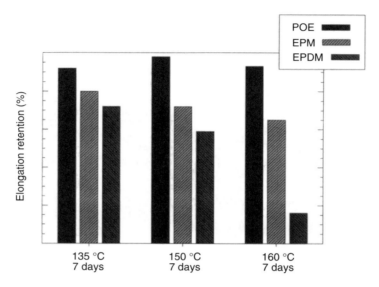

Figure 2 Heat-aging performance for cross-linked low-voltage flexible cable compound

indeed, poly(vinyl chloride) (PVC) is the most widely used polymer for FR cable insulation and jacketing globally. The incorporation of FR additives into flammable polymers has also proven to be effective; generally, such additives contain chlorine or bromine atoms. However, during combustion one of the major problems with such materials is the generation of corrosive gases that can severely damage electrical and computer systems in critical telecommunication, power and defense installations. In addition, there are also the environmental issues of disposal and recycling. Based on these concerns, end-users continue to request halogen-free systems.

An alternative halogen-free FR technology involves the use of metal hydrates such as aluminum trihydrate (ATH). At high temperatures ATH decomposes and generates water, thereby creating a vapor barrier between the flame and the polymer surface. The use of ATH is particularly attractive from an economic standpoint. From the formulators viewpoint, ATH is inexpensive relative to other FR additives. In addition, during combustion ATH-filled systems have low flame spread, and low smoke and corrosive acid gases are not generated. A typical 60 wt% ATH flame-retardant formulation is summarized in Table 4. In general, as the ATH loading increases, ignition resistance increases while processability and end-use physical properties decrease.

Flame-retardant ATH formulations are frequently based on lower melting poly-olefins such as poly(ethylene–co-vinyl acetate) (EVA). Lower melting polyolefins

Table 4 Flame-retardant (FR) cable jacket (halogen-free) formulation

Material	phr
Copolymer (EVA or POE)	100
Treated aluminum trihydrate	155
VulCup 40KE	4.5
Antioxidant	1.7

are desirable, since the formulation can be processed at lower temperatures to prevent ATH decomposition. Such FR formulations can also be prepared from CGC technology POEs.

In Table 5, one can compare the physical properties of fire-retardant POE and EVA formulations. Even at high ATH loadings, the tensile properties of these formulations were excellent. The POE formulations were softer (more flexible), as indicated by the Shore D durometer data. Water absorption was significantly lower for the POE compounds.

Table 5 Comparison of flame retardant (FR) cable jacket (halogen-free) formulations based on ENGAGE POEs and EVA (see Table 4 for formulation details)

	ENGAGE CL8002	ENGAGE CL8003	EVA
Comonomer	Octene	Octene	Vinyl acetate
Comonomer content (mol%)	24	18	11
Melt index (dg/min)	1	1	3
Density (g/cm^3)	0.870	0.885	0.951[a]
Tensile strength (MPa)	9.35	10.2	14.2
Elongation (%)	519	464	406
Heat-age, 121 °C/168 h:			
Retention tensile (%)	96	106	97
Retention elongation (%)	95	92	101
Water absorption, 82 °C/168 h (mg/in^2)	14.3	16.9	46.5
Durometer, Shore D	32	41	47
Limiting oxygen index (%)	32	33	43
NBS smoke chamber:			
Flaming (D_{max})	79	n/a	163
D_s at 4 min	5	n/a	15
Smoldering (D_{max})	168	n/a	193
D_s at 4 min	31	n/a	20

[a] Densities for EVA copolymers are not comparable to densities of ethylene–octene copolymers, owing to the presence of oxygen in the EVA.

The limiting oxygen index (LOI) and NBS smoke chamber data were used to quantify flame-resistance behavior. LOI is a measure of the amount of oxygen required in an atmosphere to sustain burning in a specified experimental set-up. Higher LOI values imply that more oxygen is required for sustained burning. Since the atmosphere contains approximately 20 % oxygen, polymers with LOI < 20 (e.g. polyethylene, LOI = 18.3) burn readily in air. Therefore, in general, higher LOI values are equated with better FR performance. As shown in Table 5, the LOI of the POE compounds were about 73 % of the EVA value. LOI values of 22.3 have been reported for PVC compounds [6].

Obviously, a low smoke density is desirable from both visibility and breathability points of view. The NBS smoke chamber was used to measure the amount of smoke produced using two standard experimental conditions. More specifically, vertically mounted samples were exposed to a propane burner (flaming condition) or a radiant heat flux (smoldering condition). The attenuation of a vertical light beam incident on a photomultiplier tube at the top of the smoke chamber was used to quantify the amount of smoke. The level of smoke is reported in terms of specific optical density (D_s), which is related to the percentage light transmission. As shown in Table 5, two values are reported for each combustion mode: D_s, the specific optical density after 4 min, and D_{max}, the maximum specific optical density observed. Excellent smoke generation results were observed for the POE based compounds relative to the EVA compounds. D_{max} values of 400 (flaming and smoldering) have been reported for PVC compounds [6].

In summary, relative to EVA-based flame retardant compounds, POE-based compounds are more flexible and have reduced water absorption and reduced smoke generation (flaming NBS smoke chamber). In addition, POEs more readily accept fillers. In other words, higher ATH loadings are possible with POEs that will result in lower formulation costs and improved FR performance.

2.3 MEDIUM-VOLTAGE INSULATION

Cross-linked high-pressure low-density polyethylene (LDPE) and ethylene–propylene rubber (EPR) have been the most commonly used materials in medium voltage cables (5–69 kV). Relative to the low-voltage applications discussed above, medium-voltage applications are more demanding, since a long cable life is desired under more severe conditions of temperature and voltage stress. As a result, polymer performance attributes such as wet electrical stability and high temperature stability become critical. In general, peroxide cross-linked compounds are preferred, since wet electrical stability and long-term cable life is improved relative to the more economic sulfur curing technologies. A typical medium-voltage compound [7] is summarized in Table 6.

Frequently, the cross-linking agent is dicumyl peroxide, and surface-treated calcined kaolin is used as the filler. A surface treatment of an organovinylsilane has proven effective; specifically, this treatment improves the mechanical properties

Table 6 Typical medium-voltage compound

Component	phr
Polymer	100
Mineral filler and coupling agent	50–120
Peroxides and co-agents	2–5
Lead oxide	0.5
Zinc oxide	0–20
Antioxidants	1–3
Process aids	0–10

and increases cable life by reducing the formation of water-trees [8]. Metallic oxides are used to improve hydrolytic stability. The medium-voltage compound can be prepared in a batch or continuous process. The pellet form of polyolefins makes them well suited for the continuous process.

During the cable production process, the compound shown in Table 6 is applied as an external jacket to a cable core and subsequently cured in a continuous vulcanization line. An extruder equipped with a crosshead die is used to apply the jacket; die geometry and jacket thickness are a function of cable design. Extrusion temperatures must be kept low such that the cross-linking agent does not decompose prematurely, e.g. the maximum extrusion temperature for dicumyl peroxide is approximately $130\,°C$. At such low extrusion temperatures, surface flow defects such as melt fracture can be a problem. However, processing aids generally eliminate such esthetic concerns. Finally, the jacketed cable is pulled through the continuous vulcanization line which cures the jacket at $95–205\,°C$ and up to 300 psi.

CGC technology polymers, both POEs and POPs, have several unique features that make them well suited for medium-voltage applications. As shown in Figure 2, peroxide cross-linked POEs have superior oxidative stability relative to peroxide cross-linked EPR and EPDM. One would expect this improved oxidative stability to increase cable life as well as the upper service temperature. CGC technology polymers have a narrow molecular weight distribution (MWD), thus low molecular weight species are minimized, and the cross-linking rate and state improve. Although a narrower MWD implies a more Newtonian viscosity profile and thus processing difficulties, i.e. high pressure and amps, CGC technology allows one to improve processability dramatically via the incorporation of long-chain branching.

2.4   CABLE JACKETING APPLICATIONS

Historically the use of high-pressure LDPE in telephone cables was pioneered in the 1940s. The advantages of LDPE included excellent electrical properties, low moisture absorption, physical toughness and ease of application relative to incumbent materials such as lead, asphalt, paper and jute. More recently, with the development of high-efficiency Ziegler–Natta catalysts in the 1970s, linear low-density polyethylene (LLDPE) has rapidly replaced much of this LDPE, especially in the North America. In Europe, LDPE has remained the preferred product, owing to its processing ease. The advantages of LLDPE over LDPE include improved toughness, abrasion resistance and environmental stress crack resistance (ESCR). In addition, relative to LDPE, the low-temperature physical properties of LLDPE are superior, and higher melting-points lead to higher upper service temperatures.

In designing LLDPE polyethylene resins for telecommunication cable jackets, cable manufacturers (and resin producers) generally compromise between cable flexibility and toughness. More specifically, lower modulus resins are desirable because they are more flexible and easier to install; however, higher modulus resins are desirable because they are more abuse resistant. Figure 3 illustrates the typical structure for a telecommunication cable.

Jackets for power cables were traditionally made with PVC. A typical power cable can be seen in Figure 4. Due to severe requirements on abrasion and mechanical toughness, as well as the much lower water vapor transmission, medium and high density polyethylene have replaced PVC in many countries. In general, blends of LDPE and HDPE are used, with some additional EVA to improve the ESCR of such blends. As demonstrated in Table 7, Dow enhanced polyethylene (EPE) resins show the desirable mechanical performance and ESCR in power cable jacket applications. Moreover, lower extrusion temperatures will lead to less penetration of the neutral wires into the insulation screen. Reduced relaxation times yield lower die-swell values and, hence, reduced heat reversion.

Additional data which compare the performance of Dow EPE resins in a cable jacketing application are summarized in Table 8 [9]. Three EPE resins and a gas-phase LLDPE product (hereafter denoted Resin GP) were characterized. All resins in Table 8 contain 2.7 wt% carbon black. The cable jackets were produced using a

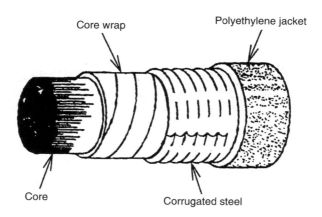

Figure 3   Telecommunication cable structure

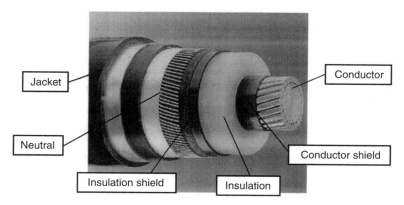

Figure 4   Schematic diagram of a power cable

Table 7   Physical properties of a black EPE compound in comparison with a standard MDPE compound for power cable jackets

Compound property	Conventional MDPE compound (black)	EPE[a] compound (black)	Method
Melt flow-rate, 190 °C/2.16 kg (dg/min)	0.2	0.7	ISO 1133/B
Density[b] (kg/m^3)	0.95	0.95	DIN 53479/C
Vicat (A) oil (°C)	113	121	ISO 306
Shore hardness (D scale)	58	59	VDE 0472 Part 631/B3sec.
Tensile strength (MPa)	23	28	VDE 0472 Part 602
Ultimate elongation (%)	800	750	VDE 0472 Part 602
Heat reversion (%)	0.8	0.6	VDE 0472
ESCR (h)	>1000	>1000	VDE 0472 Part 810

[a] Enhanced polyethylene (EPE)
[b] Density of the compound containing 2 wt% carbon black.

conventional cable jacketing line equipped with a cross-head die; a corrugated steel shield was coated which contained a PVC jacketed control cable in the core, as shown in Figure 3. Results from the jacketing line demonstrate that the EPE resins yield similar extrusion pressures, amps and specific outputs relative to Resin GP, as shown in Table 8. Surface smoothness or the degree of surface melt fracture was evaluated using a surface roughness analyzer. Based on a 3.8 mm scan of the cable surface with a diamond tipped stylus, this analyzer calculates the average roughness parameter *Ra*. *Ra* is the arithmetic mean of all departures of the roughness profile from the average mean line. The average surface roughness of EPE Resin B was much lower than Resin GP, 0.7 μm relative to 1.5 μm, respectively. Visually, there

Table 8 Summary of resin properties and key tests related to black jacketing applications (all resins contained 2.7 wt% carbon black)

	EPE resin			Gas-phase LLDPE
	A	B	C	Resin GP
Resin properties:				
I_2 (dg/min)	0.87	0.78	0.89	0.12
Density (g/cm^3)	0.952	0.958	0.957	0.958
Flexunal modulus (MPa)	1012	n/a	1147	1523
Abrasion (g lost/1000 cycles)	0.33	0.31	0.29	0.29
Resin processability at 50 rpm:				
Amps (A)	38	33	35	34
Pressure (bar)	88	74	75	101
Specific output (kg/h ph)	3.89	4.43	4.30	4.56
Average surface roughness, Ra (µm)	n/a	0.7	n/a	1.5
Melt strength	Pass	Pass	Pass	Pass
Finished cable properties:				
Flexibility: force at 5 mm deflection (kg)	6.3	7.3	6.8	8.0
Flexibility: force at 20 mm deflection (kg)	16.8	17.2	16.8	19.7
Cable tensile elongation at overlap (%)	250	280	220	40
Jacket bond strength at overlap (N/mm)	n/a	6.36	3.92	1.16
100 °C, 4 h cable jacket shrink-back (%)	0.5	1.5	0.5	1.0

was a large difference in the appearance of these two cables; the former was glossy, while the latter had a hazy appearance.

The melt strength or melt tension of cable jacketing resins is also a critical performance criterion. More specifically, the melt strength must be high enough such that the molten resin does not sag (due to the force of gravity) prior to the solidification of the cable jacket in the water-bath. The melt strength of each resin was evaluated as pass/fail as shown in Table 8. Results from the jacketing line demonstrated that EPE resins had sufficient melt strength for this application, even though these resins were significantly higher in melt index.

From the installation point of view, increased flexibility is desirable since the cables will be easier to instal. Cable flexibility was measured by clamping a piece of cable horizontally in an Instron tensiometer and measuring the force required to deflect the cable in the upward direction. In Table 8, one can compare the force required to deflect each cable 5 and 20 mm. The improved flexibility of EPE resins A (0.952 g/cm^3), B (0.958 g/cm^3) and C (0.957 g/cm^3) are evident, relative to the gas-phase produced linear low-density polyethylene control (0.958 g/cm^3).

The tensile properties of the final cable jackets were also estimated by removing the cable jacket and testing in the longitudinal (along the cable axis) and circumferential directions. For the four resins investigated there was no significant difference in the circumferential or longitudinal tensile properties. All cables passed the 400% elongation requirement. Circumferential microtensile samples were also cut from the cable jackets such that the notch produced by the steel shield overlap was centered across the waist region of the dogbone. As shown in Table 8, the EPE resins were much less notch sensitive relative to the reference GP resin. It is well known that the tensile properties of polyethylenes are sensitive to notches or surface imperfections. During the wire coating process, notches are generally produced at the steel shield overlap, as shown in Figure 3. The steel shield functions as a moisture barrier, a rodent resistant armor and greatly increases the overall strength of the cable. In the case of poor or incomplete shield overlap, severe notches are produced in the jacket, which can result in failures under relatively mild impact or tensile forces.

In general, telecommunication cables require a strong bond between the coated steel sheath and the outer polyethylene jacket. In fact, the minimum bond strength between the steal sheath and the jacket must be at least 1.7 N/mm of specimen width. Significant differences were observed in the jacket bond strength of GP and EPEs resins B and C at the sheath overlap, as shown in Table 8. Thus, in addition to superior tensile properties, reduced notch sensitivity and superior flexibility, EPE resins also formed stronger bonds with the steel sheath.

In summary, EPE resins have excellent physical properties at higher melt index, i.e. lower molecular weight, relative to traditional medium-density polyethylene (MDPE) or LLDPE. In fact, new EPE products are being developed to maximize processability by providing lower viscosity grades which consume less energy, improve extruder output and have low heat reversion in order to maximize productivity.

3 FLOORING APPLICATIONS

One of the most demanding applications for any material is in floor covering. In addition to being esthetically pleasing, the floor covering material must be capable of taking mechanical and chemical abuse and still maintain its appearance, as well as its structural integrity. Traditionally, floor covering materials have served a variety of purposes: to be pleasing to the senses, soft to walk on, durable, easily cleaned and changeable. These requirements are served by floor coverings such as carpets and rugs and the whole family of soft, fibrous coverings and, by hard durable materials such as stone, wood and synthetic polymeric materials and composites. In our current society, there is a demand to combine decorative beauty, 'softness', durability and low cost into a single floor covering. These demands are somewhat contradictory, and extremely difficult to meet adequately using natural materials. However, these demands can be reasonably met through the use of synthetic polymers. This section will focus on the benefits of using CGC ethylene copolymers in polymer flooring such as roll stock and tile flooring.

Currently, the major materials for polymeric flooring are flexible PVC, linoleum or cross-linked rubber. Globally, over one billion square meters per year of polymeric flooring are sold. In Western Europe, over 300 million square meters per year are sold, of which 80 % is PVC, 12 % linoleum and 8 % rubber. Polymeric flooring can be broadly classified into homogeneous and heterogeneous flooring. Homogeneous flooring is generally viewed as a system that is of constant composition throughout its whole structure, as shown in Figure 5. Homogeneous compositions with a thin top coat also fit into this broad category. Homogeneous flooring has the advantage of the wear layer being the full depth of the construction; consequently, deep scuffs or gouges do not generally appear as defects. Homogeneous flooring is used in industrial or commercial settings where fashionable appearance is secondary to economy and functionality. Heterogeneous flooring, as the name implies, consists of multiple components. The simplest construction consists of a reverse printed, transparent, top layer bonded to a base layer. Heterogeneous flooring may also contain additional layers, one of which is often foamed, as shown in Figure 5.

Flexible PVC, owing to a broad range of desirable properties, is an extremely important polymeric material in flooring applications. However, there has been a growing customer demand for performance and other properties that cannot be totally met by flexible PVC. In fact, the industry has tried to meet this demand through the use of materials such as polyolefins, linoleum and rubber, but there is still a desire for new materials with improved performance. The development of CGC technology is a significant development in the flooring industry because new polyolefin materials with unique physical properties are now commercially available. For example, given CGC technology's inherent ability to control precisely polyolefin

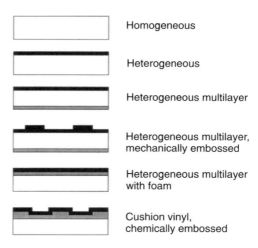

Figure 5   Polymeric flooring constructions

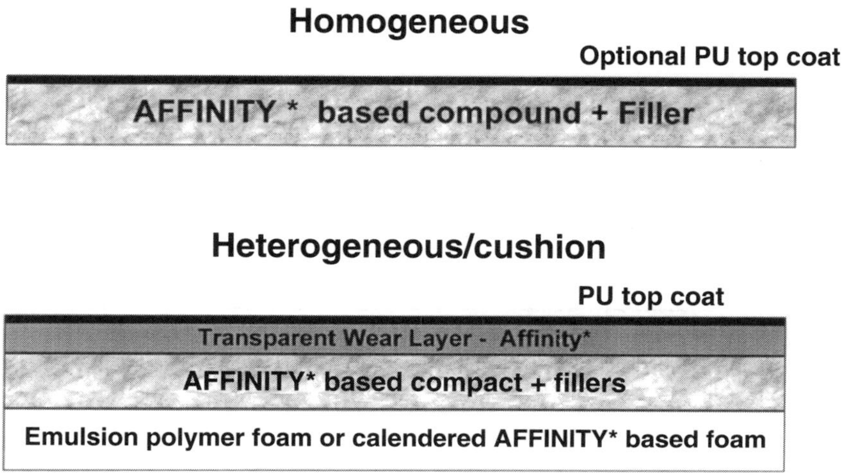

Homogeneous

Optional PU top coat

AFFINITY * based compound + Filler

Heterogeneous/cushion

PU top coat

Transparent Wear Layer - Affinity*

AFFINITY* based compact + fillers

Emulsion polymer foam or calendered AFFINITY* based foam

Figure 6 Typical flooring structures utilizing AFFINITY POP

molecular structure and morphology, one can design novel flooring systems with the desired end-use performance properties. Figure 6 shows current flooring structures that utilize these unique materials. The following sections include a brief discussion of a few key properties in homogeneous flooring which are important to both converters and end-users.

3.1 KEY PROPERTIES

3.1.1 Abrasion Resistance, Flexibility and Indentation Recovery

One of the most important functional properties of a flooring compound is its wear resistance. Table 9 shows the excellent abrasion resistance of unfilled and 50% $CaCO_3$-filled AFFINITY POP relative to commercial PVC flooring. In general, the wear resistance for filled systems is reduced relative to unfilled systems, but this can be improved by the correct choice of base resin, filler type and additives.

High flexibility of the flooring material is important for installation ease. The flexural modulus of AFFINITY POPs can be varied widely, as shown by the data for the 0.902 and 0.870 g/cm^3 POP in Table 9. The flexural modulus of the AFFINITY POP is controlled by the amount of α-olefin monomer incorporated into the backbone. Depending on the choice of base polymer, the formulated flooring can be made harder or softer than PVC without the need of plasticizers. It is important to note that the flexibility of the AFFINITY POP grades are inherent in their chemical microstructure and not the result of adding plasticizers as is the case for f-PVC.

Table 9   Comparison of AFFINITY POPs relative to PVC in flooring applications

Sample	Abrasion resistance[a] (mm^3)	Flexural modulus[b] (MPa)	Indentation[c] (mm)
POP (0.902 g/cm^3)	10	105	0.002
POP (0.902 g/cm^3) + 50 wt% CaCO$_3$	62	260	0.004
POP (0.87 g/cm^3)	50	20	0.07
POP (0.87 g/cm^3) + 50 wt% CaCO$_3$	219	27	0.057
PVC1	220	58	0.021
PVC2	350	151	0.034

[a] Zwick cylinder abrader, DIN-53516.
[b] ISO178.
[c] Remaining indentation, prEN649.

Indentation recovery is another important property for flooring. The AFFINITY POP composition performs very well both in terms of original and final indentation. Table 9 compares indentation results of POPs relative to PVC formulations. Under the conditions of the indentation test the AFFINITY POP composition creeps less than the f-PVC materials (or linoleum) giving rise to good long-term indentation performance. It is also interesting to note that recovery takes place almost instantaneously for POP compositions, with more than 70 % recovery within the first 5 mins. In contrast, f-PVC and linoleum recover more slowly.

3.1.2   Scratch and Stain Resistance

AFFINITY POP grades are of lower surface hardness than f-PVC and linoleum. Hence formulation development and design are critical in obtaining a good visual scratch resistance. In general, visual scratch ratings (ISO 458-2) show that the AFFINITY POP composition is similar to the f-PVC, and improved relative to the linoleum samples. Polyurethane (PU) acrylates can be used as top-coats to further improve the scratch resistance.

As shown in Table 10, AFFINITY POP has better stain and chemical resistance than f-PVC, which in turn has better resistance than linoleum (as defined by European DIN 51958). Similar trends were observed in the burning cigarette test (DIN 51961). More specifically, the POP was more resistance to both surface damage and blackening, relative to f-PVC and linoleum, as shown in Figure 7. Normal cleaning procedures can be used for these materials and where appropriate dry buffing can be used. Long-term wear trials have also indicated reduced traffic staining with the AFFINITY POP formulations.

3.1.3.   Processability

Calendering is a conversion technique widely used in the manufacture of homogeneous and heterogeneous floorings. The ability to be calendered is thus a

Table 10 Chemical and staining resistance, POP-based homogeneous flooring vs PVC and linoleum (higher numbers indicate more attack; top coat was a polyurethane)[a]

Treatment	POP plus top coat	POP	PVC plus top coat	Linoleum
Ammonium hydroxide, 10%				
Caustic soda, 10%	Color 1			
Iodine solution, 2%	Color 3	Color 1	Color 2	Color 2
Lactic acid, 5%				
Acetone			Softer 2	Absorb
Acetic acid			Softer 1	Matt
Amidosulfonic acid, 14%				Color 1
Sunflower oil				
Detergent solution				
Lipstick				Color 2

[a] Change of appearance: blank = none; 1 = noticeable; 2 = significant; 3 = severe.

prerequisite for any alternative material. AFFINITY POP resins show inherently good calenderability on four-roll F-calenders and two- or three-roll calenders, exhibiting good roll release properties, high thermal stability and the ability to produce high-quality products at commercially acceptable line speeds. The freedom of design and design capabilities are not altered when AFFINITY POP resins are used, although special color masterbatches may be required to prevent plate-out during calendering. Printing normally requires the use of corona treatment prior to printing. AFFINITY POP resins may also be embossed with the standard embossing rolls used for f-PVC.

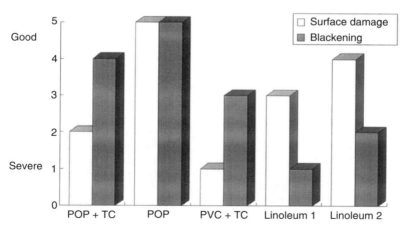

Figure 7 Resistance of POP to cigarette damage. TC = with top coat

3.2 FORMULATIONS

Formulation technology is often required to obtain the correct balance of properties. For both heterogeneous and homogeneous flooring structures AFFINITY POP-based formulations have been developed to produce floorings with comparable, and in some cases better, end-use performance relative to conventional flooring materials. AFFINITY POPs and formulations can be prepared by standard melt preparation equipment such as internal mixers, planetary extruders, Banbury-type kneaders and single- and twin-screw extruders. Although generally not required, processing agents may also be added. The addition of fillers can also be used to make an even more cost-effective product, or to improve ignition resistance. AFFINITY POP resins have excellent filler acceptance with little drop in mechanical properties; filler levels up to 60 wt% are easily accommodated. Different filler types, e.g. silica and talc, may be used where optimization of surface hardness, scratch resistance, wear resistance or coefficient of thermal expansion are required. The scratch resistance of filled systems is often a concern; however, additives may be added to improve this property. Of course, blends of AFFINITY POP with other polymers may also be used to optimize the formulation to give the required balance of properties.

In conclusion, AFFINITY POPs offer distinct advantages over traditional raw materials in flooring applications. These new materials meet, and in some cases exceed, current industry standards for flexible floor coverings.

4 GEO-MEMBRANES

One of the fastest growing markets in the polyethylene film extrusion business is geo-membranes used as landfill-liners, cap-liners, and building and construction liners. Market studies indicate that the capacity of the North American market is 200 000 MT in 1997, with growth of 62.5 % expected by the year 2000. Table 11 highlights the unique end-use properties and benefits that AFFINITY POPs bring to membrane and liner applications.

Table 11   Benefits that AFFINITY POP brings to membranes and liners

Attribute	Comments
Halogen free	Environmentally friendly
Low density	Weight savings, good flexibility
No plasticizers	Easier to handle, no plasticizer migration (embrittlement)
Narrow molecular weight distribution	Excellent toughness
Controlled long-chain branching	Excellent processability
Controlled short-chain branching	Excellent toughness and weldability

4.1 PERFORMANCE REQUIREMENTS AND PERFORMANCE

The key requirements for geo-membranes include flexibility, high load-bearing performance, excellent ESCR, chemical resistance and weldability (especially under humid conditions). In addition, the additive package must be selected such that the liner is weatherable (UV resistant) and the oxygen induction time (OIT) is greater than 100 min at 200 °C. In general, geo-membranes are produced in a blown film process with film thickness up to 3 mm; as a result, relatively low melt index, high melt strength polyethylenes are recommended. As one would expect, specific applications have their own set of unique performance requirements. For example, key requirements for tunnel liners include tensile strength, tear propagation, perforation, heat storage, ignition resistance, behavior under water pressure and water vapor transmission (as defined by European DIN 16 726 and SIA V280).

Figure 8 illustrates typical structures for 2 mm co-extruded cast-film membranes. Layer A is a LLDPE ($0.912 \, \text{g/cm}^3$, 1.0 melt index) which provides good impact strength, dart drop and puncture resistance, while layer B is a POP ($0.885 \, \text{g/cm}^3$, 1.0 melt index) which provides flexibility. Structure I is a simple two-layer structure which is used to achieve optimal toughness and flexibility. Structure II is a more complicated three-layer structure, which is frequently employed to meet country-specific compositional specifications, e.g. Germany, Austria and Switzerland.

Table 12 compares key physical properties of a monolayer membrane with the multilayer structures shown in Figure 8. Note the lower flexural modulus of the co-extruded membranes. In addition, relative to the monolayer structure, the co-extruded membranes have improved impact at −10 °C. Of course, improved puncture resistance is a highly desirable attribute in geo-membrane applications. Indeed, Table 13 compares the dart impact and puncture energy of AFFINITY POP

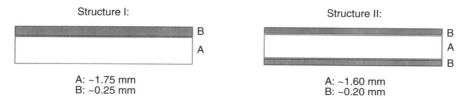

Figure 8 Typical two- and three-layer co-extruded cast membranes

Table 12 Physical properties of 2 mm thick membranes

Property	Membrane structure		
	BA	BAB	Monolayer
Flexural modulus (MPa)	77	117	185
Impact −10 °C (J/mm)	35	51	23

Table 13 Physical properties of AFFINITY POP and LLDPE 50 μm blown films

Property	AFFINITY POP	LLDPE
I_2 (dg/min)	1.5	1
Density (g/cm^3)	0.912	0.922
Dart impact (g)	710	225
Puncture energy (J)	8.1	1.3

and LLDPE blown films (50 μm). Relative to the higher density LLDPE film, the AFFINITY POP film has a threefold higher dart impact and a sixfold higher puncture energy.

In summary, AFFINITY POP multi-layer geo-membranes have been designed which are easier to instal and allow one to optimize membrane flexibility as well as overall toughness.

5 ROTATIONAL MOLDING

Rotational molding is a process which is most frequently used to produce hollow plastic products. Owing to the lower cost of equipment and molds, rotomolding is the most economic process to manufacture stress-free, large parts from a wide variety of raw materials. The rotational molding process consists of following steps: (i) the appropriate amount of resin power is added to the open mold, (ii) the mold is closed and biaxially rotated inside an oven, (iii) the resin powder becomes molten and covers the entire inside surface of the mold, (iv) the mold is cooled with air and/or a water spray and (v) the mold is opened and the hollow part removed.

Currently, polyethylene is the most commonly used raw material in the rotomolding process. More specifically, LLDPE has gained popularity because of a good balance between processing and end-use physical properties; typical resins have densities of 0.920–0.940 g/cm^3 with melt index from 3.5 to 9. Polyethylene is relatively easy to grind to the appropriate powder size (about 200–7 mesh), has good thermal stability (no major changes in properties during grinding or molding), has a well defined crystal melting-point range (with a sharp drop in viscosity allowing good fusion) and has acceptable shrinkage and good toughness, which allow easy removal from the mold. Polyolefins also have excellent chemical resistance, weatherability and low-temperature properties which make them ideal materials for many large-part applications. However, in the rotational molding industry there is a need for new polyolefin materials with improved stiffness, toughness and environmental stress crack resistance. In general, such a combination of properties is difficult to obtain through conventional Ziegler–Natta polymerization processes. However, CGC technology allows one to design new polyethylenes that are tailor-made for rotational molding. Through the control of molecular architecture one can simultaneously improve grinding, optimize rheology and increase the end-use physical properties at higher densities and/or higher melt indices. In addition, CGC technology also allows one to produce extremely flexible parts with high elasticity and good toughness.

5.1 GRINDING

Grinding and the properties of the ground resin both play an important role in rotomolding cost and part performance. The ability to produce a quality powder economically is of major importance. CGC technology can produce a polyethylene copolymer that is easier to grind than conventional LLDPE. For example, using a standard industrial grinder operating at room temperature, a CGC technology ELITE enhanced polyethylene was ground at a maximum rate of 5.5 kg/min. In contrast, the maximum grinding rate for the incumbent LLDPE was only 4.3 kg/min.

5.2 FLEXIBLE APPLICATIONS

AFFINITY POPs, in the density range 0.885–0.900 g/cm^3, have been used in applications such as road markers, crash barriers, buoys and boat fenders. Compared with other materials such as EBA (ethylene–butyl acrylate), EVA and f-PVC, AFFINITY POPs offer advantages in thermal stability, i.e. no corrosion or smell due to degradation. In addition, the AFFINITY POP was easier to demold, as shown in Figure 9.

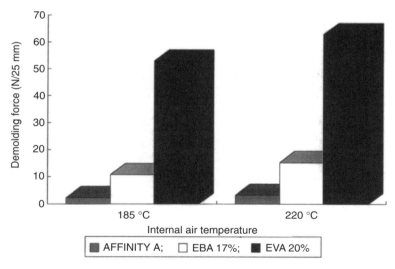

Figure 9 Demolding force comparison (AFFINITY POP A: 6 MI, 0.900 g/cm^3)

5.3 HIGH-DENSITY POLYOLEFIN RESINS

Stiff resins (generally with densities well above $0.940 \, \text{g/cm}^3$) with good low-temperature impact performance are required by manufacturers of large containers and other complex moldings, such as canoes, where either higher stiffness or lower wall thickness is desirable. Standard LLDPE resins with densities above $0.940 \, \text{g/cm}^3$ have the required stiffness, but often suffer brittle impact failures at low temperatures. In contrast, CGC technology allows a new generation of products to be produced with densities as high as $0.949 \, \text{g/cm}^3$ having excellent low-temperature impact strength. In fact, these resins exhibit similar processing and low-temperature impact properties relative to a lower density conventional LLDPE, as shown in Figure 10. Note the difference in density between the two resins shown in Figure 10. Specifically, the POP shown in Figure 10 has a 50 % higher flexural modulus (stiffness), yet this POP has the impact properties similar to those of a much lower density LLDPE. This is a distinct advantage in the above-mentioned applications.

5.4 IMPROVED PROCESSING POLYOLEFIN RESINS

Toy manufacturers and producers of items with large flat surfaces require resins that can accept dry-blended pigments, have excellent processability and shorter heating cycles, and still maintain good mechanical properties, such as toughness. Traditionally, PE resins with MI < 10 have been used; although these resins met the processability requirements, the final product often lacked toughness. In contrast,

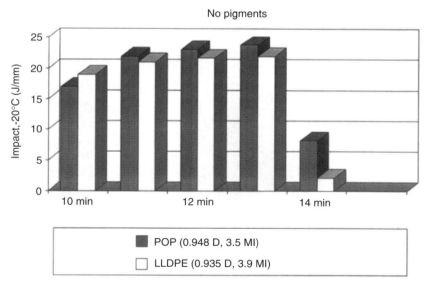

Figure 10 Impact comparison and influence of heating time

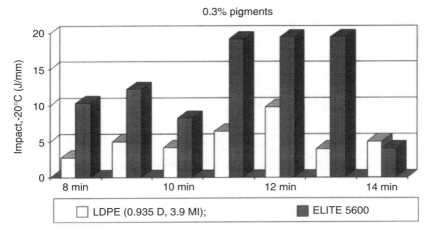

Figure 11 Cold temperature impact as a function of heating time

Table 14 Impact strength of rotomolded toys at 23, 0 and −20 °C

Product	I_2 (dg/min)	Density (g/cm^3)	Impact strength (J/mm)/no. of brittle failures		
			23 °C	0 °C	−20 °C
LLDPE-1	6	0.935	4.9/4	2.7/5	2.0/5
LLDPE-2	7	0.935	4.6/5	2.5/5	1.6/5
ELITE EPE 5600	15	0.935	10.8/0	11.2/1	5.1/5

CGC resins have improved processability, for example, melt indexes as high as 15, yet retain very high toughness. POP moldings have excellent surface finish and good mechanical properties, while offering reduced heating times to the molder, as shown in Figure 11 and Table 14.

In summary, the controlled molecular architecture of CGC technology polymers allows one to design unique polyolefins to meet the processing requirements of rotational molding and the end-use requirements of the finished parts. Given these new resins, one can explore new opportunities for the widely used rotational molding process.

6 DECORATIVE OVERLAYS FOR FURNITURE

More and more concern is being paid to our environment and our limited natural resources. This is forcing many industries to use a wider range of materials and technologies to meet performance requirements while protecting our resources. In addition, there is a global competitiveness that mandates cost reductions in both raw materials and processing. These trends also apply to the furniture industry, which has been innovative in using lower grades of wood, wood composites and new plastics to make items that, in the past, have been produced from high-quality woods. This industry has also been innovative in developing decoration technologies to make their wares more acceptable, specifically, lower cost furniture with an appearance similar to higher cost items.

Polymers continue to play an every increasing role in decoration technology owing to their extreme versatility. Plastic films can be colored, textured and made to have the touch and feel of high-quality wood. Generally, such polymeric films are laminated to the furniture piece to create the desired appearance. For example, one decoration process uses thermoplastic overlays or laminates based on f-PVC film or sheet. This overlay is either clear or colored, reverse or surface printed and embossed with a wood grain effect. Other decoration techniques are based on printed paper impregnated with thermosetting resins that enhance furniture surfaces; these systems are applied via hot-press high-pressure adhesive lamination. Both of the above systems are currently subject to environmental pressures.

More recently, polyolefin copolymers are being considered as new materials for decoration applications. Indeed, AFFINITY POPs have been developed that combine aspects of the two processes discussed above and produce a tough, flexible and environmentally friendly alternative. For example, the extrusion coating of sublimation ink printed papers (non-impregnated) with a 200–300 μm film of AFFINITY POP produces a decorative overlay with an enhanced three-dimensional print image which can be low-pressure adhesive laminated to board materials. The 3-D effect is created by the unique interaction of the ink pigments and the AFFINITY POP which occurs during the extrusion coating process. The versatility of this extrusion process allows surface texture to be modified, if desired, to give an overlay with controlled 'feel' and a matte surface finish. In addition to the esthetic constraints, these coatings must have excellent scuff and stain resistance; properties that can be optimized via CGC technology. Figure 12 illustrates a typical structure

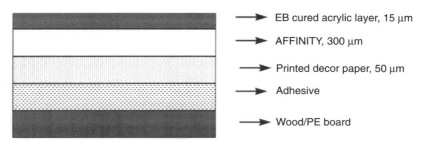

Figure 12 Typical structure used in decorative applications which includes an AFFINITY POP layer of 300 μm

Table 15 Physical property comparison

Polyolefin	Melt index (dg/min)	Density (g/cm³)	2% Secant modulus (MPa)	Tensile strength (MPa)	Haze (%)
POP	1.0	0.902	78	50	1
DOWLEX NG LLDPE	1.0	0.919	176	33	6
C₆ LLDPE	1.0	0.920	210	33	10
LDPE	1.0	0.920	200	20	10
ATTANE ULDPE	1.0	0.912	115	30	8

using an AFFINITY POP layer of 300 μm. The electron beam cured top coat is used only when an extra degree of toughness, wear and chemical resistance is required of the furniture piece. In Table 15 one can compare the relative performance of these new CGC-based copolymers with other polyolefins. Note the low modulus, high tensile strength and very low haze of the CGC copolymer; attributes the design engineer can take advantage of in decorative applications.

In summary, AFFINITY POP decorative overlays offer a more versatile, adaptable and cost-effective solution to decorative furniture laminates than conventional f-PVC and thermoset systems.

7 ACKNOWLEDGMENTS

The authors thank Maarten Aarts, Jerker Kjellqvist, Karen Katzer, Peter Schlinder and Shaun Parkinson of the Dow Chemical Co., Horgen, Switzerland, for their comments and data pertaining to the wire and cable, flooring, geo-membrane, rotational molding and decorative overlay applications, respectively.

8 REFERENCES

1. Canedo, E. and Valsamis, L., in *Proceedings of the Society of Plastics Engineers Newark Section RETEC*, Somerset, NJ, November 1993.
2. Burbank, F., in *Proceedings of the Northeast Regional Rubber and Plastics Exposition*, Mahwah, NJ, September 1994.
3. Stropoli, T. and Case, C., in *Proceedings of the Society of Plastics Engineers Newark Section RETEC*, Somerset, NJ, November 1993.
4. Kale, L. T., Hemphill, J. J., Parikh, D. R., Sehanobish, K. and Hazlitt, L. G., presented at the *American Chemical Society Rubber Division 148th Technical Meeting*, Cleveland, OH, October 17–20, 1995.
5. Barnes, M. A., Briggs, P. J., Hirschler, M. M., Matheson, A. F. and O'Neill, T. J., *Fire Mater.* **20**, 17 (1996).

6. Sawyer, D., Artingstall, L. P., Preston, J., Dell, S. and Cochard, S., in *International Wire and Cable Symposium Proceedings*, 1996, p. 520.
7. Keefe, R. J. and Hemphill, J. J., in *Proceedings of Schotland Specialty Polyolefins Business Forum SPO 96*, Houston, TX, September 1996.
8. Brown, M., *IEEE Electr. Insul. Mag.*, January/February (1994).
9. Kale, L. T., Iaccino, T. L. and Bow, K. E., presented at *Society of Plastics Engineers Annual Technical Conference, ANTEC 95*, San Francisco, CA, May 5–9, 1995.

24

Production and use of Metallocene-based Polyethylene in Rotomolding Applications

ANNE M. FATNES, I. S. MELAAEN, V. ALMQUIST AND
A. FOLLESTAD
Borealis AS, Stathelle, Norway

1 INTRODUCTION

Rotational molding (rotomolding) is a method for manufacturing hollow plastic articles [1–3]. Typical products are tanks, containers, toys, small boats and technical articles. The process was first developed in the 1940s for use with liquid poly(vinyl chloride) formulations. The first major breakthrough for this process was in the 1950s when powdered polyethylene grades were developed for rotational molding. Significant growth in the use of this technique came in the 1980s with the availability of the linear low-density polyethylene (LLDPE) types, which gave improved processing properties for the rotational molding process combined with good mechanical properties of the final product. In the last 10–15 years, LLDPEs have been the most widely used grades for rotational molding. These grades, made using Ziegler–Natta catalysts, are advantageous since they are polymers with a narrow molecular weight distribution (MWD) combined with relatively good mechanical properties.

The rotational molding process differs from all other plastic processing processes in that the heating, melting, shaping and cooling stages all occur after the plastic is placed in the mold and there is no external pressure applied during forming. This

Metallocene-based Polyolefins Edited by J. Scheirs and W. Kaminsky
© 2000 John Wiley & Sons Ltd

leads to advantages [4], some of the most important being essentially stress-free products, economic production of large containers and low mold cost.

The unique feature of the rotomolding process is that the polymer melt is not subjected to high shear forces. Because of this, polymers with a narrow molecular weight distribution (MWD) give the best combination between ease of melt flow at low shear forces and mechanical strength of the final product.

It has long been clear that with metallocene catalysts new types of polyethylenes will become available with totally new and different properties compared with the Ziegler–Natta materials. Among these properties are much narrower MWD and corresponding change in melt flow behavior. Borealis, the fourth largest producer of polyolefins in the world, has had the opportunity to develop an excellent combination of metallocene catalyst and process to satisfy the rotational molding market and its customers in this fast-growing application area [5].

2 PRODUCTION TECHNOLOGY

2.1 CATALYSTS

The traditional catalyst used for non-cross-linking rotomolding is the Ziegler–Natta catalyst family, developed in 1953 [6]. The polymer produced with such a catalyst usually has a molecular weight distribution characterized by a weight-average/number-average molecular weight ratio (M_w/M_n) of 4–6, meaning that it has substantial amounts of both long and short polymer chains. It also has a very uneven incorporation of comonomens such as butene or hexene, so that especially short chains are much richer in comonomer than others.

In the late-1970s, work was started in Kaminsky's laboratory in Hamburg with an olefin polymerization catalyst system where partially hydrolyzed TMA (trimethylaluminum) was used as an activator for Group IV metallocenes. Figure 1 shows the metallocene catalyst system. This catalyst system proved to make polymer chains that were as identical as possible using a practical catalyst. Its M_w/M_n may be close to the theoretical minimum limit of 2.

This catalyst family has been shown to be very flexible in its properties such as its ability to incorporate comonomer into the polyethylene chain, in making different levels of molecular weight and in its ability to make different kinds of stereoregular poly-α-olefins [7], mostly by suitable modification of the metallocene.

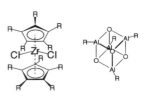

Figure 1 The metallocene catalyst system: left, metallocene; right, methylaluminoxane (MAO). R = alkyl.

Traditional catalyst

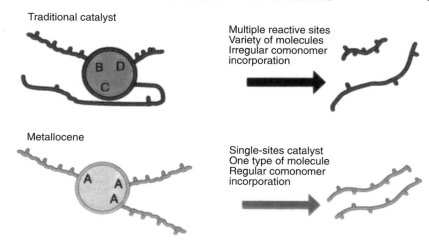

Multiple reactive sites
Variety of molecules
Irregular comonomer
incorporation

Metallocene

Single-sites catalyst
One type of molecule
Regular comonomer
incorporation

Figure 2   Major differences between traditional catalysts (Ziegler–Natta, chromium families) and single-site catalysts (metallocene family) in the way they produce polymer chains

Compared with the Ziegler–Natta catalyst family, the metallocene catalyst family gives the potential for polymers with a much narrower molecular weight distribution and more homogeneous incorporation of comonomer. Figure 2 illustrates the difference between the two catalyst families. The metallocene catalyst is the only single-site catalyst (SSC) of any importance for polyolefins.

However, achieving the combination of an acceptably high catalyst activity and good morphology of the polymer powder in a poyethylene production process, while retaining the catalyst's beneficial properties of producing identical polymer chains, is a most difficult and delicate task.

2.2   BASIC PROCESSES

Several polymer production processes are suited for the production of rotational molding grades. Process requirements are based on the process being able to handle the melt flow-rate (MFR), a measure of molecular weight, and density ranges of rotational molding products. The MFR (2.16 kg/190 °C) is most commonly 2–10 g/10 min and the density range is 930–940 kg/m^3 [8].

In 1993 it was reported that it is possible to produce polyethylene for rotational molding by all types of commercial processes such as gas-phase, slurry, liquid, bulk and high-pressure systems [9]. Exxon was the first company to apply a metallocene catalyst to polyolefin production on a commercial scale. In 1991, the 15 000 tons/year high-pressure polyethylene (HPPE) plant started producing EXACT resins [10].

Dow Chemicals has developed a new group of resins called INSITE, based on their own metallocene catalyst technology. These polymers have improved physical properties and still have good processability. They contain long-chain branches and are similar to high-pressure LDPE with respect to rheology. Dow produces their INSITE polymer in a solution process.

In 1995, Mobil produced low- and medium-density PE suited for cast film and rotational molding with a metallocene catalyst in a commercial gas-phase reactor. Exxon and Mitsui Petrochemical have been collaborating for commercialization of gas-phase technology using metallocene catalysts [11]. Exxon's supercondensed reactor phase technology makes it possible to achieve up to double capacity on gas-phase reactors. A capacity increase can be made gradually at low cost and this has been tested in Texas for producing several thousand tons of metallocene linear low-density material [12]. BP Chemicals has improved the usual gas-phase process by separating the unreacted hydrocarbon stream and injecting the liquid (up to 12 %) through a special developed nozzle into the reactor. During vaporization, the liquid removes the heat generated by the polymerization process. Theoretically, this will allow the plant to double its capacity. BPs latest generation metallocene catalyst is expected to be used for this gas-phase process [13].

In 1995, Phillips produced metallocene polymer in a pilot slurry loop reactor based on two different single-site catalysts with densities from 910 to 970 kg/m^3 and MFR from 0 to 100 [14,15]. In the same year BASF reported a successful production of VLDPE in a similar reactor using a silica-supported metallocene catalyst [16]. Borealis produced their first commercial volumes of metallocene polyethylene (mPE) in 1995 in a 130 kt low-pressure polyethylene (LPPE) plant [17,18].

3 SCALE-UP

3.1 GENERAL

New polyolefin products based on new catalysts are usually scaled up in three phases:

- laboratory scale, usually called bench scale;
- semi-commercial scale, usually called pilot-plant scale;
- full scale.

The scale-up of a new product based on a new catalyst usually takes 3–7 years. The scale-up time depends on the complexity and the need for process design changes.

3.2 BENCH SCALE

After the invention of a new catalyst or catalyst family, a large number of bench-scale polymerizations are carried out. Bench-scale polymerizations are batch experiments.

The bench-scale equipment for polymerizations consists of catalyst preparation equipment working on the grams scale and polymerization reactors of size from <1 to 10 l producing from gram to kilogram batches of polymer.

Most of the polymer properties are given by the catalyst itself. This means that the small bench-scale produced samples are very useful for predicting both rheological and mechanical properties for a scaled-up product. Bench-scale polymerizations also give valuable information for subsequent continuous pilot-plant and full-scale experiments.

Before scaling up Borealis's new mPE grades, several hundred bench-scale polymerizations in 3 l reactors had been carried out. Different metallocene compounds were tested with different metallocenes and other variations in the procedure to make a catalyst recipe for scale-up.

The main objectives were:

- high catalyst productivity (g PE/g catalyst);
- required molecular weight (measured as MFR);
- required molecular weight distribution (MWD);
- required density and density control by means of comonomer;
- convenient reactor conditions for pilot-plant and full-scale operation.

3.3 PILOT PLANT

Pilot plants are semi-commercial plants, continuously producing hundreds of kilograms of polymer per campaign. Pilot test runs have two main purposes: optimization of process conditions and resin production for application trials. Material from pilot test runs is even used for pre-marketing of new resins to customers.

At Borealis, the pilot trials with SSC took place in a slurry loop reactor producing 20–50 kg PE/h. Later several SSCs were tested in gas-phase pilot reactors with excellent results [19]. The slurry loop pilot trials went well, with high catalyst productivity at convenient reactor conditions. Vital relationships such as density versus comonomer concentration and MFR versus hydrogen/ethylene ratio were studied and the results were used in planning the first full-scale test run.

Evaluation of polymer properties and application tests showed very promising results, and it was decided to scale up and commercialize resins for the rotational molding and cast film markets.

In fact, the first SSC developed by Borealis showed excellent behavior in the slurry loop reactor. The MFR and density were even easier to control with this catalyst than with chromium and Ziegler–Natta catalysts. In addition, it gave exactly the narrow MWD targeted for the first Borecene products.

The first Borealis SSC was to be scaled up in a slurry loop reactor mainly producing chromium-based PE. SSC and chromium give different kinetics, as can be seen from Figure 3. The chromium catalyst gives a slow start and increasing activity during a polymerization period of 45 mins. Under comparable conditions, SSC gives

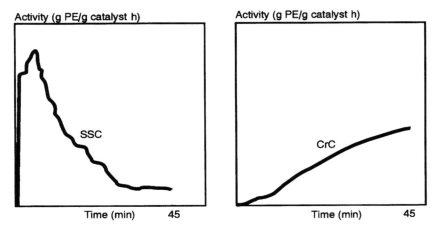

Figure 3 Comparison of polymerization kinetics for single-site catalyst and chromium catalyst systems

a very high activity at the start, slowing after a while to end at only 20–30 % of the initial activity after a polymerization period of 45 mins.

4   SSC IN ISOBUTANE SLURRY LOOP REACTOR

The metallocene catalyst developed at Borealis has the following characteristics:

(i) No induction time:
- quick response to changes in reactor control parameters;
- well suited for plant start-up from empty reactor.

(ii) Uniform incorporation of comonomers:
- lower density with the same amount of hexene compared with chromium-based polymer.

(iii) High productivity:
- 3–6 t PE/kg catalyst for rotational molding products.

(iv) High sensitivity for hydrogen on MFR:
- 100 % conversion of hydrogen.

(v) Narrow MWD:
- ideal for rotational molding.

(vi) Morphology:
- suited for slurry loop reactor system.

To produce a given resin commercially, the catalyst productivity has to be higher than 2 t/kg catalyst to satisfy customer demands, and even higher to satisfy economic demands. When reactor parameters are to be determined, all parameters need to be optimized with respect to both productivity and product properties.

SSC is much more sensitive to impurities in the raw materials than a chromium catalyst [20]. It is very important to have control over the raw materials, especially humidity and carbon dioxide.

As is already known, hexene incorporation is much more uniform in SSC than in chromium polymers. This leads to the need for smaller amounts of hexene for the same polymer density. The melt flow-rate is adjusted with hydrogen and to some degree with ethylene. An increase in hydrogen increases the density, and it is important to compensate for this with a higher hexene level.

From the operator point of view, one of the major differences between running chromium and SSC polymers is that the induction time is very short for SSC. Changes in reactor parameters give much faster response than for a chromium catalyst [21].

It is important to keep the temperature in the flash line and flash drum after the reactor as high as possible and the pressure as low as possible to remove the hydrocarbon gas from the powder particles. Steam is used in the drier to increase amount of hydrocarbons leaving the fluff in addition to using hot nitrogen in the purge tower.

Extrusion of single-site polymer is very different from extrusion of chromium polymer because of the narrow molecular weight distribution. The flow properties are more independent of shear rate and the viscosity curve as a function of shear rate is flatter. Suitable amounts of lubricants are necessary to achieve stable operation.

5 ROTATIONAL MOLDING PRODUCTS AND PROPERTIES

5.1 THE POLYMER

A good polymer for use in rotational molding must have certain features. To be able to obtain good final products it is necessary to use polymers which can be ground into a fine powder to allow even melting and building of an even wall thickness. The rheology of the melt must also be right, i.e. the polymer must have a relatively low zero shear viscosity to allow fast sintering. During cooling the crystallization must be homogeneous to give an even structure, low distortion, good stiffness and good mechanical properties.

This combination of properties has until now been achieved best by Ziegler–Natta based LLDPE polymers. However, it has recently been shown that the new family of mPE give polymers with improved properties for rotational molding, because they have an even narrower molecular weight distribution, lower zero shear viscosity and better mechanical properties, which give improvements compared with the conventional LLDPEs.

The typical resin used for rotational molding used today has a relatively narrow molecular weight distribution, which can be achieved, for instance, through using a Ziegler–Natta type catalyst in a gas-phase reactor. This is the process by which Borealis today manufactures its rotational molding resins.

The MFR and density ranges normally used for rotational molding are limited. The reasons for this are that too low an MFR limits the flow and sintering of the polymer and too high an MFR leads to dripping and splashing during processing, and high density gives too high crystallinity and too much warpage and low density results in a lack of stiffness.

5.2 POLYMER PROPERTIES OF mPE

The new metallocene catalysts are characterized by making a polymer where all chains are equal (narrow molecular weight, single-site type of polymer), and where the incorporation of the comonomer is very even and equal for each polymer chain. The Borealis metallocene catalyst has these attributes. The polymer produced has a very narrow molecular weight distribution. This is illustrated in Figure 4, where the molecular weigh distribution of a conventional Ziegler–Natta based LLDPE is compared with a Borecene mPE. It is clearly shown that the mPE has a much narrower MWD with less polymer chains in both the high and low molecular weight areas.

The narrow molecular weight distribution strongly influences the melt flow behavior of the polymer. The viscosity curves in Figure 5 clearly show that the

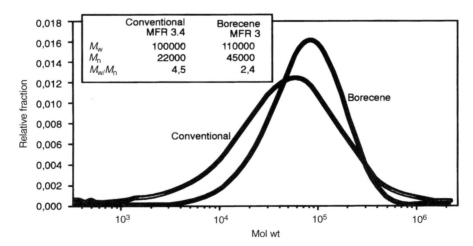

Figure 4 Molecular weight distribution for conventional Ziegler–Natta based LLDPE with MFR 3.4 compared with Borecene mPE with MFR 3.0

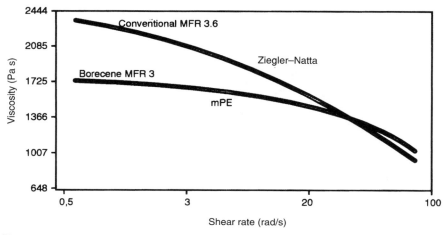

Figure 5 Viscosity curves for conventional LLDPE with MFR 3.4 compared with Borecene mPE with MFR 3.0

mPE has a lower viscosity at zero and low shear and also a lower degree of shear thinning than the comparable standard LLDPE.

5.3 PROCESSING PROPERTIES OF mPE IN ROTATIONAL MOLDING

Rheological properties are the key issue explaining the improvements in processing seen for rotational molding with these new polymers. The low zero shear viscosity makes the flow of the polymer in the zero and low shear area much easier than for a comparable Ziegler–Natta based LLDPE.

The improved flow also speeds up the sintering process, which is the process when the melted polymer particles merge into one body and trapped air/gas dissipates. It has been described [22] how the sintering follows the Frenkel equation, where it depends on the ratio between melt viscosity and surface tension [23].

Spence and Crawford [24] described the theory behind the dissipation of gas bubbles during rotational molding, showing that the same parameters as described above for the sintering are involved here.

This faster sintering means that the new mPEs provide a potential for reduced cycle times [25]. This is illustrated in Figure 6, which shows the Rotolog curve for an optimized heating cycle for mPE versus conventional material. The Rotolog measures the inner air temperature in the rotomolding machine, and it is used as a tool to follow and control the process [26].

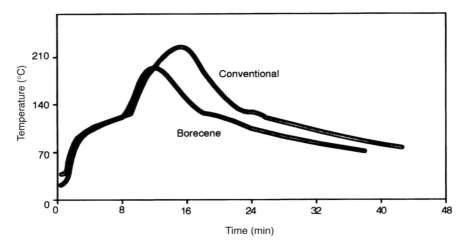

Figure 6 Potential for reduced cycle time for mPE compared with conventional LLDPE is illustated with a Rotolog curve for the two materials showing that the heating cycle can be reduced for the mPE

5.4 BROADER PROCESSING WINDOW

Rotational molding materials can be processed under a large variety of process conditions. These conditions, such as curing time, cooling time and inner air peak temperature influence the final product properties, such as impact resistance. This has been described by several authors [27]. For standard rotational molding grades there are normally a very defined peak inner air temperature and heating time where mechanical properties are at the optimum, and specifically the impact resistance is maximum. The new metallocene-based grades have a broader processing window in this respect. This is shown by the optimum mechanical properties being reached earlier, and also being retained longer during overcuring, as illustrated in Figure 7.

5.5 LONG-TERM CREEP PROPERTIES

Long-term creep properties are important for rotational molding products. This is measured as changes in tensile properties and elongation as a function of load and time. It is important that the material shows low deformation over long period at relevant loads.

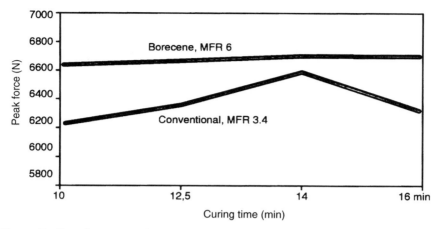

Figure 7 Dart impact resistance at −20 °C for conventional LLDPE and Borecene mPE illustrating improved impact properties and broader processing window for mPE

Measurements made for the new mPEs in comparison with conventional LLDPE materials show that the new materials have advantages, giving better long-term creep properties. The differences become more pronounced at higher temperatures.

The results are illustrated in the Figures 8 and 9. Figure 8 shows creep results at 23 °C for a conventional LLDPE compared with Borecene mPE and Figure 9 compares creep results for the same products at 45 °C.

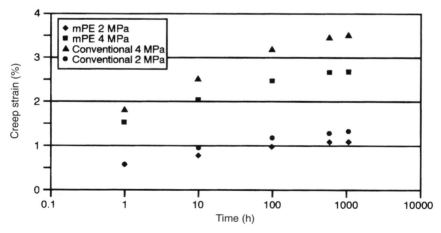

Figure 8 Creep strain curves at 23 °C for conventional LLDPE and mPE, both with density 934 kg/m^3

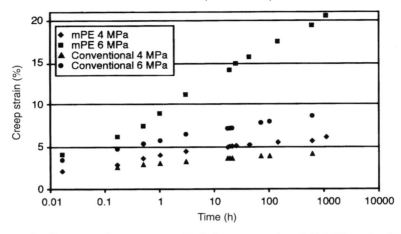

Figure 9 Creep strain curves at 45 °C for conventional LLDPE and mPE, both with density 934 kg/m^3

6 REFERENCES

1. Crawford, R. J. and Nugent, P., *Plast. Rubber Composites Process. Appl.* **1**, (17), 33 (1992).
2. Crawford, R. J., in Crawford, R. J. (ed.), *Rotational Moulding of Plastics*, Wiley, New York, 1992.
3. Rao, M. A. and Throne, J. L., *Principles Rotat. Moulding*, **4**, 237 (1972).
4. Crawford, R. J., *Mater. World*, **11**, 631 (1996).
5. Fatnes, A. M., Melaaen, I. S. and Almquist, V., presented at Metallocenes Europe '97, Düsseldorf, 1997.
6. Ziegler, K., Breil, H., Martin, H., and Holzkamp E., *Ger. Pat.*, 973 626 (1953).
7. Brintzinger, H. H, Fischer, D., Mühlhaupt, R., Rieger, B. and Waymouth, R. M., *Angew. Chem., Int. Ed. Engl.*, **34**, 1143 (1995).
8. Pu, H. T. and Habash, S. N., presented at MetCon '95, Houston, TX, May 19, 1995.
9. *Research Disclosure*, 'Rotational Moulding Resins Made from Single Site or Catalyzed Olefins.' No. 35334, 607 (1993).
10. Xie, T., McAuley, K. B., Hsu, J. C. C. and Bacon, D. W., *Ind. Eng. Chem. Res.*, **33**, 449 (1994).
11. Shut, J. H., *Plast. World*, 12 (1995).
12. Baker, J., *Euro. Chem. News*, (7), 39 (1995).
13. Goldsmith, P., *Process Eng.*, February, 27 (1995).
14. Welch, M. B., Palackal, S. J., Geerts, R. L. and Fahey, D. R., presented at MetCon '95, Houston, TX, May 1995.
15. Sinclair, K. B., presented at the Society of Plastics Engineers Polyolefins IX International Conference, Texas, February 1995.
16. Shut, J. H., *Plast. World*, August, 18 (1995).
17. Fatnes, A. M., presented at the Association of Rotational Moulders 21st Annual Fall Meeting, Vienna, 1996.
18. Melaaen, I. S., presented at Metallocenes '96, Düsseldorf, 1996.

19. Knuuttila, H. presented at Metallocenes '96, Düsseldorf, 1996.
20. Artrip, D. J., Herion, C. and Meissner, R., presented at MetCon '93, Houston, TX, 1993.
21. Almquist, V., Aastad, T., Melaaen, I. S., Hokkanen, H. and Kallio, K., *Norw. Pat.*, 301 331 (1997).
22. Bellehumeur, C. T., Bisara, M. K. and Vlachopoulos, J., *Polym. Eng. Sci.*, **17**, 2198 (1996).
23. Frenkel, J., *J. Phys.*, **9**, 385 (1945).
24. Spence, A. G. and Crawford, R. J. *Polym. Eng. Sci.*, **36**, 993 (1996).
25. Mapleston, P., *Mod. Plast. Int.*, February, 28 (1997).
26. Crawford, R. J., Oliveira, M. J. and Cramez, M. C., *Jo. Mater. Sci.*, **31**, 2227 (1996).
27. Harkin-Jones, E. and Ryan, S., presented at *ANTEC '96*, 1996.

INDEX

Contents of Volume One